Oil Palm Breeding

Oil Palm Breeding

Genetics and Genomics

Edited by

Aik Chin Soh, Sean Mayes, and Jeremy Roberts

CRC Press is an imprint of the
Taylor & Francis Group, an **informa** business

CRC Press
Taylor & Francis Group
6000 Broken Sound Parkway NW, Suite 300
Boca Raton, FL 33487-2742

First issued in paperback 2021

ISBN 13: 978-1-03-209651-3 (pbk)
ISBN 13: 978-1-4987-1544-7 (hbk)

Publisher's Note
The publisher has gone to great lengths to ensure the quality of this reprint but points out that some imperfections in the original copies may be apparent.

Library of Congress Cataloging-in-Publication Data

Names: Soh, Aik Chin, author. | Mayes, Sean, 1967- author. | Roberts, J. A. (Jeremy A.), author.
Title: Oil palm breeding : genetics and genomics / authors: Aik Chin Soh, Sean Mayes and Jeremy Roberts.
Description: Boca Raton : Taylor & Francis, 2017. | Includes index.
Identifiers: LCCN 2016055517 | ISBN 9781498715447 (hardback : alk. paper)
Subjects: LCSH: Oil palm--Breeding.
Classification: LCC SB299.P3 S592 2017 | DDC 634.9/74--dc23
LC record available at https://lccn.loc.gov/2016055517

Visit the Taylor & Francis Web site at
http://www.taylorandfrancis.com

and the CRC Press Web site at
http://www.crcpress.com

Contents

Preface vii
Acknowledgments ix
Editors xi
Contributors xiii

Chapter 1 Introduction to the Oil Palm Crop 1

Aik Chin Soh, Sean Mayes, and Jeremy Roberts

Chapter 2 The Plant and Crop 7

Aik Chin Soh, Sean Mayes, and Jeremy Roberts

Chapter 3 Genetic Resources 19

Aik Chin Soh, Sean Mayes, Jeremy Roberts, Nookiah Rajanaidu, Amiruddin Mohd Din, Marjuni Marhalil, Abdullah Norziha, Ong-Abdullah Meilina, Ahmad Malike Fadila, Abu Bakar Nor Azwani, Libin Adelinna Anak, Yaakub Zulkifli, Suzana Mustafa, Maizura Binti Ithnin, Ahmad Kushairi Din, Edson Barcelos, and Philippe Amblard

Chapter 4 Plant Genetics 57

Aik Chin Soh, Sean Mayes, and Jeremy Roberts

Chapter 5 Objective Traits 85

Aik Chin Soh, Sean Mayes, Jeremy Roberts, Kees Breure, Benoît Cochard, Bruno Nouy, Richard M. Cooper, Hubert de Franqueville, Claude Louise, and Shenyang Chin

Chapter 6 Breeding Plans and Selection Methods 143

Aik Chin Soh, Sean Mayes, Jeremy Roberts, David Cros, and Razak Purba

Chapter 7 Breeding Programs and Genetic Progress 165

Aik Chin Soh, Sean Mayes, Jeremy Roberts, Nicolas Philippe Daniel Turnbull, Tristan Durand-Gasselin, Benoît Cochard, Choo Kien Wong, Tatang Kusnadi, and Yayan Juhyana

Chapter 8 Clonal Propagation 191

Aik Chin Soh, Sean Mayes, Jeremy Roberts, Ong-Abdullah Meilina, Siew Eng Ooi, Ahmad Tarmizi Hashim, Zamzuri Ishak, Samsul Kamal Rosli, Wei Chee Wong, Chin Nee Choo, Sau Yee Kok, Nuraziyan Azimi, and Norashikin Sarpan

Chapter 9 Molecular Genetics and Breeding 225

Aik Chin Soh, Sean Mayes, Jeremy Roberts, Noorhariza Mohd Zaki, Maria Madon, Trude Schwarzacher, Pat Heslop-Harrison, Maizura Binti Ithnin, Amiruddin Mohd Din, Umi Salamah Ramli, Abrizah Othman, Yaakub Zulkifli, Leslie Eng Ti Low, Meilina Ong-Abdullah, Rajinder Singh, Chee Keng Teh, Qi Bin Kwong, Ai Ling Ong, Mohaimi Mohamed, Sukganah Apparow, Fook Tim Chew, David Ross Appleton, Harikrishna Kulaveerasingam, Huey Fang Teh, Katharina Mebus, Bee Keat Neoh, Tony Eng Keong Ooi, Yick Ching Wong, and David Cros

Chapter 10 *Elaeis oleifera* × *Elaeis guineensis* Interspecific Hybrid Improvement 283

Aik Chin Soh, Sean Mayes, Jeremy Roberts, Edson Barcelos, Philippe Amblard, Amancio Alvarado, Jeremy Henry Alvarado, Ricardo Escobar, Kandha Sritharan, Mohan Subramaniam, and Xaviar Arulandoo

Chapter 11 Commercial Planting Material Production 297

Sean Mayes, Jeremy Roberts, Choo Kien Wong, Chin Nee Choo, Wei Chee Wong, Cheng Chua Tan, Abdul Razak Purba, and Aik Chin Soh

Chapter 12 Field Experimentation 327

Rob Verdooren, Aik Chin Soh, Sean Mayes, and Jeremy Roberts

Chapter 13 Future Prospects 353

Aik Chin Soh, Sean Mayes, Jeremy Roberts, Tasren Mahamooth, Denis J. Murphy, Sue Walker, Asha S. Karunaratne, Erik Murchie, John Foulkes, Marcel de Raissac, Raphael Perez, Denis Fabre, Kah Joo Goh, Chin Kooi Ong, and Hereward Corley

Index 425

Preface

Oil palm is a remarkable crop. Combining the two oils produced ("palm oil" and "palm kernel oil"), the average hectare of oil palm in 2013 produced 4.2 tonnes oil/ha/year—over 10 times the oil production of soybean and nearly six times that of oil seed rape. In land use terms, this amounts to soybean producing around 70% as much oil as oil palm on over seven times the land surface. This is not to underplay the concerns of conservationists that new oil palm plantations lead to major loss of biodiversity—that it is a tree crop cultivated in some of the most important regions of biodiversity in the world is an important incentive to restrict further planting or at least to find ways to mitigate the effects on the biodiversity of fauna and flora. However, the figures do emphasize that oil palm (until new crops or new methods to economically manufacture oil from algae, for instance, come along) is part of the solution for producing enough oil for the growing world population, not just a problem.

In some breeders' trials, 10–12 tonnes of oil/ha/year is regularly seen for modern planting material of oil palm under best management practices. With a reduction in the yield gap by raising average yields to be closer to those obtained in breeding trials, a significant expansion of production could be possible without increasing the land under oil palm. Being a continuously cropping tree has its advantages.

However, oil palm is labor intensive, it is barely domesticated as a crop, and a single breeding cycle can easily be 10–12 years. Despite this, the average yearly farm yield increment achieved by oil palm since "scientific" breeding began in the 1920s is essentially the same as has been achieved in wheat in the United Kingdom. In many ways, oil palm has more potential to improve yields than wheat, with limited numbers of breeding cycles complete in oil palm and a large yield gap which could be addressed to raise production.

Breeding is clearly the major focus of genetic improvement in this crop and a mix of improved agronomy and management, coupled with breeding selection have quadrupled the oil yield of the crop since breeding began in earnest in the 1920s.

Since the 1970s, the development of clonal propagation techniques and marker-assisted selection have both offered promise for step changes in yield, although both have failed to live up to their potential in the past. Continuous problems with abnormal flowering have affected industry confidence in clonal material, seriously delaying its whole-hearted application in commercial fields. Marker-assisted selection within breeding and selection programs has been possible since the 1990s, but has failed to be exploited.

In recent years, next-generation sequencing and high-throughput "omics" technologies have been developed for oil palm and are being applied to research and, very recently, breeding and selection of commercial material at a large scale. The potential offered by these technologies should not be underestimated and they have already resurrected the potential for clones, through identifying the likely causal mutation which is responsible for abnormal flowering and, in the case of marker-assisted selection, removing the bottleneck of producing DNA marker data while also supplementing this with coupled information from every conceivable level of cellular activity.

The final challenge, as yet untackled in earnest, is that of "phenomics"—the generation of high-density phenotype data which complements the other available "omics" for a systems breeding approach to oil palm.

This book presents a series of thematic chapters with a clear focus on genetic improvement.

Chapter 2 provides an introduction to plant and crop genetics, Chapter 3 examines germplasm collections in the two oil palm species, while Chapter 4 focuses on methods for breeding qualitative and quantitative traits. Chapter 5 gives detailed examples of these target traits within oil palm breeding and Chapters 6 and 7 some of the applied breeding programs as examples. The current state of clonal propagation is presented in Chapter 8. The potential applications of "omics" and next-generation sequencing are presented in Chapter 9. The potential of hybrids between the two species is examined in Chapter 10. Chapter 11 goes on to examine how all this is translated through to commercial material. Chapter 12 examines how new material is assessed in breeding field trials.

Being a perennial tree crop represents a challenge for breeding and phenotyping for genetic improvement. Chapter 13 discusses issues which are major challenges for the future in this crop, using current research and approaches in other crops and systems for promising routes for phenotyping, future breeding, and genetic improvement in oil palm, with a summary of future perspectives on crop sustainability from one of the pioneers of oil palm physiology and breeding, Dr. Hereward Corley.

Chapters have been authored through a single contribution (often one of the editors) or contributions combined from research and industry experts active in the field of discussion. The overall aim of this book is to provide an up-to-date perspective on oil palm breeding and genetics from the people actively involved. The breeding of many crops has been or will be revolutionized by the new high-throughput approaches, but how they are applied and how much of an impact they make is determined by the same factors which make each crop, and the agriculture and uses of each crop, unique.

Aik Chin Soh, Sean Mayes, and Jeremy Roberts

Acknowledgments

The editors express their great gratitude and appreciation to all the authors and their supportive parent organizations namely; University of Nottingham, Crops For the Future, Malaysian Palm Oil Board, Indonesian Oil Palm Research Institute, Cirad, PalmElit, Applied Agricultural Resources, Sime Darby, ASD Costa Rica, United Plantations, PT Salim Ivomas Pratama, PT London Sumatra, EMBRAPA, University of Bath, University of Leicester, University of Sri Lanka, University of Malaya, for their invaluable contributions and efforts to the successful accomplishment of this comprehensive book.

Editors

Professor Aik Chin Soh is a leading international oil palm breeder and geneticist although currently in an advisory capacity. Currently he is an advisor (previously research theme leader in breeding & agronomy) in the new international crop research center, Crops For the Future.

Earlier, he was the CEO and research head supervising more than 30 research scientists at Applied Agricultural Resources Sdn. Bhd., a leading plantation R&D company. Over the years, he has developed semi-dwarf and improved harvest oil palm hybrid seed and clonal varieties which continue to make impact on the Malaysian oil palm industry. He earned his bachelor's and master's degrees from the University of Malaya and PhD from Oregon State University, all on industry and university scholarships and teaching assistantships. He has published about a hundred papers and book chapters, many in leading international crop journals and also served as their reviewer. For example, *Euphytica*, *Crop Science*, *Phytopathology*, *J. Agric. Sci.*, *SABRAO J.*, *J. Oil Palm Res.*, *Plant Breeding Reviews* (Wiley), *Handbook of Plant Breeding: Oil Crops* (Springer), *Encyclopedia of Fruits & Nuts* (CABI) and *Palm Oil* (CABI). He also served as a reviewer and an advisor to the oil palm breeding programs of the Malaysian Palm Oil Board, Cirad (France), and other plantation R&D companies.

He is a Fellow of the Academy of Sciences Malaysia, International Society of Oil Palm Breeders, and Malaysian Oil Scientists' and Technologists' Association. He is a winner of the Malaysia Toray Outstanding Scientist Award and the Accomplished Plant Breeder Award of the Genetics Society of Malaysia. As honorary professor, he also lectures and supervises MSc and PhD students at The University of Nottingham Malaysia Campus since 2007.

His current and previous research interests include

- Breeding, genetics, and biotechnology of tropical crops with experiences in oil palm, chilli pepper, maize, papaya, and cacao
- Breeding, agronomy, and crop physiology particularly with respect to resource (light, water, nutrients, gas, labor) use efficiencies in cropping (mono/mixed/intercropping) systems to adapt and mitigate against climate change effects
- Renewable energy production through biomass/biofuel crops
- Varietal and agronomic development of underutilized plant species
- High-tech agriculture
- Field experimentation and applied statistics

soh.aikchin@cffresearch.org/sohac28@gmail.com

Dr. Sean Mayes is currently an associate professor in crop genetics at the University of Nottingham. He started his research career with Unilever at the Plant Breeding International (Cambridge) working with Peter Jack and Hereward Corley on applying molecular genetics to the oil palm program, applying genetic fingerprinting to clonal propagation, and developing the first genetic map for oil palm. Over the years, he has worked on a range of tropical and temperate crops, with a focus on generating genetic markers and applying them in a research and breeding context.

His research interests are in the assessment of genetic variation and the use of genetic tools to dissect agricultural traits of breeding interest for marker-assisted selection. He has published more than 90 peer-reviewed papers and acts as a consultant to the oil palm industry.

Professor Jeremy Roberts is a deputy vice chancellor Research and Enterprise at the University of Plymouth. Prior to this appointment, in February 2017, he was at the University of Nottingham where he was director of the Biotechnology and Biological Sciences Research Council (BBSRC) AgriFood Advanced Training Partnership and of the University of Nottingham's BBSRC Doctoral Training Partnership. He has been a long-standing member of the Malaysian Palm Oil Board (MPOB) Programme Advisory Committee and for a number of years was chair of FELDA's R&D Advisory Panel.

Jerry's primary research interests are focused on understanding the molecular and cellular mechanisms responsible for regulating cell separation processes in plants. He has published more than 100 papers in international peer-reviewed journals and his work has led to the application and granting of a number of patents.

Contributors

Amancio Alvarado
Oil Palm Breeding Director
ASD Costa Rica SA
San José, Costa Rica

Jeremy Henry Alvarado
Oil Palm Breeding Program
ASD Costa Rica SA
San José, Costa Rica

Philippe Amblard
Oil Palm Breeder
Cirad/PalmElit
Montpellier, France

Libin Adelinna Anak
Malaysian Palm Oil Board (MPOB)
Kajang, Selangor, Malaysia

Sukganah Apparow
Biotechnology and Breeding Department
Sime Darby Plantation R&D Centre
Selangor, Malaysia

David Ross Appleton
Biotechnology and Breeding Department
Sime Darby Plantation R&D Centre
Selangor, Malaysia

Xaviar Arulandoo
United Plantations Research Department
Perak, Malaysia

Nuraziyan Azimi
Malaysian Palm Oil Board (MPOB)
Kajang, Selangor, Malaysia

Abu Bakar Nor Azwani
Malaysian Palm Oil Board (MPOB)
Kajang, Selangor, Malaysia

Edson Barcelos
EMBRAPA (Brazilian Agricultural Research Corporation)
Manaus, Amazonas, Brazil

Kees Breure
Bakrie Sumatra Plantations (BSP)
Kisaran, North Sumatra, Indonesia

Fook Tim Chew
Department of Biological Sciences
National University of Singapore
Singapore

Shenyang Chin
Applied Agricultural Resources Sdn. Bhd.
Petaling Jaya, Selangor, Malaysia

Chin Nee Choo
Advanced Agriecological Research Sdn. Bhd.
Petaling Jaya, Selangor, Malaysia

Benoît Cochard
PalmElit
Montferrier-sur-Lez
Montpellier, France

Richard M. Cooper
Department of Biology and Biochemistry
University of Bath
Bath, United Kingdom

Hereward Corley
Plantation Crop Physiology
Great Barford, United Kingdom

David Cros
AGAP Research Unit
Cirad
Montpellier, France

Hubert de Franqueville (deceased)
PalmElit
Montferrier-sur-Lez
Montpellier, France

Marcel de Raissac
AGAP Research Unit
Cirad
Montpellier, France

Ahmad Kushairi Din
Malaysian Palm Oil Board (MPOB)
Kajang, Selangor, Malaysia

Amiruddin Mohd Din
Malaysian Palm Oil Board (MPOB)
Kajang, Selangor, Malaysia

Tristan Durand-Gasselin
PalmElit
Montferrier-sur-Lez
Montpellier, France

Ricardo Escobar
General Manager
ASD Costa Rica SA
San José, Costa Rica

Denis Fabre
AGAP Research Unit
Montpellier, France

Ahmad Malike Fadila
Malaysian Palm Oil Board (MPOB)
Kajang, Selangor, Malaysia

John Foulkes
Plant and Crop Sciences, Biosciences
Sutton Bonington Campus
University of Nottingham
Leicestershire, United Kingdom

Kah Joo Goh
Advanced Agriecological Research Sdn. Bhd.
Petaling Jaya, Selangor, Malaysia

Ahmad Tarmizi Hashim
Malaysian Palm Oil Board (MPOB)
Kajang, Selangor, Malaysia

Pat Heslop-Harrison
Department of Genetics
University of Leicester
Leicester, United Kingdom

Zamzuri Ishak
Malaysian Palm Oil Board (MPOB)
Kajang, Selangor, Malaysia

Maizura Ithnin
Malaysian Palm Oil Board (MPOB)
Kajang, Selangor, Malaysia

Yayan Juhyana
Genetics and Statistics Division
Bah Lias Research Station (BLRS)
Sumatra Bioscience (SumBio)
London Sumatra Indonesia
Medan, North Sumatra, Indonesia

Asha S. Karunaratne
Crop Production Systems and Modelling
Sabaragamuwa University of Sri Lanka
Belihuloya, Sri Lanka

and

Crops For the Future
and
University of Nottingham Malaysia Campus
Semenyih, Selangor, Malaysia

Sau Yee Kok
Malaysian Palm Oil Board (MPOB)
Kajang, Selangor, Malaysia

Harikrishna Kulaveerasingam
Sime Darby Plantation R&D Centre
Selangor, Malaysia

Tatang Kusnadi
Oil Palm Breeding Division
PT Salim Ivomas Pratama (SIMP)
Riau, Indonesia

Qi Bin Kwong
Biotechnology and Breeding Department
Sime Darby Plantation R&D Centre
Selangor, Malaysia

Claude Louise
PalmElit
Montferrier-sur-Lez
Montpellier, France

Leslie Eng Ti Low
Malaysian Palm Oil Board (MPOB)
Kajang, Selangor, Malaysia

Maria Madon
Malaysian Palm Oil Board (MPOB)
Kajang, Selangor, Malaysia

Tasren Mahamooth
Advanced Agriecological Research Sdn. Bhd.
Petaling Jaya, Selangor, Malaysia

Marjuni Marhalil
Malaysian Palm Oil Board (MPOB)
Kajang, Selangor, Malaysia

Sean Mayes
Plant and Crop Sciences, Biosciences Sutton Bonington Campus
University of Nottingham
Leicestershire, United Kingdom

and

Crops For The Future
and
University of Nottingham Malaysia Campus
Semenyih, Selangor, Malaysia

Katharina Mebus
Centre for Research in Biotechnology for Agriculture (CEBAR)
University of Malaya
Kuala Lumpur, Malaysia

Ong-Abdullah Meilina
Malaysian Palm Oil Board (MPOB)
Kajang, Selangor, Malaysia

Mohaimi Mohamed
Biotechnology and Breeding Department
Sime Darby Plantation R&D Centre
Selangor, Malaysia

Erik Murchie
Plant and Crop Sciences, Biosciences
Sutton Bonington Campus
University of Nottingham
Leicestershire, United Kingdom

Denis J. Murphy
Genomics and Computational Biology Research
University of South Wales
Pontypridd, Wales

Suzana Mustafa
Malaysian Palm Oil Board (MPOB)
Kajang, Selangor, Malaysia

Bee Keat Neoh
Biotechnology and Breeding Department
Sime Darby Plantation R&D Centre
Selangor, Malaysia

Abdullah Norziha
Malaysian Palm Oil Board (MPOB)
Kajang, Selangor, Malaysia

Bruno Nouy
PalmElit
Montferrier-sur-Lez
Montpellier, France

Ai Ling Ong
Biotechnology and Breeding Department
Sime Darby Plantation R&D Centre
Selangor, Malaysia
and
University of Nottingham Malaysia Campus
Semenyih, Selangor, Malaysia

Chin Kooi Ong
Crops For The Future
Semenyih, Selangor, Malaysia

Siew Eng Ooi
Malaysian Palm Oil Board (MPOB)
Kajang, Selangor, Malaysia

Tony Eng Keong Ooi
Biotechnology and Breeding Department
Sime Darby Plantation R&D Centre
Selangor, Malaysia

Abrizah Othman
Malaysian Palm Oil Board (MPOB)
Kajang, Selangor, Malaysia

Raphael Perez
AMAP Research Unit
Cirad
Montpellier, France

Abdul Razak Purba
Plant Breeding and Biotechnology Research Unit
Indonesia Oil Palm Research Institute (IOPRI)
Medan, North Sumatra, Indonesia

Nookiah Rajanaidu
Malaysian Palm Oil Board (MPOB) and International Society of Oil Palm Breeders (ISOPB)
Kajang, Selangor, Malaysia

Umi Salamah Ramli
Malaysian Palm Oil Board (MPOB)
Kajang, Selangor, Malaysia

Jeremy Roberts
Office of the Vice Chancellor
University of Plymouth
Plymouth, United Kingdom

Samsul Kamal Rosli
Malaysian Palm Oil Board (MPOB)
Kajang, Selangor, Malaysia

Norashikin Sarpan
Malaysian Palm Oil Board (MPOB)
Kajang, Selangor, Malaysia

Trude Schwarzacher
Department of Molecular Cytogenetics
University of Leicester
Leicester, United Kingdom

Rajinder Singh
Malaysian Palm Oil Board (MPOB)
Kajang, Selangor, Malaysia

Aik Chin Soh
Crops For the Future
and
University of Nottingham Malaysia Campus
Semenyih, Selangor, Malaysia

Kandha Sritharan
United Plantations Research Department
Perak, Malaysia

Mohan Subramaniam
United Plantations Research Department
Perak, Malaysia

Cheng Chua Tan
Applied Agricultural Resources Sdn. Bhd.
Petaling Jaya, Selangor, Malaysia

Chee Keng Teh
Biotechnology and Breeding Department
Sime Darby Plantation R&D Centre
and
University of Nottingham Malaysia Campus
Semenyih, Selangor, Malaysia

Huey Fang Teh
Biotechnology and Breeding Department
Sime Darby Plantation R&D Centre
Selangor, Malaysia

Nicolas Philippe Daniel Turnbull
PalmElit
Montferrier-sur-Lez
Montpellier, France

Rob Verdooren
Danone Nutricia Research
Utrecht, The Netherlands

Sue Walker
Agricultural Meteorology
University of the Free State
Bloemfontein, South Africa

and

Crops For the Future
and
School of Biosciences
University of Nottingham Malaysia Campus
Semenyih, Selangor, Malaysia

Choo Kien Wong
Crop Improvement Division Applied Agricultural Resources Sdn. Bhd.
Petaling Jaya, Selangor, Malaysia

Wei Chee Wong
Biotechnology Section
Advanced Agriecological Research Sdn. Bhd.
Petaling Jaya, Selangor, Malaysia

Yick Ching Wong
Biotechnology and Breeding Department
Sime Darby Plantation R&D Centre
Selangor, Malaysia

Noorhariza Mohd Zaki
Malaysian Palm Oil Board (MPOB)
Kajang, Selangor, Malaysia

Yaakub Zulkifli
Malaysian Palm Oil Board (MPOB)
Kajang, Selangor, Malaysia

1 Introduction to the Oil Palm Crop

Aik Chin Soh, Sean Mayes, and Jeremy Roberts

CONTENTS

1.1 Historical Development ... 1
1.2 World Trade in Palm Oil ... 3
1.3 Economic Uses of Oil Palm Products ... 3
 1.3.1 Palm Oils and Products ... 3
 1.3.2 Secondary/Complementary Products ... 5
  1.3.2.1 Field Waste ... 5
  1.3.2.2 Mill Waste ... 5
1.4 Conclusions and Aims of This Book ... 5
References ... 6

1.1 HISTORICAL DEVELOPMENT

Oil palm historically is endemic to the humid tropical regions of West Africa and occurs as semiwild groves close to settlements. Traditionally, the pulp and oil from the fruit constitute a major part of the native diet, providing needed dietary fat, fiber, and vitamins (A, E). Early colonial interest in palm oil to substitute tallow in the manufacture of soap, candle, and margarine led to the development of oil palm plantations in Indonesia and Malaysia to provide a steady supply of the oil (Hartley 1988). Out of botanical interest and curiosity, four seedlings of the thick shell *dura* (D) type presumably derived from the same fruit bunch were sent to Bogor Botanic Gardens in Indonesia via the botanic gardens in Amsterdam and Reunion or Mauritius. Subsequent (out)-crossing, selection, and distribution of the progenies from these four Bogor Palms to Deli province in Sumatra (hence known as Deli D) and Malaysia led to the early establishment of commercial oil palm plantations in these two areas. Commercial D × D plantations persisted until the early 1960s, when breeding progress became limited by the thick shell *dura* fruit types. Beirnaert and Vanderweyen (1941) elucidated the monogenic inheritance of the shell gene from their work in the Congo and that crossing the homozygous *dura* (D) with the homozygous female sterile *pisifera* (P) would give 100% *tenera* (T) palms having fruits with thicker oil-bearing mesocarp due to a thinner shell (incomplete dominance). The switch to the D × P or T hybrid planting material was very rapid especially in Malaysia and Indonesia with the availability of P pollen derived from West African Ts, having been imported and bred earlier by their researchers (Hartley 1988). Rapid expansion of oil palm plantations followed, driven particularly by the Far East. By

2014, this had reached 16.5 million hectares worldwide; 10.3 million hectares in Indonesia, 5.39 million in Malaysia, and the rest mainly from Thailand, Colombia, Nigeria, Papua New Guinea, Ecuador, and Côte d'Ivoire (Review of the Malaysian Oil Palm Industry 2014). Yields of 6–9 t/ha oil have been achieved in some plantation areas as compared to less than 0.1 t/ha in the original semiwild groves. These yield gains have arisen from concomitant improvements in varieties and agromanagement practices, derived from public and private investments in breeding and agronomic research.

The earliest oil palm breeding efforts in West Africa were in the Congo by INEAC (Institut National pour l'Etude Agronomique du Congo) where Beirnaert and Vanderweyen (1941) did their seminal fruit inheritance study; in La Mé, Côte d'Ivoire by IRHO (Institut de Researches pour les Huiles Oleagineux), which later became Cirad (French Agricultural Research Center for International Development); and in Nigeria and Ghana by WAIFOR (West African Institute for Oil Palm Research), which later separated into NIFOR (Nigerian Institute For Oil Palm Research) and the Oil Palm Research Institute Ghana. In the Far East, they were the Department of Agriculture in Malaya (later also in Sabah) and AVROS (Algemeene Vereniging van Rubberplanters ter Oostkust van Sumatra) in Indonesia. With the formation of MARDI (Malaysian Agricultural Research and Development Institute), oil palm research was included as part of its remit. With the fast-growing importance of the palm oil industry, PORIM (Palm Oil Research Institute of Malaysia) was inaugurated as the principal custodian of oil palm and palm oil research, including breeding, tissue culture, and biotechnology. PORIM subsequently became MPOB (Malaysian Palm Oil Board). In Indonesia, AVROS became RISPA (Research Institute of the Sumatra Planters Association), the latter subsequently teamed up with Marihat Research Station to form IOPRI (Indonesian Oil Palm Research Institute), amalgamating both their breeding programs. Private plantation research companies played a key role in oil palm breeding, driven not only by the need to supply superior planting material to support their plantation expansion efforts but also to derive lucrative revenue from commercial seed sales. In Malaysia, the early players were Guthrie Chemara, Harrisons and Crosfield (H&C, later Golden Hope), Pamol, Dunlop, and Socfin, the first four later formed the Oil Palm Genetics Laboratory (1964–1974), which contributed significantly to oil palm breeding research methodology. Subsequent players were United Plantations (UP) and Applied Agricultural Resources or AAR, which evolved from HRU (Highlands Research Unit) and also made significant breeding research contributions. Dunlop and Pamol eventually became IOI Plantations and Guthrie and Golden Hope were amalgamated with Ebor Research Sime Darby to become Sime Darby Research. In Indonesia, Socfindo and Lonsum (London Sumatra) have been the leading private oil palm breeding companies, the former linked to the breeding programs of Cirad and the latter to H&C. The breeding program of Dami Research Station, Papua New Guinea was also derived from H&C and was recently acquired by Sime Darby Plantations. Unilever Plantations were active in breeding research in Zaire (now the Democratic Republic of the Congo) and Cameroon. ASD de Costa Rica amassed a wide range of Deli and West African breeding materials from exchanges with breeding programs in the Old World using its American oil palm (*Elaeis oleifera*) genetic materials. A number of newer breeding programs set up in Indonesia were based on these ASD-derived materials.

1.2 WORLD TRADE IN PALM OIL

Crude palm oil (CPO) and palm kernel oil (PKO) obtained from the mesocarp and kernel of the palm fruit, respectively, and their refined olein and stearin derivatives are traded at the Kuala Lumpur Commodity Exchange and, previously (before 1980), at Chicago Board of Trade. The production of 59.3 million tons of CPO and 6.6 million tons of PKO worth around US$57 billion in 2014 from the 16.5 million hectares of oil palm in the world, dominated the international edible oil trade at 61% (Review of the Malaysian Oil Palm Industry 2014, MPOB 2015). The success story of the oil palm in terms of its high productivity and profitability has attracted many developing countries in the humid tropics as an instrument for agricultural and economic development. Currently, the main markets are India, China, European Union, Pakistan, the United States, and Bangladesh, with increasing interest shown by other developing countries such as eastern European and central Asian countries. Owing to its cheaper price compared to other vegetable oils and with increasing population and per capita dietary fat intake in these countries, the demand for palm oil will rise. New oil palm plantation development is also on the rise in the traditional oil palm areas in West Africa and Latin America with investments from large plantation companies from the Far East trying to ensure palm oil supply to meet the increasing demand. Palm oil will further enhance its position as the dominant vegetable oil in the international oil and fats market, primarily due to the significant yield advantage (per year per hectare) it has over oil seed crops.

1.3 ECONOMIC USES OF OIL PALM PRODUCTS

1.3.1 Palm Oils and Products

Eighty percent of the oils from the oil palm are used as food and 20% as oleochemicals, replacing those traditionally derived from mineral oil as the latter become increasingly more expensive and concerns grow for their effects on the environment. The crude oils extracted from the sterilized mesocarp and kernel are bleached and deodorized to give refined oils used solely or in blends with other oils in cooking oil, salad oil, vanaspati or vegetable ghee, margarines, spreads, and shortenings. The refined oil is fractionated to give olein and stearin and with further fractionation gives fatty acids and alcohols, intermediate commodities traded and used in food and oleochemical industries (toiletry, cosmetic, lubricant, packaging, upholstery, etc.). Carotene extracted from CPO is processed into vitamin A supplements and a natural dye for snack foods, for example, instant noodles. Likewise, tocopherols and tocotrienols are extracted for industrial vitamin E production (Choo 2000; Jalani et al. 1997), tocotrienols being more abundant and a more desirable form of vitamin E nutritionally. Palm oil can also be made into biodiesel and has been blended with mineral diesel, for example, B5, B10, in Europe, Malaysia, and Indonesia. Figure 1.1 is a summary chart of the palm oil fractionated products and their uses.

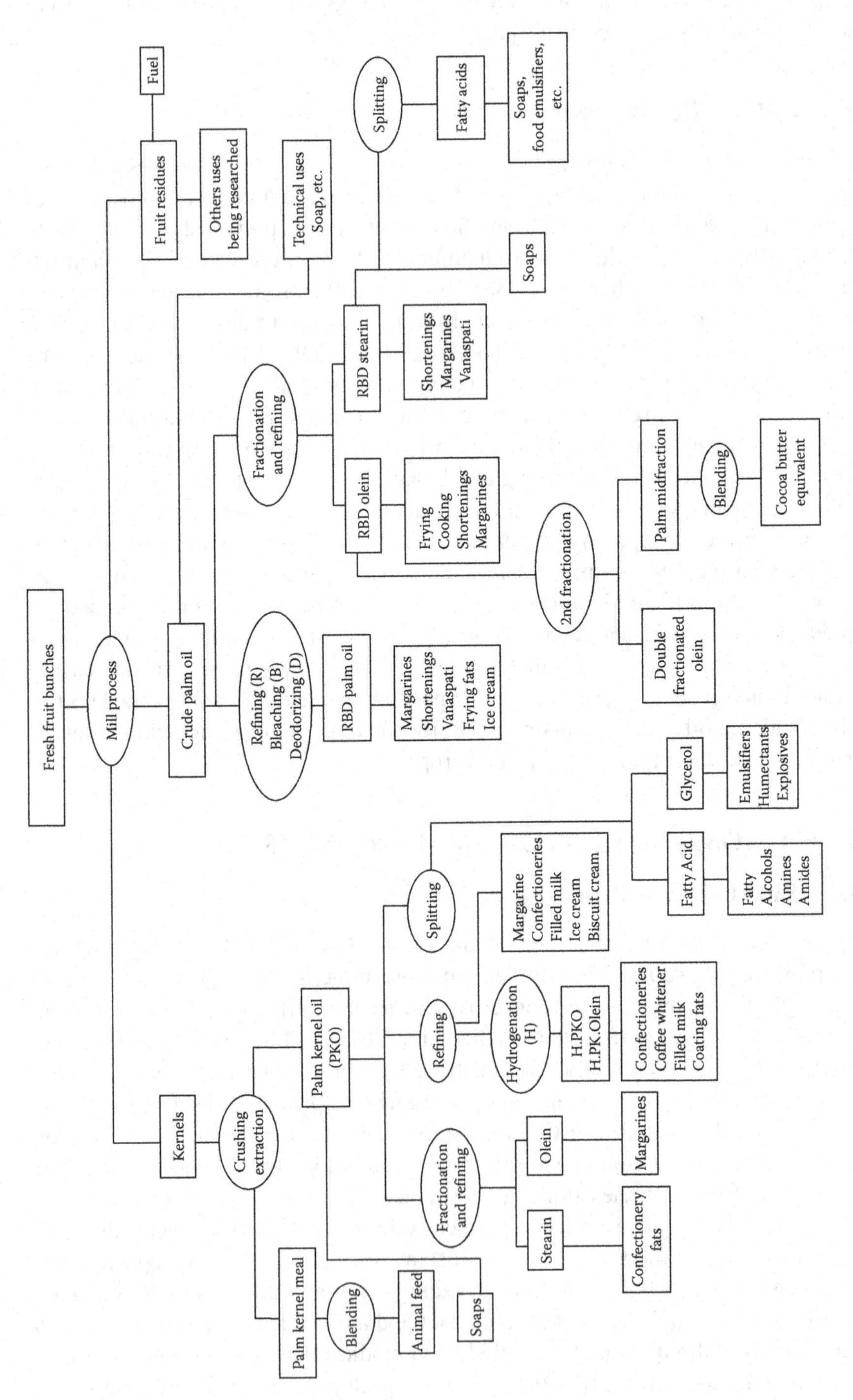

FIGURE 1.1 Uses of palm oil. (Adapted from Pantzaris, T.P. 2000. *Pocketbook of Palm Oil Uses*. Palm Oil Res. Inst. Malaysia, Kuala Lumpur.)

1.3.2 Secondary/Complementary Products

The oil palm generates a lot of waste with only 10% of the dry matter ending up as oil.

1.3.2.1 Field Waste

Palm fronds or leaves are pruned in daily cycles during harvesting and at special rounds. The fronds are usually left to rot in the field acting as a mulch and returning valuable organic matter and nutrients to the soil. However, the upper two thirds of the frond contain most of the valuable nutrients, the lower petiole part comprising mainly lignocellulose biomass material suitable for conversion into second-generation biofuels. During replanting, with the existing ban on field burning, the whole tree is chipped into small fragments and also left to rot in the field. The felled trunk contains enough nutrients to support the growth of the replanted seedling for the next 3–5 years; however, the nutrient release from the chips is very rapid and neither the root system of the young replant nor that of the legume cover is well established enough to retain the nutrients, which are consequently lost through leaching and soil wash. The trunk is also a potential source of biomass for biofuel and biomaterial production.

1.3.2.2 Mill Waste

Secondary and waste products from the palm oil industry are assuming economic importance prompted largely by health and environmental concerns. The stripped fruit bunch stalk fiber or EFB (empty fruit bunch) waste can be used directly in the plantation as an organic mulch or processed into an organic compost, dried or processed, and used as a biofuel feedstock for local use and export. Kernel cake and sludge cake, wastes from the palm and kernel oil extraction processes, respectively, have use as animal feed while the shell previously used in the mill furnace, road paving, and nursery seedling mulching has added value as a pelleted biofuel feedstock and activated carbon (for medical and industrial purposes) with the biochar (waste from biofuel production process) having good use as a soil ameliorant and also having other industrial applications. The mill effluent also contains valuable phytonutrients (antioxidants), which can be extracted prior to discharge into the effluent ponds, with the methane emitted from the latter trapped and used as biogas.

1.4 CONCLUSIONS AND AIMS OF THIS BOOK

The oil palm can be considered the "miracle crop" of modern times. From a wild plant in the "backwaters" of West Africa, within a century, it has become the most important and highest-yielding oil crop. Besides the high revenue obtainable from palm and kernel oils, it has a host of value additions from the secondary products and crop processing wastes as discussed. The demand for palm oil will continue to rise to meet the per capita intake of the increasing world population and many developing countries in the humid tropics are also venturing into oil palm plantation development, trying to emulate the successes of their Southeast Asian counterparts, despite concerns of negative impacts on biodiversity, environment, and climate change, often championed by the NGOs.

The success of the oil palm crop to date has been attributed to both breeding and agronomy with the former usually blazing the trial. Recent developments, particularly associated with sequence analysis and oil palm genomics, have the potential to continue the highly successful breeding history of this crop, helping to confirm it as a mainstay of food security as the world population grows. There are also significant challenges ahead, most notably from the predicted effects of climate change; however, we believe that tools and genetic resources exist to address these concerns. The current volume has contributions from authors who are directly involved in the attempt to ensure that oil palm plays its part in the future of agriculture, while recognizing that sustainability and protection of the environment must be balanced with the potential of the oil palm to produce more oil per unit area than any other oil crop.

REFERENCES

Beirnaert, A. and Vanderweyen, T. 1941. *Contribution a l'Etude Genetique et Biometrique des Varieties d'Elaeis guineensis Jacq.* Publ. INEAC, Serie Scientifique 27.

Choo, Y.M. 2000. Specialty products. pp. 1036–1060. In: Y. Basiron, B.S. Jalani, and K.W. Chan (eds.), *Advances in Oil Palm Research*. Malaysian Palm Oil Board, Kuala Lumpur.

Hartley, C.W.S. 1988. *The Oil Palm (Elaeis guineensis* Jacq.*)*. Longman Scientific and Technical Publication, Wiley, New York.

Jalani, B.S., Cheah, S.C., Rajanaidu, N., and Darus, A. 1997. Improvement of oil palm through breeding and biotechnology. *Journal of the American Oil Chemicals Society* 47: 1451–1455.

Pantzaris, T.P. 2000. *Pocketbook of Palm Oil Uses*. Palm Oil Res. Inst. Malaysia, Kuala Lumpur.

Review of the Malaysian Oil Palm Industry. 2014. *MPOB 2015*. Malaysian Palm Oil Board, Kuala Lumpur.

2 The Plant and Crop

Aik Chin Soh, Sean Mayes, and Jeremy Roberts

CONTENTS

2.1 Plant Morphology and Reproductive Biology 7
2.1.1 Vegetative Traits 7
2.1.2 Reproductive Traits 8
2.1.3 Crop Physiological Traits 9
2.2 Ecological and Agronomic Requirements of the Crop 10
2.2.1 Ecological Requirements 10
2.2.1.1 Rainfall 10
2.2.1.2 Relative Humidity and Vapor Pressure Deficit 11
2.2.1.3 Temperature 11
2.2.1.4 Solar Radiation 11
2.2.1.5 Soil 11
2.2.1.6 Wind 11
2.2.1.7 Flood 12
2.2.1.8 Salinity 12
2.2.2 Agronomic Requirements 13
2.2.2.1 Light 13
2.2.2.2 Fertilizer 13
2.2.2.3 Water 14
2.3 Abiotic and Biotic Stress 15
2.4 Resource Use Efficiency 15
2.5 Conclusions 15
References 15

2.1 PLANT MORPHOLOGY AND REPRODUCTIVE BIOLOGY

2.1.1 VEGETATIVE TRAITS

The oil palm is a monocot and a typical mono-stem palm with a single apical meristem. In the field, the stem emerges from the ground at about 3 years (Corley and Tinker 2016). During the pre-competition phase (3rd to 5th year) stem growth is about 30–50 cm per year. Then at the palm-to-palm competition phase palm height increases and can reach 60–100 cm per year. Palm height to a large extent determines the economic life span of the palm; attaining 10–12 m at 25–30 years old, which is really the maximum limit for economic harvesting. Slower growing and dwarf genotypes are available, for example, Dumpy E206, Yangambi 16R, Pobé Dwarf, and MPOB's Population 12 (Soh 2004). In conventional material the palm girth remains constant, that is, cylindrical at 20–75 cm. Stem constrictions represent periods of stress (moisture, nutrients). However, dwarf palms tend to have thicker trunks. Genetic variability exists for palm height and girth.

At pre-competition phase, the palm produces three fronds (leaves) per month reducing subsequently to two fronds. A standing mature palm would have 30–50 fronds. A frond of about 3–4 m length bears around 30 pinnae in two ranks on each side of its petiole. Genetic variability in frond and petiole length and size, pinna width and thickness, and their subtending angles are evident. The fronds are borne in a one-eighth spiral parastichy. Left- and right-handed palms are reported to exist even within clones (hence the trait is not likely to be genetically controlled) but have no reported consequences on their yields (Arasu 1970; Corley and Tinker 2016). By marking the first fully open new frond at two different periods, usually half-yearly or yearly, the number of fronds produced over the period can be easily counted. Genetic variability in frond production rate also exists.

The oil palm has a fibrous root system comprising primary, secondary, tertiary, and quaternary roots. The rooting system is found to be concentrated in the top 50 cm of the soil with the actively absorbing roots (quaternary) occupying the top 25 cm. Under flooded conditions, pneumatodes or breathing roots are formed on the soil surface and under severe drought conditions, the primary roots can travel many meters in distance and depth. Understandably, root growth and biomass are seldom estimated. Consequently, the palm and crop biomass productivity is usually underestimated. Root structures will also involve maintenance respiration costs, so are a legitimate, if difficult, target for breeding.

2.1.2 Reproductive Traits

The oil palm is monoecious, bearing male and female inflorescences on the same palm. Hermaphrodite inflorescences occur infrequently and usually with the first inflorescences produced in young palms. Inflorescence initiation and sex differentiation occur around 24 months prior to emergence from the frond axils. Prevailing favorable growing conditions favor female inflorescence production while stress conditions favor male inflorescences. Under severe stress, the inflorescences (even after pollination) can abort. Pisifera palms being often female sterile are used as male or pollen parents in seed production, but do produce male and female inflorescences, although the latter often abort. Under favorable conditions they will keep producing and aborting female inflorescences at the expense of male inflorescence and pollen production. In such situations in commercial hybrid seed production, induced stress treatments, for example, frond and root pruning, withdrawal of fertilizer and moisture inputs are performed to trigger male inflorescence production. The inflorescence is a compound spike with thousands of flowers borne on the spikelets. The flower is tricarpelloid. In the male flower the ovary is vestigial while in the female flower, the stamens become staminodes.

Pollen viability and stigma receptivity periods persist for around 3–4 days under natural conditions. The pollination agent is the weevil, *Elaeidobius kamerunicus*, in most oil palm growing countries. Prior to the introduction of the pollinating weevil into Malaysia and South East Asia in the early 1980s, thrips and wind were the attributed agents. They were inefficient, thus requiring laborious assisted-pollination (with pollen collected from other palms) to achieve satisfactory bunch and fruit set. The pollinated bunch requires four and a half to five and a half months to ripen. The ripe bunch for harvest is judged by the change in fruit color from dark purple to orange red and the number of loose fruits on the ground. Standard practice in many plantations for harvesting is the presence of five detached fruits on the ground below the bunch. However, it is likely that most bunches reach maximal oil content when around 100 fruits have detached. Premature (under-ripe, unripe) bunch harvesting will result in lower oil content, while delayed harvesting will result in loss of loose fruits from poor collection practices and limiting labor availability (in Malaysia, for instance) as well as poor quality oil with high FFA or free fatty acid content. The fruit is a drupe with a pulpy mesocarp containing the crude palm oil (CPO) of commerce and the kernel inside the nut providing the palm kernel oil (PKO). Usually only one seedling emerges from the germinating seed although two to four emergent seedlings from multiple kernels do occasionally occur. Variability in bunch and fruit and their component traits tend to be more heritable.

2.1.3 Crop Physiological Traits

Crop physiological traits are extensions of plant physiological traits taking into consideration the resources (light energy, space, soil, moisture, nutrients) utilized for productivity in a crop situation. Plant vegetative growth measurement, for example, frond production rate; frond dry weight; leaf area; trunk dry matter production; reproductive growth measurement, for example, bunch production, bunch dry weight; bunch analysis, for example, oil to dry bunch, kernel to dry bunch measurement procedures; have been standardized and are well documented (Blaak et al. 1963; Corley et al. 1971; Corley and Breure 1981).

From these the following crop physiological parameters can be derived:

Leaf area index (*LAI*) = leaf area/land area
Leaf area ratio (*LAR*) = leaf area/leaf weight
Vegetative dry matter production (*VDM*) = trunk dry matter production + leaf dry matter production
Bunch dry matter production (*Y*)
Total dry matter production (*TDM*) = $VDM + Y$
Bunch index (*BI*) = $Y/(Y + VDM)$
Harvest index (*HI*) = Oil yield ($O/(Y+VDM)$

The actual procedures in computing these are given in Chapter 5.

Leading from these measurements are the concepts and studies on the efficiency of the photosynthetic process relating to the fraction (*f*) of solar energy (*S*) captured (from the photosynthetic active range or PAR 400–700) by the canopy, its conversion to dry matter or conversion efficiency (*e*) and the partitioning (*HI*) of the latter to vegetative and reproductive (*Y*) growth, that is, $TDM = S.\ f.\ e.$

$Y = S.\ f.\ e.\ HI$ (Squire 1985; Squire and Corley 1987; Henson 1992; Corley and Tinker 2016).

Integral with these crop physiological parameter measurements are the following concepts:

Potential yield—the yield of a theoretical or hypothetical genotype possessing plausible optimal physiological attributes and grown under the best growing conditions.
Yield potential—yield realized by a cultivar growing in its adapted environment that is abiotic (soil, water, nutrient) and biotic (pest, disease, weed) stress free.
Site yield potential (*SYP*)—the yield of the current genotype grown under good agro-management but with site limitations (Tinker 1984; Evans and Fischer 1999; Corley 2006).
Plant ideotype and *ideotype breeding*—ideotype (form denoting an idea) breeding was the concept proposed by Donald (1968) to seek morphological and physiological traits that will enhance the efficiency of breeding to achieve high yields.

Adoption of these concepts by the plant breeders since the 1960s and 1970s strongly influenced the Green Revolution, which led to the development of superior semi-dwarf, erect leafed, high *HI* and nitrogen responsive, grain crop varieties. A similar approach has since been pursued in other crops including tree crops such as oil palm (Dickmann et al. 1994; Soh 2004). Although the emphasis earlier was on morpho-physiological traits, the approach is still very relevant and complementary with current molecular breeding approaches.

2.2 ECOLOGICAL AND AGRONOMIC REQUIREMENTS OF THE CROP

2.2.1 Ecological Requirements

Existing as natural and semi-wild stands, the oil palm is indigenous to tropical Africa stretching from latitude 16° N (Senegal in the west) to latitude 20° S (Madagascar in the east). The bulk of the wild oil palm distribution is concentrated between latitudes 10° S and 10° N, especially along the riverine areas of humid West and Central Africa, down to Angola. There is limited wild material in eastern Africa because of low moisture and high altitude restrictions, although oil palms can be found in the Cameroon Mountains at an altitude of 1300 m with adequate moisture from rainfall and even up to 1700 m (Henson 2012; Corley and Tinker 2016).

Cultivated oil palm performs best under the following environmental conditions (Lim et al. 2011).

2.2.1.1 Rainfall

A rainfall of 2000–2500 mm per year evenly distributed with no month having less than 100 mm is desirable. Severe drought stress would result in reduced young leaf expansion, older leaf snapping and desiccation, and inflorescence and bunch abortion culminating in reduced crop growth and yield.

2.2.1.2 Relative Humidity and Vapor Pressure Deficit

Relative humidity (RH) above 75% is favorable for palm growth and development, reflecting its origin in high humidity areas. Vapor pressure deficit (VPD) is the difference between the actual vapor pressure (VP) and the saturated vapor (SVP) pressure, the latter being a function of the temperature. RH is VP/SVP. The VPD directly affects stomata conductance and photosynthesis. RH or VPD varies during the day depending on air temperature and soil moisture. High VPD adversely affects gross photosynthesis and therefore dry matter production. This is more significant if soil moisture deficit is high. Thus high RH is needed at high temperature to maintain low VPD, otherwise the stomata may close, limiting carbon dioxide availability and reducing photosynthetic activity.

2.2.1.3 Temperature

A mean temperature of 24–28°C is the normal range requirement. Prolonged higher temperatures above 32°C reduce dry matter production. Prolonged lower temperatures of less than 21°C can cause delayed palm maturity, increased inflorescence abortion, and delayed fruit bunch ripening. Palm growth is inhibited below 15°C temperatures.

2.2.1.4 Solar Radiation

The oil palm requires 16–17 MJm^{-2} per day or more than 5 hours of sunlight per day and more than 1825 hours per year. A 1 MJm^{-2} drop could translate into 4.8 t^{-1} ha^{-1} per year decrease in bunch yield (Chan, 1991; Lim et al. 2011). Higher sunshine hours would increase bunch yield. Given the limited suitability range around the equator (±10°), it is unlikely that the oil palm has undergone direct selection for photoperiod effects.

2.2.1.5 Soil

Well-structured clay, sandy clay, clay loam, and silty loam soils with effective soil depth of up to 50 cm (the most actively feeding roots are found within the top 25 cm) which are well drained with good nutrient and moisture supplies would be most suitable for oil palm. Oil palm is very adaptable and can grow on soils with pH 3.5–7.5 and on lateritic, sandy to organic muck and shallow peat soils. Bunch production on such marginal areas would not be acceptable without appropriate agro-management inputs. For example, for acid sulfate and peat soils, the water table has to be maintained at less than 50 cm depth; in the former to prevent formation of a sulfate or jarosite due to oxidation and a consequent pH reduction to below pH 2 which is toxic to the palm, in the latter to prevent shrinkage and subsidence of the peat and also an increase in acidity due to oxidation of the peat.

2.2.1.6 Wind

Light winds are desirable especially during hot dry seasons as they help to cool the palm leaves by enhancing transpiration. Tropical wind storms, hurricanes, and typhoons which occur particularly in plantations in the Pacific and Caribbean islands and coastal areas can destroy the canopy and uproot the trunk and have been responsible for the loss of a number of important historical palms.

2.2.1.7 Flood

With pneumatodes, the oil palm can withstand wet soils. It can also withstand periodic flooding with running water for about a couple of weeks provided the apical meristem is not submerged (Siburat et al. 2003; Corley and Tinker 2016).

2.2.1.8 Salinity

Palm planted on coastal areas can be affected by salinity from seawater. Soil salinity will impair the palm's ability to absorb water and nutrients. Agronomic amelioration of salinity involves the construction and maintenance of a drainage system that allows the flushing out and replacement of the seawater with fresh water (Mahamooth et al. 2011). With the predicted sea level rise due to climate change, oil palm varieties tolerant of salinity would be useful for large areas of coastal plantings.

As evident judging from palms growing in the extreme boundaries of the distribution of natural and cultivated stands of palm (latitude and altitude), genetic variability for tolerance and adaptability to stress-related traits does exist and potentially could be used for genetic improvement. MPOB and oil palm breeding companies have already collected germplasm from such regions. While many researchers have studied the effects of individual climatic/ecological factors, in reality there are significant interactions and interplay among these factors and the palm's internal physiological processes. This was summarized by Henson (2003), see Figure 2.1, with this article forming the framework for crop physiological and agrometeorological modeling and climate change studies. With a better understanding of the interacting

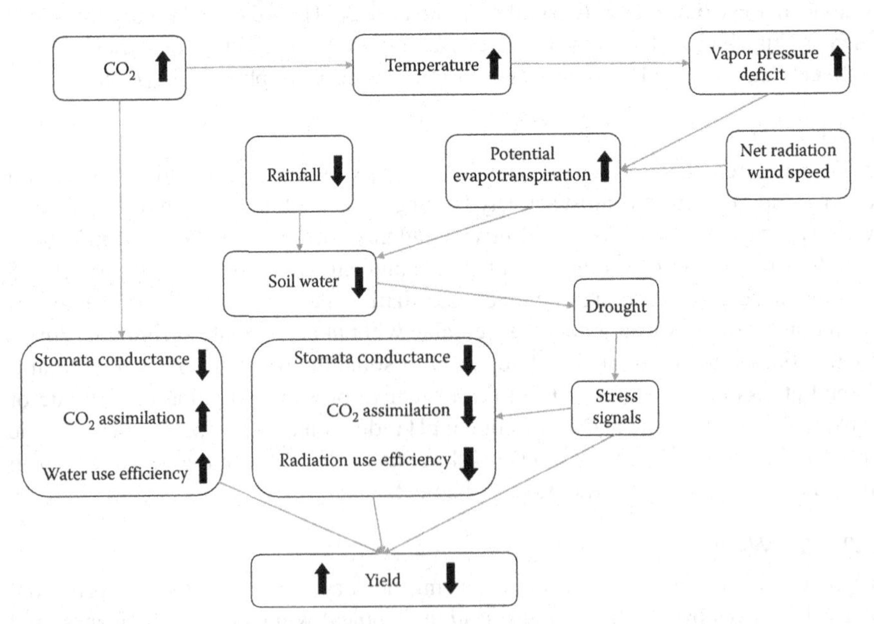

FIGURE 2.1 Probable global warming effects and their relationships with physical and physiological factors affecting oil palm yield. (Adapted from Henson, I.E. 2003. *Probable global warming effects and their relationships with physical and physiological factors affecting oil palm yield*, Malaysian Palm Oil Board. Unpublished report.)

factors and relationships, gaps and bottlenecks can be identified where breeding (conventional and molecular) could intervene.

2.2.2 Agronomic Requirements

The oil palm is one of the highest yielding C_3 oil crops (per hectare, per year) and requires and is responsive to substantial inputs of nutrients, water, as well as the high levels of solar energy required to support yield production.

2.2.2.1 Light

Photosynthetically active radiation or PAR (400–700 nm) is over half that of the total radiant energy wavelength (300–3000 nm). Although it is generally known that the oil palm requires full sunlight for maximum photosynthesis, the exact amount in terms of solar energy or sunshine hours for optimum requirements has yet to be established (von Uexkull and Fairhurst, 1991), With a defined circular canopy, triangular-spaced planting would ensure maximum canopy cover per unit land area. At the normal planting density of 138–148 palms per hectare, canopy closure occurs at between year 6–10 with an LAI of about 5–6 depending on the growth vigor. The 40–50 green leaves present per palm at maturity remain photosynthetically active throughout their 2-year life span, although the shaded lower leaves receive less than half of the light levels of the upper leaves (Squire, 1986; Henson 2009).

2.2.2.2 Fertilizer

The oil palm continues to be largely planted on highly weathered soils with poor nutrient content. Consequently, large amounts of fertilizer are applied to achieve the high yields and to cater for erosion and run-off losses especially during the juvenile phase. During replanting, old palm debris is generally chipped and returned to the soil. However, nutrient release occurs over the first 1–2 years, long before the new crop is able to capture and fix the released nutrients. The cover legume can reduce soil erosion and may capture some of the initial nutrient release from the palm debris. In the peak mature phase much of the under canopy vegetation has died out and does not contribute to fertilizer capture. Perhaps not surprisingly, fertilizer costs thus constitute more than 35% of the total production cost of the oil palm and this is likely to increase with the rising cost of petroleum. Fuel is a major requirement for the Haber-Bosch process which is a major contributor to the Haber–Bosch process of artificially fixing atmospheric nitrogen (N; applied as ammonium chloride/nitrate, urea). There are also dwindling and dwindling mined sources of potassium (K; applied as muriate of potash) and phosphate (P; applied as rock phosphate). Consequently, the major agronomic research area in oil palm has been on site-specific fertilizer use and further efficient fertilizer management (Goh et al. 1994).

The approach in drawing up fertilizer recommendations for a "manuring block" (10–100 ha) commonly used by some companies involves the following considerations and steps:

Site Yield Potential (SYP) determination. This is the Yield Potential (of a cultivar) discounted by non-easily amenable factor limitations, for example, light utilization, local climate, topography, soil, palm spacing, water availability.

Expected yield (EY) is the discounted yield (from SYP) from amenable agronomic factor limitations, for example, nutrition, canopy damage, drainage.

Actual yield (AY) is discounted yield (from EY) from management inefficiency factor effects, for example, poor harvesting practice, limited labor, etc.

The agronomist will first determine the growth, yield, and nutrition status of the block of palms as well as its historical agronomic background. Next the nutrient status of the soil will be determined. Using the SYP, the yield target is established and the nutrient requirements (based on experimental data) computed to get the existing palms to achieve the yield target using the *Nutrient Balance* approach:

Demand—nutrients needed for growth and production, immobilized in the plant and lost via run-off and leaching.

Supply—nutrients from the atmosphere, recycled plantation waste (fronds, EFB or empty fruit bunch, effluent), soil nutrient, and fertilizer nutrients.

The fertilizer (required) nutrients are obtained by subtracting the demand (without fertilizer) from the supply requirement.

The usual strategy is to maintain or build up the soil component so as not to mine it and to provide a reserve for periods of unexpected shortfall in fertilizer application due to, for example, fertilizer unavailability or bad weather complications. The agronomist will then draw up the most economical fertilizers, rates, and combinations (straight, mixed, compound) and recommend when and where best to apply them together with other best agro-management practices.

2.2.2.3 Water

Water is a very important "element" or resource for the crop, both functionally, to use in photosynthesis, but also as a critical part of the physiological processes in the palm. Water is essential for the roots to absorb nutrients and transport them through the vascular system for diffusion to various parts of the plant for essential metabolic processes, for transpiration to cool the plant, support the metabolic processes, and provide turgidity to maintain the cell, tissue, structure, and form. Transpiration is affected by solar radiation, temperature, rainfall, and RH and VPD and thus their deviations from the optimum affect the plant's growth or yield. As mentioned earlier, although drought-tolerant palms do exist, the oil palm is a crop of the humid tropics requiring high soil moisture and high humidity for maximum production. Good regular rainfall throughout the year without a distinct dry season maximizes oil palm bunch yield (>40 tonnes FFB per hectare per year) with nutrients and sunlight non-limiting. Irrigation studies in oil palm areas with a distinct dry season and even short periodic drought stress have shown positive yield responses. High irrigation infrastructure costs and the lack of available water in times of drought have discouraged expansion of irrigated oil palm areas.

2.3 ABIOTIC AND BIOTIC STRESS

Palms that are grown in agroecological environments or agro-management regimes suboptimal for their growth and productivity are said to undergo abiotic stress. Observations from commercial material or progeny trials planted in such situations often reveal differential responses suggesting genetic variability exists in resistance or tolerance to the abiotic or physical environment stress. Similarly, oil palms planted in different parts of the world experience different biotic (pests and diseases) stresses. For example, Ganoderma disease is serious in South East Asia while it is Fusarium Wilt in West Africa and Bud Rot in Latin America. Again progeny trials in these areas reveal differential susceptibility to their local disease suggesting that genetic variability for tolerance and resistance exists for breeding improvement. Although there are some important pests (e.g., rhinoceros beetle, leaf eating caterpillar, rat) causing periodic chronic outbreaks, none has warranted a resistance breeding approach as integrated pest management (IPM) or the combination of cultural, biological, and chemical method approaches appears adequate and expedient.

2.4 RESOURCE USE EFFICIENCY

Resource use efficiency or R_eUE (radiation, nutrients, and water) can be defined simply as crop output/resource input. The crop output can be in the form of the economic product (biomass, grain, protein, oil, energy, other bio-products) and the resource input can be those supplied externally or available *in situ*. The R_eUE concept arose from the application of agricultural inputs which proved to be damaging to the environment (Parry and Hawkesford 2010) thus affecting the sustainability of crop production. Genetic (progeny, clone) differences in canopy architecture, photosynthetic rates, differential palm nutrient accumulation under the same fertilizer regime, and differential responses to irrigation treatments have been demonstrated (Donough et al. 1996; Lamade and Setiyo 1996; Lamade et al. 1998; Rao et al. 2008; Lee et al. 2011) suggesting breeding for R_eUE traits is feasible.

2.5 CONCLUSIONS

This chapter has introduced the concepts and definitions of the oil palm plant and crop vegetative, reproductive, morpho-physiological, and agroecological traits of breeding interest. These will be discussed in further detail in subsequent chapters.

REFERENCES

Arasu, N.T. 1970. Foliar spiral and yield in oil palms (*Elaeis guineensis* Jacq.). *Malaysian Agriculture Journal* 47: 409–415.

Blaak, G., Sparnaaij, L.D., and Menendez, T. 1963. Breeding and inheritance in the oil palm (*Elaeis guineensis* Jacq.). Part II. Methods of bunch quality analysis. *Journal of West African Institute of Oil Palm Research* 4: 146–155.

Chan, K.W. 1991. Predicting oil palm yield potential based on solar radiation. In: Second National Seminar on Agrometeorology. Malaysian Meteorological Service, Petaling Jaya, Malaysia, pp. 129–139.
Corley, R.H.V. 2006. Potential yield of oil palm—An update. In: *Proceedings of ISOPB Seminar*, Phuket, Thailand, 2006.
Corley, R.H.V. and Breure, C.J. 1981. *Measurements in oil palm experiments.* Internal Report, Unilever Plantations, London.
Corley, R.H.V., Hardon, J.J., and Tan, G.Y. 1971. The analysis of the growth of the oil palm (*Elaeis guineensis* Jacq.) 1. Estimation of growth parameters and application in breeding. *Euphytica* 20: 307–315.
Corley, R.H.V. and Tinker, P.B. 2016. *The Oil Palm.* Wiley-Blackwell, Oxford.
Dickmann, D.I., Gold, M.A., and Flore, J.A. 1994. The ideotype concept and the genetic improvement of tree crops. *Plant Breeding Reviews* 12: 163–193.
Donald, C.M. 1968. The breeding of crop ideotypes. *Euphytica* 17: 385–403.
Donough, C.R., Corley, R.H.V., Law, I.H., and Ng, M. 1996. First results from an oil palm clone x fertilizer trial. *The Planter* 72: 69–87.
Evans, L.T. and Fischer, R.A. 1999. Yield potential: Its definition, measurement, and significance. *Crop Science* 39: 1544–1551.
Goh, K.J., Chew, P.S., and Teo, C.B. 1994. Maximising and maintaining oil palm yields on commercial scale in Malaysia. In: Chee, K.H. (ed.) *Management of Enhanced Profitability in Plantations.* Incorp. Soc. Planters, Kuala Lumpur, pp. 121–141.
Henson, I.E. 1992. Carbon assimilation, respiration and productivity of young oil palm. (*Elaeis guineensis* Jacq). *Elaeis* 4: 51–59.
Henson, I.E. 2003. *Probable global warming effects and their relationships with physical and physiological factors affecting oil palm yield*, Malaysian Palm Oil Board. Unpublished report.
Henson, I.E. 2009. Ecophysiology of oil palm biodiversity. In: Singh, G., Lim, K.H., Teo, L., and Chan, K.W. (eds.) *Sustainable Production of Palm Oil.* MPOA, Kuala Lumpur, pp. 1–52.
Lamade, E. and Setiyo, I.E. 1996. Variation of in maximum photosynthesis of oil palm in Indonesia: Comparison of three morphologically different clones. *Plantations, Recherche, Developpment* 3: 429–435.
Lamade, E., Setio, I.E., Muluck, C., and Hakim, M. 1998. Physiological studies of three contrasting clones in Lampung (Indonesia) under drought in 1997. In: Paper Presented at International Conference "Developments in Oil Palm Plantation Industry for the 21st Century", September 21–22, Bali, Indonesia.
Lee, C.T., Zaharah, A.R., Chin, C.W., Mohamed, M.H., Mohd, S.N., Tan C.C., and Wong, M.K. 2011. Variation of leaf nutrient concentrations in oil palm genotypes and their implication on oil yield. In: *Paper presented at Int. Soc, Oil Palm Breeders International Seminar on Breeding for Sustainability in Oil Palm*, November 18, 2011, Kuala Lumpur.
Lim, K.H., Goh, K.J., Kee, K.K., and Henson, I.E. 2011. Climatic requirements of oil palm. In: Goh, K.J., Chiu, S.B., and Paramanathan, S. (eds.) *Agronomic Principles & Practices of Oil Palm Cultivation.* Agricultural Crop Trust, Kuala Lumpur, pp. 1–37.
Mahamooth, T.N., Gan H.H., Kee K.K., and Goh K.J. 2011. Water requirements and cycling of oil palm. In: Goh, K.J., Chiu, S.B., and Paramanathan, S. (eds.) *Agronomic Principles & Practices of Oil Palm Cultivation.* Agricultural Crop Trust, Kuala Lumpur, pp. 89–131.
Parry, A.J. and Hawkesford, M.J. 2010. Food security: Increasing yield and improving resource use efficiency. In: *Proceedings of the Nutrition Society* 69: 592–600.
Rao, V, Palat T., Chayawai, N., and Corley R.H.V. 2008. The Univanich oil palm breeding programme and progeny trial results from Thailand. *The Planter* 84: 519–531.

Siburat, S., Dusimin, H., and Sim S.T. 2003. Flood mitigation measures for oil palm planting in Segama floodplain—A PBB oil palm's experience. *The Planter* 76: 15–28.

Soh, A.C. 2004. Selecting the ideal oil palm: What you see is not necessarily what you get!. *Journal of Oil Palm Research* 16: 121.

Squire, G.R. 1985. A physiological analysis of oil palm trials. *PORIM Bulletin* 12: 12–31.

Squire, G.R. A physiological analysis for oil palm trials, Palm Oil Research Institute Malaysia Bulletin, pp. 12–31.

Squire, G.R. and Corley, R.H.V. 1987. Oil palm. In: Sethuraj, M.R. and Raghavendra, A.S. (eds.) *Tree Crop Physiology*. Elsevier, Amsterdam, pp. 141–167.

Tinker, P.B. 1984. Site-specific yield potentials in relation to fertilizer use. In: von Peter, A. (ed.) *Nutrient Balances and Fertilizer needs in Temperate Agriculture*. International Potash Institute, Berne, pp. 193–208.

Von Uexkull, H.R. and Fairhurst, T.H. 1991. The oil palm: Fertilising for high yield and quality. *IPI Bulletin* 12: 12–15.

3 Genetic Resources

Aik Chin Soh, Sean Mayes, Jeremy Roberts, Nookiah Rajanaidu, Amiruddin Mohd Din, Marjuni Marhalil, Abdullah Norziha, Ong-Abdullah Meilina, Ahmad Malike Fadila, Abu Bakar Nor Azwani, Libin Adelinna Anak, Yaakub Zulkifli, Suzana Mustafa, Maizura Binti Ithnin, Ahmad Kushairi Din, Edson Barcelos, and Philippe Amblard

CONTENTS

3.1 Introductory Overview (*Sean Mayes, Aik Chin Soh, and Jeremy Roberts*)....20
3.1.1 General Principles: Origins and Diversity....20
3.1.2 Prospection and Conservation....22
3.1.3 Genetic Base Broadening (Other Exemplar Crops Compared to Oil Palm)....23
3.1.4 Genetic Base of Current Breeding Programs....24
3.1.4.1 *Deli*....25
3.1.4.2 AVROS....25
3.1.4.3 Yangambi....26
3.1.4.4 La Me....26
3.1.4.5 Binga....26
3.1.4.6 Ekona....26
3.1.4.7 Calabar....27
3.1.4.8 Derived and Recombinant BPROs....27
3.1.5 Future Needs and Traits....27
3.2 Prospection, Conservation and the Broadening of the Genetic Base in Oil Palm (*Nookiah Rajanaidu, Amiruddin Mohd Din, Marjuni Marhalil, Abdullah Norziha, Ong-Abdullah Meilina, Ahmad Malike Fadila, Abu Bakar Nor Azwani, Libin Adelinna Anak, Yaakub Zulkifli, Suzana Mustafa, Maizura Binti Ithnin, and Ahmad Kushairi Din*)....27
3.2.1 Background: Exploiting Interesting Traits from Germplasm....27
3.2.2 Germplasm Collection and Sampling....32
3.2.3 MPOB's Genetic Resources Program....32
3.2.4 Centers of Distribution of Natural Oil Palm Populations....33
3.2.5 Collections....33
3.2.5.1 *Elaeis guineensis*....33
3.2.5.2 *Elaeis oleifera*....36
3.2.6 Introduction of Exotic Germplasm and Quarantine Procedures....37

3.2.6.1 Phytosanitary Measures in the Country of Origin 37
3.2.6.2 Intermediate Quarantine Station.. 37
3.2.6.3 Phytosanitary Measures in Malaysia.................................. 37
3.2.7 Evaluation 38
3.2.7.1 Evaluation of Oil Palm Genetic Material 38
3.2.7.2 Methods Used to Evaluate Oil Palm Genetic Material........ 38
3.2.8 Utilization 39
3.2.8.1 Germplasm Utilization.. 39
3.2.8.2 Introgression Programs... 39
3.2.9 Conservation 40
3.2.9.1 *Ex Situ* Conservation.. 40
3.2.9.2 *In Situ* Conservation.. 41
3.2.9.3 Development of Core Collection in Oil Palm 42
3.2.10 Germplasm Management, Characterization, and Conservation Using Molecular-Based Assays 42
3.3 *Elaeis oleifera* Genetic Resources (*Edson Barcelos and Philippe Amblard*) 45
3.3.1 Background... 45
3.3.2 Prospections and Collections.. 46
3.3.2.1 Central America and Northern Region of S. America 46
3.3.2.2 South America 46
3.3.2.3 Complementary Collections (Peru, Ecuador, French Guyana)... 47
3.3.3 Characteristics and Genetic Diversity of Each Origin 47
3.3.3.1 Phenotypic Diversity.. 47
3.3.3.2 Genetic Diversity 48
3.4 Conclusion (*Sean Mayes, Aik Chin Soh, and Jeremy Roberts*) 50
References... 51

3.1 INTRODUCTORY OVERVIEW

Sean Mayes, Aik Chin Soh, and Jeremy Roberts

3.1.1 General Principles: Origins and Diversity

Harlan (1992) reports that an estimated 7000 plant species may have been used throughout history for food, feed, fuel, and industrial purposes. How (and where) a crop becomes adopted by people is dependent on many factors. A recent analysis by the Global Crops Diversity Trust (https://www.croptrust.org/blog/how-much-do-countries-benefit-from-one-anothers-crop-diversity/) summarized the wide geographical spread of current crop use, with extensive use of some crops beyond their initial areas of use and domestication.

The so-called "major crops" are those which often have extremely wide cultivation, with maize, rice, and wheat accounting for around 60% of all calories consumed by people on the planet. For oil crops, there are a number in use, both perennial (e.g., oil palm and coconut) and annual (e.g., soybean, oil seed rape/canola, etc.).

The process of crop adoption and development has often been a long story, with natural variation or spontaneous mutation having been selected by ancient farmers in the field to develop the crops we see today, in a process of domestication. Adaptation traits will also have been selected for when a crop has been introduced into a new environment, such as is the case with the spread of wheat cultivation from the Fertile Crescent to the United Kingdom, a process which took several thousand years with coupled selection for major gene effects on vernalization and photoperiod requirement. Indeed, it is likely that for crops which have been grown for thousands of years in the same location, some further selection for adaptation has occurred as the climate has changed, even if the geographical location has not. Oil palm is a good example of a crop derived from one region (probably the region around Angola in Africa) but which has had most impact in another region, South East Asia.

For peanut (*Arachis hypogaea* L.) and bread wheat (*Triticum aestivum* L.) the current cultivated form is a polyploid (auto- and allo-, respectively) which represents a genetic block to continued natural gene exchange with the lower ploidy ancestor species. Such a problem does not exist (at least overtly) in oil palm, with both main species being primarily diploid, although with an ancient tetraploid event in evolutionary history (Singh et al., 2013b). A better understanding and more data on recombinational patterns between genetically diverse breeding populations of restricted origins (BPROs; Rosenquist, 1986) is needed to confirm that the reported regions of segmental genome duplication observed in the African oil palm genome do not have a major effect on fertility or viability.

Geographically distinct/isolated populations may show differences in allele frequencies through stochastic processes, such as new mutation and genetic drift in small populations, as well as selection. If the number of individuals which form the basis of the population is small, then there is also the danger of a founder effect, which can restrict the available genetic variation for future breeding based on crossing lines from within that location or region. These processes, along with selection both by environment and farmers are some of the factors which drive the genetic differentiation of geographically isolated populations. Perhaps one of the more remarkable cases of isolation by geographical distance is the situation that exists between African and American oil palm, which is believed to have been separated some 60 million years ago with the continental plate separation of what is now Africa from South America. Crosses between the two species are viable (although do have some problems with fertility, probably associated with developmental processes, but repeated backcrossing can restore fertility) and such crosses have already made a major contribution to the oil palm industry, particularly where disease resistance is concerned (e.g., for bud-rot resistance). In the future, they also have the potential to introduce beneficial traits, such as less saturated oil, into African germplasm (see Chapter 7).

For conventional breeding, the first point of call is to evaluate the available germplasm for traits of potential economic value with a view to introducing those traits into current elite material. Molecular markers can be used to estimate the overall genetic distance of individuals to each other, and have often been used to develop core collections, where a subset of the population are chosen to represent the majority of the allelic variation present in the complete population (this chapter). However, it is really trait variation which is of most interest. For crops where multiple-gene heterosis is an important component of yield, optimizing genetic distance between the hybrid parents can be an important consideration. In maize, contrasting inbreds can produce extensive hybrid vigor, but there is an optimal genetic distance for exploiting this, as well as an optimal combination of traits (e.g., Stuber, 1995). For oil palm, it seems very likely that multigenic heterosis is captured in the crossing of *dura* × *pisifera* populations, although different breeding companies and research institutes use a range of different origins. The single gene heterosis of the SHELL gene itself has had the biggest effect on breeding improvement in the last century, with a 30% increase in oil yield per hectare per year, when comparing the thin-shelled *tenera* with the thick-shelled *dura* (Singh et al., 2013a).

One significant problem with the introduction of new traits into elite material is the loss of breeding progress which can occur in the hybrid and the generations immediately afterward. This can also lead to extensive segregation in the offspring (where the same final ideotype has been achieved through selection for alleles of different genes in a pathway for a trait) or simply it can produce low yielding palms. Repeated backcrossing (or intercrossing to other elite palms) is slow in oil palm, with even marker-assisted selection to facilitate the process of transferring the trait of interest into the elite background taking between 3 and 4 years per generation.

An awareness of (and access to) a diverse germplasm collection for a species is probably the most important resource for breeding programs and for oil palm access to germplasm is the foundation of both commercial and research institute breeding programs.

3.1.2 Prospection and Conservation

Prospection and conservation efforts have been made for many crops of importance and most of the UN Consultative Group on International Agricultural Research (CGIAR) centers have extensive germplasm collections, *ex situ* (e.g., CIMMYT in Mexico has extensive collections of wheat and maize germplasm worldwide; www.cimmyt.org/). Oil palm is no exception and the very extensive collections made by MPOB in Africa and South America are perhaps the best known (Section 3.2). There have also been breeding germplasm exchanges, such as the industry-based combined breeding programs.

However, prospections can be physically difficult, expensive, and if there is an intention for one country to utilize the germplasm collected from another, there can be political issues to be overcome. Quite often a bilateral agreement to share the germplasm collected or a reciprocal exchange can be agreed. The Convention on Biological Diversity (CBD) covers such processes for crops which do not fall within the International Treaty for Genetic Resources for Food and Agriculture Annex 1 list of crops—oil palm does not, although coconut is included. The IPTGRFA establishes a multilateral system of exchange and standard protocols for germplasm exchange and benefit sharing. Oil palm, not being on the Annex 1 list, is automatically covered by the Nagoya Protocol of the CBD, which has recently come into force. While the intentions of the Protocol are well meaning, it could make such prospections more difficult in future.

For countries with important existing industries based on the crop species, such as oil palm in Malaysia and Indonesia, phytosanitary considerations must be paramount. The accidental introduction of crop diseases into the country with the collection could have very significant effects on the existing industry and phytosanitary screening in a third country where there is no industry (or where the crop will only grow under controlled environment conditions) is preferable, but not always practical. For oil palm, genetic fingerprinting evidence suggests that *Fusarium oxysporum* fsp has been introduced on imported seed from Africa to South America, although subsequently contained and now being eliminated through the use of resistant germplasm. Interestingly, Fusarium Wilt is not a major problem in South East Asia, despite repeated official (and almost certainly, unofficial) introductions of oil palm material from endemic regions into South East Asia. This either argues for good phytosanitary screening, luck, or perhaps the presence of a preexisting nonpathogenic species in the ecological niche the pathogen usually occupies in Malaysia.

Ex situ (gene bank) collections are extensive for many species, but *in situ* collections or even farmer-based collections (working with farmers to preserve their "active" germplasm) are being increasingly considered for many crops. For tree species, even relatively small growing collections can take extensive land, so maintenance of germplasm in tissue culture systems and through cryopreservation are actively pursued. In practice, most plantations will maintain material from different origins within their active breeding programs, with crossing programs to evaluate traits of potential breeding interest and to begin the trait introgression process.

3.1.3 Genetic Base Broadening (Other Exemplar Crops Compared to Oil Palm)

The records suggest that the *Deli dura* BPRO is derived from four palms originally planted in the Bogor botanical garden in Indonesia in 1848. This is the most important composite BPRO on the *dura* side. On the *pisifera* side, the most widely used origin is the AVROS (Algemeene Vereniging van Rubberplanters ter Oostkust van Sumatra) and its derivatives, which are, according to pedigree, 75% derived from a single introduction to Indonesia of offspring from the "Djongo" palm from the Congo (now the Democratic Republic of Congo). In theory, this makes the two most important breeding BPROs in world oil palm extremely narrow genetically, despite considerable breeding progress having been made historically using these populations, their derivatives, and commercial hybrids. In practice, there have been other introductions into these lineages (both deliberate and accidental, with oil palm naturally being open-pollinated), but concerns with breeding progress decreasing in recent years within these elite pools has led to an increasing desire to introduce new sources of genetic and trait variation into breeding programs. The loss of breeding progress, when introducing unimproved germplasm with beneficial traits into elite lineages is a major concern in all crop species, even annual species.

Recent work in wheat is a good example of different levels at which variation can be introduced and where there is also concern with potential limitations imposed by a narrowing genetic base which may be limiting breeding progress. Most breeding programs in wheat work on the basis of "crossing the best with the best" and selecting for incremental changes in a range of quantitative characters. Making wide crosses can introduce potentially useful new traits (and a major driver in wheat has always been introducing disease-resistance genes), but can require a number of generations of backcrossing to recover the elite recurrent parent background. More ambitious approaches recently have revisited the ancestral genomes and a number of synthetics have been created based on re-crossing the tetraploids wheats (AB genomes) with the diploid D genome grasses, with marker analysis suggesting current bread wheat only includes genes from a limited section of the *Triticum tauchii* (D) germplasm. More ambitious still is the development of germplasm introgression lines containing chromosomes from alien relatives of bread wheat, which could bring desirable traits, such as salt and drought tolerance, aiming to produce structured combinations of genomes which are novel and have not existed before. Similar attempts to widen the genetic base in the other two major cereals, rice and maize are also being made. However, the individual domestication and adaption history and the modern breeding systems of these crops differ, with consequences for the genetic variation present and the levels of variation accessible to modern breeders. Both maize and rice are also essentially diploids (although maize is genetically an ancient tetraploid and rice shows duplication of some regions of the chromosomes (e.g., on rice 11 and 12), with maize derived from probably a single lineage from Teosinte. Rice, however, has a number of sister species with indica and japonica being the better known cross-fertile subspecies, but other relatives, such as *Oryza glaberrima* (African, non-paddy, rice) which hold important traits for stress resistance. Crosses with the two main groups have developed so-called Nerica rice types, which hold great potential as sources of genes for both African and Asian rice improvement (http://www.africarice.org/).

In oil palm, there are many parallels, although the timescales involved are potentially significantly longer and the fact that oil palm is a naturally out-crossing species, adds further complications. The reciprocal recurrent selection/family and individual (palm) selection (RRS/FI(P)S) approaches adopted in most programs often involve the improvement of *dura* and *pisifera* sides separately, followed by testing for the best combinations in the hybrid. In essence, this is "crossing the best with the best," but with added complications. Many potential traits of interest exist in non-elite germplasm (and occasionally in elite, but unrelated germplasm), such as long stalk, *virescens*, low lipase activity, etc., which could be introgressed into elite material, bearing in mind the genetic control of the trait and the need for the final product to be a *tenera* hybrid. The combination of such traits to produce a new ideotype for oil palm will take a number of generations, although the genetics of the trait and the penetrance and expressivity of the trait can be determined in earlier generations. At the widest level, crosses and derived material from *E. guineensis* × *E. oleifera* have the potential for introduction of substantially different traits (very low height increment, more liquid oil, potential disease resistances, etc.) and some of these uses have already been realized. Whether crosses or introduction of material from non-*Elaeis* species is possible is unknown, but certainly individual genes might be introduced through genetic modification in the future and potentially rapid changes made through Genome Editing approaches to existing germplasm, which could remove the needs for extensive (and slow) introgression programs, for some traits.

3.1.4 Genetic Base of Current Breeding Programs

The breeding populations which form the genetic base and diversity of most, if not all, of the current oil palm breeding programs have been derived from a limited number of progenitor palms prospected by early oil palm breeders at the various centers of origin and diversity in W. Africa, for example, Nigeria (Calabar, Ufuma), Ivory Coast (La Me), and Zaire (Sibiti, Yangambi), especially after World War 2 (Rosenquist, 1986; Corley and Tinker, 2015) or through chance planting in ornamental avenues in South East Asia. Rosenquist (1986) coined the term "breeding populations of restricted origins" or BPROs to describe these genetic materials. The application of different selection criteria on different plantations has led to the development of additional BPROs from the same original germplasm, with *Deli dura* (D) being a good example. In genetic terms, the breeding approach adopted will also have a strong influence on the genetic structure of breeding programs. Some of these BPROs have since remained intact and been improved. Many of them have been crossed out in recombinant crossing programs, subsequent to the exchange of genetic material among the various oil palm breeding centers. Brief descriptions of some of the major BPROs are given below.

3.1.4.1 *Deli*

This thick-shelled *Dura* BPRO is reportedly derived from the original four palms planted in the Bogor Botanical Garden in Java in 1848. Progenies of these four palms were planted in Deli province in Sumatra (hence are known as "*Deli dura*") and thence to other countries followed by local selection, which led to the development of the Elmina (E), Serdang Avenue (S), and Ulu Remis (UR) Deli D subpopulations in Malaysia and the Dabou (Dab) and Socfin (Soc) D subpopulations in Ivory Coast, named after the places where they were further bred. The rather uniform high-yielding Deli population led to the speculation of a common progenitor for the four Bogor palms, although significant genetic variation has been observed within this origin, consistent with a wider genetic base or the intention (or accidental) introduction of additional genetic variation into the Deli D lineages. The UR Deli Ds which resulted from the Guthrie Chemara/OPGL (Oil Palm Genetics Laboratory) breeding programs selected for high oil yield and high bunch number (although there were lines selected for large bunches as well) are the most widely distributed. They appeared to combine well with the AVROS (AVR) pisiferas (Ps) which confer moderate bunch number and size. Dab Deli has presumably been selected for fewer but larger bunches which combine well with the high bunch number and small bunch size characteristics conferred by the La Me (LM) Ps. Deli Ds provide the mother palms for almost all major oil palm commercial hybrid seed production programs worldwide. The Dumpy (Dy) (E206) and Gunung Malayu (GM) palms are dwarf variants of the Deli, the former was discovered in a commercial field in Elmina Estate, Selangor, Malaysia by Jagoe (1952), a research officer from the Department of Agriculture, and the latter in Gunung Melayu Estate in Sumatra. The dwarfing traits of these two progenitors have been exploited in the breeding programs of AAR (Dy) in Malaysia and IOPRI (Dy) and SUMBIO (GM) in Indonesia. The last of the four Bogor palms died about 10 years ago, but were unfortunately not sampled for DNA.

3.1.4.2 AVROS

Researchers from Algemeene Vereniging van Rubberplanters ter Oostkust van Sumatra (AVROS) collected seeds from the Djongo (best) palm at Eala Botanical Garden in Zaire and planted them at Sungai Pancur (SP), Sumatra, in 1923. These gave rise to the well-known SP540 *tenera* (T) palm. Crossing with the T at Bangun Bandar Experimental Station and subsequent backcrossing to the SP540 selfed palms resulted in the BM119 or AVR population. AVR Ps are noted for their vigorous growth, precocious bearing, thin shell, thick mesocarp, and high oil yield conferring attributes. Major commercial hybrid seed production programs in W. Africa (UNIPALMOL), Colombia (UNIPALMA), Indonesia (SUMBIO, SMART), Malaysia (Sime Darby, IOI, Kulim), and Papua New Guinea (Dami), which were associated with the original Unilever/Harrisons and Crosfield breeding programs, are based on Deli D × AVR P lineages. RISPA (now part of IOPRI) has maintained the pure SP540T derived breeding population and still uses them in its breeding and seed production programs. A top selling variety of RISPA is based on the Dy D × RS P (SP540T derived). The original SP540T also died recently although there was an effort made to clone it.

3.1.4.3 Yangambi

INEAC or Institut National pour l'Etude Agronomique du Congo started the breeding program at Yangambi, Zaire, with open-pollinated seeds from the Djongo T palm and from Ts in Yawenda, N'gazi, and Isangi. Subsequent breeding led to the development of the Yangambi (Ybi) population characterized by its excessive vigor, bigger fruit, and high oil yield conferring attributes. The Ybi population also features as male parents in many breeding and seed production programs and its lineage is represented in a number of breeding populations worldwide, for example, Malaysia (AAR, United Plantations, FELDA), Indonesia (SOCFINDO, IOPRI), W. Africa (La Me). Ybi 16R is a dwarf variant and its lineage is also featured in a number of breeding populations.

3.1.4.4 La Me

The Institut de Recherches pour les Huiles et Oleagineux's (IRHO, now Cirad) original breeding program was based on the La Me (LM) population derived from seeds collected in the wild groves of Ivory Coast from 21T palms, notably the Bret 10 palm. Pisiferas and Ts derived from the L2T palm are used in breeding and seed production programs in W. Africa (Ivory Coast, Benin, Cameroon) Indonesia, Thailand, and Latin America. LM Ts and their progenies are characteristically smaller palms with smaller leaves and pinnae borne on thinner trunks, producing many smaller bunches and fruits. They appear to be more tolerant of sub-optimal growing conditions.

3.1.4.5 Binga

This subpopulation is derived essentially from the F_2 and F_3 Ybi progenies planted in Binga Plantation, Yangambi, in Zaire (now the Democratic Republic of Congo) belonging to the Unilever Plantations Group. Palms Ybi 69 MAB and Bg 312/3 are the parent palms of breeding interest featured in the breeding programs of Unilever Plantations Group and their ex OPGL/CBP (combined breeding program) partners.

3.1.4.6 Ekona

The Ekona (EK) BPRO originated from the wild palms in Ekona, Cameroon and was bred further in the Unilever plantations of Cowan Estate, Ndian Estate, and Lobe Estate. Progenies from palm Cam 2/2311, noted for its high bunch yield, good oil content, and wilt resistance, have been distributed to Costa Rica and Malaysia.

Besides the ex OPGL/CBP partners, EK derived breeding materials are also featured prominently in the newer breeding programs in Indonesia based on the ASD Costa Rica supplied breeding materials.

3.1.4.7 Calabar

The breeding populations of the Nigerian Institute for Oil Palm Research (NIFOR) are much broader based with accessions from Aba, Calabar, Ufuma, and Umuabi. Progenies from the Calabar selections were of most interest and have been distributed to Costa Rica, Ghana, Ivory Coast, and Malaysia. Palm NF 32.3005 was featured in most of the distributions. In Ghana this progenitor is known as GHA851 and this breeding material also found its way into Indonesia via ASD Costa Rica.

3.1.4.8 Derived and Recombinant BPROs

A number of breeding programs have made significant progress in the recombinant phase of intercrossing or introgressing parent palms from the various traditional BPROs to form new BPROs with mixed lineages or recombinant BPROs, for example, URT (Ulu Remis Ts) by Guthrie Chemara/Sime Darby; Dumpy.AVROS (Dy.AVR), Dumpy.Yangambi.AVROS (Dy.Ybi.AVR), La Me × Yangambi (LM.Ybi), Dumpy × AVROS × LM (Dy.AVR.LM) by AAR (Soh et al., 2006); Deli × EK/Bg/etc (SUMBIO).

Many of these materials have been introduced into the commercial seed market.

3.1.5 Future Needs and Traits

Oil palm breeding goals to date have focused on maximum yield improvement through selection on bunch and oil yield components, improved crop yield traits based on physiological analysis, and improved oil quality and resistance to diseases (Ganoderma Basal Stem Rot, Fusarium Wilt). In the future as a consequence of climate change and the rising cost of inputs (such as fertilizer, energy, and labor), a new suite of traits, for example, resource use efficiency and abiotic and biotic stress adaptation traits will be sought. Some of these may have been lost due to a breeding focus that used the best land for trait assessment and has targeted yield alone. Such traits may still exist within the undeveloped BPROs or wild and semi-wild accessions. This topic will be discussed further in subsequent chapters.

3.2 PROSPECTION, CONSERVATION AND THE BROADENING OF THE GENETIC BASE IN OIL PALM

Nookiah Rajanaidu, Amiruddin Mohd Din, Marjuni Marhalil,
Abdullah Norziha, Ong-Abdullah Meilina, Ahmad Malike Fadila,
Abu Bakar Nor Azwani, Libin Adelinna Anak, Yaakub Zulkifli,
Suzana Mustafa, Maizura Binti Ithnin, and Ahmad Kushairi Din

3.2.1 Background: Exploiting Interesting Traits from Germplasm

Considerable progress in yield enhancement through breeding was achieved despite the extremely narrow genetic base of oil palm (*Elaeis guineensis*) breeding

TABLE 3.1
MPOB Oil Palm Germplasm Collections from Center of Origins

Variety and Country	Year	No. of Accessions
Elaeis guineensis		
Nigeria	1973	919
Cameroon	1984	95
Zaire	1984	369
Tanzania	1986	60
Madagascar	1986	17
Angola	1991/2010	54/127
Senegal	1993	104
Gambia	1994	45
Sierra Leone	1996	56
Guinea	1994	61
Ghana	1996	58
Elaeis oleifera		
Honduras	1982	14
Nicaragua	1982	18
Costa Rica	1982	61
Panama	1982	27
Colombia	1982	41
Suriname	1982	6
Ecuador	2004, 2006	5

Source: Adapted from Rajanaidu, N. et al. 2013. Breeding for oil palm for strategic requirement of the industry. *MPOB International Palm Oil Congress (PIPOC 2013)*, November 19–21, 2013, Kuala Lumpur.

populations. The necessity to broaden the genetic base for continuous yield improvement was generally recognized by the industry in Malaysia. It was for this reason that several expeditions were mounted to the centers of diversity in West Africa and South and Central America to collect oil palm germplasm (Table 3.1). The germplasm collections were carefully evaluated, and as expected they contained potentially interesting traits, such as; long stalk for ease of harvesting, low lipase to help maintain oil quality, high oleic acid to help improve levels of unsaturation of the oil, to name a few. The availability of the germplasm with these traits allowed MPOB and the industry to identify and prioritize 10 traits that could be incorporated for further improvement in the advanced breeding lines (Table 3.2). The selected germplasm has been disseminated to the industry to exploit the palms containing these traits either through conventional breeding or tissue culture (Table 3.3). The dissemination of the materials to various members of the industry allows for effective exploitation of this important resource. However, it cannot be denied that in order to hasten the introgression of the desirable traits from germplasm into advanced breeding lines, modern genomic tools will have to be applied in conventional breeding programs (Table 3.4). This will help Malaysia reach the objectives of the National Key Economic Area

TABLE 3.2
Ten Priority Traits in Oil Palm

No.	Trait	Current	Benchmark
1	High oil yield	3.70 t/ha/year	9.00 t/ha/year
2	Ganoderma tolerance	70%	90%
3	High bunch index	0.40	0.60
4	Low height/compactness	45–75 cm/year	30 cm/year
5	Long stalk	10–15 cm	25 cm
6	Low lipase	22%–73% FFA level	Half of the current level of FFA
7	High oleic acid	22%–40%	65%
8	Large kernel	5%	20%
9	Vitamin E	600 ppm	1000–1500 ppm
10	High carotene content	500 ppm (*E. guineensis*); 1500 ppm (*E. oleifera*)	2000 ppm (*E. guineensis*); 3000 ppm (*E. oleifera*)

Source: Adapted from Mohd Din, A., Kushairi A., Rajanaidu, N., Noh, A. and Isa, Z.A. 2005. Performance of various introgressed oil palm populations at MPOB. In: *Proceedings of the International Palm Oil Congress—Technical Break Throughs and Commercialisation—The Way Forward.* pp. 111–143, September 25–29, 2005, Malaysian Palm Oil Board, Kuala Lumpur.

TABLE 3.3
Commercial Oil Palm Planting Materials and Breeding Populations Released to the Industry by MPOB

No.	PORIM Series No.	Special Trait	Year of Release	Intended Use of Material
1	PS1	Dwarf	1992	Planting material
2	PS2	High iodine value	1992	Planting material
3	PS3	Large kernel	1996	Breeding population
4	PS4	High carotene (*E. oleifera*)	2002	Breeding population
5	PS5	Thin-shell *tenera*	2003	Breeding population
6	PS6	Large-fruit *dura*	2003	Breeding population
7	PS7	High bunch index	2004	Breeding population
8	PS8	High vitamin E	2004	Breeding population
9	PS9	*Bactris gasipaes*	2004	Economic palm (non-oil palm)
10	PS10	Long stalk	2006	Breeding population
11	PS11	High carotene (*E. guineensis*)	2006	Breeding population
12	PS12	High oleic	2006	Breeding population
13	PS13	Low lipase	2008	Breeding population

(NKEA), which is to significantly enhance national oil palm productivity by 2020. More details of the prospection will be given in the following sections.

Rajanaidu and Jalani (1994c) and Zohary (1970) have documented the centers of origin/variation for various crops. Zeven (1967) studied natural oil palm populations

TABLE 3.4
Characteristics of Germplasm Collections

			Dura				*Tenera*			
No.	**Country**		**S/F**	**M/F**	**MFW**	**O/B**	**S/F**	**M/F**	**MFW**	**O/B**
1	Nigerian	n	5947	5947	5947	5947	2117	2117	2117	2117
		Mean	42.37	44.90	9.32	14.64	14.91	74.67	7.53	22.58
		Min	22.13	26.33	2.83	6.73	3.88	57.34	3.23	11.16
		Max	57.43	62.93	26.60	25.31	28.33	91.32	20.07	36.51
2	Cameroon	n	1894	1894	1894	1894	517	517	517	517
		Mean	46.03	41.42	8.18	12.32	18.71	67.88	6.97	19.18
		Min	23.62	23.9	2.62	5.11	6.76	54.27	3.26	9.52
		Max	62.87	64.81	18.8	22.53	29.17	86.49	16.8	31.14
3	Zaire	n	3439	3439	3439	3439	556	556	556	556
		Mean	43.23	44.57	10.38	14.21	16.99	70.34	8.80	21.5
		Min	21.21	26.4	3.38	7.16	4.04	55.08	4.24	12.45
		Max	57.99	65.95	28.16	25.05	28.71	89.44	17.00	31.51
4	Tanzania	n	2098	2098	2098	2098	612	612	612	612
		Mean	39.54	47.11	9.65	13.92	16.33	70.54	8.40	20.16
		Min	19.55	32.25	3.26	6.71	2.80	55.54	3.73	11.22
		Max	53.10	65.46	27.90	23.44	28.87	92.30	21.45	30.89
5	Madagascar	n	16	16	16	16	–	–	–	–
		Mean	47.52	36.91	3.02	7.91	–	–	–	–
		Min	42.86	31.65	2.42	3.83	–	–	–	–
		Max	57.02	43.11	4.18	11.79	–	–	–	–
6	Angola	n	1935	1935	1935	1935	434	434	434	434
		Mean	39.17	49.54	12.56	15.38	14.18	74.32	10.85	22.70
		Min	20.14	28.02	4.63	6.16	6.11	60.73	5.30	14.08
		Max	60.24	67.19	33.02	24.33	24.39	88.86	24.55	31.31
7	Senegal	n	503	503	503	503	–	–	–	–
		Mean	48.94	35.58	2.46	9.10	–	–	–	–
		Min	33.65	23.56	1.19	2.84	–	–	–	–
		Max	60.07	57.09	6.2	16.69	–	–	–	–
8	Gambia	n	151	151	151	151	–	–	–	–
		Mean	44.73	37.9	2.53	11.07	–	–	–	–
		Min	21.62	26.64	1.26	4.35	–	–	–	–
		Max	59.73	51.84	2.96	18.58	–	–	–	–
9	Sierra Leone	n	805	805	805	805	44	44	44	44
		Mean	51.39	35.41	6.1	10.21	15.69	75.27	9.60	19.92
		Min	23.29	23.60	2.58	5.02	6.64	47.98	3.46	7.98
		Max	64.03	67.43	17.81	22.72	40.82	88.50	16.65	29.94
10	Guinea	n	613	613	613	613	39	39	39	39
		Mean	44.88	40.30	6.42	11.08	17.43	72.66	6.30	19.78
		Min	23.56	17.61	2.28	5.01	6.68	51.91	2.71	12.95
		Max	62.63	67.84	14.83	23.31	33.14	88.73	15.36	27.78

(*Continued*)

TABLE 3.4 (*Continued*)
Characteristics of Germplasm Collections

			Dura				*Tenera*			
No.	**Country**		**S/F**	**M/F**	**MFW**	**O/B**	**S/F**	**M/F**	**MFW**	**O/B**
11	Ghana	n	1417	1417	1417	1417	80	80	80	80
		Mean	44.18	41.96	5.42	11.88	18.19	69.79	7.54	18.74
		Min	19.28	23.81	2.38	3.19	4.27	52.29	2.51	7.63
		Max	63.29	72.91	16.03	24.26	32.02	92.17	15.86	31.81

in Africa and concluded that W. Africa was the center of origin of the oil palm. In 1971, Charles Hartley, consultant to MARDI, strongly recommended the systematic sampling of oil palm germplasm in W. Africa. The MARDI collection team comprised Drs. Arasu and Rajanaidu. Before embarking on the collection, the MARDI team held lengthy discussions with the population geneticists at the University of Birmingham to formulate a sampling strategy for collection in Nigeria. Meetings with leading oil palm breeders such as Drs. Zeven, Sparnaaij, and others were also made to familiarize the team with the distribution of natural populations and the customs and traditions of the local communities. The first collection in Nigeria was carried out in collaboration with Nigerian Institute for Oil Palm Research (NIFOR) with Mr. C.O. Obasola as the counterpart. The Department of Agriculture (DOA), Malaysia, requested CABI UK to be the intermediate quarantine station for the collection.

The main objective of the germplasm collection was to broaden the genetic base of current breeding materials. It has been well documented that the genetic base of oil palm breeding materials is extremely narrow, originating from only four Deli D palms planted in Bogor Botanical Gardens in 1848 (Hartley, 1988) and a limited number of P genotypes. It has been generally recognized that the restriction in effective gene pools has been a major obstacle to rapid selection progress in oil palm (Jagoe, 1952; Hardon and Thomas, 1968; Thomas et al., 1969; Ooi et al., 1973; Arasu and Rajanaidu, 1975; Ooi and Rajanaidu, 1979; Ahiekpor and Yap, 1982; Hardon et al., 1985; Rajanaidu and Abdul Halim Hassan, 1986) although this was before the advent of molecular markers. Nevertheless this concern provided the initial impetus for MPOB to carry out extensive prospections of oil palm genetic materials in the centers of origin of W. Africa for *E. guineensis* (*EG*) and Latin America for *Elaeis oleifera* (*EO*). Thus, the Malaysian Palm Oil Board (MPOB) has carried out extensive oil palm germplasm collections in Africa and Latin America.

From 1973 to 2010, the MPOB team led by Rajanaidu collected *EG* germplasm from the centers of origin in 11 countries, namely Nigeria, Cameroon, Zaire, Tanzania, Madagascar, Angola, Senegal, Gambia, Sierra Leone, Guinea, and Ghana. For *EO*, collection covered seven countries: Honduras, Panama, Costa Rica, Suriname, Ecuador, Peru, and Colombia. In addition, MPOB also collected palms of economic interest such as *Bactris*, *Jessenia*, *Oenocarpus*, *Euterpe*, and *Babassu*. By bringing in these genetic materials into the country, it is anticipated that new traits of economic interest will be discovered and subsequently introgressed into current breeding materials to broaden the genetic base.

3.2.2 Germplasm Collection and Sampling

The objective of collection is to gather the maximum amount of genetic variability in the minimum number of samples (Marshall and Brown, 1975). The basic sampling strategy is to collect a few individuals per site from as many sites as possible to cover various ecological niches. In the case of *EG*, 2–32 palms were collected per site and the number of sites visited in a country depended on the distribution and density of natural oil palm groves. For each bunch sampled, data on bunch weight, bunch length, bunch breadth, bunch depth, fruit diameter, nut diameter, kernel diameter, mesocarp to fruit (%), shell thickness, fruit weight, and nut weight were recorded *in situ* (Rajanaidu et al., 1979). In addition, the local name for oil palm, fruit forms, and uses for palm oil were also recorded.

CBD is an international legally binding treaty that was adopted in Rio de Janeiro in June 1992. Among other functions, it sets principles for the fair and equitable sharing of the benefits arising from the use of genetic resources, notably those destined for commercial use. Before CBD, the negotiation was normally on a bilateral basis between the countries involved. These include prior informed consent (PIC), sharing the cost involved in the prospection, sharing the germplasm collected, providing training, and the transfer of technology to the country of origin.

3.2.3 MPOB's Genetic Resources Program

It involves four major steps:

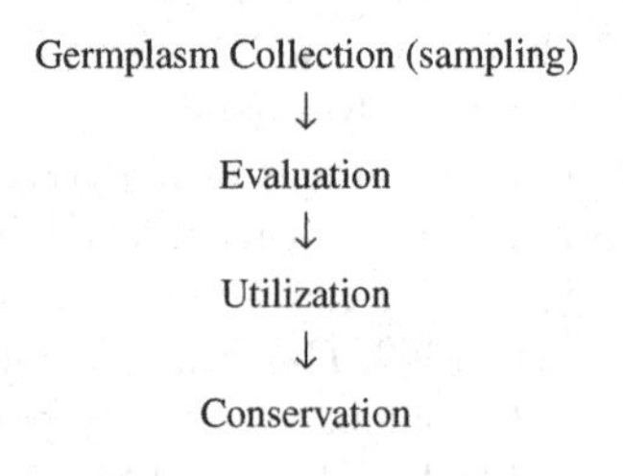

The major objectives are

- To find new sources of genes to broaden the genetic base of current oil palm breeding program, which is currently dependent on an extremely narrow genetic base.
- To conserve a cross-section of oil palm germplasm for future use as the natural palm groves in Africa are rapidly disappearing due to development and human population pressure.
- To introduce novel genes, for example, fatty-acid composition resulting in a high iodine value or IV (more unsaturated oil), dwarfness, and high bunch index (BI). Current breeding materials possess limited variation for these traits.
- To create an entirely new oil palm breeding population based on the new introduction.

- To understand the level of natural variation for traits within the oil palm species.
- To study the organization of variation between and within natural oil palm groves. This information is extremely useful to devise a sampling strategy for germplasm collection in the wild and for long-term conservation.

3.2.4 Centers of Distribution of Natural Oil Palm Populations

The centers of distribution of oil palm can be broadly classified into two areas

1. *E. guineensis* is endemic to West and Central Africa. Extensive natural or semi-wild palm groves are distributed along the west coast of Africa from Senegal to Angola (Zeven, 1967).
2. *E. oleifera* is found in Honduras, Nicaragua, Costa Rica, Panama, Colombia, Venezuela, Suriname, Ecuador, and Brazil (Meunier, 1975; Escobar, 1982; Ooi et al., 1981; Rajanaidu, 1983). Recently, new oleifera populations were discovered in Peru (Julian Barba, 2012).

3.2.5 Collections

3.2.5.1 *Elaeis guineensis*

E. guineensis germplasm from the centers of origin in 11 countries in Africa was collected by the MPOB team as shown in Figure 3.1; whereas for *EO* involving seven countries in South America is illustrated in Figure 3.2, see later in Section 3.2.5.2.

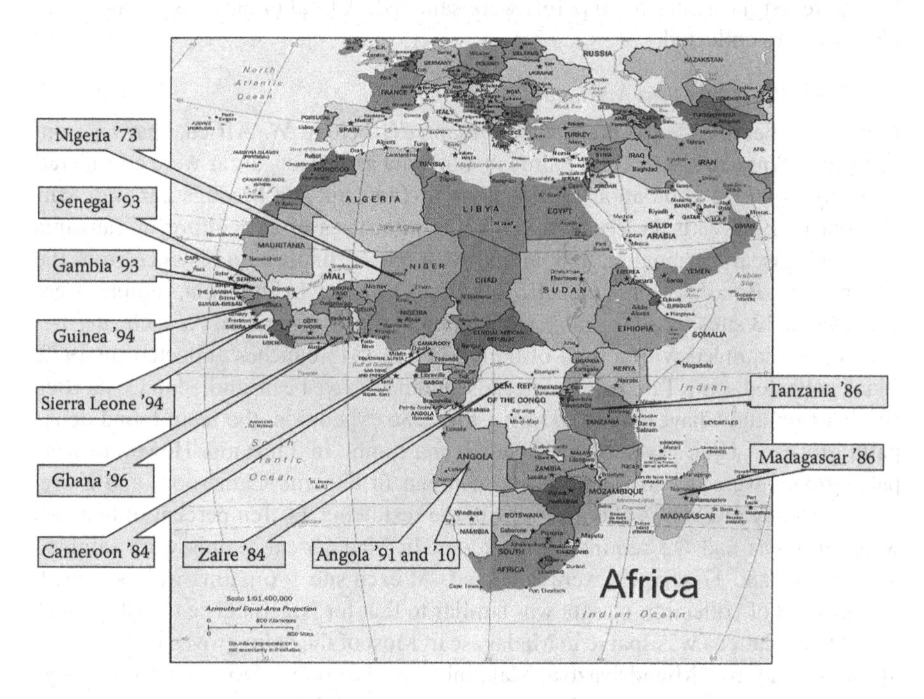

FIGURE 3.1 **(See color insert.)** Germplasm collection (*E. guineensis*) site in Africa.

3.2.5.1.1 Nigeria

In 1973, MARDI and NIFOR collected *OG* genetic materials at 45 sites from an average of 20 palms per site with 200 seeds per palm. A total of 919 (595 D and 324T) bunches were harvested during this prospection. One bunch was harvested from each of the sampled palms and the fruits from each bunch were kept separate until field planting. For the sampled palms, data on bunch weight, bunch length, bunch breadth, bunch depth, fruit diameter, nut diameter, kernel diameter, mesocarp-to-fruit (%), shell thickness, fruit weight, and nut weight were recorded *in situ* (Ooi and Rajanaidu, 1979; Rajanaidu et al., 1979; Obasola et al., 1983).

3.2.5.1.2 Cameroon

In 1984, PORIM with the cooperation of Unilever, made collections in both the western and eastern parts of Cameroon. Samples were collected at 32 different sites distributed throughout the country. One to 15 palms per site were chosen at random; the objective was to cover the whole country as far as possible. A total of 95 (58 D and 37T) palms were sampled during the prospection. The method of collection and characteristics studied were similar to those adopted in Nigeria.

3.2.5.1.3 Zaire

PORIM collected oil palm genetic material in Zaire from April to July 1984 also with the cooperation of Unilever. In the past, Belgian workers had prospected extensively for *EG* germplasm in Africa. During PORIM's expedition, palms were sampled at 56 different sites distributed throughout Zaire; Equator, Kivu, Kikwit—Kwango, and Bas Zaire. At most sites 5–10 palms were sampled. A total of 369 (283 D and 86T) bunches were collected.

3.2.5.1.4 Tanzania and Madagascar

Previous collections were made in the main *EG* belt in W. Africa, that is, from Senegal to Angola. The natural palms in Tanzania and Madagascar were considered basically as fringe populations. Collections in Tanzania and Madagascar were carried out in 1986 with the cooperation of the ministries of agriculture in Tanzania and Madagascar (Rajanaidu, 1986a) and with partial financial assistance from the International Board for Plant Genetic Resource (IBPGR). In Tanzania, samples were collected at 13 sites located near Kigoma along Lake Tanganyika. At each site, 1–7 palms were sampled with a total of 60 (42 D and 18T) bunches collected. In 1979, Blaak collected some Tanzanian materials and these have been studied in Costa Rica (Richardson and Chavez, 1986). During the 1986 expedition, the team found dense palm groves at Ujiji, Mwandiga, Kiganza, and Simbo in Tanzania. However, these palm groves were not as dense as those found in Nigeria, Cameroon, and Zaire. The frequency of Ds was about 90% with the rest being Ts. Ten percent of bunches were *virescens* and the reminder *nigrescens.* In Madagascar, palms were sampled at four sites and 17 samples were collected. At each site 1–6 palms were sampled. The method of collection of data was similar to that for Nigeria. The distribution of natural populations was sparse in Madagascar. Most of the palm groves were noticed along the road from Miandrivazo to Malaimbandy and confined to sandy river valleys intermingled with forest trees. In Madagascar, limited data were scored in the field due to lack of harvesters to obtain intact bunches for measurements. It was also not possible to identify the fruit forms because they were extremely small and the mesocarp of the fruit was generally damaged by birds. In general, the palm growth, bunch and fruit traits were poorer than those collected elsewhere in Africa. This could be due to the poor environment and low rainfall in the western part of Madagascar.

3.2.5.1.5 Angola

Oil palm germplasm was sampled at 8 sites. At each site 2–14 samples were collected. A total of 54 (42 D and 12T) bunches were collected in Angola (Rajanaidu et al., 1991). Bunch and fruit data were obtained similarly. The palm groves at Cabinda were dense and actively exploited by the farmers for the oil. The natural palm groves at Sumbe and Benguela were sparse. The palms at Caxito and Funda were moderate in density (Rajanaidu et al., 1991). A second collection expedition to Angola was carried out in 2010 and 127 accessions were collected. The second expedition included some of the sites that were not covered in the earlier trip due to civil war.

3.2.5.1.6 Senegal

A collection was carried out in July/August 1993 with the cooperation of the Ministry of Agriculture, Senegal. The palms were sampled at 13 sites. At each site 5–10 palms were sampled. A total of 104 samples were collected. Only D palms were encountered. This could be due to low rainfall and differential survival of Ds and Ts in this harsh environment (Rajanaidu and Jalani, 1994a).

3.2.5.1.7 Gambia

Gambia is a narrow strip of land which is wedged within Senegal. During the course of oil palm germplasm collection in Senegal, a limited number of accessions were sampled in Gambia. The oil palm germplasm collection in Gambia was carried out with the cooperation of the Ministry of Agriculture and Forestry, Gambia. Collections were made at six sites. At each site 5–10 palms were sampled. As in Senegal, only D palms were encountered in Gambia. A total of 45 palms (bunches) were sampled. Isolated palm groves were noticed and the natives regularly harvested the bunches and extracted oil using crude village methods (Rajanaidu and Jalani, 1994b).

3.2.5.1.8 Sierra Leone

In April/May 1994, collections were made in Sierra Leone with the cooperation of the Ministry of Agriculture. The team visited 14 sites covering York, Waterloo, Mckorowo, Mayira, Bonyeya, Kabaiama, Matopie, Rotifunk, Rogbene, Mayonkoli, Mamanka, Kambia, Masenie, and Rokupr. Extensive natural palm groves were encountered around Kambia and Rokupr. At each site 2–6 palms were sampled. A total of 56 samples (52 Ds, 3 Ts, 1 P) were collected. In terms of fruit color, 54 samples were *nigrescens*, 1 *virescens*, and I *albescens*. The collected seeds were divided equally between the Ministry of Agriculture, Sierra Leone, and MPOB.

3.2.5.1.9 Guinea

Collections were made in Guinea in May 1994. The plant exploration was carried out jointly with the Ministry of Agriculture, Guinea. The collection was made at 14 sites including Kissidougou, Geukedou, Macenta, Seredou, Somota, Golowe, Bomati, Kindia, Boffa, Boke, Kamsar, and Coyah. Dense palm groves were noticed around the Boffa and Boke areas.

3.2.5.1.10 Ghana

Ghana germplasm was collected in July/August 1996. Bunches were collected at 14 sites and a total of 58 (49 D and 9T) samples were gathered. The characteristics of the Ghana samples collected at various sites were studied. Table 3.1 provides the list of germplasm collection and the characteristics of collections is listed in Table 3.4.

3.2.5.2 *Elaeis oleifera*

E. oleifera was collected in Colombia, Panama, Costa Rica, Honduras, Brazil and Suriname in 1981–1982 (Figure 3.2). The palms were screened for fatty acid composition. The collections from Colombia, Panama, and Costa Rica had IVs (iodine values, indicating percent oil unsaturation) of more than 90. Their oleic acid (C18:1) level ranged from 52% to 66% and linoleic (C18:2) levels varied between 15% and 23%. The IV in the Brazilian *EO* ranged from 76 to 81 and the level of C18:1 was lower than the accessions from Colombia, Panama, Costa Rica, and Honduras. The fatty acid composition of Surinam *EO* is rather unique. It has the highest level of C18:1 and the lowest C18:2 when compared to other populations. The mean IV of the Surinam population is the lowest, that is, 67.5. The lowest level of C16:0 (palmitic acid) in the *EO* collections is 13% (Rajanaidu et al., 1994).

FIGURE 3.2 Germplasm collection (*E. oleifera*) site in South America.

The mean fruit weight (MFW) varied from 3.7 g (Colombia) to 4.6 g (Panama) and the highest value (6.5 g) was found in the Panama population. For M/F, the mean values ranged from 28.6% (Honduras) to 36.0% (Colombia) and the highest value (53.4%) was found in fruits collected from Costa Rica. Escobar (1980) has given values for MFW for Costa Rica as 3.2 g, for Panama, 3.3 g and for Colombia, 3.6 g. In the case of M/F, the same author gave values for Costa Rica as 37.2%, Panama, 36.4%, and Colombia 35.6%. The mesocarp content of *EO* in C. America was lower than that of *EO* Suriname, Brazil and the KLM (Kuala Lumpur Melanococca) palm which was introduced by the DOA, Malaysia in 1952 from the Congo. In the case of MFW, Brazil and KLM fruits were bigger than those found in C. America. However, the Brazilian population had a lower BW. It may be possible to intercross C. or S. American populations with Brazilian or KLM in order to obtain more desirable *EO* populations for further crossbreeding with *EG*. Even though the Brazilian material has superior fruit qualities, the level of unsaturation of mesocarp oil seems to be lower than that of the C. American population. Recently, Rajanaidu and Kushairi (2003, 2006) collected *EO* in Ecuador at Taisha. The fruit size and mesocarp content are superior to previous *E. oleifera*. The hybrids between Taisha *oleifera* and AVROS pollen are being evaluated in the field in Palmar del Rio in Ecuador (Julian Barba, 2012).

3.2.6 Introduction of Exotic Germplasm and Quarantine Procedures

The oil palm industry aims at preventing the introduction and spread of harmful organisms into Malaysia. So far, Malaysia is fairly free of major diseases such as Fusarium Wilt from Africa and Bud-rot from South America. Obviously, Malaysia has adopted very strict quarantine measures to prevent these diseases. The DOA of Malaysia has adopted a number of quarantine measures (Kang, 1986).

3.2.6.1 Phytosanitary Measures in the Country of Origin

The seeds or pollen are collected from healthy palms and the genetic material inspected by the local quarantine officer and certified free from any harmful diseases. The seeds must be free of mesocarp before dispatch to intermediate quarantine stations, normally in a temperate country. At present, the intermediate quarantine station used is CABI, United Kingdom.

3.2.6.2 Intermediate Quarantine Station

A representative sample of the seed is examined for *Fusarium oxysporum* and *Cercospora elaeidis*. If any serious pathogen is detected, the consignment is not transmitted to the importing country. If clean, the seeds and pollens are re-packed before despatching to the recipient country.

3.2.6.3 Phytosanitary Measures in Malaysia

In Malaysia, samples of the seeds are examined for *Fusarium oxysporum*, *Cercospora elaeidis*, and other pathogens. The seeds are also fumigated with methyl bromide. The seeds are germinated and planted in a post-entry quarantine nursery and are observed closely by pathologists for 1 year. After the nursery stage, the seedlings are released for field planting. The pathologists and breeders are required to report t DOA if any exotic disease is detected in the materials.

3.2.7 Evaluation

3.2.7.1 Evaluation of Oil Palm Genetic Material

The oil palm genetic materials collected in the wild were planted in the form of "open-pollinated" families at the MPOB Research Station Kluang, Johor. Experimental designs such as the cubic lattice, randomized complete block (RCBD), and completely randomized design (CRD) were used to study the performance of the material. It takes more than 20 years from the collection of germplasm in the wild to the release of oil palm planting materials. During this time, detailed data are collected on an individual palm basis. They comprise

1. Fresh fruit bunch (FFB) (kg/palm/year)
2. Bunch number (BNO) (bunches/palm/year)
3. Average bunch weight (ABW) (kg/palm/year)
4. Mean nut weight (MNW) (g)
5. Fruit/bunch (F/B) (%)
6. Mesocarp/fruit (M/F) (%)
7. Oil/dry mesocarp (O/DM) (%)
8. Oil/wet mesocarp (O/WM) (%)
9. Oil/bunch (O/B) (%)
10. Kernel/fruit (K/F) (%)
11. Shell/fruit (S/F) (%)
12. Mean fruit weight (MFW) (g)
13. Frond production (FP) (no./palm/year)
14. Rachis length (RL) (m)
15. Leaflet number (LN) (no./palm/year)
16. Height (HT) (m)
17. Physiological parameters, for example, leaf area index (LAI), bunch index (BI)
18. Fatty-acid composition, for example, iodine value (IV)
19. Oil yield (OY) (kg/palm/year)
20. Kernel yield (KY) (kg/palm/year)
21. Total economic product (TEP) (kg/palm/year)

Most of the vegetative and physiological traits were computed as described by Corley and Breure (1981) and Squire (1986).

3.2.7.2 Methods Used to Evaluate Oil Palm Genetic Material

3.2.7.2.1 Yield and Its Components

Data collection for FFB and its components; BNO and ABW on individual palms were made 36 months after field planting.

3.2.7.2.2 Bunch Quality Characters

Bunch quality characters were determined using the bunch analysis method developed by Blaak et al. (1963). Sampling started after 4 years of field planting with a minimum of three bunches from each palm and randomly sampled between intervals of at least 3 months from the previous sampling. Samples were then taken to the laboratory for bunch analysis. The basic parameters such as F/B, M/F, O/DM, O/WM, O/B, and K/B were computed.

3.2.7.2.3 Vegetative Traits

The vegetative measurements using the nondestructive method (Corley et al., 1971) and one shot method (Breure and Powell, 1988) were recorded. The vegetative measurements included RL, LN, LL, LW, petiole cross-section (PCS), and HT. Vegetative measurements were taken at the 8th-year after field planting.

MPOB has developed a database known as breeding information system (BIS) to manage the data collected on nearly 100,000 palms of germplasm collections (Mohd Din et al., 2012).

3.2.8 Utilization

3.2.8.1 Germplasm Utilization

The main objective of oil palm germplasm collection at MPOB is to utilize the elite palms with interesting attributes such as dwarfness, high bunch number, high iodine value, high kernel, and high vitamin E to introgress into current breeding materials. Some Cameroon × Zaire crosses showed tolerance to Ganoderma strains. High palmitic and linoleic acids, low stearic and oleic acids were found in the Madagascar collection. Some palms from Tanzania gave thin-shelled T and high BI. The Angolan palms exhibited large fruit with bunch and fruit characteristics similar to Deli Ds. Individuals from both the Tanzanian and Angolan collections are used in the long-stalk breeding program. Some palms collected from Guinea are found to be high in vitamin E. *E. oleifera* is an important source of high iodine value and high carotene oil. Palms that showed interesting traits were selected and distributed to the members of industry within Malaysia. Table 3.3 shows the germplasm materials which have been developed and distributed by MPOB.

3.2.8.2 Introgression Programs

3.2.8.2.1 Nigerian

The Nigerian oil palm germplasm was prospected in 1973 and field planted in 1975/76 at MPOB Kluang Station. Extensive evaluation for yield, bunch traits, fatty acid composition, physiological parameters, and vegetative characters were carried out between 1982 and 1987. Elite Nigerian palms for high oil yield, high IV, and low stature were distributed to members of the Malaysian industry for progeny-testing, introgression into the current breeding material, and to initiate new breeding lines for seed production. After undergoing extensive progeny-testing, the PS1 and PS2 composite varieties were then released as commercial planting material. New generations of these materials, that is, PS1.1 and PS2.1 materials are in the pipeline as future shorter planting materials (Rajanaidu et al., 2013).

3.2.8.2.2 Cameroonian

Cameroonian oil palm germplasm collected in 1994 were evaluated on inland soil of predominantly the Rengam Series at Bukit Lawiang, MPOB Kluang Research Station, Johore. After undergoing systematic selection, individuals were then introgressed and progeny-tested with AVR Ps. These progenies together with the selfs of the selected Ds are currently being evaluated there.

3.2.8.2.3 Zairean

After undergoing systematic evaluation, a few individual palms were selected and introgressed with AVR Ps. The progenies are currently being evaluated in a trial at MPOB Keratong Station while the intercrossed progenies of some selected Ds were planted in trials Keratong and Sessang (Sarawak).

3.2.9 Conservation

Oil palm germplasm can be conserved *ex situ* (outside their natural habitats) and *in situ* (in their natural habitats). *Ex situ* conservation involves methods such as seed storage, field genebanks, and botanical gardens. Seed, tissue culture, pollen, and DNA storage also contribute to *ex situ* conservation of genetic resources (Rajanaidu, 1982).

3.2.9.1 *Ex Situ* Conservation

3.2.9.1.1 Field Genebank (Living Collection)

This method is used for conserving perennial tree crops such the oil palm with long life cycle and species producing recalcitrant seeds (Jain, 2011). However, field genebanks requiring large land areas are expensive and labor intensive to maintain. Currently, the MPOB field genebank comprises 19 germplasm sets collected from W. Africa and Latin America planted in an area covering nearly 500 ha. For these reasons, germplasm collections in the field are augmented with other conservation methods. However, field genebanks are readily available for utilization (e.g., Figure 3.3).

3.2.9.1.2 Seed Storage

Oil palm seed is now considered an orthodox seed not a recalcitrant seed as previously thought, although it displays intermediate storage behavior (Engleman, 1991). Nevertheless the preservation of oil palm collections through storage of seeds is not practical as the seeds can be stored for only 2 years (Rajanaidu, 1980).

3.2.9.1.3 In Vitro Storage

The cryopreservation technique is used as a back-up conservation strategy to field genebanks (living collections). MPOB routinely stores zygotic embryos in liquid nitrogen for long-term conservation and to date 33,250 accessions have been stored in the cryotank. These embryos were derived from MPOB African collections (Rajanaidu and Ainul, 2013). Oil palm pollen, somatic embryos, and DNA are also similarly preserved for conservation and breeding purposes.

FIGURE 3.3 Comparative palms: (a) a good *E. guineensis* D × P (AVROS) hybrid, (b) a good *E. guineensis* D (Nigerian Prospection) × P (AVROS) hybrid, (c) a good pure *E. oleifera*, (d) a good *E. oleifera* × *E. guineensis* hybrid.

3.2.9.2 *In Situ* Conservation

The natural palm groves in the centers of distribution in Africa are disappearing at a rapid rate due to development and human population growth. Under the guidance of FAO, it is recommended to preserve a sample of natural palm groves with genetic diversity for future generations.

3.2.9.3 Development of Core Collection in Oil Palm

Brown (1989) and Odong et al. (2013) reviewed extensively the concept of core collection. *Ex situ* collections increased tremendously in the last 40 years. MPOB's collections occupy nearly 500 ha. In order to increase the efficiency of conservation, characterization, use, and regeneration of germplasm, the development of core collection is cost effective and efficient. Basically various approaches are pursued to preserve as much genetic diversity as possible with a minimum number of accessions. Odong et al. (2013) recommended the use of genetic distance-based criteria. At MPOB genetic distance of collections were studied using both molecular and phenotypic data. It is worthwhile to note that, while genetic distance studies provide gross difference between populations, as breeders, we have included some of the elite germplasm in the core collection.

3.2.10 Germplasm Management, Characterization, and Conservation Using Molecular-Based Assays

Genetic evaluation of the MPOB oil palm genetic materials was carried out using both phenotypic traits and molecular markers. The results generated are useful in identifying duplications, unique populations, or individuals as well as populations that are either of high or low variability. Such information is valuable in developing core collections, where maximum diversity is preserved in the reduced population size. The establishment of core collection helps to optimize the use of land resources and decrease the cost as well as number of workers required to maintain the field genebank, while ensuring the greatest possible genetic variation within a single trial.

Most of the molecular markers available currently, for example, restriction fragment-length polymorphism (RFLP), amplified fragment-length polymorphisms (AFLPs), simple sequence repeats (SSRs), and single-nucleotide polymorphism (SNPs) have been applied in characterizing the oil palm populations. Based on 16 SSRs, Bakoume et al. (2015) revealed high genetic diversity among 45 *E. guineensis* natural populations. Other researchers have also analyzed oil palm populations using RFLPs (Maizura et al., 2006), AFLPs (Kularatne et al., 2000), SSRs (Zulkifli et al., 2012), and isozymes (Hayati et al., 2004). For each marker, the allelic frequencies and other genetic variability parameters such as the mean number of alleles per locus (A), the percentage of polymorphic loci (P), and the expected heterozygosity (He) were reported. Table 9.6 summarizes the number of rare alleles (alleles with frequency less than 0.05), the mean number of alleles per locus, and the percentage of polymorphic loci for each oil palm population analyzed using RFLPs, isozymes, SSRs, and AFLPs, respectively. These results are further illustrated in Figure 3.4.

In general, the number of rare alleles detected by SSR markers was relatively higher than that observed for RFLPs and isozymes indicating their usefulness in diversity studies. Most of the populations from Nigeria exhibited rare alleles except for populations 40 and 42, with populations 44 and 45 having the most. Rare alleles were also detected in populations from other countries (13 and 22 of Cameroon; 7 and 36 of Zaire; 5 and 7 of Tanzania; 7 and 8 of Angola; 1, 5, and 13 of Sierra Leone;

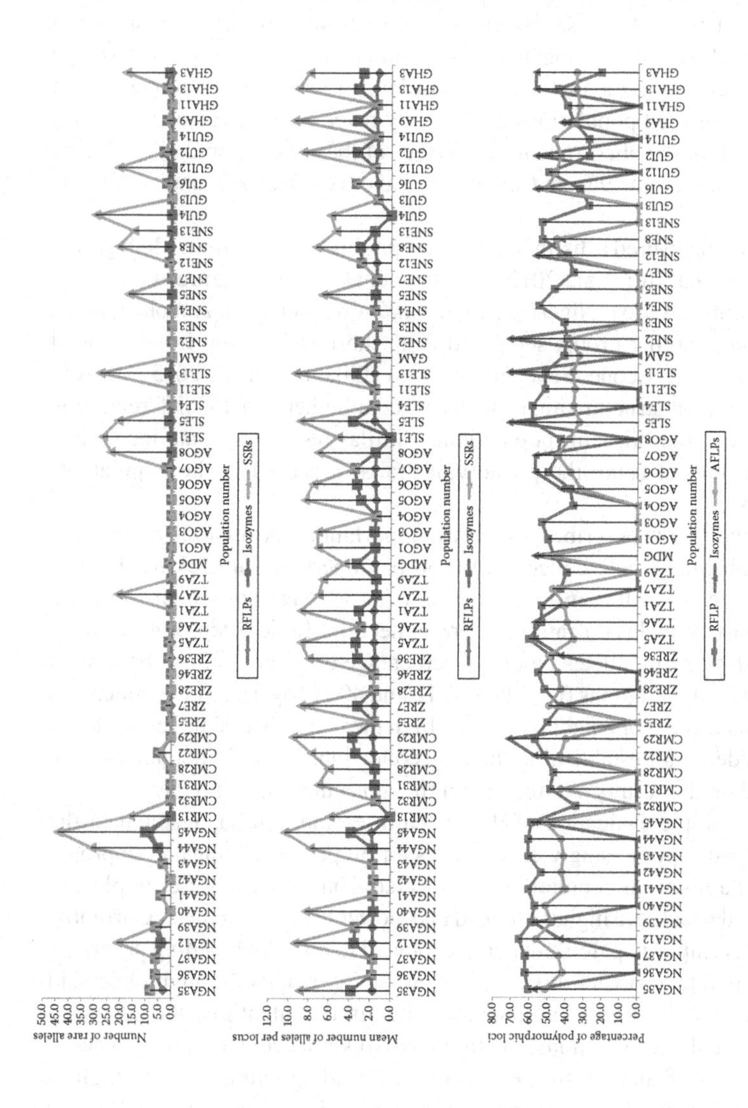

FIGURE 3.4 The levels of the number of rare alleles, the mean number of alleles per locus and the percentage of polymorphic loci estimated from RFLPs, isozymes, SSRs and AFLPs are shown for selected oil palm natural populations. The illustration above helps in selecting populations for future conservation efforts. From 56 populations, 30 populations exhibited considerably high values in terms of the estimated parameters. These include NGA35, NGA12, NGA39, NGA40, NGA44, NGA45, CMR22, CMR29, CMR31, CMR32, ZRE7, ZRE36, TZA1, TZA5, TZA7, AGO5, AGO6, AGO8, SLE1, SLE5, SLE13, SNE2, SNE8, SNE13, GUI2, GUI4, GUI6, GUI12, GHA3 and GHA13. Conserving these populations would results in higher diversity than conserving at random. NGA—Nigeria, CMR—Cameroon, ZRE—Zaire, TZA—Tanzania, AGO—Angola, SLE—Sierra Leone, GUI—Guinea Conakry, GHA—Ghana.

5, 8, 12, and 13 of Senegal; 2, 4, 6, and 12 of Guinea Conakry; 3, 9, and 13 of Ghana). No rare allele was observed among palms from Gambia.

Among the molecular markers, SSRs also revealed the highest mean number of alleles per locus across the oil palm populations, ranging from 1.1 (among palms from Madagascar) to as high as 6.7 (palms from the Nigerian populations). The range for this parameter detected by isozymes and RFLPs was relatively narrower, varying from 1.4 to 2.1 and 1.2 to 1.8, respectively. From the SSR analysis, the majority of the populations analyzed showed mean number of alleles per locus above the average (5.0) signifying their value for further conservation. In terms of the percentage of polymorphic loci, selected populations consistently revealed high values for at least two of the marker types applied. The most obvious of these were populations 35, 39, and 45 from Nigeria; 22 and 29 from Cameroon; 5 and 6 of Tanzania, and population 5 of Sierra Leone.

Cluster analysis uncovered a distinct group comprising of palms from Madagascar (Hayati et al., 2004; Zulkifli et al., 2012; Bakoume et al., 2015; Maizura et al. 2015). Palms from the central Africa (Nigeria, Cameroon, Zaire, Tanzania, Angola, Ghana) were classified into another group, separated from those originating from the west (Guinea Conakry, Sierra Leone, Senegal, Gambia) (Bakoume et al., 2015; Maizura et al., 2015). Population structure analysis disclosed a higher genetic differentiation within than between the oil palm populations. Similar results were attained when analysis was carried out using the phenotypic traits measured in the populations (Rajanaidu, 1985).

Collectively, the results described above are important in formulating a sampling strategy for establishing a core collection for oil palm germplasm. Despite the limited population size (15 palms in total), palms from Madagascar which formed a distinct cluster justify inclusion into the core collection. In addition, a total of 28 populations exhibiting rare alleles and high percentage of polymorphism (indicated above) and another 6 populations (populations 12 and 40 of Nigeria; 31 of Cameroon, 1 of Tanzania, and 5 and 6 of Angola) having a high mean number of alleles per locus can also be considered for establishing the core collection. These 35 populations are shortlisted based on the results obtained from the molecular marker assay.

Similarly, phenotypic evaluation of the oil palm germplasm has resulted in the identification of palms possessing valuable traits for oil palm breeding and improvement programs. Palms from population 12 collected from Nigeria, for example, are actively being used to achieve higher oil yield and slower height increment. Currently, as many as 1000 germplasm palms are being exploited in oil palm breeding programs in order to incorporate various traits such as low lipase, long stalk, high oleic acid and iodine value as well as high carotene and vitamin E content into the advanced breeding lines. In addition, germplasm with improved kernel contents are also desirable by some sectors of the industry, especially in breeding palms with oil high in medium-chain fatty acids for industrial applications. The palms that are actively being exploited in the breeding program should be further conserved for posterity.

By combining information obtained both from the molecular marker and phenotypic analyses, selection of populations for core collections can be done more effectively. This will increase the efficiency of conserving the oil palm germplasm collection to a manageable size that represents the diversity of the collection, which in fact allows for a more efficient access to the genetic materials. Van Hintum et al. (2000) recommended that a core collection should include between 5% and 20% of the entire collection. If a minimum figure of 5% is taken into consideration, the oil palm core collection should accommodate approximately 20,000 entries to be preserved in 135 ha of land. In actual practice, a standard oil palm breeding experiment evaluates 64 palms per populations that are divided into four replicates. Thus, conserving 35 populations identified based on the molecular-based results presented above, would involve approximately 2240 palms. If the selected 1000 palms currently being used in oil palm breeding programs are also taken into account, about 16 ha of land is needed to house this initial set of genetic materials for the establishment of the oil palm core collection. This would result in a more manageable hectarage considering the scarcity of good arable land, especially in Peninsular Malaysia.

3.3 *ELAEIS OLEIFERA* GENETIC RESOURCES

Edson Barcelos and Philippe Amblard

3.3.1 BACKGROUND

The genus *Elaeis* described as occurring in the African continent, also contains an American species, *E. oleifera* (Kunth, Cortés), known as the American oil palm. The center of *EO* origin is not well defined, being broadly dispersed in C. America and the northern region of S. America. Native small and dense populations growing along the riverbanks have been found in Brazil, Colombia, Costa Rica, Ecuador, French Guiana, Honduras, Nicaragua, Panama, Peru, Suriname, and Venezuela. *E. oleifera* populations, growing or tolerating both shaded and flooding conditions, suggest a broader environmental adaptability compared to the African oil palm (Corley and Tinker, 2015). In Brazil, many populations of *EO* are found on anthropogenic soils called Amazonian Dark Earths or Terra Preta de Índio, formed due to occupation by pre-Columbian populations, over sites that are highly fertile soils and different from adjacent soils (Lima et al., 2014). The use of the American oil palm by the indigenous people can only be speculated on, as for many American species, pre-Columbian societies shaped both the biogeographic distribution, and most likely introduced genetic bottlenecks by establishing local populations, with strong founder effects (Ting et al., 2014). Despite lower yields, *EO* is a source of many economically desired characteristics, of which the most important are

1. Shorter stature, due to a slow trunk growth, which facilitates harvesting and ensures a longer economic life span
2. Higher proportion of unsaturated fatty acids in the oil
3. Lower lipase activity in the mature fruit mesocarp, extending the time between harvest and fruit processing
4. Higher vitamins A and E content, improving the oil nutritional value
5. High levels of pest and disease tolerance which are of particular importance

The F1 *EO* × *EG* or *OG* hybrids are becoming widely planted in Latin America due to their disease tolerance and a number of the *EO* traits have been the object of introgression into high yielding *EG* varieties. Due to a poorly understood cytogenetic problem (Schwendiman et al., 1983), F_1 *OG* hybrids still have some reproductive limitations, particularly lower natural fertility that results in reduced pollen production with lower viability and poorer dispersion. Consequently, hybrids exhibit fruit abortion and lower oil production. Assisted pollination is used to overcome this limitation in the plantation with consequent high costs and labor inconvenience. In spite of the highly promising characteristics of *EO*, it is only in the last 40 years that natural populations have been thoroughly sampled to create and research *ex situ* germplasm collections (Meunier, 1975; Escobar, 1982; Ooi et al., 1981; Rajanaidu, 1985; Barcelos, 1998; Barcelos et al., 2002).

3.3.2 Prospections and Collections

3.3.2.1 Central America and Northern Region of S. America

Systematic studies and collections of *EO* from Central America (Mexico, Nicaragua, Guatemala, Costa Rica, Panama), and from the northern region of South America (Colombia) were started in 1968 by IRHO (now Cirad) (Meunier, 1975). At Cirad and its partners' research stations in Africa, an important collection was established from 1975 comprising various accessions from the following origins: Monteria, Cerete, Turbo, and San Alberto.

At ASD Costa Rica's (ex-CBCR) oil palm breeding program in Coto Oil Palm Plantation since 1967 (Escobar, 1982) they have included *EO* genetic resources originating from 43 regions covering seven countries: Colombia, Suriname, Panama, Brazil, Costa Rica, Nicaragua, and Honduras. Panama was the most important source represented in this collection, with 12 regions and 88 accession sites totaling 365 individual palm accessions. Currently, the breeding program in Costa Rica has achieved remarkable success in exploiting the *EO* gene pool toward commercialization of F_1 hybrids (Amazon) and backcross hybrids (Compactas) as seeds and clones (Alvarado et al., 2010).

In 1968–1969, 24 groups of *EO* (collected at Panamá and Costa Rica) and five *EO* F_1 hybrid seeds were sent to Lobe, Cameroon (Richardson, 1995). Malaysian oil palm breeding programs also received or collected *EO* germplasm during 1967 to 1982 from Colombia, Panama, Costa Rica, Suriname, Nicaragua, and Honduras (Mohd Din et al., 2000).

3.3.2.2 South America

In 1968 Cirad began systematic studies and collections of *EO* in S. America (Colombia, Brazil, Suriname, Guyana) (Meunier, 1975). Collections were planted by Cirad in a number of African oil palm research stations (La Mé/Ivory Coast, Pobé/Benin and La Dibamba/Cameroon) and in partner with private plantations in Colombia (Indupalma, La Cabaña), resulting in an important collection with accessions representing the main occurrences and origins of this species (Meunier, 1975).

Currently, the breeding program of Cirad has also made remarkable progress in evaluating and exploiting the *EO* gene pool, to the commercial stage for F_1 hybrids (Coari) initially and subsequently for backcross hybrids.

Preliminary observations and studies on *EO* in Brazil started in 1940, when 30 two-year-old plants of pure Caiaué origin were planted at Belém. Fruiting had started by November 1941. These plants originated from a natural population located at Paraná da Eva region, Manaus, Amazonas state. These same accessions were supposedly also planted at Museu Paraense Emílio Goeldi (Homma and Furlan Júnior, 2001).

By 1948/1949, the first F_1 *OG* hybrid was obtained by George O'Neill Adison working at Instituto Agronômico do Norte (IAN), from Caiaué trees planted at 1940 (Homma and Furlan Júnior, 2001). In 1960, Professor Mangenot, introduced to La Mé Experimental Station/IRHO/Ivory Coast, seeds of plants from Museu Emilio Goeldi. Hybrids derived from these plants were tested on site and also in Colombia without any exceptional results being obtained (Genty and López, 2013). In 1977, Dr. Ollagnier/IRHO travelling through the Brazilian Amazon identified a small population at Mamiá Lake border, near Coari, Amazonas state. Some seeds were collected and transferred to Indupalma in Colombia giving rise to the current commercial Coari hybrids (Genty and López, 2013). From 1980 to 1984, EMBRAPA (Empresa Brasileira de Pesquisa Agropecuária) jointly with Cirad, undertook a major *EO* prospection, covering the main concentrations of this species in the Brazilian Amazon, aiming to support the Oil Palm Breeding Program, launched by the Brazilian government. As a result of this effort, the EMBRAPA *EO* collection currently has 237 accessions originating from 16 areas and more than 53 populations dispersed mainly along the rivers of the Brazilian Amazon (Barcelos, 1986; Rios et al., 2012).

3.3.2.3 Complementary Collections (Peru, Ecuador, French Guyana)

Seeds collected from Peru constituting Genarro Herrera's population were introduced into Africa and Colombia. In Ecuador, *E. oleifera* seeds were collected at Taisha (Amazonian area) and planted in various estates. In French Guyana, bunches were harvested and the seeds added to Colombian, Brazilian, and African collections. This last population is closely related to the Surinam populations.

3.3.3 Characteristics and Genetic Diversity of Each Origin

3.3.3.1 Phenotypic Diversity

Plant breeders constantly need to introduce desirable new characteristics and new alleles to refresh breeding stocks. This requires an assessment of the phenotypic characteristics and potential of new accessions to identify the most promising genotypes to be exploited to improve the existing material.

With respect to several *EO* populations of Latin America, a number of important traits were characterized (Meunier, 1975; Barcelos et al., 1985; Rajanaidu, 1985; Santos et al., 1985; Barcelos, 1986; Mohd Din et al., 2000; Rey et al., 2004; Rios et al., 2011):

- Bunch and fruit traits: stalk-to-bunch, normal and parthenocarpic F/B, M/F, MFW, O/DM, nut weight (normal fruit) were presented by several authors. The highest M/F was obtained for the Surinam and Brazilian populations while the Honduras' populations gave the lowest.
- Leaf and bunch traits: RL, leaf petiole length, leaflet numbers.

- Fatty acid composition: IV. *EO* oil is mainly unsaturated with IVs ranging from 70 to 86.
- Although all the accessions have yet to be not fully characterized, some gave very interesting phenotypes: Surinam and French Guyana populations exhibit slow growth and are compact palms; Taisha and Genaro Herrera populations have long flower peduncles.

3.3.3.2 Genetic Diversity

Genetic diversity studies have been undertaken using many different marker types (phenotypic, enzymatic, and molecular).

3.3.3.2.1 Isozymes

Isozyme studies were started based on results from other crops and on *EO* undertaken earlier by Ghesquière (1985). Analysis on *EO* populations from the Amazonian basin (Ghesquière et al., 1987; Moretzsohn, 1995) revealed relatively low genetic diversity. The genetic diversity appeared to be similar within and between groups with an organization strongly affected by the hydrographic network of the region (Amazon, Madeira, and Negro) and with a distinct group originating from the Caracaraí/Roraima region.

3.3.3.2.2 Random Amplified Polymorphic DNA

A sample of 175 accessions obtained along the Amazon River Basin was analyzed and compared to 17 accessions of *EG* from Africa. Ninety-six random amplified polymorphic DNA (RAPD) marker bands were used in the analysis, of which 14 were shown to be specific to *EG*, while 12 were specific to *EO*. The Brazilian Caiaué accessions had moderate levels of genetic diversity as compared to *EG* accessions. The data allowed the establishment of similarity groups for *EO* accessions. Cluster analysis showed that, in general, genetic similarities were not correlated to geographical distances, but consistent with geographical dispersal along the Amazon River network, showing that most of the genetic variation was found within populations (Moretzsohn et al., 2002).

3.3.3.2.3 Restriction Fragment-Linked Polymorphisms

Analysis was applied to over 241 accessions of *EO* originating from different countries in Latin America, using 37 selected cDNA probes and produced 248 RFLP polymorphic fragments, coded as dominant markers, and used for a factorial analysis of correspondences (FAC) statistical analysis. For all 248 polymorphic markers, the percentage of polymorphism was 86%, with the Brazilian accessions alone revealing polymorphism in 65% of marker bands, while other populations had between 8% and 15% polymorphic bands. Nei's (1973) genetic diversity was $H = 0.225$ for the whole of America (Brazil, Colombia, Costa Rica, Nicaragua, Panama, French Guyana, Peru, and Suriname) (Barcelos et al., 2002).

In this study, four geographically distinct *EO* groups were identified: Brazil, Surinam/French Guyana, north of Colombia/C. America, and Peru. The Brazilian group was clearly separated from all others. The genetic distances (Nei, 1978) among these groups revealed divergence among *EO* (mean = 0.330, varying 0.114–0.425).

Cluster analysis showed a close relationship between the French Guyana and Surinam accessions. RFLP analysis on 241 accessions of *EO* previously reported by Barcelos et al. (2000), revealed that 74% of the genetic variability was due to inter-group divergence, while for the 32 populations in the Brazilian accession only 35% of variability was due to inter-populations divergence. The genetic structure revealed by RFLP within *EO* accessions from the Brazilian Amazon region was linked to the fluvial network of this region (Barcelos et al. 1997, 2002).

3.3.3.2.4 Amplified Fragment-Length Polymorphisms

Analysis of 40 accessions of *E. oleifera* revealed 169 polymorphic fragments.

AFLP polymorphic products were scored for presence or absence, and were used in FAC. Four genetic groups were distinguished: Brazil, Surinam/French Guyana, north of Colombia/C. America, and Peru. This result was similar to that based on RFLPs. The genetic diversity within each of the 32 studied Brazilian populations was higher than those in the non-Brazilian groups (Barcelos et al. 1997, 2002). This suggests a substantial genetic drift and bottleneck effect among the American groups other than the Brazilian.

As with previous markers, the structure of the *EO* Amazonian population was associated with the Amazon River network.

3.3.3.2.5 Simple Sequence Repeat

Ten markers were developed and characterized from 1500 sequences of an *EO* genomic library. These markers were used to assess the genetic diversity of *EO* germplasm collections from four S. American countries (Colombia, Costa Rica, Panama, and Honduras). The number of alleles per locus varied from 2 to 11. These SSR markers are expected to become useful tools to determine the population structure for conservation of *E. oleifera* populations (Zaky et al., 2010).

3.3.3.2.6 Other Markers

Genome DNA Quantity. To provide an understanding of the fertility problems manifested by F_1 *OG* hybrids the genome sizes of both species were evaluated by flow cytometry (Rival et al., 1997; Barcelos, 1998; Barcelos et al., 1999; Srisawat et al., 2005; Madon et al., 2008; Camillo et al., 2012). This was based on the hypothesis that a large difference in DNA quantity between species could be the underlying basis for fertility problems in the hybrids. The genome size of *EG* varied between 3.24 and 4.32 pg/2n while for *EO* the variation was between 2.08 and 4.43 pg/2n. However for the *OG* hybrid, the genome size was quite homogeneous between 4.16 and 4.40 pg/2n. It was concluded that differences in DNA quantity between these two species could not simply account for fertility problem in the hybrids.

Pollen morphology. With the aim toward using this trait for population discrimination, pollen morphology of 10 Amazonian populations (Acajatuba, Amatari, Autazes, BR-174, Careiro, Manicoré, Maués, Moura, Novo Aripuanã, and Tefé) were also studied. The acetolysis method was used to eliminate cell contents to enable observations on pollen size, shape, aperture, and exine surface. Pollen grains from the Novo Aripuanã population had the largest mean grain size (49 μm) and populations from Amatari, BR-174 and Moura the smallest (42.5–44 μm). Pollen is elliptical or piriform, aperture type is mono-colpate and exine surface is micro-reticulate. Pollen grains with morphological abnormalities were also found. Pollen morphology was able to discriminate three groups within Amazonian populations and offers an important tool for population discrimination, especially when associated with other analyses (Martins et al., 2012)

The utilization of *EO* genetic resources in oil palm breeding programs is given in Chapter 8.

3.4 CONCLUSION

Sean Mayes, Aik Chin Soh, and Jeremy Roberts

Currently, for oil palm, as for most crop species, access to germplasm is the one fundamental requirement to improve yields, to cope with disease pressure, and also to adapt crops to new environments—whether new countries or rapid changes in climate, as are now being experienced. Perhaps in the future it will be possible (and acceptable) to introduce genes from other species or directly engineer novel traits into crop species using our understanding of the genetic architecture of these traits in other species and tools such as genome editing. However, for now conventional breeding forms the bedrock of genetic improvement programs and of adaptation to specific environments. The available genetic diversity is the starting place to screen for improved traits.

Collecting, conserving, and understanding the genetic and trait diversity present in a species is the first step to exploiting it for breeding gain. In the oil palm, as detailed in Sections 3.2 and 3.3, the research and breeding community have been well served by the activities of a number of key players, alongside numerous bilateral exchanges of material between industry members. The geographical separation of *E. guineensis* from *E. oleifera* has already had practical breeding consequences in terms of finding genes for resistance to disease in the sister species and many other traits of interest that may be transferable, despite the initial difficulties with fruit fertility seen in the hybrids.

Molecular markers can help to make this process more efficient, by helping to inform collection strategies and also developing core collections, but it is important that general genetic distances are seen as a tool, with the trait phenotypes guiding the collection.

Alternative methods to preserve material will become more important in the future, particularly pollen and embryo cryopreservation have the potential to store very large numbers cost effectively and can help to preserve the existing diversity in oil palm as natural populations come under increasing pressure from expanding populations and changing climates.

Access to material will always be a point of contention, with the originating countries having a legitimate claim to ownership and certainly in other crops there has been significant historical development of the crop through farmer-based selection in those countries. This must be true to some extent in oil palm, but is less clear than for some other crops. CBD and the recently enacted Nagoya Protocol have a direct impact on such collection activities and set down conditions for the use of such material, particularly for commercial exploitation.

It is also important that the focus of germplasm collections is toward their use and exploitation in breeding programs, rather than collecting material simply to preserve it. The latter is important, but a collection which is never used adds no new value to the crop species.

REFERENCES

Ahiekpor, E.K.S. and Yap, T.C. 1982. Heritability, correlation and path coefficient analyses of some of oil palm breeding populations in Malaysia. In: Pushparajah, E. and Chew, P.S. (eds.) *The Oil Palm is Agriculture in the Eighties*, Volume 1. ISP, Kuala Lumpur, Malaysia. pp. 47–54.

Alvarado, A., Escobar, R., and Peralta, F. 2010. ASD's oil palm breeding program and its contributions to the oil palm industry. *ASD Oil Palm Papers* 34: 1–16.

Arasu, N.T. and Rajanaidu, N. 1975. Conservation and utilization of genetic resources in the oil palm (*Elaeis guineensis*, Jacq.). In: Williams, J.T. et al. (eds.) *South East Asia Genetic Resources*. pp. 182–186.

Bakoume, C., Wickneswari, R., Siju, S., Rajanaidu, N., Kushairi, A., and Billotte, N. 2015. Genetic diversity of the world's largest oil palm (*Elaeis guineensis* Jacq.) field genebank accessions using microsatellite markers. *Genetic Resources and Crop Evolution* 62: 349–360.

Barcelos, E. 1986. Características Genético-ecológicas de Populações Naturais de Caiaué (*Elaeis oleifera* H.B.K. Cortés) na Amazônia Brasileira. *Master*, Fundação Universidade do Amazonas, Manaus, AM, Brasil.

Barcelos, E. 1998. Etude de la Diversité Génétique du Genre Elaeis (*E. oleifera* Kunth Cortés et *E. guineensis* Jacq.) par Marqueurs Moléculaires (RFLP et AFLP). *PhD*, Ecole Nationale Supérieure Agronomique de Montpellier, Montpellier, France.

Barcelos, E., Amblard, P., Berthaud, J., and Seguin, M. 2002. Genetic diversity and relationship in American and African oil palm as revealed by RFLP and AFLP molecular markers. *Pesquisa agropecuaria brasileira* 37: 1105–1114.

Barcelos, E., Cunha, R.N.V., and Nouy, B. 2000. Recursos Genéticos de Dendê (*Elaeis guineensis*, Jacq.) *E. oleifera* (Kunth), Cortes) Disponíveis na Embrapa e Sua Utilização. In: Ocidental, E.A. (ed.) Seminario Internacional "Agronegocio de Dende: Uma Alternativa SocialL, Economica e Ambiental Para Desenvolvimento Sustentavel da Amazonia" Belem, PA. Brasil. Embrapa Amazonia Ocidental. pp. 39–40.

Barcelos, E., Lebrun, P., and Barre, P. 1999. Variação na Quantidade de DNA Nuclear no Gênero Elaeis (*E. guineensis* Jacq. *E. oleifera* (Kunth) Cortés) Determinado Pela Técnica de Citometria De Fluxo. In: Simposio de Recursos Genetics Para America Latina e Caribe—Sirgealc, 2., 1999, Brasília. Embrapa Recursos Genéticos e Biotecnologia, Anais... Brasília. 1 CD-ROM.

Barcelos, E., Santos, M.M., and Vanconcellos, M.E.C. 1985. Phenotypic variation in natural populations of Caiaué (*Elaeis oleifera* (H.B.K., Cortés) in the Brazilian Amazon. In: *Proceedings of the International Workshop "Oil palm Germplasm and Utilisation"*. Palm Oil Research Institute of Malaysia, Kuala Lumpur, Bangi. Selangor, Malaysia.

Barcelos, E., Second, G., Kahn, F., Amblard, P., Lebrun, P., and Seguin, M. 1997. Molecular markers applied to the analysis of genetic diversity and to the biogeography of Elaeis (Palmae). *Conference on Evolution, Variation and Classification of Palms*. Botanical Garden, Bronx, NY.

Blaak, G., Sparnaaij, L.D., and Menendez, T. 1963. Breeding and inheritance in the oil palm (*Elaeis guineensis* Jacq.) Part II. Methods of bunch quality analysis. *Journal of the West African Institute for Oil Palm Research* 4: 146–155.

Breure, C.J. and Powell, M.S. 1988. The one-shot method of establishing growth parameters in oil palm. In: *Proceedings of the 1987 International Palm Oil Conference on Progress and Prospects*. Palm Oil Res. Inst. Malaysia, Kuala Lumpur. pp. 203–209.

Brown, A.D.H. 1989. Core collection: A practical approach to genetic resources management. *Genome* 31: 818–824.

Camillo, J., Leão, A.P., Alves, A.A., Formighieri, E.F., Azevedo, A.L., Nunes, J.D., Capdeville, G., Mattos, J.K., and Souza Junio, M.T. 2012. Reavaliação do Tamanho do Genoma de *Elaeis guineensis* Jacq., *Elaeis oleifera* (Kunth) Cortés Seus Híbridos Interespecíficos. *Tese de doutorado*. Universidade de Brasília/Faculdade de Agronomia e Medicina Veterinária, Brasília, p. 137.

Corley, R.H.V. and Breure, C.J. 1981. *Measurements in Oil Palm Experiments*. Internal Report, Univeler Plantation Group, London.

Corley, R.H.V., Hardon, J.J., and Tan, G.Y. 1971. Analysis of growth of the oil palm (*Elaeis guineensis* Jacq.). 1. Estimation of growth parameters and application in breeding. *Euphytica* 20: 307–315.

Corley, R.H.V. and Tinker, P.B. 2015. *The Oil Palm*. 5th Edition. Wiley Blackwell, Chichester.

Engleman, F. 1991. Current development of cryopreservation for oil palm somatic embryos. In: *Proceedings of the XV11Ith International Refrigeration Congress*, vol. 1V, August 10–17, 1991, Montreal, pp. 1676–1680.

Escobar, R. 1980. An improved oil palm fruit depulper for single bunch lots. *Planter*, 56: 540–542.

Escobar, R. 1982. Preliminary results of the collection and evaluation of the American oil palm *Elaeis oleifera* (H.B.K.) Cortes in Costa Rica, Panama and Colombia. In: Pushparajah, E. and Chew, P.S. (eds.) *The Oil Palm in Agriculture in the Eighties*, vol. 1, Incorp Soc Planters, Kuala Lumpur, Malaysia, pp. 79–93.

Genty, P. and López, M.R.U. 2013. *Relatos Sobre el Híbrido Interespecífico de Palma de Aceite OxG-Coari x La Mé: Esperanza Para el Trópico*. FEDEPALMA, Bogotá. ISBN 78-958-8616-53-7. 209p

Ghesquière, M. 1985. Polymorphisme Enzymatique Chez le Palmier à Huile (*Elaeis guineensis*, Jacq.). II – Variabilité et Structure Génétique de Sept Origines de Palmiers. *Oléagineux* 40: 529–540.

Ghesquière, M., Barcelos, E., Santos, M. D., and Amblard, P. 1987. Enzymatic polymorphism in *Elaeis oleifera* H.B.K. Cortés (*E. melanococca*). Analysis of Populations in the Amazon Basin. *Oleagineux* 42: 151–153.

Hardon, J.J., Rao, V., and Rajanaidu, N. 1985. A review of oil palm breeding. In: Russell, G.E. (ed.) *Progress in Plant Breeding*. Butterworths, U.K., pp. 139–163.

Hardon, J.J. and Thomas, R.L. 1968. Breeding and selection of the oil palm in Malaya. *Oleagineaux*, 23: 85–90.

Harlan, J.R. 1992. *Crops and Man*. 2nd ed. American Society of Agronomy, Madison, WI. 284pp.

Hartley, C.W.S. 1988. *The Oil Palm (Elaeis guineensis Jacq.)*. Longman Wiley, New York.

Hayati, A., Wickneswari, R., Maizura, I., and Rajanaidu, N. 2004. Genetic diversity of oil palm (*Elaeis guineensis* Jacq.) Germplasm collections from Africa: Implications for improvement and conservation of genetic resources. *Theoretical and Applied Genetics* 108: 274–1284.

Homma, A.K.O. and Furlan Júnior, J. 2001. Desenvolvimento da Deindeicultura na Amazônia: Cronologia. In: Muller, A.A. and Furlan Junior, J. (eds.) *Agronegócio do Dendê: Uma Alternativa Social, Econômica e Ambiental Para o Desenvolvimento Sustentável da Amazônia*. Embrapa Amazônia Ocidental, Belém. pp. 193–207.

Jagoe, R.B. 1952. The "dumpy" oil palm. *Malaysian Agricultural Journal* 35: 12.

Jain, S.M. 2011. Prospects of *in vitro* conservation of date palm genetic diversity for sustainable production. *Emirates Journal of Food Agriculture* 23(2): 110–119.

Julian Barba, R. 2012. *Ecuadorian Oleiferas* minimize losses caused by the bud-rot in Latin America. In: *ISOPB Proceedings International Seminar Proceedings on "Breeding for Oil Palm Disease Resistance and Field Visits"*. September 21–24, 2012, Bogota, Colombia.

Kang, S.M. 1986. Plant quarantine in the international transfer of oil palm genetic materials. In: *Proceedings of the International Workshop on "Oil Palm Germplasm and Utilisation"*. Palm Oil Res. Inst. Malaysia, Kuala Lumpur. pp. 206–215.

Kularatne, R.S., Shah, F.H., and Rajanaidu, N. 2000. Investigation of genetic diversity in African Natural Oil palm populations and Deli using AFLP markers. In: *Proceedings of the International Symposium on Oil Palm Genetic Resources and Their Utilisation,* Malaysian Palm Oil Board, Bangi, Malaysia, pp. 497–555.

Lima, A.B., Cannavan, F.S., Navarrete, A.A., Teixeira, W.G., Kuramae, E.E., and Tsai, S.M. 2014. Amazonian dark earth and plant species from the Amazon region contribute to shape Rhizosphere bacterial communities. *Microbial ecology* 1–12.

Madon, M., Phoon, L.Q., Clyde, M.M., and Mohd Din, A. 2008. Application of flow cytometry for estimation of nuclear DNA content in Elaeis. *Journal of Oil Palm Research* 20: 447–452.

Maizura, I., Chan, P.L., Norhalida, M.S., Low, E.T., and Singh, R. 2015. Genetic structure of selected oil palm accessions as revealed by genotyping by sequencing (GBS) method. In: *Poster Presented at PIPOC 2015.* KLCC Conventional Center, Kuala Lumpur.

Maizura, I., Rajanaidu, N., Zakri, A.H., and Cheah, S.C. 2006. Assessment of genetic diversity in oil palm (*Elaeis guineensis* Jacq.) using restriction fragment length polymorphism (RFLP). *Genetic Resources and Crop Evolution* 53: 187–195.

Marshall, D.R. and Brown, A.H.D. 1975. Optimum sampling strategies for genetic conservations. In: Frankel, O.H. and Hawkes, J.G. (eds.) *Crop Genetic Resources for Today and Tomorrow.* Cambridge Press, Cambridge. pp. 53–80.

Martins, L.H.P., Miranda, I.P.A., and Nunes, C.D. 2012. Morfologai Polonica de Populacoess Amazonica de *Elaeis oleifera*. *Acta Amazônica* 33: 159–166. INPA Manaus. Amazonas. Brasil.

Meunier, J. 1975. Le Palmier à Huile Américain *Elaeis melanococca*. *Oléagineux* 30: 51–61.

Mohd Din, A., Rajanaidu, N., and Jalani, B.S. 2000. Performance of *Elaeis oleifera* from Panama, Costa Rica, Colombia and Honduras in Malaysia. *Journal of Oil Palm Research* 12: 71–80.

Mohd Din, A., Rajanaidu, N., Kushairi, A., Marhalil, M., and Zaharah, R. 2012. MPOB Breeding Information System (MPOB-BIS). MPOB TT No. 512.

Mohd Din, A., Kushairi, A., Rajanaidu, N, Noh, A., and Isa, Z.A. 2005. Performance of various introgressed oil palm populations at MPOB In: *Proceedings of the International Palm Oil Congress—Technical Break Throughs and Commercialisation—The Way Forward.* pp. 111–143, September 25–29, 2005, Malaysian Palm Oil Board, Kuala Lumpur.

Moretzsohn, M.C. 1995. *Estudo de Padrões Eletroforéticos e Análise da Variabilidade Genética de Dendê (Elaeis guineensis), Caiaué (E. oleifera), e Gerações F1 e 1025 RC1.* Universidade de São Paulo, Ribeirão Preto, São Paulo, Brasil.

Moretzsohn, M.C., Ferreira, M.A., Amaral, Z.P.S., Coelho, P.J.A., Grattapaglia, D., and Ferreira, M.E. 2002. Genetic diversity of Brazilian oil palm (*Elaeis oleifera* H.B.K.) germplasm collected in the Amazon Forest. *Euphytica* 124: 35–45.

Nei, M. 1973. Analysis of gene diversity in subdivided populations. In: *Proceedings of the National Academy of Sciences of the United States of America* 70: 3321–3323.

Nei, M. 1978. Estimation of average heterozygosity and genetic distance from a small number of individuals. *Genetics.* 89: 583–590.

Obasola, C.O., Arasu, N.T., and Rajanaidu, N. 1983. *Collection of oil palm genetic material in Nigeria. I. Method of collection.* MARDI Report No. 80. p. 38.

Odong, T.L., Jansen, J., van Eauwijk, F.A., and van Hintum, T.J.L. 2013. Quality of core collections for effective utlization of genetic resources: Review, discussion and interpretation. *Theoretical and Applied Genetics* 126: 289–305.

Ooi, S.C., Barcelos, E., Muller, A., and Nascimento, J. 1981. Oil palm genetic resources—Native *E. oleifera* populations in Brazil offer promising sources. *Pesquisa Agropecuária Brasileira* 16: 385–395.

Ooi, S.C., Hardon, J.J., and Phang, S. 1973. Variability in the Deli *dura* breeding population of the oil palm (*Elaeis guineensis* Jacq) I. Components of bunch yield. *Malaysian Agricultural Journal* 49: 112–121.

Ooi, S.C. and Rajanaidu, N. 1979. Establishment of oil palm genetic resources: Theoretical and practical considerations. *Malay. Appl. Biol.* 8(1): 15–28.

Rajanaidu, N. 1980. Oil palm genetic resources: Current methods of conservation. Paper presented at International Symposium on Conservation Inputs from Life Sciences. MPOB, Bangi. pp. 25–30.

Rajanaidu, N. 1982. Oil palm genetic resources—Current methods of conservation. *FAO Plant Genetic Resources Newsletter,* 48: 25–30.

Rajanaidu, N. 1983. Elaeis oleifera collection in South and Central America. *Plant Genetic Resources–Newsletter.* MPOB, No 56. F.A.O., Rome, pp. 52–51.

Rajanaidu, N. 1985. *Elaeis oleifera* collection in Central and South America. In: *Proceedings of the International Workshop on Oil Palm Germplasm and Utilisation*, Palm Oil Research Institute of Malaysia (PORIM), Bangi, Malaysia, March 26–27, pp. 84–94.

Rajanaidu, N. 1986a. Collection of oil palm (*Elaeis guineensis*) genetic material in Tanzania and Madagascar. *ISOPB Newsletter* 3(4): 2–6.

Rajanaidu, N. and Abdul Halim, Hassan. 1986. Oil palm genetic resources programme in relation to other perennial tree crops. In: Banpot Napompeth and Suranant Subhadrabandhu (eds.). *New Frontiers in Breeding Researches.* Kasetsart University Bangkok, Thailand, pp. 505–521.

Rajanaidu, N. and Ainul, L. M. 2013. Conservation of oil palm and coconut genetic resources. In: Normah, M.N. et al. (eds.) *Conservation of Tropical Plant Species.* Springer Science, New York. pp. 189–212.

Rajanaidu, N., Arasu, N.T., and Obasola, C.O. 1979. Collection of oil palm (*Elaeis guineensis,* Jacq.) genetic material in Nigeria II. Phenotypic variation of natural population. *MARDI Res Bull* 7(1): 1–27.

Rajanaidu, N. and Jalani, B.S. 1994a. Oil palm germplasm collections in Africa I. Senegal. *ISOPB Newsletter* 10(1): 1–18.

Rajanaidu, N. and Jalani, B.S. 1994b. Oil palm germplasm collections in Africa II. Gambia. *ISOPB Newsletter* 10(2): 1–17.

Rajanaidu, N., Jalani, B.S. 1994c. Oil palm genetic resources collection, evaluation, utilization, and conservation. In: *Paper Presented at Colloquium Oil Palm Genetic Resources.* Palm Oil Research Institute of Malaysia, Bangi, Malaysia.

Rajanaidu, N., Jalani, B.S., and Domingos, M. 1991. Collection of oil palm germplasm in Angola. *ISOPB Newsletter* 8(2): 2–3.

Rajanaidu, N. and Kushairi, A. 2003. *Negotiation of oil palm (Elaeis oleifera) germplam collections in the Amzonian Regions of Peru, Ecuador, Colombia and Brazil,* July 12 to August 6, 2003. Report OP (111) 2003.

Rajanaidu, N. and Kushairi, A. 2006. Oil palm planting materials and their yield potential. On International Society of Oil Palm Breeders Symposium. "Yield Potential in Oil Palm II". Phuket, Thailand, Nov. 27–28.

Rajanaidu, N., Kushairi, A.D., Jalani, B.S., and Tang, S.C. 1994. Novel oil from exotic palms. *Malaysian Oil Science and Technology Journal* 3(2): 22–28.

Rajanaidu, N., Kushairi, A., Marhalil, M., Mohd Din, A., Fadila, A.M., Noh, A., Meilina, A., Raviga, S., and Isa, Z.A. 2013. Breeding for oil palm for strategic requirement of the industry. *MPOB International Palm Oil Congress (PIPOC 2013)*, November 19–21, 2013, Kuala Lumpur.

Rey, L., Gomez, P., Ayala, I., Delgado, W., and P. Rocha. 2004. Coleciones Geneticas de Palma de Aceite *Elaeis guineensis* (Jacq.) y *Elaeis oleifera* (H.B.K.) de Cenipalma: Caracteristicas de Importancia para el sector Palmicultor. *Palmas* 25: 39–48.

Richardson, D.L. 1995. The history of oil palm breeding in the United Fruit Company. *ASD Oil Palm Papers (Costa Rica),* 11: 1–22.

Richardson, D.L. and Chavez, C. 1986. Oil palm germplasm of Tanzania origin. *Turrialba* 36(4): 493–498.

Rios, S. A., Cunha, R. N. V., Lopes, R., and Barcelos, E. 2012. *Recursos Genéticos de Palma de Óleo (Elaeis guineensis, Jacq.) e Caiaué (E. oleifeira (H.B.K.), Cortés).* Embrapa Amazônia Ocidental, Manaus. 39p. Ddocumentos, 96. ISSN 1517-3135.

Rios, S. A., Cunha, R. N. V., Lopes, R., Barcelos, E., Teixeira, P. C., Lima, W. A. A., and Abreu, S. C. 2011. Caracterização Fenotipica e Diversidade Genética em Subamostras de Caiaué (*Elaeis oleifera*) de Origem Coari. In: Congresso Brasiliero de Melhoramento de Plantas, 6. 2011 Buzios. RJ. Panorama Atual e Perspectivas do Nelhoramento de Plants no Brasil. SBMP. 2011. 1 CD RROM.

Rival, A., Beule, T., Barre, P., Hamon, S., Duval, Y., and Noiret, M. 1997. Comparative flow cytometric estimation of nuclear DNA content in oil palm (*Elaeis guineensis* Jacq.) tissue cultures and seed-derived plants. *Plant Cell Reports* 16: 884–887.

Rosenquist, E.A. 1986. The genetic base of oil palm breeding populations. In: Soh, A.C., Rajanaidu, N., and Mohd Nasir (eds.) *Int. Workshop on Oil Palm Germplasm and Utilization.* Palm Oil Research Institute of Malaysia. Malaysia, Kuala Lumpur. pp. 16–27.

Santos, M. M., Barcelos, E., and Nascimento, J. C. 1985. Genetic resources of *Elaeis oleifera* (H.B.K.), Cortés) in the Brazilian Amazon. In: *Proceedings of the International Workshop "Oil Palm Germplasm and Utilisation".* Palm Oil Research Institute of Malaysia, Bangi, Selangor, Kuala Lumpur, Malaysia.

Schwendiman, J., Pallares, P., Amblard, P., and Baudouin, L. 1983. Analysis of fertility during bunch development in the interspecific oil palm hybrid *Elaeis melanococca* × *E. guineensis. Oléagineux* 38(7), 411–420.

Singh, R., Low, E.-T. L., Ooi, L. C.-L., Ong-Abdullah, M., Ting, N.C., Nagappan, J., Rajanaidu, N. et al. 2013a. The oil palm SHELL gene controls oil yield and encodes a homologue of SEEDSTICK. *Nature* 500: 340–344.

Singh, R., Ong-Abdullah, M., Low, E-T. L., Abdul Manaf, M.A., Rosli, R., Rajanaidu, N., Ooi, L.C-L. et al. 2013b. Oil palm genome sequences reveals divergence of interfertile species in old and new worlds. *Nature* 500: 335–339.

Soh, A.C., Wong, G., Hor, T.Y., Tan, C.C., and Chew, P.S. 2003. Oil palm genetic improvement. In: Janick, J. (ed.). *Plant Breeding Reviews*, vol. 22. John Wiley & Sons, New Jersey, pp. 165–219.

Squire, G.R. 1986. A physiological analysis for oil palm trials. *PORIM Bulletin.* 12: 12–31.

Srisawat, T., Kanchanapoom, K., Pattanapanyasat, K., Srikul, S., and Chuthammathat, W. 2005. Flow cytometric analysis of oil palm: A preliminary analysis for cultivars and genomic DNA alteration. *Songklanakarin Journal of Science and Technology*, 27(Suppl. 3): 645–652.

Stuber, C.W. 1995. Mapping and manipulating quantitative traits in maize. *Trends in Genetics*, 11(12): 477–481.

Thomas, R.L., Watson, I., and Hardon, J.J. 1969 Inheritance of some components of yield in the "Deli dura variety" of oil palm. *Euphytica* 18: 92–100.

Ting, N.C., Jansen, J., Mayes, S., Massawe, F., Sambanthamurthi, R., Ooi, L.C., Chin, C.W. et al. 2014. High density SNP and SSR-based genetic maps of two independent oil palm hybrids. *BMC Genomics* 15: 309.

van Hintum, T. J. L., Brown, A. H. D., Spillane, C., and Hodgkin, T. 2000. *Core Collections of Plant Genetic Resources.* IPGRI Technical Bulletin No. 3. International Plant Genetic Resources Institute, Rome, Italy.

Zaky, N. M., Ismail, I., Rosli, R., Chin, T. N., and Singh, R. 2010. Development and characterization of *Elaeis oleifera* microsatellite markers. *Sains Malaysiana* 39: 909–912.

Zeven, A.C. 1967. *The Semi-World Oil Palm and Its Industry in Africa.* Wageningen University, Wageningen.

Zohary, D. 1970. Centers of diversity and centers of origin. In: Frankel, O.H., and Bennett, E. (eds.) *Genetic Resources of Plants—Their Exploration and Conservation.* Blackwell, Oxford. pp. 33–42.

Zulkifli, Y., Maizura, I., and Singh, R. 2012. Evaluation of MPOB oil palm (*Elaeis guineensis*) germplasm using EST-SSR. *Journal of Oil Palm Research*, 4: 1368–1377.

4 Plant Genetics

Aik Chin Soh, Sean Mayes, and Jeremy Roberts

CONTENTS

4.1 Introductory Overview ... 58
4.2 Qualitative Trait Inheritance and Applications in Oil Palm Breeding ... 60
4.2.1 Inheritance of Qualitative Traits ... 60
4.2.2 Applications ... 61
4.2.2.1 Shell Marker ... 62
4.2.2.2 Dwarfing Markers ... 62
4.2.2.3 Long Bunch Stalk Markers ... 62
4.2.2.4 *Virescens* Fruit Marker ... 62
4.2.2.5 Delayed Fruit Shedding with Low Lipase (Reduced Undesirable Free-Fatty Acid or FFA Production) Markers ... 62
4.2.2.6 Crown Disease Markers ... 62
4.3 Quantitative Trait Inheritance and Applications in Oil Palm Breeding ... 63
4.3.1 Inheritance of Quantitative Traits ... 63
4.3.2 Estimation of Genetic Variance and Heritability ... 63
4.3.2.1 Genetic Assumptions for Estimating Genetic Variance and Heritability ... 65
4.3.2.2 Applications ... 65
4.3.2.3 Appropriate Use of Heritability Estimation Methods ... 66
4.3.3 Combining Ability and Breeding Value Estimates ... 67
4.3.4 Genetic Covariance and Correlation ... 68
4.3.5 Genotype–Environment Interaction ... 68
4.4 Conclusions ... 71
Appendix 4A.1 Estimation of Genetic Variance and Heritability ... 71
Appendix 4B.1 Selection Response and Realized Heritability ... 75
Appendix 4C.1 Estimation of Genetic Covariance, Correlation, and Correlated Response to Selection ... 76
Appendix 4D.1 Estimation of Genotype–Environment Interaction Effects ... 76
Appendix 4E.1 Estimates of Within-Family Genetic Variability for Clonal Selection in Oil Palm ... 76
References ... 79

4.1 INTRODUCTORY OVERVIEW

Plant genetics is the core science behind plant breeding. Farmers for thousands of years have been selecting for traits that meet their needs by saving the seeds or vegetative propagules from favored plants after each season's harvest. Such desirable phenotypes have included non-shedding, lack of spines, larger and better tasting seeds, tubers, and fruits. The process of farmer-based selection can often be separated into two, often overlapping, phases—domestication and adaptation.

Domestication is the selection of variants which make crop plants more suitable for human agricultural practices. Many domestication genes cause significant changes in morphology (such as changes in the DNA in the upstream region of the teosinte branched 1 gene (tb1; Doebley et al. 1995) which transformed the morphology of the many branched teosinte into the single stem maize) or harvestability traits (such as non-shattering, allowing the grain to remain on the ear during harvest; Lin et al. 2012). In selecting such changes, early farmers have been altering the genetic make-up of the crops they grew and have dramatically changed them compared to their wild relatives, without any knowledge of plant genetics and breeding. Many of the early crop varieties or landraces arose in such a manner. Once a crop has been domesticated (often with similar selected changes in different crop species, which have been termed Domestication Syndrome Factors; Purugganan and Fuller 2009) crops may also have been further selected for adaptation to their growing environment. Adaptation has been critical to allow the geographical and climatic spread of the cultivation of particular crops. Photoperiod sensitivity and vernalization requirements are classic examples of adaptation in wheat which modify the phenological development of the crop so that it fits better with the farming year in a particular location and more reliably produces a good yield in the new environment.

It is often said that plant genetics began with Gregor Mendel, the amateur plant breeder monk who elucidated the inheritance (Mendelian) of flower, pod, and seed color (qualitative traits) in the pea plant, giving rise to the famous Laws of Segregation and Independent Assortment which remain relevant today (Mendel 1865). Plant breeders however are more concerned about economic traits, for example, yield and these often exhibit continuous variation. It is worth noting that a number of Mendelian genes have had significant effects on such quantitative traits in the past, such as the Norin-10 allele of the *Reduced Height* genes in wheat (*Rht* genes on the homeologous group 4 chromosomes) which significantly increased yields (with increased artificial nitrogen) and kick-started the Green Revolution (Fick and Qualset 1973). However, when the best available version of such genes have been incorporated into elite germplasm, the remaining variation to be used in breeding is often quantitative, often representing many genes of small effect scattered through the genome. In these cases, statistical descriptions (means, variances) are more appropriate for analysis of genetic control of these traits than frequencies of discrete genotype classes. Fisher (1918) in his seminal work "The correlation between relatives on the supposition of Mendelian inheritance" essentially married statistics with genetics, to give rise to quantitative genetics. Quantitative genetics has been very popular with plant and animal breeders and many breeding, selection, and analysis methodologies were developed based on quantitative genetic principles, for example, reciprocal recurrent selection (RRS), selection index (SI), best linear unbiased prediction (BLUP) breeding values (BV), and genome-wide/genomic selection (GS).

Molecular genetics is essentially the study of the inheritance of genes or DNA markers linked to genes which affect traits. This can range from a detailed understanding of the structure and function of specific genes at the DNA/RNA level, through to studies of the inheritance of DNA markers of unknown function. Studies can focus on serial or parallel approaches, with single genes in pathways of known function explored in detail one at a time or multiple genes (or individuals) genotyped to understand global genetic variation and develop markers linked to specific traits of breeding interest. While it is useful to know the identity of a causative gene for a phenotype, as it allows allele mining and a better understanding of the interactions between the genes and genes and environments that contribute to the phenotype, it is not necessary to clone a gene variant (allele) to be able to use it in a breeding program. Where the gene is known and the genetic (or epigenetic) variant that causes the phenotype understood, so-called "perfect" markers can be developed, where the marker test used in breeding evaluates the presence of the causative variant in the gene. Where linkage allows the development of tests based on DNA markers which are close (physically) to the causative gene, such associated markers can be used to infer the presence of the desired gene form in the offspring, without any understanding of the nature of the causative gene. Associative tests always risk recombination occurring between the associated marker and the causative gene, which would lead to an incorrect test result. However, as marker density increases, the likelihood of undetected recombination events decreases. With the development of "omics" technologies, allowing massively parallel testing of markers, transcripts, proteins, metabolites, and other cellular components, and also a rapidly increasing number of sequenced crop genomes made possible by third and fourth generation sequencing technologies, there is now more convergence between these two approaches. Molecular breeding is the application of molecular genetic tools to plant breeding, covering both gene discovery and also selection of the offspring genome for desired combinations of often functionally unknown genes. The applications of molecular breeding are broad and range from quality control in breeding programs and confirmation of hybrids using genetic fingerprinting, through germplasm diversity analysis, genetic mapping, qualitative and quantitative trait locus (QTL) analysis in bi-parental populations, association genetics in breeding germplasm, through to the current area of most research interest, genomic selection.

Another potential option in the toolkit of molecular breeding is genetic modification, which effectively bolts-on genetic modification to a number of the well-developed tissue culture procedures that have been used for clonal propagation in crop plants. However, there are clearly public acceptability issues with the current technologies for genetic modification in a number of countries which may (or may not) partly be addressed by the recent development of genome editing tools.

Oil palm genetics and breeding has essentially followed the same historical development as other crops.

4.2 QUALITATIVE TRAIT INHERITANCE AND APPLICATIONS IN OIL PALM BREEDING

4.2.1 Inheritance of Qualitative Traits

Qualitative traits are traits that can be scored into distinct classes. They are usually controlled variants of one (monogenic) or a few (oligogenic) genes and are inherited in a Mendelian fashion, exhibiting dominant, recessive, or epistatic (gene–gene interaction) segregation patterns. Classically, the mode of inheritance is usually experimentally inferred from the segregation patterns of the derived generations: parental (P_1, P_2); filial, first (F_1), second (F_2), sometimes third (F_3), and backcross generations to both parents ($BC_{1.1}$ and $BC_{1.2}$) of a cross between two inbred parents with contrasting traits. The P_1, P_2, F_1, $BC_{1.1}$, $BC_{1.2}$, and F_2 are often described as the "basic generations" in classical genetic analysis. Large sample sizes and the availability of BC (and F_3) segregation data besides the F_2 data are beneficial to obtain a convincing goodness-of-fit chi-square test, where a specific hypothesis to explain the inheritance of the quantitative trait can be tested against a null hypothesis (Sinnot et al. 1956; Soh et al. 1977). However, alternative explanations can be evaluated (e.g., the inheritance of testa color variants in beans; e.g., Sax 1923) with quite limited numbers of offspring showing trait segregation at the F_2 generation.

Most of the earliest presumed Mendelian inheritance traits in oil palm were inferred from observations from natural or breeding populations rather than from planned genetic experiments (Soh and Tan 1983; Hardon et al. 1985; Corley and Tinker 2016) and hence these unstructured observations could be open to misinterpretation because of structure within the observed stands which was unknown to the observer, with the observation effectively being *posthoc* experiments. Beirnart and Vanderweyen (1941) carefully designed and controlled experiment led to the revelation of the simple Mendelian inheritance of the fruit shell-thickness trait and provided the basis of modern oil palm breeding and hybrid seed production programs: the thick-shelled *dura* (D) being homozygous dominant; the shell-less pisifera (P) homozygous recessive, and the thin-shelled *tenera* (T) heterozygous. The gene responsible for the inheritance of the shell has been elucidated recently (Singh et al. 2013b). Although the presence of the shell is under monogenic control, variation in shell-thickness in the D and T could be attributed to multiple alleles of the *SHELL* gene (although all variants identified so far seem to be nonfunctional, rather than partly functional), an epigenetic/epigenomic basis or attributed to the presence of "modifying genes" (Van der Vossen 1974), which could also be related to Okwuagwu and Okolo's (1992, 1994) postulated maternal inheritance of kernel size. Fruit color (*nigrescens*—black unripe to red ripe, *virescens*—green unripe to orange ripe, *albescence*—whitish to yellow) are presumed simply inherited traits from observations and genetic analyses (Hartley 1988). Singh et al. (2014) recently unraveled the genomic basis of the *virescens* trait. Although *virescens* is a dominant trait, curiously nigrescens palms appear to be more prevalent in the natural populations. The genomic basis of the natural variant of the mantled fruit palm, which was observed to be transmitted as a dominant trait (Hartley 1988) is unknown although the epigenetic/epigenomic basis of the mantled fruit somaclonal variant has now been uncovered (Ong Abdullah et al. 2015).

The P phenotype and genotype is strongly linked to female sterility. Fertile Ps do exist although very infrequently in West African (Menendez and Blaak 1964) and Malaysian breeding populations especially related to the SP29/36 pedigree (Tang 1971; Chin and Tang 1979). Monogenic inheritance of female sterility has been suggested by Wonky-Appiah (1987) from his studies. From sib-crosses of fertile Ps derived from SP29/36, T segregants were obtained, suggesting that some of the fertile Ps were "paper" thin-shelled Ts or illegitimate contamination or the inheritance of shell thickness may not be that simplistic (Singh et al. 2013b) and may involve multiple alleles. Fertile Ps can confer a number of advantages to oil palm breeding and seed production. It allows direct determination of the genotype without resort to auxin induction of fertile fruit set or progeny-testing. It also allows initial phenotypic selection of bunch and yield traits prior to progeny-testing and also for vegetative growth trait selection without the confounding effects of sterility (i.e., enhanced vegetative vigor) and as potential cultivars, presumably higher mesocarp oil content in the absence of the shell. The earlier belief that fertile Ps would give rise to thicker-shelled D × P or Ts (Menendez and Blaak 1964; Chin 1982) has been debunked (Rosenquist 1990; Nelson 1993). Although oil yields of fertile Ps comparable to those of commercial D × P are obtainable (Chin 1982), none have surpassed those of the better D × P progenies, mainly due to the variable fertility levels in the fertile Ps. The other disadvantage is the mixing of mesocarp oil (palm oil) with kernel oil (palm kernel oil) of the shell-less P fruit in the mechanical fruit pressing mill oil extraction process, the latter oil having a different fatty-acid composition and consequent preferred product uses, for example, oleochemicals (Figure 1.1, Chapter 1; Pantzaris 2000) and fetching a premium price (about 20%) over the former. Nevertheless, breeding for fertile P persists in some programs.

Crown disease (CD) causes young leaf curling in young palms and can lead to arrested growth and even death when severe. It is more prevalent in Deli × AVROS/Yangambi hybrid materials. The genetic basis of the susceptibility to CD exemplifies the classical case of epistatic gene action: *C/c* genes involved in CD and *S/s* in disease expression, with *ccss* genotype exhibiting susceptibility while *c_ S_* genotypes appear unaffected (Blaak 1970). Long bunch stalk palms are available in inbred populations of both Deli D and W. African T (e.g., Yangambi) suggesting a simply inherited trait or a major QTL trait (Priwiratama et al. 2010, Chookien Wong personal communication 2016), the former suggesting a dominant trait and a recessive trait suggested by the latter. Likewise, dwarf palms or dwarfing genes (as major QTL) are available in the Dumpy Deli, Yangambi 16R, Pobe Dwarf, Gunung Melayu, MPOB's Population 12 (Soh et al. 1981; Rosenquist 1986, 1999). The genomic basis of CD, long bunch stalk, and dwarf stature should be elucidated over the coming years.

4.2.2 Applications

All the above traits have economic values directly or indirectly. The availability of molecular markers for marker-assisted selection or MAS will facilitate (time, space, and accuracy) selection early in the nursery stage before the expression of the trait later in the field or after palm maturity.

4.2.2.1 Shell Marker

Marker kits such as the SureSawit Shell Gel Kit™ (Orion™ Biosains) have been marketed as an early diagnostic tool to detect *dura* contamination as a consequence of illegitimacy of commercial D × P seeds. This would supplant the previous tedious method of cutting the first ripe fruits for examination of fruit type. Legitimacy testing using DNA fingerprinting would be more powerful (Wong et al. 2015; see also Chapter 10). Nevertheless, the shell marker would be helpful in the palm breeding T × T/P crosses to separate out the Ds (usually of no interest) and the Ps (to be planted as a "pollen garden" for commercial seed production), and the Ts (to be planted in the trial plots unencumbered by the more vigorously growing female sterile Ps).

4.2.2.2 Dwarfing Markers

Dwarf or short palms facilitate bunch harvesting and prolong the economic life of the palm. Old palms are generally replanted when they have grown too tall for economic harvesting. Nursery screening for dwarf material with attenuated sensitivity to gibberellic acid (GA) has been demonstrated and molecular markers are being developed (Roberts and Zubaidah 2013).

4.2.2.3 Long Bunch Stalk Markers

Long thin bunch stalks facilitate the bunch-cutting process, potentially speeding up the harvesting or allowing less-skilled harvesters to be used. In addition, it is standard to cut the frond below the bunch, to allow access to harvest the bunch. However, the subtending frond is still photosynthetically active, so its preservation may influence carbon assimilation.

4.2.2.4 *Virescens* Fruit Marker

Virescens fruit could facilitate identification of ripe bunches for harvesting of taller palms for unskilled harvesters (Singh et al. 2014). For this to be confirmed, the timing of color accumulation needs to be investigated, to see how it corresponds to oil accumulation.

4.2.2.5 Delayed Fruit Shedding with Low Lipase (Reduced Undesirable Free-Fatty Acid or FFA Production) Markers

These could extend the harvesting rounds and reduce labor requirement while still maintaining the oil quality. Current plantation practice almost certainly harvests bunches before peak oil content is achieved, to minimize loose fruit collection and reduce labor input. Loose fruit usually accumulate FFA due to the induction of lipases on bruising with high levels of FFA in the final oil reducing the potential market value. Major genes or major QTL may be involved and appropriate molecular markers could be developed (Roongsattham et al. 2012; Morcillo et al. 2013).

4.2.2.6 Crown Disease Markers

Currently crown disease palms can only be identified in the young plantings and are culled and replaced. To "weed out" the genes involved from the seed and breeding parents, the breeder could remove both parent palms of the susceptible progeny from commercial seed production (with consequently have reduced capacity for seed production) and also from further use in the breeding program. The other breeding parents would still be carriers for the *c* and *S* genes. Molecular markers can assist to breed away these undesirable genes, without simply discarding the best germplasm lines, which happen to contain these deleterious recessive alleles.

4.3 QUANTITATIVE TRAIT INHERITANCE AND APPLICATIONS IN OIL PALM BREEDING

4.3.1 Inheritance of Quantitative Traits

Quantitative traits are traits that are controlled by many genes (polygenes) usually with cumulative small effects with environmental effects superimposed on them giving a continuous normal curve distribution (Falconer and Mackay 1996; Bernardo 2002).

The traits are measured rather than scored and the parameters studied are means and variances (genetic and environmental) where

P = phenotypic effect or value of an individual or effect, G = genotypic effect, E = environmental effect.

P = G + E; the phenotypic value of an individual is the sum of the effects of its genotype and environment.

G = A + D + I; the genotypic effect is the sum of the additive (A), the non-additive or dominance (D) and epistatic or non-allelic interaction effects (I) which comprises AA + AD + DD.

E = E + GE; genotype × environment interaction effect (GE) is confounded with the environmental effect if the phenotypes are evaluated in one environment.

The corresponding variances for a group of individuals are

$V_P = V_G + V_E$

$V_P = V_A + V_D + V_I$ and

Broad-sense heritability, $h_B^2 = V_G/V_P$; the proportion of the phenotypic variation attributed to the genotype which can only be captured by cloning

Narrow sense heritability, $h_N^2 = V_A/V_P$; the proportion of the phenotypic variation attributable to the additive effects of genes which can be transmitted to the next generation by breeding

4.3.2 Estimation of Genetic Variance and Heritability

The development of quantitative genetic theory and principles owes much of its origin to the need to provide the scientific basis and methods to improve plant and animal breeding. Estimation of genetic variance and heritability is needed to provide answers to the following questions (Dudley and Moll 1969):

- Is there sufficient genetic variability in the desired trait within the germplasm pool for breeding improvement?
- Which germplasm holds the best promise for improved breeding material and the extent of testing (replications, locations, seasons, or years) needed to identify the superior genetic populations and their superior parents?
- Which breeding procedure is most efficient to achieve the desired improvement in the selected individual and for the combination of traits and which cultivar type (hybrid, pure line, synthetic, clone, etc.) that is the most appropriate to achieve this goal.

Additionally, the estimates can help the breeder to

- Predict the response to recurrent selection ($R = ih^2V$, Figure 4.1 and Appendix 4B.1) which aims to sustain steady breeding progress over a longer period for which a large V_A is helpful
- Allocate limited resources in field performance trials
- Construct SIs and BLUP, and predict single-cross performance

Methods for estimating genetic variance and heritability are illustrated in Appendix 4A.1

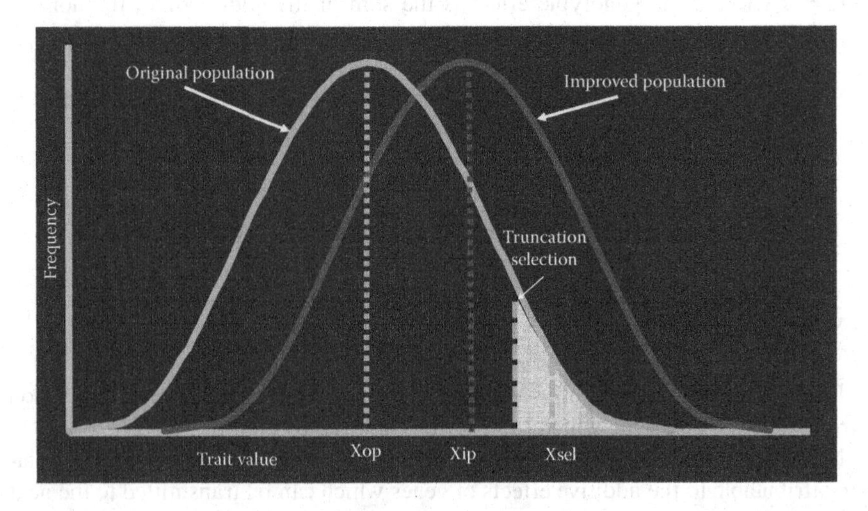

FIGURE 4.1 Response to Selection (Xip – Xop), $RS = i\ h^2\ V_p$, where i = selection intensity i.e. selection differential ((Xsel – Xop) in standard deviation units (SD), h^2 = heritability (narrow sense for breeding, broad sense for cloning), and V_p = square root of phenotypic variance; where Xop, Xsel, and Xip are the means of the original, selected and improved populations respectively.

4.3.2.1 Genetic Assumptions for Estimating Genetic Variance and Heritability

Quantitative genetic theory and principles are developed from a random mating population in Hardy–Weinberg equilibrium and inferences made from genetic experimental populations derived from this population refer back to it or a future reconstituted random mating population (Falconer and Mackay 1996).

The assumptions needed for valid estimations of the genetic variance and heritability are

- The parents have not been selected and are random members of the genetic population and that experimental errors are independent
- The genetic population is assumed to be at random mating equilibrium with linkage equilibrium
- Normal diploid and solely Mendelian inheritance
- No environmental correlations among progenies
- The progenies are not inbred and can be considered random members of some non-inbred population

4.3.2.2 Applications

Estimations of genetic variance and heritability have been attempted by oil palm breeders to provide information to guide them to increased breeding efficiency (Soh and Tan 1983). North Carolina Model 1 (hierarchal) and Model 2 (factorial) mating designs and the parent–offspring method have been most commonly used. Heritability estimates for various yield, component, and crop physiological traits have been compiled by Soh and Tan (1983), and Hardon et al. (1985) and recently in greater detail by Corley and Tinker (2016). Expectedly, heritabilities for yield (FFB or fresh fruit bunch, OY or oil yield, and KOY or kernel oil yield) and percent fruit to bunch (F/B%) tended to be very low, low to moderate for bunch weight (BW) and bunch number (BNo), and higher for percent mesocarp to fruit (M/F%), kernel to fruit (K/F%), and shell to fruit (S/F%).

With the use of data from breeding experiments derived from selected parents from restricted populations, the assumptions of random choice of parents, no correlation of genotypes at separate loci, and a definite reference population, necessary for valid quantitative genetic analyses could not be met and the biases incurred could not be estimated. Commonly, the estimates have been derived from inter-population Deli D × African P progeny tests of selected parents. The estimates would not apply to the Deli D nor the Africa P random mating parental populations. At best they would reflect the parental estimates in the inter-population hybrid combination which would be biased by selection and linkage disequilibrium. Heritability is estimated for the purpose for selection and selection response $\left(R = ih\delta_A^2\right)$, and thus based on the appropriate selection unit which is usually the individual (see Figure 4.1). Sometimes heritability estimates were obtained from family mean data rather than from individual data and would thus be higher as $h^2 = V_G/(V_G + V_E/N)$, where N is the number of individuals in the family. Family heritability would perhaps be useful only in combined family and individual selection (FIPS; see Chapter 6).

TABLE 4.1
Covariance of Relatives and Estimation of Genetic Variance and Heritability

Relationship/ Covariance	Genetic Variance	Regression (b) or Intraclass Correlation (t)	Heritability (h^2)
Between—within individuals	$V_G + V_{Eg}$	$r = V_G + V_{Eg}/(V_G + V_{Eg} + V_{Es})$ or V_p	$r = h_B^2$ (upper estimate)
Parent/Clone—clone		$b = 1$	$b = h_B^2$
Offspring—one parent	$1/2\ V_A + 1/4\ V_{AA}$	$b = 1/2\ V_A/V_P$	$b = 1/2\ h_N^2$
Offspring—mid-parent	$1/2\ V_A + 1/4\ V_{AA}$	$b = 1/2\ V_A/V_P$	$b = h_N^2$
Half-sibs	$1/4\ V_A + 1/16\ V_{AA}$	$t = 1/4\ V_A/V_P$	$t = 1/4\ h_l^2$
Full-sibs	$1/2\ V_A + 1/4\ V_D + V_{Ec}$[a]	$t = (1/2\ V_A + 1/4\ V_D + V_{Ec})/V_p$	$t >/= 1/2\ h_N^2$

Source: Adapted from Falconer, D.S. and Mackay, T.F.C. 1996. *Introduction to Quantitative Genetics*. Longman, England.

Note: V_{Eg} = general environmental variance, V_{Es} = special environmental effects, and V_{Ec} = common environment variance. In computations of b and t, interactions are confounded in the main genetic effects.

[a] Includes all the epistatic interactions V_{AA}, V_{AD}, V_{DD}, and their product coefficients.

4.3.2.3 Appropriate Use of Heritability Estimation Methods

Some examples are given below

1. Potier et al. (2006): Broad-sense heritabilities were appropriately computed from ortet (clonal parent)—ramet (clonal offspring) correlation or regression, that is, $h_B^2 = r = COV_{PO}/(V_P \times V_O)^{1/2}$
2. Lawrence and Rajanaidu (1986): The experimental materials were open-pollinated-derived palms from prospections made in different restricted or isolated locations in Nigeria. The open-pollinated progenies would range from half-sibs (full open-pollination) to full-sibs (full self-pollination). Consequently, heritability was computed using intraclass correlation t = 1/4 h^2 and t = 1/2 for both half-sibs and full-sibs (Falconer and Mackay 1996; see Table 4.1) to the range.
3. Soh et al. (2003): Used a combination of methods to estimate the within-family genetic (genotypic) variability in D × P crosses for ortet (clonal parent) selection: between (genetic) and within-family variance components, between (genotypic) and within clone (environmental) variance components, difference between pooled within family (genotypic) and pooled within clone (environmental) variance, from a set of three genetically connected trials. This is detailed in Appendix 4E.1.

Genetic variance and heritability estimates merely serve as a guide for selection decisions as any estimate is unique, that is, dependent on the population and environment sampled and the estimation method used. For prediction of selection response for a particular population and environment, specific genetic variance and heritability estimates obtained from the population concerned should be used. In lieu of reliable specific estimates or if the responses are for different environments, average heritability (from literature or from various estimation methods as in Soh et al. 2003) may be substituted (Soh 1994). In routine oil palm breeding, heritability estimates are not available for the target population, the oil palm breeder relies on his experience and intuition. For example with the more heritable yield (BNo), fruit (M/F, K/F), and vegetative growth (height, girth, and petiole size) component traits can be confidently selected by eye. Breeders also need to know the history of the palms under selection, whether they have been derived from highly selected or inbred that will affect the expression of the quantitative traits, for example, inbreeding or heterotic effects and could only be revealed upon subsequent progeny-testing. This then brings us to the concept of combining ability which is the relevant and much applied principle in oil palm breeding and plant breeding in general.

4.3.3 Combining Ability and Breeding Value Estimates

General combining ability (GCA) is the mean progeny performance (corrected from the mean of all crosses) of a parent or line in all its crosses (Sprague and Tatum 1942; Griffing 1956; Falconer and Mackay 1996).

The specific combining ability (SCA) is the deviation of the cross mean, X (mean of total crosses) from the GCAs of the two combining parents (M, F) of the cross, that is,

$X = GCA_F + GCA_M + SCA_{FM}$ and their variances:

$\delta^2_X = \delta^2_{GCA(F)} + \delta^2_{GCA(M)} + \delta^2_{SCA(FM)} = 2\,\delta^2_{GCA+}\delta^2_{SCA}$ with no differentiation between the parents.

If the two parents have been derived from the same random mating population then:

$$\delta^2_{GCA} = V_A \text{ and } \delta^2_{SCA} = V_D$$

The breeding value (BV) of a parent is twice the deviation of the mean of its progenies from the population mean, that is, BV = 2GCA. If however the parents have not been derived from a random mating population but from a breeding population, estimation of genetic variance is not appropriate, but GCA and SCA effects are and they refer to the breeding population under study and would provide information on the relative importance of the additive (GCA) and nonadditive (SCA) effects for the development of breeding and selection strategies. This is revealed by computing: MS_{GCA}/MS_{SCA} ratio or $SS_{GCA}/SS_{Total} = r^2$ (coefficient of determination of total variation in the progeny due to GCA) (Baker 1978; Soh et al. 1984; Simmonds and Smartt 1999). When GCA or additive genetic effects predominates, mass selection, simple recurrent selection, and recurrent selection for GCA (based on top-cross test with a broad (genetic based) tester are effective (Allard 1960). When SCA effects predominate, however, hybrid breeding through inbreeding of the parents or recurrent selection for SCA (using inbred tester in top-cross test) is advisable.

The BV of the breeding parent is loosely equated to 2GCA as the former is couched in random mating population terms. The computation methods for GCA and SCA estimates are illustrated in Appendix 4A.1 (Tables 4A.6 and 4A.7)

Combining ability is the most useful tool to an oil palm breeder. It is straightforward to compute the GCA or SCA effects or just compare the average values of crosses from each parent (GCA) and outstanding cross means (SCA). The breeder can also deduce these "intuitively" from field observations prior to confirmation from data analyses.

4.3.4 Genetic Covariance and Correlation

Traits especially quantitative traits tend to be genetically correlated, for example, yield component traits. Genetic correlation can arise from multiple or associated effects of genes or pleiotropy, for example, yield components or genetic linkage. Genetic linkage effect is considered to be transient especially with wide crosses and can theoretically be removed by repeated hybridization. If two desirable traits are genetically correlated positively selection in one trait can result in favorable response in the other and if negative then it is the converse. Selection for a correlated response is useful if selection in the target trait is less efficient than its correlated trait due to lower heritability, less precise or costly measurement, inapplicable sex-related trait in the former, and genetic correlation for the two traits high. The same trait expressed in two different environments can be considered as two different traits and indeed may be so because of a different physiological and genetic basis. If there is a high genetic correlation, that is, genotype–environment interaction is low, then essentially the two traits are the same and selection needs only be conducted in one environment.

Estimation methods of genetic covariance and correlation and correlated response to selection are illustrated in Appendix 4C.1 (Tables 4C.1 and 4C.2).

Estimates of genetic covariance and correlations are seldom attempted in oil palm unless needed in the development of SI and BLUP (see Chapter 6).

Again oil palm breeders practice correlated response selection intuitively or indirectly by selecting the more heritable component traits, for example, BNo, O/M, HT for OY, and HI (harvest index). Marker-assisted selection is essentially for correlated response where each molecular marker linked to a QTL could account for a certain portion of the quantitative trait variation, that is, coefficient of determination or r^2.

4.3.5 Genotype–Environment Interaction

When different genotypes perform differently in different environments (e.g., location, season, year, planting density, irrigation, and fertilizer application) in the expression of traits (e.g., yield, crop quality, and stress resistance) GE interaction is said to exist. In such a situation:

P = G + E + I (or GE) for an individual evaluated in a single environment

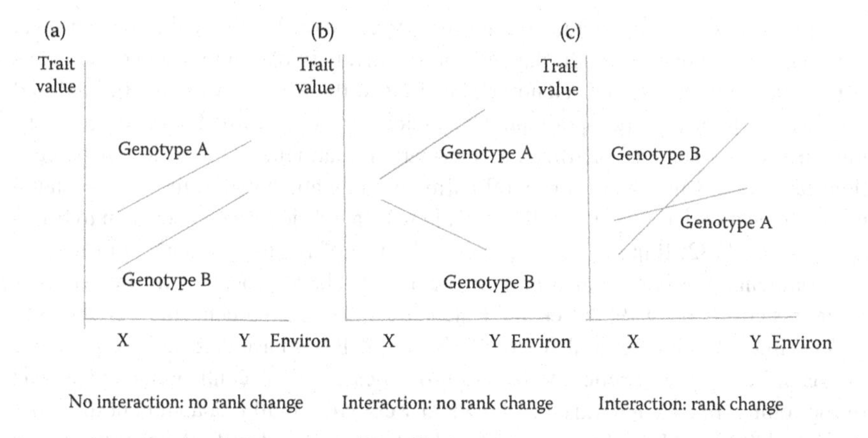

FIGURE 4.2 Theoretical examples of G × E interactions. (a) No interaction: no rank change, (b) interaction: no rank change, and (c) interaction: rank change.

$V_P = V_G + V_E + V_I$ for different individuals or groups of individual tested in multiple environments. The patterns of GE interaction are illustrated in Figure 4.2.

Studies in GE are important in plant breeding to ascertain whether varieties selected in one environment will perform equally well in another with respect to themselves and in relation to others. Interaction may or may not involve a change in ranking of the varieties. They also bring forth the concept of breeding for varietal stability and varieties with general adaptation versus varieties with specific adaptation. There may be biological mechanisms underlying GE interactions which may show phenotypic plasticity or sensitivity within a genotype or through genes differentially expressed. Statistically V_I comprises two parts, the differential genetic variances under differential environments, that is, scaling effect, and the deviation of the genetic correlation of the trait among the genotypic groups under the different environments, that is, ranking effect (Yates and Cochran 1938; Finlay and Wilkinson 1963; Eberhart and Russel 1966; Perkins and Jinks 1968).

The classical 2-way ANOVA will only detect the presence or absence of GE effects (Appendix 4D.1, Table 4D.1) (Allard 1960). If the variety or genotype trials are replicated over many environments, the mean of each variety can be regressed onto the trial mean for each environment. The "average" variety will have a slope or b = 1 passing through 0. For the interacting varieties; a "stable" or unresponsive variety will have $b < 1$, and "unstable" or responsive variety $b > 1$ (Yates and Cochran 1938; Finlay and Wilkinson 1963). Adaptation or adaptability of a variety is measured by its overall mean. Eberhart and Russel (1966) suggested inclusion of the residual deviation from regression besides slope as measure of varietal stability.

Classical analysis of GE by 2-way ANOVA which can be followed-up with joint regression analysis is illustrated in Appendix 4D.1 (Table 4D.2)

Most of the GE studies made in oil palm were based on the classical 2-way ANOVA method. The expression of GE depends on the diversity of genotypes and the diversity of environments tested. Genotypes can be in the form of varieties (mixed and single-cross hybrids), open-pollinated accessions or clones while E can be agro-ecological zones or spacing, fertilizer, and irrigation treatments. The earlier misperception that GE was not important in oil palm could be attributed to the use of related mixed hybrid varieties (genetically variable within hybrids, less so between hybrids) (Rosenquist 1982; Rajanaidu et al. 1986); the use of related progenies in different environments (Cochard et al. 1993), or the use of related progenies in similar environments (Rosenquist 1982; Lee and Rajanaidu 1999). Subsequent studies (Obesisan and Parimoo 1985; Ong et al. 1986; Corley et al. 1993) have detected the presence of GE effects using genetically diverse progenies but their contributions (3%–4%) to the total experimental variance for yield were still small (Rajanaidu et al. 1993; Rafii et al. 2000). Most of these papers did not attempt to identify the characteristics that conferred stability in the progenies or provide a biological explanation of the GE effects (Caligari 1993; Lee and Rajanaidu 1999). Absence of statistical significance for GE does not imply that a few progenies (especially from a large set) do not exhibit them. Corley et al. (1993), using a simple analytical approach found that progenies with few bigger bunches rather than many smaller bunches tended to yield poorly in stressful environments because the abortion of a single large bunch would result in a considerable loss in yield. Plasticity of bunch weight was also found to vary between progenies. In terms of stability analysis, Lee and Donough (1993) used the Francis and Kannenberg (1996) genotype-grouping technique which subdivides the genotypes into four groups based on their yield and coefficient of variation (CV). Genotypes with high mean yield and low CV which fall into Group 1 are considered to be stable genotypes.

Currently, with the advent of near true hybrid, clonal hybrid, and clonal cultivars, GE effects with respect to location, spacing, fertilizer, and climate change induced abiotic and biotic stress factors, will assume increased importance (Corley and Donough 1992; Lee and Donough 1993; Corley et al. 1995; Soh et al. 1995; Donough et al. 1996; Kushairi et al. 2001; Palat et al. 2006; Corley and Palat 2013; Soh 2016) and may warrant consideration in the development of cultivars or ideotypes suited for different situations, in line with the trend toward precision farming or precision plantation practices (Chew 1998). Also previously for logistic and economic reasons, GE or adaptability trials tended to be done at the end of the breeding program prior to cultivar release (Allard 1960; Simmonds and Smartt 1999). In the current context as discussed, such trials should strategically be done at the start of a breeding program as adaptability traits may be lost in the intensive process of selecting for high yields (see Future Prospects—Chapter 13).

In terms of selection, an attractive alternative approach used in animal breeding to handle GE effects is to treat a trait in different environments as separate but genetically correlated traits (Falconer and Mackay 1996). They can then be incorporated into SI to predict selection response for the trait in the second environment as a correlated response to selection for the trait in the first environment (White and Hodge 1989; Yamada 1993; Soh 1999).

4.4 CONCLUSIONS

Phenotypic selection presumably based on the assumption of "what you see is what you get" has served the plant breeder well and will continue to do so even in the current age of molecular or genomics-assisted breeding. For qualitative traits this is largely true although sometimes the phenotype may be obscured or modified by gene interaction, polygenic, multiallelic, and epigenetic (gene expression) effects. For quantitative traits, for example, yield, vegetative growth traits, some resistance and adaptation traits, this is seldom the case. Quantitative genetic theory and tools, for example, genetic variance and covariance, heritability, selection methods, selection differential and genetic progress, developed are based on random mating populations derived from unselected parents, a situation rarely achieved in applied breeding programs especially for perennial tree crops such as the oil palm. Hence, the estimates obtained from breeding programs would be imprecise and biased. Nevertheless, average values from such reported estimates would serve as a guide to the plant breeder complemented by knowledge of the selection history of the breeding populations to derive an appropriate breeding program cum selection method.

The concepts of GCA and SCA are routinely applied by plant breeders as they do not have strict genetic assumptions, their estimates being easily computed and directly applicable to the population under study. BLUP BV's estimations which also has less genetic and statistical assumption restrictions and coupled with readily available software programs are now in routine use in animal and tree breeding programs (Henderson 1984; White and Hodge 1989) and this is likely to follow suit in oil palm (Soh 1994; Purba et al. 2001; see Chapter 6). BLUP has also found use in GS (which is MAS for quantitative traits) in oil palm (Wong and Bernardo 2008; Cros et al. 2015).

APPENDIX 4A.1 ESTIMATION OF GENETIC VARIANCE AND HERITABILITY

4A.1.1 Estimation Methods

Estimation of genetic variance and heritability is based on the resemblance or coefficient of relationship or covariance and expressed as genetic variances (Becker 1975; Hallauer and Miranda 1981; Falconer and Mackay 1996; Simmonds and Smartt 1999) (see Table 4.1).

The regression approach is appropriate for computing heritability for linear (parent–offspring) relationship and intraclass correlation for collateral (sib) relationships.

Full-sibs are not preferred for estimating V_A and h_N^2 due to confounding of V_D and V_{Ec} (common environmental variance).

4A.1.2 Mating Designs and Genetic Analyses

Mating designs and genetic experiments are needed to obtain the types of relatives for genetic analysis to derive the genetic variance and heritability estimates.

TABLE 4A.1
Offspring–Male Parent Regression Data Structure and Analysis

Male Parent S	Individual Progeny P
S_1	P_1
S_2	P_2
S_3	P_3
S_4	P_4
S_n	P_N

Note: Offspring–parent covariance, Cov SP = 1/2 V_A.
Offspring–parent regression, b_{op} = Cov SP/Vs.
Heritability, $h_N^2 = 2\ b_{op}$.

The common mating designs and their genetic analyses are illustrated (see Tables 4A.1–4A.7).

Parent–offspring regression (Table 4A.1). This is intuitively the best approach.

Repeatability (R, see Tables 4A.2 and 4A.3). This is obtained easily by repeated (in time or space) measurements within and between individuals, sets the upper limit of heritability for a trait. It is also used to decide on the limit on the number of repeated measurements needed for a trait:

$$V_{Pn}/V_P = 1 + r(n-1)/n$$

where V_{Pn} = variance of multiple measurements, V_P = variance of a single measurement, n = number of measurements, $r = V_G + V_{Eg}/(V_G + V_{Eg} + V_{Es})$ or V_p, and V_{Eg} = general environmental variance contributing to permanent difference between individuals, for example, developmental defects and V_{Es} = special environmental effects occurring on an individual due to localized or temporary circumstances. The inverse C (hyperbolic) graph of V_{Pn}/V_P against n will indicate the point whereby additional measurements would result in negligible gain in precision (Falconer and Mackay 1996). With a highly heritable trait this will be reached in a couple of measurements whereas for lowly heritable traits a higher number of additional

TABLE 4A.2
Repeatability Data Structure

	Measurements	M1	M2	M3	Individual Mean
Individuals					
I_1		I_1M_1	I_1M_2	I_1M_3	I_1
I_2		I_2M_1	I_2M_2	I_2M_3	I_2
I_3		I_3M_1	I_3M_3	I_3M_3	I_3

TABLE 4A.3
ANOVA for Estimation of Repeatability

SV	df	MS	EMS
Between individuals	n–1	MS_B	$\delta_w^2 + m\delta_B^2$
Between measurements within individuals	n(m–1)	MS_W	δ_w^2

Note: $\delta_B^2 = (MS_B - MS_W)/m$
$\delta_w^2 = MS_W$
Repeatability, $R = \delta_B^2/\left(\delta_B^2 + \delta_w^2\right)$

measurements would be needed. This approach would be useful in determining the number of years of bunch yield recording and number of bunches for bunch analysis needed to characterize a palm in breeding. This presumably has been attempted although it is not so evident in published literature (Hardon et al. 1985) as the correlation between early and later periods of recording approach (Blaak et al. 1963; Blaak 1965; Corley et al. 1988).

North Carolina Model 1 (NCM1)/Nested/Hierarchal Design (see Tables 4A.4 and 4A.5). This is perhaps the most common mating design practiced in oil palm breeding especially those adopting the modified recurrent selection (MRS) or FIPS schemes where each P is progeny-tested with a random sample of D mother palms (Chapter 7). The GCA means and variance of the Ps are estimated.

In the *poly-cross* mating design where a number of plants from all the lines to be tested are planted close together for natural cross-pollination, the seeds of one line would be a mixture of random crosses with other lines. The mean performances and variance of the lines computed will be the GCA and V_{GCA} of the female parents. A similar case is when open-pollinated seeds are collected from randomly selected

TABLE 4A.4
North Carolina Model 1 (NCM1)/Nested/Hierarchal Mating Design

Female (F)	Male (M)	Cross	Mean (Effect)	General Combining Effect (GCA)
F_1	M_1	$F_1 \times M_1$	M_1	GCA_{M1}
F_2	M_1	$F_2 \times M_1$		
F_3	M_1	$F_3 \times M_1$		
F_4	M_2	$F_4 \times M_2$	M_2	GCA_{M2}
F_5	M_2	$F_5 \times M_2$		
F_6	M_2	$F_6 \times M_2$		

Note: The GCA of a parent is the average performance of the parent in hybrid combinations reflecting its additive genotype. The SCA or nonadditive genotype of a cross is the difference between its mean (corrected for the population mean) and the sum of the GCAs of its parents. With this mating design (top-cross test) only the GCA or additive genetic variance and effects of the male parents can be estimated.

TABLE 4A.5
ANOVA for NCM1 Design

Source of Variation (SV)	Degree of Freedom (df)	Sum of Squares (SS)	Mean Square (MS)	Expected Mean Square (EMS)
Between males	m−1	SS_M	MS_M	$\delta_w^2 + p\delta_f^2 + pf\delta_m^2$
Between females within males	m(f−1)	$SS_{F/M}$	$MS_{F/M}$	$\delta_w^2 + p\delta_f^2$
Between progeny within females	mf(p−1)	SS_W	MS_w	δ_w^2

Note: m = number of males, f = number of females per male, and p = number of progeny plants per female.
Computations:
$\delta_m^2 = (MS_M - MS_{F/M})/pf = \text{Cov HS} = 1/4\,V_A,\ V_A = V_{GCA} = 4\,\delta_m^2$
$\delta_f^2 = (MS_{F/M} - MSw)/p = \text{Cov FS} = 1/2\,V_A + 1/4\,V_D$
$\delta_m^2 + \delta_f^2 + \delta_w^2 = V_P,\ h_N^2 = V_A/V_P$

plants from an open-pollinated or natural population (Lawrence and Rajanaidu 1986). *Top-cross* mating design is when a line is crossed with random individuals from the base population and gives GCA estimates as well.

North Carolina Model 2 (NCM2)/Factorial Design (see Tables 4A.6 and 4A.7). In this mating design, a set of parents is crossed *inter se* (factorial) combinations with another set of parents. This design allows the estimation of SCA effects and variances besides those of GCA. This design is preferred by the breeding programs which aim to exploit both GCA and SCA, that is, modified reciprocal recurrent selection (MRRS) schemes in oil palm.

Incomplete factorial/connected mating designs, for example, alpha Incomplete factorial design is frequently used to estimate BV or GCA of a large number of parents with a reduced number of factorial combinations to allow detection of SCA if it exists. Incomplete factorial can also arise due to the inability to achieve the full set of factorial crosses in time. Balanced incomplete factorial designs are available to

TABLE 4A.6
North Carolina II (NCM2)/Factorial Mating Design

Parents	M1	M2	M3	F Mean/Effect	F GCA (Corrected Mean from U)
F_1	$F_1 \times M_1$	$F_1 \times M_2$	$F_1 \times M_3$	F_1	GCA_{F1}
F_2	$F_2 \times M_1$	$F_2 \times M_2$	$F_2 \times M_3$	F_2	GCA_{F2}
F_3	$F_3 \times M_1$	$F_3 \times M_2$	$F_3 \times M_3$	F_3	GCA_{F3}
M Mean/Effect	M_1	M_2	M_3		
M GCA	GCA_{M1}	GCA_{M2}	GCA_{M3}	U (overall) mean	

Note: Cross mean, $X_{MF} = U + GCA_M + GCA_F + SCA_{MF}$.
$SCA_{MF} = X_{MF} - U - GCA_M - GCA_F$.

TABLE 4A.7
ANOVA for NCM2 Design

SV	df	MS	EMS
Between males	m–1	MS_M	$\delta^2_w + p\delta^2_{mf} + pf\delta^2_m$
Between females	f–1	MS_F	$\delta^2_w + p\delta^2_{mf} + pm\delta^2_f$
Males × females	(m–1)(f–1)	MS_{MF}	$\delta^2_w + p\delta^2_{mf}$
Between progeny within females	mf (p–1)	MS_W	δ^2_w

Note: m = number of males, f = number of females per male, p = number of progeny plants per female.
Computations:
Random Model
$\delta^2_m = (MS_M - MS_{MF})/pf = \text{Cov HS} = 1/4\ V_A, V_{Am} = V_{GCAm} = 4\ \delta^2_m$
$\delta^2_f = (MS_F - MS_{MF})/pm = \text{Cov HS} = 1/4\ V_A, V_{Af} = V_{GCAf} = 4\ \delta^2_f$
$\delta^2_{mf} = \text{Cov FS} - \text{Cov HS}_m - \text{Cov HS}_f = 1/4\ V_D, V_{SCA} = 4\ V_D$
$\delta^2_m + \delta^2_f + \delta^2_{mf} + \delta^2_w = V_P,$
$h^2_N = V_A/V_P$
Fixed Model
Ratios MS_M/MS_{MF}, MS_F/MS_{MF} indicate relative importance of GCA and SCA effects.
$(SS_M + SS_F)/SS_{Total} = r^2_{GCA}$ = coefficient of determination (contribution of GCA or additive effects) to the total variation among the crosses evaluated.

achieve better and balanced precision of the estimates. General linear model (GLM) or general least-squares method is adequate to analyze and estimate GCA and SCA effects. Maximum likelihood (ML), in particular ReML (residual ML) and mixed model methods, for example, BLUP are more versatile and robust in handling highly unbalanced data and can integrate information from other relatives and sources (Henderson 1975, 1984; Soh 1994; Bernardo 1996; Chapter 6 in this book).

APPENDIX 4B.1 SELECTION RESPONSE AND REALIZED HERITABILITY

The response to selection is given by the formula:

$$R = h^2 S$$

where R is the response or the difference between the phenotypic mean of the offspring of the selected parents and the mean of the original unselected parental population, and S, the selection differential, is the deviation of the selected parental mean from its population mean. If the trait can be assumed to be randomly distributed, S can be standardized to give selection intensity, that is, $i = S/\delta_P$, consequently $R = ih^2\delta_P$ or $R = ih\delta^2_A$

Figure 4.1 illustrates this. The expected selection responses with respect to the different selection methods, that is, individual, family, sib, within family and combined selection, and their heritabilities are given by Falconer and Mackay (1996).

It can be easily seen that by manipulating the equation $R = ih^2\delta_P$ we can obtain realized heritability, that is, $h^2_{(realized)} = R/i\delta_P$ if the values for the right-hand side of the equation are available. Realized heritability is the true heritability operating during the selection process.

APPENDIX 4C.1 ESTIMATION OF GENETIC COVARIANCE, CORRELATION, AND CORRELATED RESPONSE TO SELECTION

The estimation methods for genetic covariance, correlation, and correlated response to selection are merely extensions of the genetic variance methods to multiple traits. Two examples, parent–offspring and NCM1 methods are illustrated (see Tables 4C.1 and 4C.2).

APPENDIX 4D.1 ESTIMATION OF GENOTYPE–ENVIRONMENT INTERACTION EFFECTS

The ANOVA for GE experiments and its incorporation of joint regression analysis for stability analysis are illustrated in Tables 4D.1 and 4D.2.

APPENDIX 4E.1 ESTIMATES OF WITHIN-FAMILY GENETIC VARIABILITY FOR CLONAL SELECTION IN OIL PALM

Soh et al. (2003) illustrated the genetic assumptions needed to compute the various genetic variance estimates and to compare the results of the various methods used.

Within-family genetic and environmental variabilities and heritability were estimated from a D × P (NCM1 mating design) progeny test trial, a clonal trial, and a clonal cum progeny test (NCM1 mating design) trial linked genetically. The clones in the second were derived from seedling embryos that were progeny reproductions

TABLE 4C.1
ANACOVA and Estimation of Genetic Correlation from Offspring–Male Parent Relationship Data Structure and Analysis

Male Trait X	Parent (S) Trait Y	Individual Trait X	Progeny (P) Trait Y	Cross Products S_XP_Y	Cross Products S_YP_X
S_{1X}	S_{1Y}	P_{1X}	P_{1Y}	$S_{1X}P_{1Y}$	$S_{1Y}P_{1X}$
S_{2X}	S_{2Y}	P_{2X}	P_{2Y}	$S_{2X}P_{2Y}$	$S_{2Y}P_{2X}$
S_{3X}	S_{3Y}	P_{3X}	P_{3Y}	$S_{3X}P_{3Y}$	$S_{3Y}P_{3X}$
S_{4X}	S_{4Y}	P_{4X}	P_{4Y}	$S_{4X}P_{4Y}$	$S_{4Y}P_{4X}$
S_{nx}	S_{nY}	P_{nX}	P_{ny}	$S_{nx}P_{ny}$	$S_{nY}P_{nX}$

Note: Genetic correlation, $r_G = (COVS_XP_Y + COVS_YP_X)/\sqrt{(COVS_XP_X + COVS_YP_Y)}$.

TABLE 4C.2
ANACOVA and Estimation of Genetic Correlation from NCM1 Mating Design

Source of Variation (SV)	Degree of Freedom (df)	Sum of Cross Products (SCP)	Mean Cross Products (MCP)	Expected Mean Cross Products (EMCP)
Between males	m–1	SCP_M	MCP_M	$cov_w + pcov_f + pfcov_m$
Between females within males	m(f–1)	$SCP_{F/M}$	$MCP_{F/M}$	$cov_w + pcov_f$
Between progeny within females	mf(p–1)	SCP_W	MCPw	cov_w

Note: $cov_w = MCPw$
$cov_f = (MCP_{F/M} - MCPw)/p = 1/4\ COV_A + 1/4\ COV_D$
$cov_m = (MCP_M - MCP_{F/M})/pf = 1/4\ COV_A$
$r_A = 4\ cov_m / \sqrt{(4\ \delta^2{}_m x + \delta^2{}_{mY}}$
Correlated response of trait Y to direct selection on trait is given by the regression equation:
$CRy = b_{(A)YX} Rx$
where
$Rx = ihx\delta^2_{AX}$, the direct response to selection of the primary trait X
$b_{(A)YX} = COV_A / \delta^2_A x = r_A \delta^2_{AY} / \delta^2_{AX}$ consequently
$CRy = i\ hxh_Y r_A \delta_{PY}$
where $hxh_Y r_A$ is co-heritability, that is, the response of a correlated trait can be predicted if the heritabilities of the two traits and their genetic correlation are known.

TABLE 4D.1
ANOVA of Data from Genotypes Planted in Replicated Trials at Different Sites and Recorded Over Different Periods

SV	df	MS	EMS
Genotypes	g–1	MS_g	$\delta^2_e + r\delta^2_{gys} + rs\delta^2_{gy} + ry\delta^2_{gs} + rsy\delta^2_G$
Genotype × Year/Period	(g–1)(y–1)	MS_{gy}	$\delta^2_e + r\delta^2_{gys} + rs\delta^2_{gy}$
Genotype × Site	(g–1)(s–1)	MS_{gs}	$\delta^2_e + r\delta^2_{gys} + ry\delta^2_{gs}$
Genotypes × Year × Site	(g–1)(y–1)(s–1)	MS_{gys}	$\delta^2_e + r\delta^2_{gys}$
Error	(r–1)(gys–1)	MS_e	δ^2_e

Note: g, y, s refer to the respective number of genotypes, years, and sites and δ^2_G, δ^2_{gys}, δ^2_{gy}, δ^2_{gs}, their corresponding variance and interaction variance components.

TABLE 4D.2
ANOVA and Joint Regression Analysis of a Genotype × Environment Experiment

SV	df	MS	EMS
Environment (E)	k–1	MS_k	
Genotype (G)	g–1	MS_G	$\delta_e^2 + n\delta_I^2 + nk\delta_G^2$
G × E	(k–1)(g–1)	MS_I	$\delta_e^2 + n\delta_I^2$
Heterogeneity Regression	k–1	MS_{HR}	
Deviations	(k–1)(g–1)n–(k–1)	MS_D	
Error (e)	gk (n–1)	MS_e	δ_e^2

Note: n = number of replication, δ_I^2 = G × E variance components.

of the families tested in the first trial. The progenies in the third trial were the source families of the clones planted in the same trial.

From the clonal trial: $h_B^2 = \delta^2_{\text{between clone}} / \delta^2_{\text{between clone}} + \delta^2_{\text{within clone}} = V_g / V_g + V_e$, that is, between clone variance is completely genetic while within clone variance is completely environmental. The h_B^2 computed here is a precise estimate of the within-family heritability.

For the D × P progeny test data:

The statistical model:

$$Y = \mu + P + D + B + C + R,$$

where P = pisifera effect, D = dura effect, C = plot effect, and R = residual or error.

$$\delta^2_{\text{Psifera}} \text{ or } \delta^2_{\text{Dura}} = \text{Cov FS} - \text{Cov HS} = \frac{1}{4}\delta_A^2 + \frac{1}{4}\delta_D^2$$

where δ^2_{Ap} is the additive genetic variance when P from P population is crossed to D from D population and δ^2_{Ap} vice versa. If the P and D parents are from the same source population, $\delta^2_{Ap} = \delta^2_{Ad}$, dominance variance $\delta_D^2 = 4(\delta^2_{\text{Dura}} - \delta^2_{\text{Pisifera}})$ and $h_B^2 = 4\delta^2_{\text{Dura}} / (\delta^2_{\text{Pisifera}} + \delta^2_{\text{Dura}} + \delta^2_{\text{Resudual}} + \delta^2_{\text{Block}} + \delta^2_{\text{Plot}})$

Since the Ps and Ds are from separate populations,

δ^2_{Ap} is not equal to δ^2_{Ad} and δ_D^2 cannot be estimated unless two extreme assumptions or hypotheses are made: $\delta_D^2 = 0$ or $\delta^2_{Ad} = 0$

Hypothesis 1: $\delta_D^2 = 0, h_B^2 = 2\left(\delta^2_{\text{Dura}} + \delta^2_{\text{Pisifera}}\right) / \left(\delta^2_{\text{Pisifera}} + \delta^2_{\text{Dura}} + \delta^2_{\text{Resudual}} + \delta^2_{\text{Block}} + \delta^2_{\text{Plot}}\right)$

Hypothesis 2: $\delta^2_{Ad} = 0, h_B^2 = 4(\delta^2_{\text{Dura}}) / \left(\delta^2_{\text{Pisifera}} + \delta^2_{\text{Dura}} + \delta^2_{\text{Resudual}} + \delta^2_{\text{Block}} + \delta^2_{\text{Plot}}\right)$

Hypothesis 2 was chosen on the basis that with an inter-population cross, the confounded dominance, epistasis, and linkage disequilibrium would assume importance.

The computed h_B^2 is the between-family heritability.

The within-family heritability (appropriate for selecting palms for cloning) is

$$h^2_{Bwf} = h^2_B \times \frac{(1-r)}{(1-t)}$$

The within-family broad-sense heritabilities estimated would be similar but not as precisely estimated as those using the clonal data earlier.

Estimates were found to be generally similar with the various methods used.

REFERENCES

Allard, R.W. 1960. *Principles of Plant Breeding*. Wiley, New York.

Baker, R.J. 1978. Issues in diallel analysis. *Crop Science* 18: 533–536.

Becker, W.A. 1975. *Manual of Quantitative Genetics*. Third Edition. Washington State University, Pullman, Washington.

Beirnart, A. and Vanderweyen, T. 1941. *Contribution a l'Etude Genetique et Biometrique desVarieties d'Elaeis guineensis Jacq.* INEAC, *Serie Scientifique* 27.

Bernardo, R. 1996. Best linear unbiased prediction of maize single-cross performance. *Crop Science* 36: 50–56.

Bernardo, R. 2002. *Breeding for Quantitative Traits in Plants*. Stemma Press, Woodbury, Minnesota.

Blaak, G. 1965. Breeding and inheritance in the oil palm (*Elaeis guineensis* Jacq.). Part III. Yield selection and inheritance. *Journal of West African Institute of Oil Palm Research* 4: 262–283.

Blaak, G. 1970. Epistasis for crown disease in the oil palm (*Elaeis guineensis* Jacq.). *Euphytica* 19: 22–24.

Blaak, G., Sparnaaij, L.D., and Menendez, T. 1963. Breeding and inheritance in the oil palm (*Elaeis guineensis* Jacq.). Part II. Methods of bunch quality analysis. *Journal of West African Institute of Oil Palm Research* 4: 146–155.

Caligari, P.D.S. 1993. G × E studies in perennial tree crops: Old familiar friend or awkward unwanted nuisance. In: *Proceedings of 1991 International Society of Oil Palm Breeders Workshop on Genotype–Environment Interaction Studies in Perennial Tree Crops*. pp. 1–11, Palm Oil Research Institute of Malaysia, Kuala Lumpur.

Chew, P.S. 1998. Prospects for precision plantation practices in oil palm. *The Planter* 74: 661–684.

Chin, C.W. 1982. Segregation of fertility, fruit forms, and early yield performance of selected sibbed and outcrossed fertile pisifera progenies. In: *The Oil Palm in Agriculture in the Eighties*. Eds. E. Pushparajah and P.S. Chew, Vol. 1, pp. 63–68, Palm Oil Research Institute of Malaysia, Kuala Lumpur.

Chin, C.W. and Tang, T.L. 1979. The oil palm—Fertile pisifera. *The Planter* 55: 64–77.

Cochard, B., Noiret, J.M., Baudouin, L., Flori, A., and Amblard, P. 1993. Second cycle reciprocal recurrent selection in oil palm, *Elaeis guineensis* Jacq. results of Deli × La Mé hybrid tests. *Oléagineux* 48: 441–451.

Corley, R.H.V., Boonrak, T., Donough, C.R., Nelson, S., Dumortier, F., Soebagio, F.X., and Vallejo, G. 1995. Yield of oil palm clones in different environments. In: *Recent Developments in Oil Palm Tissue Culture and Biotechnology*. Eds. V. Rao, I.E. Henson and N. Rajanaidu, pp. 145–157. Palm Oil Research Institute of Malaysia, Kuala Lumpur.

Corley, R.H.V. and Donough, C. 1992. Potential yield of oil palm clones—The importance of planting density. In: *Proceedings of the Workshop on Yield Potential in the Oil Palm*. Eds. V. Rao, I.E. Henson and N. Rajanaidu, pp. 58–70. Int. Soc. Oil Palm Breeders, Kuala Lumpur.

Corley, R.H.V., Lee, C.H., Law, I.H., and Cundall, E. 1988. Field testing of oil palm clones. In: *Proceedings of 1987 International Oil Palm Conference on Progress and Prospects*. Ed. A. Hassan et al., pp. 173–185. Palm Oil Research Insitute of Malaysia, Kuala Lumpur.

Corley, R.H.V. and Palat, T. 2013. Maximising lifetime yield for greater economic sustainability. International Palm Oil Congress 'Green Opportunities for the Golden Crop'. Kuala Lumpur, November 19–21.

Corley, R.H.V., Tan, Y.P., Timti, I.N., and de Greef, W. 1993. Yield of oil palm progenies in Zaire, Cameroun and Malaysia. In: *Proceedings of the International Society of Oil Palm Breeders Workshop Genotype—Environment Interaction Studies I Perennial Tree Crops*. Eds. V. Rao, I.E. Henson, and N. Rajanaidu. pp. 46–54. International Society for Oil Palm Breeders, Kuala Lumpur.

Corley, R.H.V. and Tinker, P.B. 2016. *The Oil Palm*. Wiley Blackwell, Chichester, West Sussex.

Cros, D. et al. 2015. Genomic selection prediction accuracy in a perennial crop: Case study of oil palm (*Elaeis guineensis* Jacq.). *Theoretical and Applied Genetics* 128: 397–410.

Doebley, J., Stec, A., and Gustus, C. 1995. *Teosinte branched1* and the origin of maize: Evidence for epistasis and the evolution of dominance. *Genetics* 141: 333–346.

Donough, C.R., Corley, R.H.V., Law, I.H., and Ng, M. 1996. First results from an oil palm clone × fertiliser trial. *The Planter* 72: 69–87.

Dudley, J.W. and Moll, R.H. 1969. Interpretation and use of estimates of heritability and variances in plant breeding. *Crop Science* 9: 257–262.

Eberhart, S.A. and Russel, W.A. 1966. Stability parameters for comparing varieties. *Crop Science* 6: 36–40.

Falconer, D.S. and Mackay, T.F.C. 1996. *Introduction to Quantitative Genetics*. Longman, England.

Fick, G.N. and Qualset, C.O. 1973. Genes for dwarfness in wheat, *Triticum aestivum* L. *Genetics* 75: 531–539.

Finlay, K.W. and Wilkinson G.N. 1963. The analysis of adaptation in a plant breeding programme. *Australian Journal of Agricultural Research* 14: 742–754.

Fisher, R.A. 1918. The correlation between relatives on the supposition of Mendelian inheritance. *Transactions of the Royal Society (Edinburgh)* 52: 399–433.

Francis, T.R. and Kannanberg, L.W. 1996. Yield stability studies in short season maize I. A descriptive method of grouping genotypes. *Canadian Journal of Plant Science* 58: 1029–1034.

Griffing, B. 1956. Concept of general and specific combining ability in relation to diallel crossing systems. *Australian Journal of Biological Science* 9: 465–493.

Hallauer, A.R. and Miranda, J.B. 1981. *Quantitative Genetics in Maize Breeding*. Iowa State University Press, Ames, Iowa.

Hardon, J.J., Rao, V., and Rajanaidu, N. 1985. A review of oil palm breeding. In: *Progress in Plant Breeding*. Ed. G.E. Russell, pp. 139–163. Butterworths, UK.

Hartley, C.W.S. 1988. *The Oil Palm (Elaeis guineensis* Jacq.*)*. Longman Scientific and Technical Publication, Wiley, New York.

Henderson, C.R. 1975. Best linear unbiased estimation and prediction under a selection model. *Biometrics* 31: 423–447.

Henderson, C.R. 1984. *Applications of Linear Models in Animal Breeding*. University of Guelph, Ontario, Canada.

Kushairi, A., Rajanaidu, N., and Jalani, B.S. 2001. Response of oil palm progenies to different fertiliser rates. *Journal of Oil Palm Research* 13: 84–96.

Lawrence, M.J. and Rajanaidu, N. 1986. The genetical structure of natural populations and sampling strategy. In: *Proceedings of the International Workshop Oil Palm Germplasm and Utilisation*. pp. 15–26. Palm Oil Research Institute, Kuala Lumpur, Malaysia.

Lee, C.H. and Donough, C.R. 1993. Genotype–environment interaction in oil palm clones. In: *Proceedings of the 1991 International Society Oil Palm Breeders Workshop on Genotype—Environment Interaction Studies in Perennial Tree Crops*. pp. 106–117, September 9–14, 1991, Kuala Lumpur, Malaysia.

Lee, C.H. and Rajanaidu, N. 1999. Genotype–environment interaction in oil palm. In: *Proceedings of the Symposium of the Science of Oil Palm Breeding*. Eds. N. Rajanaidu and B.S. Jalani. pp. 96–111. Palm Oil Research Institute, Kuala Lumpur, Malaysia.

Lin, Z. et al. 2012. Parallel domestication of the shattering1 genes in cereals. *Nature Genetics* 44: 720–724.

Mendel, G. 1865. Experiments in plant hybridization. *Verhandlungren nuturforschender Verein in Brunn*, Abhandlungen.

Menendez, T. and Blaak, G. 1964. *Plant breeding division*. In: 12th Annual Report of Western African Institute for Oil Palm Research, pp. 45–75, Benin, Nigeria.

Morcillo, F. et al. 2013. Improving oil palm quality through identification and mapping of the lipase gene causing oil deterioration. *Nature Communications* 4: 2160.

Nelson, S.P.C. 1993. Production de semillas clonal y sexual en Dami—Futuros desarollos. *Palmas* 14 (Special Issue): 40–48.

Obesisan, I.O. and Parimoo, R.D. 1985. Stability and adaptation responses of dumpy deli palms (*Elaeis guineensis* Jacq.) to different environments. *Malaysian Applied Biology* 14: 84–88.

Okwuagwu, C.O. and Okolo, E.C. 1992. Maternal inheritance of kernel in the oil palm (*Elaeis guineensis* Jacq.). *Elaeis* 4: 72–73.

Okwuagwu, C.O. and Okolo, E.C. 1994. Genetic control of polymorphism for kernel to fruit ratio in the oil palm (*Elaeis guineensis* Jacq.). *Elaeis* 6: 75–81.

Ong, E.C., Lee, C.H., Law, I.H., and Ling, A.H. 1986. Genotype-environment interaction and stability analysis of bunch yield and its components, vegetative growth and bunch characters in the oil palm (*Elaeis guineensis* Jacq.). In: *New Frontiers in Breeding Research. Proceedings of the 5th International Congress on SABROA*. Eds. B. Napompeth and S. Subhadrabandhu. pp. 523–532. Bangkok.

Ong-Abdullah, M. et al. 2015. Loss of karma transposon methylation underlies the mantled somaclonal variant of oil palm. *Nature* 525: 533–537.

Palat, T., Chayawat, N., Rao, V., and Corley, R.H.V. 2006. The Univanich oil palm breeding programme and recent progeny trials in Thailand. International Society on Oil Palm Breeders Symposium 'Yield Potential in Oil Palm II', Pukhet, Thailand, November 27–28.

Pantzaris, T.P. 2000. *Pocketbook of Palm Oil Uses*. Palm Oil Res. Inst. Malaysia, Kuala Lumpur.

Perkins, J.M. and Jinks, J.L. 1968. Environmental and genotype-environmental components of variability. III. Multiple lines and crosses. *Heredity* 23: 339–356.

Potier, F., Nouy, B., Flori, A., Jacquarmard, J.C., Edyana Suryna, H., and Durand-Gasselin, T. 2006. Yield potential of oil palm (*Elaeis guineensis* Jacq.) Clones: Preliminary results observed in the Aek Loba genetic block in Indonesia. *International Society on Oil Palm Breeders Symposium "Yield Potential in Oil Palm II,"* Pukhet, Thailand, November 27–28.

Priwiratama, H., Djuhjana, J., Nelson, S.P.C., and Caligari, P.D.S. 2010. Progress of oil palm breeding for novel traits: Late abscission, virescence and long bunch stalks. *International Oil Palm Conference*, Yogyakarta, Indonesia, June 1–3.

Purba, A.R., Flori, A., Baudouin, L., and Hamon, S. 2001. Prediction of oil palm (*Elaeis guineensis* Jacq.) Agronomic performances using the best linear unbiased predictor (BLUP). *Theoretical and Applied Genetics* 102: 787–792.

Purugganan, M.D. and Fuller, D.Q. 2009. The nature of selection during plant domestication. *Nature* 457: 843–848.

Rafii, M.Y., Rajanaidu, N., Kushairi, A., Jalani, B.S., and Din, M.A. 2000. Evaluation of cameroon and zaire genetic materials. Paper presented at International Symposium on Oil Palm Genetic Resources and Utilization. Malaysian Palm Oil Board, Kuala Lumpur.

Rajanaidu, N., Tan, Y.P., Ong E.C., and Lee, C.H. 1986. The performance of inter-origin commercial D × P planting material. In: *Proceedings of the International Workshop on Germplasm and Utilization*. pp. 155–161. Palm Oil Research Institute of Malaysia.

Rajanaidu, N., Jalani, S., Rao, V., and Kushari, A. 1993. Genotype-environment interaction (GE) studies in oil palm (Elaeis guineensis) progenies. In: *Proceedings of the International Society of Oil Palm Breeders Workshop Genotype—Environment Interaction Studies I Perennial Tree Crops*. Eds. V. Rao, I.E. Henson, and N. Rajanaidu. pp. 12–32. International Society for Oil Palm Breeders, Kuala Lumpur.

Roberts, J.A. and Zubaidah, R. 2013. Genetic strategies to regulate height in oil palm. In: *Proceedings of 2013 PIPOC Paper A18*. Malaysian Palm Oil Board, Kuala Lumpur.

Roongsattham, P. et al. 2012. Temporal and spatial expression of galacturonase members reveals divergent regulation during fleshy fruit ripening and abscission in the monocot species oil palm. *BMG Plant Biology* 12: 150.

Rosenquist, E.A. 1982. Performance of identical oil palm progenies in different environments. In: *The Oil Palm in Agriculture in the Eighties*. Eds. E. Psuhparajah and P.S. Chew. Incorp. Soc. Planters, Kuala Lumpur. Vol.1. pp. 131–143.

Rosenquist, E.A. 1986. The genetic base of oil palm breeding populations. In: *Proceedings of the International Workshop on Oil Palm Germplasm and Utilization*. Eds. A.C. Soh, N. Rajanaidu and M. Nasir. pp. 16–27. Palm Oil Res. Inst. Malaysia, Kuala Lumpur.

Rosenquist, E.A. 1990. An overview of breeding technology and selection in *Elaeis guineensis*. In: *Proceedings of 1989 International Oil Palm Developmental Conference in Agriculture*. Eds. B.S. Jalani et al. pp. 5–25. Palm Oil Research Institute of Malaysia, Kuala Lumpur.

Rosenquist, E.A. 1999. Some ancestral palms and their descendants In: *Proceedings of the Seminar "Science of Oil Palm Breeding"*. Eds. N. Rajanaidu and B.S. Jalani, pp. 8–36. Palm Oil Research Institute of Malaysia, Kuala Lumpur, Malaysia.

Sax, K. 1923. The association of size difference with seed coat patterns and pigmentation in *Phaseolus vulgaris*. *Genetics* 8: 552.

Simmonds, N.W. and Smartt, J. 1999. *Principals of Crop Improvement*. Blackwell Science, Oxford.

Singh, R. et al. 2013a. Oil palm genome sequence reveals divergence of interfertile species in Old and New Worlds. *Nature* 500: 335–339.

Singh, R. et al. 2013b. The oil palm *SHELL* gene controls oil yield and encodes a homologue of SEEDSTICK. *Nature* 500: 340–344.

Singh, R. et al. 2014. The oil palm *VIRESCENS* gene controls fruit colour and encodes a R2R3-MY. *Nature Communications* 5: 4106, doi: 0.1038/ncomms5106.

Sinnot, E.W., Dunn, E.W., and Dobzhanzsky, T. 1956. *Principles of Genetics*. McGraw-Hill Book Company Inc., New York.

Soh, A.C. 1994. Ranking parents by best linear unbiased prediction (BLUP) breeding values in oil palm. *Euphytica* 76: 13–21.

Soh, A.C. 1995. Commercial potential of oil palm clones: Early results of their performance in several locations. In: *Recent Developments in Oil Palm Tissue Culture and Biotechnology*. Eds. V. Rao, I.E. Henson and N. Rajanaidu, pp. 134–14. International Society for Oil Palm Breeders, Selangor.

Soh, A.C. 1999. Breeding plans and selection methods in oil palm. In: *Proceedings of the Symposium on the Science of Oil Palm Breeding*. Eds. N. Rajanaidu and B.S. Jalani. pp. 65–95. Palm Oil Research Institute of Malaysia, Kuala Lumpur.

Soh, A.C. 2016. Breeding for climate change mitigation and adaptation in oil palm. *Paper Presented at International Conference on Oil Palm and the Environment (ICOPE) 2016*, Bali, Indonesia, March 15–18.

Soh, A.C., Frakes, R.V., Chilcote, D.O., and Sleper, D.A. 1984. Genetic variation in acid detergent fiber, neutral detergent fiber, hemicellulose, crude protein and their relationship with in vitro dry matter digestibility in tall fescue. *Crop Science* 24: 721–727.

Soh, A.C., Gan, H.H., Wong, G., Hor, T.Y., and Tan, C.C. 2003. Estimates of within family genetic variability for clonal selection in oil palm. *Euphytica* 133, 147–163.

Soh, A.C. and Tan, S.T. 1983. Estimation of genetic variance, and combining ability in oil palm breeding. In: *Proceedings of the 4th International SABRAO Congress Crop Improvement Research*. Eds. T.C. Yap and K.M. Graham. pp. 379–388. SABRAO, Kuala Lumpur.

Soh, A.C., Vanialingam, T., Taniputra, B., and Pamin, K. 1981. Derivatives of the dumpy palm—Some experimental results. *Planter* 57: 227–239.

Soh, A.C., Yap, T.C., and Graham, K.M. 1977. Inheritance of resistance to pepper veinal mottle virus in chilli. *Phytopathology* 67: 115–117.

Sprague, G.F. and Tatum, L.A. 1942. General vs specific combing ability in single crosses of corn. *Journal of the American Society of Agronomy* 34: 923–932.

Tang, T.L. 1971. The possible use of fertile pisifera in plantation. *Malaysian Agriculture Journal* 48: 57–68.

Van der Vossen, H.A.M. 1974. Towards More Efficient Selection for Oil Yield in the Oil Palm. (*Elaeis guineensis* Jacq.). *PhD thesis*. University of Wageningen, The Netherlands.

White, T.L. and Hodge, G.R. 1989. *Predicting Breeding Values with Applications in Forest Tree Improvement*. Kluwer Academic Publishers, Boston.

Wong, C.K. and Bernardo, R. 2008. Genomewide selection in oil palm: Increasing selection gain per unit time and cost with small populations. *Theoretical and Applied Genetics* 116: 815–824.

Wong, W.C., Teo, C.J., Wong, C.K., Mayes, S., Singh, R., and Soh, A.C. 2015. Development of an effective SSR-based fingerprinting system for commercial planting materials and breeding applications in oil palm. *Journal of Oil Palm Research* 27: 113–127.

Wonky-Appiah, J.B. 1987. Genetic control of fertility in the oil palm (*Elaeis guineensis* Jacq.) *Euphytica* 36: 505–511.

Yamada, Y. 1993. Interaction: Its analysis and interpretations. In: *Proceedings of the 1991 International Society of Oil Palm Breeders Workshop on Genotype—Environment Interaction Studies in Perennial Tree Crops*. pp. 131–139, Kuala Lumpur, Malaysia.

Yates, F. and Cochran, W.G. 1938. The analysis of groups of experiments. *Journal of Agricultural Science* 28: 556–580.

5 Objective Traits

Aik Chin Soh, Sean Mayes, Jeremy Roberts, Kees Breure, Benoît Cochard, Bruno Nouy, Richard M. Cooper, Hubert de Franqueville, Claude Louise, and Shenyang Chin

CONTENTS

5.1 Introductory Overview (*Aik Chin Soh, Sean Mayes, and Jeremy Roberts*) 86
5.2 Selection for Physiological Traits in Oil Palm (*Kees Breure*) 87
5.2.1 Background 87
5.2.2 Gross CO_2 Assimilation 88
5.2.3 Canopy Efficiency (e) 88
5.2.4 Harvest Index 88
5.2.5 Height 89
5.2.6 Selection for Auxiliary Traits 89
5.2.7 Parent Selection 89
5.2.8 Traits for Parent Selection 91
5.2.8.1 Traits of D Palms in Dura Lines and T Palms in Tenera Lines 91
5.2.8.2 Traits of P Parents 92
5.2.9 Recording Techniques 92
5.2.9.1 Leaf Measurements 92
5.2.9.2 Dry Matter Production 93
5.2.9.3 Bunch Index 93
5.2.9.4 One-Shot Method of Growth Recording 94
5.3 Harvestable Yield (*Jeremy Roberts, Aik Chin Soh, and Sean Mayes*) 94
5.3.1 Background 94
5.3.2 Identification and Manipulation of Phenotypic Traits That Might Enhance Harvestable Yield 96
5.3.2.1 Tree Architecture 96
5.3.2.2 Fruit Ripening 97
5.3.2.3 Fruit Abscission 98
5.4 Adaptability and Abiotic Stress Tolerance 99
5.4.1 Genotype × Environment Interaction (*Aik Chin Soh, Sean Mayes, and Jeremy Roberts*) 99
5.4.2 Abiotic Stress Tolerance and Resilience (*Aik Chin Soh, Sean Mayes, and Jeremy Roberts*) 100
5.4.3 Oil Palm Resistance to Drought (*Benoît Cochard and Bruno Nouy*) 101
5.4.3.1 Distribution and Main Culture Conditions 101

5.4.3.2 Symptoms and Impact of Drought on Oil Palm 101
5.4.3.3 Water-Deficit Environments 102
5.4.3.4 Methods of Breeding Drought-Resistant Crosses 103
5.4.3.5 Physiological Impact of Water Deficit 103
5.4.3.6 Prospects 104
5.5 Biotic Stress Resistance 105
5.5.1 Background (*Aik Chin Soh, Sean Mayes, and Jeremy Roberts*) 105
5.5.1.1 Diseases 105
5.5.1.2 Pests 105
5.5.1.3 Physiological Disorders 105
5.5.2 Aspects of Host–Pathogen Interactions as Applied in Breeding for Disease Resistance against *Ganoderma* Stem Rots, Fusarium Wilt, and Spear/Bud Rots (*Richard M. Cooper*) 106
5.5.2.1 Background 106
5.5.2.2 Fusarium Wilt 108
5.5.2.3 *Ganoderma* Stem Rots 110
5.5.2.4 Spear Rot (Lethal Bud Rot; Pudricion de Cogollo [PC]; Fatal Yellowing Amerelecimento Fatal [AF]) 112
5.5.2.5 Innate Immunity Genes as Candidates for Genetic Modification for Disease Resistance? 115
5.5.2.6 Conclusions 119
5.5.3 Breeding for Resistance to Oil Palm Diseases (*Hubert de Franqueville and Claude Louise*) 119
5.5.3.1 Main Oil Palm Diseases 119
5.5.3.2 Fusarium Vascular Wilt 120
5.5.3.3 *Ganoderma* Basal Stem Rot 121
5.5.3.4 Bud Rot Diseases: Pudricion Del Cogollo (PC) 124
5.5.4 Some Aspects of Screening for Resistance to *Ganoderma* sp. in Malaysia (*Shenyang Chin*) 125
5.5.4.1 Background 125
5.5.4.2 Early Nursery Screening Methods 126
5.5.4.3 Physiological Age 127
5.5.4.4 Disease Assessment 128
5.5.4.5 Alternative Screening Methods 129
5.5.4.6 Field Resistance and Nursery–Field Screening Relationships 129
5.5.4.7 Current Molecular Understanding of Oil Palm Defense Response toward *Ganoderma* sp. 129
5.5.4.8 Conclusions 130
References 131

5.1 INTRODUCTORY OVERVIEW

Aik Chin Soh, Sean Mayes, and Jeremy Roberts

The oil palm is a versatile crop with a myriad of uses for its oil. Depending on the disciplinary and plantation cultural biases of the growers, crop researchers, oleo-chemists and technologists, engineers, and marketers, many suggestions for the various desirable selection traits to improve the palm and its oil have been made. In 2003, Malaysian Palm Oil Board (MPOB) (Basri 2003) held a brainstorming session among the key stakeholders to reconcile, rationalize, and prioritize the selection traits for improvement as it is cumbersome and inefficient for breeders to include all the desirable useful traits in their breeding program at any one time and there may also be obvious or hidden negative trade-offs between pairs of traits.

Ten traits were listed in order of priority: (1) high oil yield, (2) resistance to *Ganoderma*, (3) high bunch index (BI), (4) slow height increase, (5) long bunch stalk, (6) low lipase fruit, (7) high oleic acid, (8) large kernel, (9) high vitamin E, and (10) high carotene. The top three priority traits are agronomic traits relating to yield while the fourth and fifth are for ease of harvesting.

This is understandable as palm oil is still essentially a commodity crop where high yield ensures lower production costs and competitiveness against other vegetable oils, primarily soybean oil and rapeseed oil. The highest-ranking oil quality trait was oleic acid at position 7. Presumably, increased oil yield would concurrently increase the relative production of the other oil quality components, although it is unlikely to lead directly to major changes in overall composition. Perhaps for this reason, breeding improvement for these traits has not been conferred a high priority and any production of "novelty" oil may require dedicated plantations of the novel germplasm and dedicated milling and extraction facilities.

The "current" priority traits are still relevant in the near future and in fact should be more urgently, if not desperately, pursued as the increased *Ganoderma* incidence and the inability to harvest the bunches are undermining the profitability and sustainability of the crop.

The traits required for the future are those relating to environmental sustainability: adaptability to marginal environments, resilience to extreme climate or weather volatility, and resource use efficiencies or efficient use of natural resources and agronomic inputs (light, soil, water, nutrients, chemicals). They should begin to be pursued now as breeding improvements require decades to reach fruition, even with the latest technologies.

5.2 SELECTION FOR PHYSIOLOGICAL TRAITS IN OIL PALM

Kees Breure

5.2.1 Background

In oil palm, as in other plants, dry matter is formed by the process of photosynthesis (CO_2 assimilation). In this process, CO_2 from the air is converted into carbohydrates. Only 25% of the carbohydrates produced are used for structural dry matter production of the vegetative growth and fruit bunches (Breure 2003). The main goal of breeding is to develop planting material that maximizes the conversion of available solar radiation into bunch production. Breeding therefore implies selection for increased gross CO_2 assimilation, and hence the increased supply of carbohydrates; increased canopy efficiency, that is, the rate of conversion of photosynthetically active radiation (PAR) into carbohydrates; and increased harvest index (HI), that is, the proportion of total carbohydrates used for economic products (palm oil and kernels).

5.2.2 Gross CO_2 Assimilation

This is first dependent on the amount of PAR available and the size of the light-intercepting leaf surface. In oil palm, total leaf surface is calculated as the product of the number of green leaves on the palm and the mean area of a single leaf. The total leaf surface per hectare is defined as the leaf area index (LAI). LAI increases with the expansion of the crown until it reaches an optimum for obtaining the highest bunch yield per hectare, which for oil palm is estimated to be about 4.5. To optimize early yield, optimum LAI should be reached as soon as possible after planting. The increase in leaf area with age fits a logistic growth curve from which the maximum leaf area (L_{max}) and the time to reach 95% of L_{max} ($t_{0.95}$) can be estimated (Breure 1985). There were pronounced and significant differences between general combining ability (GCA) values for L_{max} of three ASD (ASD Costa Rica S.A.) varieties tested in South Sumatra and, more importantly, considerable differences between the GCA values of individual *pisiferas* [Ps] per variety. It is of particular interest that the latter L_{max} values are not significantly related to the corresponding GCA values for bunch yield. This paves the way for selecting the most favorable combination of high bunch yield and low L_{max} (Breure 2010).

Light interception is hampered by the incidence of crown disease (CD). This appears as the bending of newly opened leaves and occurs during the period from field planting until canopy closure. Breure and Soebagyo (1991) showed, as expected, that CD affects yield during the early years of production, especially for the first year.

5.2.3 Canopy Efficiency (e)

Squire (1983) concluded from a detailed recording of fertilizer experiments that fertilizers increased light interception by about 4%, but increased canopy efficiency by about 25%. This is because an adequate amount of mineral nutrients in the leaf benefits the rate of photosynthesis and, hence, increases e. Breure and Bos (1992) showed that the Mg status of *pisifera* (P) and *dura* (D) in a D × P experiment in Papua New Guinea accounted for 80% of the variance in yield of their *tenera* (T) offspring. Another method of increasing e is to improve light distribution over the leaf surface, particularly during the final stage of canopy closure (Breure 2003). Diminished light distribution, therefore, explains why increasing LAI beyond the optimum reduces bunch yield.

5.2.4 Harvest Index

Corley and Lee (1992) showed that HI has increased considerably due to breeding in the last 50 years. This has been obtained by increasing the bunch yield. There is further scope for increasing HI through reducing vegetative dry matter production. This reduction mainly concerns those parts that do not contribute to photosynthesis: increasing leaf area/leaf weight ratio, reducing leaf production, and reducing height increment (Squire 1984; Breure and Bos 1992).

5.2.5 Height

Selection for slow height increment without sacrificing yield is feasible, as Soh and Chow (1993) concluded by using index selection. A progeny test in South Sumatra showed that there was no relationship between GCA values for height and yield of the *tenera* offspring of Ps from five origins. The Ps of the Nigerian origin even showed a pronounced inverse relationship. A similar effect occurred in wheat breeding, where the major step change in yield increase during the last 100 years was obtained by introduction of reduced height genes (*RHT*) reducing the stem height and also increasing the fertility of the ear (Austin et al. 1989). The reduced height also allowed dwarf and semidwarf plants to utilize greater artificial nitrogen without the danger of lodging.

5.2.6 Selection for Auxiliary Traits

It makes sense to select for auxiliary traits that are indirectly associated with high oil yield per hectare. How these traits can be incorporated in the selection procedure will be described for an ongoing breeding program in North Sumatra. The ultimate goal of this breeding strategy is to produce heterozygous but relatively homogeneous T palms, generated by pollinating female palms from D inbred lines by unrelated P palms selected from a series of interconnected D × P test crosses. As *pisifera* palms are female-sterile, homozygous, genetically uniform P palms cannot be developed by repeated self-pollination ("selfing") using saved pollen. To circumvent this drawback, two approaches may be considered:

One could select *pisifera* palms in families obtained by T × P full-sib (FS) mating. With FS mating, the frequency of heterozygous palms, as a proportion of the frequency in the previous generation, is 0.809. This approach thus gives only a small decline in heterozygosity.

One might, instead, consider (continued) selfing T palms, where this decline is 0.5. As 0.5 is about equal to $(0.809)^3$, three generations of T × P FS mating result in the same decrease of heterozygosity as one generation selfing.

Because of the difficulty of selfing *pisifera* palms, a strategy has been adopted to develop P pure lines by continued selfing of T palms occurring in T-derived lines. Selection in and among the T-derived lines is mainly based on the average phenotypic values for relevant traits across the palms within a line. (In passing, it is noted that this average estimates the genotypic values of the T parents of the lines.)

The *pisifera* palms used in the test crosses are selected in the above-described T-derived lines on the basis of the GCA/breeding values estimated from a connected design of D × P crosses. Only proven P palms are used as pollen parents for commercial seed production. As the *dura* lines and test crosses are planted at about the same time, this breeding strategy allows improvement of both D and P parents of commercial T planting material every 8 years.

5.2.7 Parent Selection

Earlier, parents were selected for yield per palm, with recording during the early years of production. Later, it was realized that this strategy would result in producing vigorous off-spring that would not perform well when surrounded by other competitive palms, and consequently selection for increasing yield per hectare was proposed (Corley et al. 1971).

Breure and Corley (1983) subsequently found that young palms selected for high BI (the proportion of dry matter used for bunch production) performed better later on than palms selected for yield per palm. Breure (1985, 1986) showed that selection for high leaf area ratio (LAR) led to improved BI. Smith et al. (1996) showed that during the early years in the field, palms with a shorter rachis competed with each other less than those with a long rachis. Later in plantings, when the leaves of all palms were overlapping, leaf area mainly becomes a more important parameter related to competition for light among palms. These traits were therefore included in selection programs in addition to yield per palm and extraction rate.

A more direct method of evaluating interpalm competition is through recording the light penetrating through the canopy with a light meter under the oil palm canopy. An alternative, but less precise, method of measuring light penetration is to score the vigor of the (legume) cover crop (cc score) (Breure 2003). This novel method for indirectly evaluating the effect of competition for light among palms on yield per palm was confirmed by a multiple regression analysis of the GCA values for traits recorded in test crosses planted in North Sumatra. This showed that 65% of the total variation in yield *per palm* was significantly explained by, in decreasing order of importance, cover crop (cc) score, LAR, and height. The cc score itself accounted for 42% of the variation in yield per palm. In ranking order of importance, leaf area, frond production, and rachis length (RL) in combination accounted for 58% of the variation of cc score. As will be explained below, selection for some of the above-mentioned growth parameters that are indirectly associated with light competition (BI, RL, and LAR) should be better restricted to the female parents. But, since selection of individual female parents for these traits is hampered by low heritability, they should be exclusively selected for in the *dura* lines, which will be the source of mother palms for commercial seed production—in addition to yield per palm and components of extraction rate.

Using the GCA values for L_{max} of P parents and adopting 40 green leaves in mature palms as a standard complement per palm, as reported by Gerritsma and Soebagyo (1999), offers the opportunity to estimate the optimal planting density of the D × P offspring.

$$\text{Optimal planting density} = \frac{4.5 \times 10{,}000}{L_{max} \times 40}$$

Along with an estimate of yield per palm, this would then give a direct estimate of yield per hectare. But this still needs a fair estimate of early yield per palm that is not biased by the size of the light-intercepting leaf surface and, hence, gross CO_2 assimilation during the period of canopy closure. An option is to select for yield per palm during year 6–9 of production, as adopted by the Cirad system of selection. This considerably delays, however, the period of selecting a new generation of parent palms for seed production.

An alternative approach to account for light competition is by applying an independent culling system and selecting only those P palms that transmit high yield per hectare and a relatively high cc score, as shown in Figure 5.1. This gives a more valid

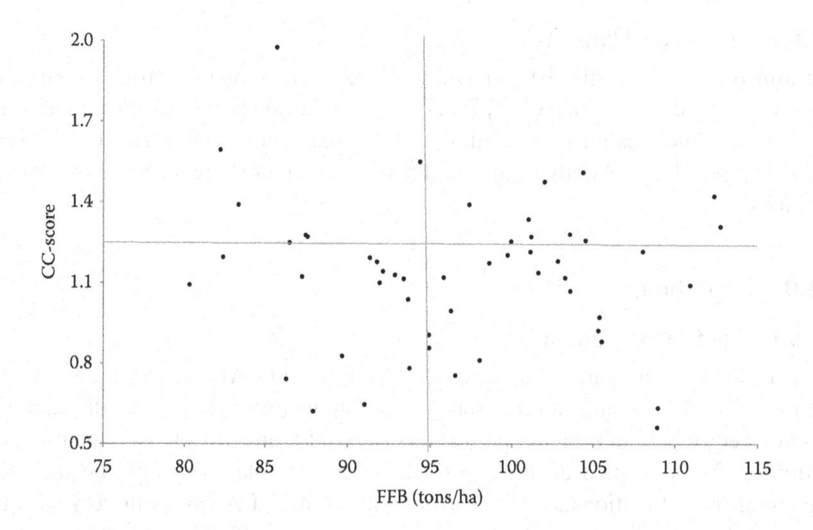

FIGURE 5.1 Relationship between cumulative FFB (tons/ha), based on the breeding value for bunch yield (kg per palm) and estimated optimal density, and vigor of the Mucuna cover (CC-score) under the palm canopy of 52 *pisiferas* tested in North Sumatra.

selection of P palms, since the yield per hectare—in contrast to yield per palm—is not related with cc score.

It may further be difficult to entirely combine selection of P parents for offspring with high yield per hectare with, for example, slow height increment. To correct this defect, we split the seed palms into three height groups—without sacrificing the yield per palm or extraction rate.

Pisiferas that transmit excessive height are used exclusively to pollinate seed palms from the group with low height. Similarly, we have grouped the seed palms according to components of extraction rates to improve this defect in the P parent, if needed.

5.2.8 Traits for Parent Selection

To formulate a schedule of recording for auxiliary traits, we assume that yield recording of D and T lines as well as D × P test crosses is restricted to only the first 5 years of bunch production.

5.2.8.1 Traits of D Palms in Dura Lines and T Palms in Tenera Lines

In addition to the target traits (yield and components of oil and kernel extraction), the following traits are thus relevant:

Bunch index (BI), the proportion of aboveground dry matter production used for bunch production
Height and annual height increment (Ht, HInc)
Rachis length (RL)
Leaf area (LA)/leaf weight (LW) ratio (LAR)
Crown disease (CD)

5.2.8.2 Traits of P Parents

The aim is to estimate the BVs or GCAs of the P male parents from a connected D × P crossing design (Chapters 4, 7, and 12). In addition to the parameters used for the D lines, a more extensive recording of leaf area is needed to estimate the maximum leaf area (L_{max}). Another important trait is the vigor score of the below-canopy vegetation.

5.2.9 Recording Techniques

5.2.9.1 Leaf Measurements

These measurements aim to estimate LA, LW, RL, and LAR. Leaves are marked at opening 48 and 72 months after planting and then measured some months later. For a more precise estimate of the parameters derived from leaf measurements, measuring the frond marked at opening 60 months after planting could be included. Logistic growth functions estimated from eight annual LA values fit very well the leaf area plotted against years after planting (Breure 2010). This set of records can thus be obtained during the usual period of 5 years yield recording. Leaf area has been estimated from the mean length×mid-width of a sample of six leaflets and the number of leaflets, using regression parameters reported by Hardon et al. (1969) and more recently those by Henson (1993). However, the parameters of the regression line in these two reports differ considerably. This is not surprising considering the very small samples being used for determining true leaf area. Breure recorded actual leaf area from annual measurements of one palm sampled from each 400 plots of test crosses. Samples were taken during the first 4 years after planting from all six distinct P origins (varieties) that were tested, planted at high and standard planting densities.

He found no significant difference between varieties, planting density, and the slope of the regression equations. More importantly, there was no statistical difference in the slope between ages, as claimed in the previous two estimates. By fitting a common regression line through all 3100 records assembled during 4 years in all varieties, the following formula for estimating leaf area was established:

$$\text{Leaf area} = -0.245 + 0.455\,n^{*}(\text{mean } L^{*}W)$$

where n = number of leaflets and L*W = mean length × mid-width for a sample of six of the largest leaflets.

This formula gives a more accurate estimate of the true leaf area than that estimated from the formula of Hardon et al. (1969) and Henson (1993). In particular, the widely used formula of Hardon et al. grossly overestimates the true leaf area. Using this more precise formula implies that previous physiological analyses of productivity that were based on estimated leaf area may need reviewing. For example, by recalculating the leaf area with this new formula, Breure's 2010 estimate of LAI of 5.6 for optimum yield becomes 4.5. The leaf area values of the logistic growth curve are estimated as $A\,[1 + B \times \exp(-C \times T)]$, where T = years after planting and A, B, and C are constants. The asymptotical maximum leaf area attained (L_{max})

and the time to reach 95% of L_{max} (t0.95) are estimated as described by Breure and Verdooren (1995).

5.2.9.2 Dry Matter Production

Our interest is restricted to the above-ground parts: *bunch dry matter* (Y) and *vegetative dry matter* (VDM). The latter is the sum of *leaf dry matter* (LDM) and *trunk dry matter* (TDM).

5.2.9.3 Bunch Index

This is estimated from annual values of Y and VDM:

$$Y/(Y + VDM)$$

Bunch dry matter production is 0.53*weight of fresh fruit bunches (FFB).

Leaf dry matter production is the product of leaf production and mean leaf weight.

Leaf production is recorded from 48 to 72 months after planting by using a specially designed form. This gives the ranking of the (annually) marked fronds on a spiral as well as the corresponding order of the leaf after opening (Breure and Verdooren 1995). Leaf production is then obtained by the difference between the orders of the leaves.

Leaf weight is estimated from the petiole cross section (PCS) using the formula proposed by Corley et al. (1971):

$$\textit{Leaf dry weight} = 0.1023\ \text{PCS} + 0.2062$$

As with estimating leaf area, this formula is based on measuring the true weight from a small sample and consequently various different formulae have been reported. Establishing a more precise formula is very laborious, but less important than for leaf area.

Trunk dry matter is estimated from the trunk diameter and the annual height increment, using the formula of Corley et al. (1971).

Height is measured from the trunk base to the insertions of a leaf base of known opening date, which corresponds to the level of the growing point. A moveable bar with pointer and attached measuring tape fitted on a supporting pole is brought to the base of the marked frond. The height is then directly read from the attached sliding measuring tape. Because the reference leaf is usually above a point with different ground level, the moveable base of the pole is fitted with a fluid leveller, which is brought under the trunk base. This ensures that height is accurately measured to the base of the trunk.

Annual height increment (HInc) is calculated as half the difference in height between the base of leaves opening at 72 and at 48 months after planting.

Trunk diameter is measured at about 150 cm above the ground, using a large pair of calipers with a sliding measuring tape that allows a direct reading of the diameter. To ensure that it is measured at the widest diameter, three spirals are left between two measuring points. At these points, the frond bases are removed by pruning to expose the trunk.

Crown disease severity is scored on the eight youngest opened leaves at 4-month intervals, using score 0 (absence of symptoms) to 3 (severe symptoms). Recording starts about 10 months after planting and continues until CD is no longer observed (usually after the seventh round). This is more precise than solely recording the number of affected palms.

Vigor score is recorded under individual palms, using a score 0 (no vegetation) to 3 (vigorous vegetation), and then averaged per plot. For a reliable score, it is essential to establish and maintain a pure legume cover of, for example, Mucuna, which is moderately shade tolerant.

5.2.9.4 One-Shot Method of Growth Recording

Breure and Powell (1988) showed that in case only yield records and components of oil extraction rate are available, a reliable estimate of growth records could still be obtained from one single round of measurements. This method aims mainly to estimate vegetative dry matter production and, hence, BI as well as LA/LW ratio and annual HInc. Since leaves measured at a certain age give a reliable estimate for comparative purposes, this one-shot method only needs an estimate of frond production and annual height increment.

Annual frond production can be estimated from counting the visible frond bases on two opposite spirals, multiplied by four gives the total visible fronds. From long-term actual frond production, it was estimated that 16 months of production are concealed under the trunk base. Total fronds produced per month at a certain age (t in months after planting) is then estimated as t – 16.

Annual height increment at a certain age must be corrected for by the period of expansion of trunk base with little internodal elongation during the early years after planting.

Table 5.1 shows true values of height increment and height recorded at research units of Bah Lias and Selapan Jaya. This revealed that a precise estimate of annual height increment could be obtained as t – 3.

Note that Breure and Powell (1988) concluded to subtract 2 years from the palm age (due to a misprint, the formula in their paper incorrectly shows to subtract 1 year). Since this widely used adjustment grossly underestimates height increment, small values of annual height increment as, for example, reported for the Nigeria collection by Rajanaidu and Jalani (1994) should be considered with suspicion are probably underestimates of the true value.

5.3 HARVESTABLE YIELD

Jeremy Roberts, Aik Chin Soh, and Sean Mayes

5.3.1 Background

While the concept of potential yield (Evans and Fischer 1999) is helpful in attempts to aim toward achieving the theoretical physiological limit, the concept of harvestable yield is important for a perennial tree crop. This is particularly true where the actual harvesting (cutting) operation is still largely manual and the prospect of a

TABLE 5.1
Estimated versus True Annual Height Increment of Tenera Palms at Bah Lias and Selapan Jaya

	Height (cm) at Leaf Base of Frond Marked at Opening[a]		Height Increment (cm/year)			
			True		Estimated	
Research Station	Age at Marking (Years after Planting)	Height (cm)	Age (Years after Planting)	Annual Height Increment (cm)	Age (t Years after Planting) Divided by (t – 3)	Annual Height Increment (cm)
Bah Lias	7	290.2	5–7	74.3	$\frac{\text{Year 7}}{4}$	72.6
Selapan Jaya	8	334.0	6–8	64.0	$\frac{\text{Year 8}}{5}$	66.8
	Height (cm) at Leaf Base of Frond 25					
Bah Lias	13.5	766.7	10.5–13.5	70.2	$\frac{\text{Year 13.5}}{10.5}$	73.0

[a] Level of the growing point.

feasible and cost-effective mechanized alternative—especially for tall palms—is still elusive. Harvestable or recoverable yield relates to ease of harvesting, harvesting at optimal fruit ripeness, and minimal loss of ripe fruits. Dwarf palms with longer bunch stalks will facilitate harvesting, manual or mechanical (Soh et al. 2003). Palms with ripe fruits exhibiting a distinct color change could ensure harvesting at optimum ripeness, if color change coincides with optimal oil content (highest oil content and quality), while those with a non- or delayed-abscising fruit habit could reduce the need for frequent harvesting rounds and also minimize loose ripe fruit loss in the below palm undergrowth. Overripe fruits also result in poorer oil quality due to high free fatty acid (FFA) production arising from activated lipase enzyme activity, particularly from bruising of fruit (Kushairi et al. 2011). These beneficial trait versions are generally found in less advanced breeding material and wide/exotic germplasm, including *Elaeis oleifera* material. Breeding programs that combine breeding and genetic engineering approaches have been proposed earlier as one way to combine such traits in elite material (Osborne et al. 1992; Rao 1998). However, breeding programs combining high oil yield with one and two of these ease of harvesting traits are well advanced already using conventional breeding approaches (Bah Lias Research Station Annual Report 2010 [See Chapter 7]; Tan et al. 2015). Semidwarf varieties exploiting the dwarfing genes of the Dumpy, Yangambi 16R, Pobe dwarf, and MPOB's Population 12 germplasm are already available (Kushairi et al. 2011). A molecular marker for the *virescens* trait is available at MPOB (Singh et al. 2013) while molecular markers for the non/delayed abscission and dwarfing genes (including a nursery screening technique) are being actively pursued (Tranbarger et al. 2011; Roberts and Zubaidah 2013).

5.3.2 Identification and Manipulation of Phenotypic Traits That Might Enhance Harvestable Yield

5.3.2.1 Tree Architecture

Domestication of crops has led to the selection of phenotypes with architectural features that contribute to an enhancement in harvestable yield. The "green revolution" of the 1960s and 1970s was primarily fuelled by the development of shorter varieties of wheat that increased ear fertility and partitioned assimilates into the grain rather than into the stem, as a major gene effect. The height of most annual crops is now routinely reduced by the use of dwarfing genes, the application of growth-retarding chemicals, or in the case of tree crops such as apple and pear, the exploitation of dwarfing root stocks. Although some semidwarf oil palm germplasm is available, developed from material collected from the wild, there has been limited dedicated research effort to understanding the genetic and functional basis of the observed variation in height or trying to develop/select for novel dwarfing genes analogous to the situation in wheat and other cereals. Such developments could lead to focused efforts to produce dwarf ideotypes with defined architectural characteristics.

Gibberellins (GAs) play a key role in regulating plant growth (Claeys et al. 2014). The biosynthetic pathway of this group of plant growth regulators is now well established and the signaling events downstream of the binding of GAs to the GID1 receptor have been characterized in a range of plant species (Yamaguchi 2008; Santner and Estelle 2009; Colebrook et al. 2014). Furthermore, the specific "reduced height" (Rht) loci in wheat that have been associated with dwarfing phenotypes have been identified as encoding nuclear transcription factors that play a key role in gibberellin signaling (Peng et al. 1999). The availability of the oil palm genome sequence (Singh et al. 2013) provides the community with the resource necessary to identify the putative genetic components of the gibberellin biosynthesis and signaling pathways and explore how this knowledge might be exploited to develop germplasm with reduced height characteristics. Some progress is already being made in this direction by MPOB, which isolated a range of germplasm from oil palm seedling nurseries that have a spectrum of dwarfing phenotypes. Preliminary analysis has revealed that some of the material appears to be GA deficient in that it exhibits enhanced growth after application of GA_3 and the level of specific gibberellins in leaf and internode tissues is attenuated. The next step in this program is to identify the genes in the biosynthetic pathway that are mutated and how these modifications might provide an explanation for these observations. Once the genetic characteristics of this dwarfing material have been identified, it should be possible to determine the phenotypic consequences of defined lesions in the GA pathway and ascertain whether impacts on growth can be restricted to specific vegetative tissues without impinging on the size of the fruit tissues themselves or their chemical constituents. The ultimate goal of this research will be to generate a range of elite germplasm with reduced height characteristics that would require less fertilizer and less water, and with fruit bunches that would be easier to harvest. Suppression of gibberellin signaling has also been associated recently with abiotic stress responses (Colebrook et al. 2014) and therefore the generation of germplasm with altered GA characteristics might prove of value in the quest to develop oil palm material that can tolerate exposure to extremes of temperature or water availability.

5.3.2.2 Fruit Ripening

Oil palm is a climacteric fruit whose ripening is associated with a spectrum of physiological and biochemical events leading to changes in the accumulation of specific metabolites involved in pigmentation, respiration, hormone biosynthesis, and cell wall remodeling (Tranbarger et al. 2011). As the fruit matures, there is an increase in the synthesis of certain oils and up to 90% of these accumulate in the mesocarp tissues. The high oil content is associated with an increase in transcript levels for all fatty acid synthesis enzymes (Bourgis et al. 2011) and these may be temporally coordinated by an ortholog of the *Arabidopsis* seed oil transcription factor WRINKLED1 suggesting a common regulatory mechanism for oil synthesis in seeds and fruit (Tranbarger et al. 2011). By carrying out a detailed analysis of transcript accumulation at specific days after pollination (DAP) and correlating this with the accumulation of metabolites, it has been possible to correlate specific changes in gene expression during fruit development (Tranbarger et al. 2011). This analysis has identified a number of genes that might be key candidates for the manipulation of fruit ripening characteristics and could be used as specific molecular markers in breeding programs. One such gene is EgLIP1, which encodes a lipase that is restricted to the mesocarp tissues (Morcillo et al. 2013). Lipase activity within the mesocarp tissues increases over 100-fold during ripening and this enzyme has been proposed to enhance the palatability of the fruit to animals that then consume it and as a consequence facilitate seed dispersal. High lipase activity in the fruit causes a loss in quality as the enzyme releases FFAs that can render the oil unfit for human consumption. Lipase activity invariably increases when fruits become bruised due to the processes associated with harvesting. The identification of *EgLIP1* may prove a breakthrough as it is anticipated that this could be used as a molecular marker to identify low lipase genotypes (Morcillo et al. 2013).

It has long been proposed that the changes associated with fruit ripening are coordinated by the gaseous plant hormone ethylene. The study by Tranbarger et al. (2011) has revealed that by 140 DAP there is a substantial increase in transcripts encoding the ethylene biosynthetic enzymes 1-aminocyclopropane-1-carboxylic acid (ACC) synthase (ACS) and ACC oxidase (ACO) in addition to a rise in the expression of ethylene response factor (*ERF*) genes. Further work will be necessary to determine the consequence of manipulating the expression of such genes on the time course of metabolite production and cell wall remodeling. However, it is likely that future breeding strategies will exploit probes for these genes to identify new elite germplasm with enhanced ripening characteristics.

Fruit color is an important trait for the optimization of harvesting. Although fruit shedding is routinely used as an indicator of bunch maturity, pigment accumulation is another valuable characteristic. The majority of oil palms produce fruit that are either almost black throughout their development (*nigrescens*) or change from green to yellow when they are ripe (*virescens*). The deep purple coloration in *nigrescens* fruit is due to the accumulation of anthocyanins, which obscures the degradation of chlorophyll and accumulation of carotenoids that accompanies ripening in *virescens* fruit. Recently, the *VIR* gene has been identified as an R2R3-MYB transcription factor with homology to the *Arabidopsis PAP1* gene (Singh et al. 2014). Mutations of the VIR gene lead to a dominant negative phenotype resulting in the development of *virescens* fruit. This discovery has paved the way for the development of a molecular toolkit that can be utilized to identify *virescens* genotypes in oil palm seedlings rather than having to wait 5 years until phenotypic characterization of fruit can be carried out on mature plants. The ability to identify stages of ripening in *virescens* fruit, using devices that can measure spectral composition associated with chlorophyll degradation/carotenoid accumulation, will prove a valuable tool to aid harvesting and increase oil yield in future years, if the development of maximal carotenoid coincides with maximal oil content or a management regime could permit it to be used, even if it does not coincide directly.

5.3.2.3 Fruit Abscission

Harvesting of fruit bunches is traditionally "triggered" by the identification of loose fruit underneath a tree. While this is a useful indicator of the state of ripeness of an individual bunch, it also means that the fruit shedding process has been initiated to varying degrees and since bunches are normally excised and left to drop, under gravity, to the floor, then a consequence of this is that individual fruit breaks free. As yet no effective way has been developed that can collect the abscinded fruit other than brushing them up, which is both time consuming and laborious. Now the *VIRESCENS* gene has been identified (see the previous section), the color of the fruit may become a more effective way of determining the optimal time for bunch harvesting and therefore approaches to prevent fruit shedding are of interest. The manipulation of organ shedding was one of the first characteristics that was selected for by farmers and now many domesticated crops no longer abscind their seeds while their wild ancestors continue to do so. Only a limited amount of research has been undertaken to understand the mechanisms responsible for fruit shedding and one of the first observations that was made was that the fruit abscission zone comprises two sites where wall loosening takes place (Henderson and Osborne 1990). Recently, a research group led by Tim Tranbarger at Institut de Recherches en droit Des Affaires (IRDA) in Montpelier has been undertaking a detailed anatomical, biochemical, and molecular characterization of the events associated with oil palm fruit abscission in *Elaeis guineensis*. Their work has revealed that both ethylene- and excision-promoted fruit shedding is associated with a substantial increase in the expression of the polygalacturonase EgPG4, particularly in the abscission zone. While elevated expression is also seen in both mesocarp and pedicel tissue exposed to ethylene, this is one thousand times less than that observed at the site of organ loosening (Roongsattham et al. 2012). EgPG4 groups with a clade containing two other *Arabidopsis* PGs (At2g43880 and At2g43890) have been reported to be upregulated during floral organ shedding (Kim and Patterson 2006). It is unlikely that manipulating the expression of this PG would be sufficient to attenuate the timing of fruit shedding; however, it should be possible to use abscission-related transcripts to identify gene hubs that may play a key regulatory role in cell wall remodeling and that this information might be exploited to control the timing of fruit loss.

The ability to regulate the timing of shedding raises the question as to whether the ideal abscission ideotype in oil palm would be a tree that failed to shed its fruit or one where individual fruit could be readily freed from the bunch in the field so that surplus biomass could be retained primarily *in situ* and composted to provide a more sustainable growing environment for the trees. Complete abscission of fruitlets may allow the subtending leaf (usually removed prior to bunch cutting) to be retained, as it is still photosynthetically active.

5.4 ADAPTABILITY AND ABIOTIC STRESS TOLERANCE

5.4.1 Genotype × Environment Interaction

Aik Chin Soh, Sean Mayes, and Jeremy Roberts

Most oil palm breeding programs have been developed on their own plantation site. Consequently, the varieties developed are likely to be more adapted or suited to their own plantation site and agro-management practices. Thus, it is not surprising that in comparative trials of different sources of commercial materials at the different source sites, the host source tended to perform best at its own site. As the plantation group owner of the breeding stations spread its plantation interest to other countries and environments, especially those with less favorable growing conditions, the adaptability of the proprietary variety to the new environments assumes importance and the setting up of variety × location or genotype × environment (G × E) experiments becomes mandatory. The first "International Experiment" set up aimed to compare the adaptability of the different major planting materials at the different producer sites and evaluate the importance of G × E effects. The sources of planting materials tested were (1) Deli × LaMe (IRHO), (2) Deli × Yangambi (IRHO), (3) African D × 2/2311 (Pamol), (4) African D × ex Yangambi (Yaligimba, Pamol), (5) Deli × ex Yangambi (Marihat), and (6) Deli × 27B + 233B + local unknown (UP), Deli × AVROS (HMPB), Deli × (Deli + Congo + Nigeria + Serdang). The 8 testing sites were: Binga (Zaire), Yaligimba (Zaire), Pamol (Cameroon), Marihat Research Station (Indonesia), Chemara (Malaysia), United Plantations (Malaysia), HMPB (Malaysia), and San Alberto (Colombia) (Corley et al. 1993). Genotype (Origin) × environment interaction effects for yield were found to be small and non-significant (Rajanaidu et al. 1986). Studies by Rosenquist (1982) and Cochard et al. (1993) also confirmed this. The presence or absence of G × E effects depends on the variabilities of the genotypes and environments tested (Corley et al. 1993; Soh et al. 2003). The misperception that G × E is not important could be attributed to the lack of diversity in the variety/genotype and/or environmental treatments used in the trials (Rosenquist 1982; Chan et al. 1986; Rajanaidu et al. 1986; Cochard et al. 1993; Lee and Rajanaidu 1999). Other researchers (Obisesan and Fatunla 1983; Ong et al. 1986; Corley et al. 1993) have detected the presence of G × E effects using genetically diverse progenies although their contributions (3%–4%) to the total experimental variance for yield or coefficient of determination remained small (Rajanaidu et al. 1992; Rafii et al. 2000). Although purported stable progenies have been identified

(Ong et al. 1986; Lee and Donough 1993; Rafii et al. 2000), their exploitation as cultivars depends on the ability to reproduce these *tenera* hybrid progenies in large quantities (clonal seeds, clones) as only limited hybrid seeds can be produced from a pair of *dura* and *pisifera* sexual parents. While the G × E effects may be found to be small and nonsignificant in the analysis of variance of an experiment with a large number of entries, a few progenies may in fact exhibit differential response to different environments. This may be of interest to the breeder of developing varieties for specific adaptation and understanding the physiological basis of the adaptability seldom emphasized in the reported studies (Caligari 1993; Lee and Rajanaidu 1999). Corley et al. (1993), using a simple analytical approach, found that progenies with few bigger bunches rather than many smaller bunches tended to yield poorly in stressful environments because the abortion of a single large bunch would result in a considerable loss in yield, confirming observations by others. Although plasticities of bunch weight and number apparently do exist and vary between progenies, it has been found that some progenies producing high numbers of small bunches did not yield so well under good growing conditions because of their limiting bunch sizes. Conversely, progenies producing few large bunches did not yield well under poorer condition because of their limiting bunch numbers.

With the advent of near true hybrid cultivars from inbred parents or cloned parents and clones, G × E effects with respect to location, spacing, fertilizer, soils, and other factors will likely assume increased importance (Lee and Donough 1993; Corley et al. 1995; Soh et al. 1995) and may warrant consideration in the development of cultivars suited for different situations, in line with the trend toward precision farming or precision plantation practices (Chew 1998).

In terms of selection, an attractive approach used in animal breeding to handle G × E effects is to treat a trait measured in two different environments as two separate but genetically correlated traits (Falconer and Mackay 1996). They can then be incorporated into a selection index to predict selection response for the trait in the second environment as a correlated response to selection for the trait in the first environment (Yamada 1962, 1993; White and Hodge 1989; Soh 1999).

5.4.2 Abiotic Stress Tolerance and Resilience

Aik Chin Soh, Sean Mayes, and Jeremy Roberts

As the available good agricultural land areas have been fully exploited by the oil palm in the current palm oil–producing countries and with oil palm being a profitable and easily grown crop, oil palm plantations have expanded into suboptimal and marginal areas (e.g., dry sandy areas, podsols, peat, highlands) within these countries and in new countries. In such areas, abiotic or physical (water, temperature, nutrient) environmental stresses, which impede or reduce production, are likely to be encountered. These would be exacerbated with the onset of extreme weather patterns with climate change predicted by IPCC (International Panel on Climate Change) (IPCC 2014). Drought resistance or tolerance is reviewed here while the other aspects of abiotic stress tolerance and resilience related to sustainability and climate change are discussed later in this chapter.

5.4.3 Oil Palm Resistance to Drought

Benoît Cochard and Bruno Nouy

5.4.3.1 Distribution and Main Culture Conditions

Oil palm (*E. guineensis* Jacq.) is a plant originating from the humid tropical regions located between Guinea and north of Angola (11°N to 10°S). However, isolated palm groves can be found in neighboring regions further north in Senegal (16°N) and further south of Angola (15°S). This plant is commonly found on edges of swamps, river banks, gallery forest, and forest edges. Oil palm does not exist in primary forest but is present in secondary forests and in fallows around villages (Zeven 1967).

The minimum temperatures must be higher than 20°C with a maximum between 28°C and 34°C. Below 20°C, bunch abortion is more frequent; slower growth as well as slower bunch maturation is observed. This plant grows mainly at low altitude, between 0 and 400 m. However, if there is enough and well-distributed rainfall, oil palm can be found in altitudes as high as 1000 m. The two main limiting factors are the lack of sun and the lack of rainfall. Oil palm needs at least 1800 hours of sunshine per year and an average global radiation greater than 12 MJ/m^2/day. This plant demands a lot of water: 1800–2000 mm annual rainfall well distributed, with a water deficit less than 250 mm per year. The air humidity should be high.

In Benin and Togo, oil palm is traditionally farmed although it is a region with limiting rainfall conditions. During the nineteenth century, under the leadership of the kings of Abomey, plantation areas reached 500,000 ha. Since the 1990s, oil palm cultivation has strongly increased. Over time, people have started planting oil palm in less favorable conditions, with reduced precipitation and higher water deficits. Breeding for oil palm better suited to these limiting conditions becomes a necessity. Therefore, the effect of drought has to be better studied, breeding criteria for this trait to be developed, and best origins, crosses, and parents must be identified.

5.4.3.2 Symptoms and Impact of Drought on Oil Palm

Oil palm can survive relatively long periods of drought due to the efficiency of its stomata (Maillard et al. 1974). Any deficit between perspiration and water absorption causes stomata closure, allowing the maintenance of tissue hydration for several months. An average water deficit of less than 600 mm affects the growth and production potentials. Beyond a water deficit of 600 mm, vegetative disorders can appear and can lead to death. Drought impacts first affect the FFB yield. The immature period is extended. In Benin, harvesting begins 2 years later compared to North Sumatra (Indonesia). Nouy et al. (1999) showed the effect of drought on the same cross, with three different water deficit levels: 0–100 mm (A), 300–400 mm (B), and 400–700 mm (C). Using case A as the FFB yield reference with 100%, there is a decrease of the yield of 46% in case B and 76% in case C. In addition, drought induced a seasonality effect on FFB production, with erratic yield distribution. In case A, monthly peak yield represented 13%–16% of total production; in case B, monthly peak yield was 17%–23%; and in case C, monthly peak yield rose to 35%–43% of the total production. Drought has little impact on the oil extraction rate and its components, except in most severe cases at the end of the dry season. With more than 2 months of severe deficit, oil to mesocarp percentage is greatly reduced.

Besides the loss in yield, visual symptoms related to drought can be seen. These visual symptoms have been well described (Maillard et al. 1974; Nouy et al. 1999). The first symptoms are an accumulation of unopened leaves. Thereafter, green leaves break and leaves at the base of the crown dry out. Next, bunches and immature fruits also dry out. Finally, the crown bends over and the palm dies. However, the oil palm has very high resilience; surviving palms will recover and resume growth rapidly when the rainy season returns.

5.4.3.3 Water-Deficit Environments

In environments with high water deficit, for example, Pobè (Benin), percentage mortality is one of the most relevant criteria for assessing susceptibility to drought. However, this only occurs when water deficits are higher than 700 mm and exacerbated by high bunch load and young age. This criterion is therefore irrelevant for lower water deficits (Maillard et al. 1974). The salient results, obtained under severe water-deficit conditions, for the first cycle crosses of the reciprocal recurrent selection (RRS) were:

Progenies with high drought resistance can be among the most productive progenies.
Progenies highly susceptible to drought also carry high bunch load.

More detailed results of these first cycle progenies revealed that Deli × Yangambi crosses seemed more resistant to mortality link to drought as there were lower yields and a better distribution of the production throughout the year, whereas Deli × La Mé crosses seemed more susceptible to drought because mortality was higher linked to higher production during the peak season. With respect to the oil content components, drought affected Deli × La Mé more than Deli × Yangambi in the percentage of oil to fresh mesocarp. Nevertheless, Deli × La Mé crosses had an average FFB yield of 65.8 kg/palm/year, well above that of the Deli × Yangambi FFB with 43.5 kg/palm/year.

In the 1999 study, Nouy et al. (1999) gave the details of a specific experimental design planted in Benin (Obéké-Obéké) and Indonesia (North Sumatra), which had 50 common crosses. Unfortunately, in Benin, an exceptional drought in 2000 led to high mortalities. Very few common crosses remained available. Despite this, the extraction rate components were observed to be relatively unaffected except the percentage of the fruit to bunch in the Deli × Yangambi crosses. The increased water deficit affected the FFB components, more particularly in the number of bunches, especially for Deli × Yangambi crosses. These results are even more interesting since Deli × La Mé and Deli × Yangambi crosses had the same FFB yield in Indonesia but in Benin, oil and FFB yields were 50% higher for Deli × La Mé compared to Deli × Yangambi. Furthermore, Deli × Yangambi production in Benin had a very high year to year variability. In some cases, Deli × Yangambi crosses had a year without any production.

In environments with medium water deficit, for example, La Mé (Ivory Coast), there have not been any proper impact studies on yield or any behavioral studies on different crosses, under medium water-deficit conditions. Moreover, no trait related to water-deficit levels has been identified. However, behavioral comparisons of the same crosses planted in different environments have been done. Thus, between Aek Kwasan in Indonesia, with a water deficit estimated at 0 mm, and La Mé in Ivory Coast, with a water deficit of 350 mm, the cross LM2T × DA10D produced 6 t/ha of oil in Indonesia against 3 t/ha in Ivory Coast (Cochard et al. 1993). Nouy et al. (1999) presented the FFB yield profile of the LM2T × DA10D cross in Indonesia and Ivory Coast and showed that, apart from a lower FFB production in Ivory Coast, the production was also less well distributed over the year. Likewise, the eight Deli × La Mé crosses studied had the same oil yields when expressed in percentage of the control cross in Indonesia and in Ivory Coast. This confirmed that oil production of all these eight crosses was twice as low in the La Mé environmental conditions (Cochard et al. 1993). These results showed that within the Deli × LM materials, the highest producers in low-water-deficit area were also the highest producers in medium-water-deficit area (Nouy et al. 1999).

5.4.3.4 Methods of Breeding Drought-Resistant Crosses

5.4.3.4.1 Offset Production Peak

In the absence of water deficit, there exists a significant genetic diversity for yield distribution during the year (Nouy et al. 1996) with productions varying between crosses and origins. However, the more the water deficit increases, the less this genetic diversity is expressed. Thus, production profiles become similar (Corley 1977). Having crosses with shifted peak production seems illusory.

5.4.3.4.2 Distribution of Production

The graphic representation of annual production spread over years enables the estimation of the variability of the distribution of that production. We can see the peak sizes, peak frequencies, and monthly yields that characterize a cross. Thus, production peaks are significantly greater for Deli × La Mé than Deli × Yangambi. Deli × Yangambi FFB production is more homogeneously distributed across the year (Nouy et al. 1999). Another way to estimate and present the FFB production distribution is to use the Gini coefficient and illustrate it by translating it into Lorenz curves (Cros et al. 2013). The closer to 0 the Gini coefficient is, the more homogeneous is the distribution of the FFB production. Conversely, when the Gini coefficient is close to 1, the FFB production is not well distributed. As was seen for crosses of the first RRS cycle, the second RRS cycle crosses of Deli × Yangambi gave a better distribution of the FFB production with a Gini coefficient varying between 0.382 and 0.614, whereas for the Deli × La Mé crosses, the coefficient varied between 0.645 and 0.699 (Cros et al. 2013). But if one considers crosses of different origins having an equivalent overall FFB production, therefore, no major differences in the distribution of FFB production could be identified.

5.4.3.5 Physiological Impact of Water Deficit

Apart from FFB production, which is the main trait affected by water deficit, secondary traits that are easier to select could become interesting indicators of drought resistance or susceptibility.

5.4.3.5.1 Root Development

Houssou et al. (1989) and Cornaire et al. (1994) showed that the most drought-resistant crosses, based on the lowest mortality rates, had the highest root densities. However, these palms were generally also the worst for FFB production. Nevertheless, they also showed that it was possible to find crosses with good FFB production while having high root densities. Based on this observation, and in order to have a better understanding of the oil palm performance in water-deficit conditions, other physical parameters (root length, root surface area, and root biomass) were studied in three crosses: resistant, intermediate, and susceptible. The resistant cross had more fine roots (tertiary and quaternary roots) either in terms of dry matter, root surface area, or length of the roots (Nodichao et al. 2011). In order to highlight the significant differences between the three crosses for the studied traits, it was necessary to take into account many sources of variation for total root production in trees. This helped to discriminate genetic variation in rooting depth, lateral extent of roots around the palms, and their interaction. These results confirmed the importance to select for plant vigor, especially roots, without compromising oil palm yield (Nodichao et al. 2011).

5.4.3.5.2 Other Traits

Stomata regulation was also studied. Susceptible crosses seemed to keep their stomata open longer during drought periods (Reis de Carvalho 1991). Functional traits, such as the potential root extraction rate for water and the soil moisture extraction efficiency, displayed differential behavior between resistant and susceptible crosses to drought. The presumed resistant cross had the best soil exploration ratio for water uptake, due to a higher root length density and better distribution throughout the soil (Chopart 1999; Da Matta 2004). Water uptake expressed as the volume of water taken up per unit root surface area per day was lower for this cross than the susceptible ones. In this study, it seemed that the tolerant cross was able to access a larger volume of soil and with a slightly slower drying out of the soil around the roots (Nodichao et al. 2011).

5.4.3.6 Prospects

Most of these results were obtained from experiments not specifically designed for drought studies but from comparing common crosses planted in different environments, on different continents, and during the same periods. To correct this deficiency, it was decided a few years ago to set up specifically designed experiments that would enable proper studies on drought effects, alongside other objectives. Such trials were conducted in Indonesia, Nigeria, and Benin between 2010 and 2012. A range of crosses and environmental conditions were chosen and with different approaches agreed upon to study diverse parameters, for example, carbon metabolism with biomass assessments, measures of photosynthetic activity, changes in carbon reserves, etc.; running water with water flow measurements, water and osmotic potential, root studies, etc.; yield and yield components: sex ratio, number of bunches, number of spikelets, parthenocarpic fruit, fruit weight, oil content, etc.; palm architecture, by monitoring the emission of leaves, inflorescences, stage of flowering and maturity of bunches, etc.

5.5 BIOTIC STRESS RESISTANCE

5.5.1 Background

Aik Chin Soh, Sean Mayes, and Jeremy Roberts

5.5.1.1 Diseases

Basal stem rot (BSR) disease (caused by *Ganoderma boninense*) is the most important disease in the Far East. The disease occurs in increasingly younger palms at each consecutive replanting, with casualties reported of 5% at year 3% to 40% by year 15, the important yielding period of the palms (Gurmit 1991; Tan 2015). With no foolproof preventive and therapeutic control methods, plant resistance is the only hope, and breeding has been actively pursued with some partially resistant varieties becoming available (Durand Gasselin et al. 2005; Idris et al. 2004).

Fusarium (*Fusarium oxysporum*) wilt (FW) is the major disease in West Africa but resistant varieties are available (De Franqueville and Renard 1990). In Latin America, besides BSR and FW diseases, bud rot purportedly caused by *Phytophthora palmivora* seriously undermines the establishment and sustainability of *E. guineensis* (*EG*) plantations (De Franqueville 2003; Torres et al. 2010).

Breeding efforts to provide some tolerance or resistance to bud rot are focused on *EG* × *EO* (*E. oleifera*) hybrids (de Franqueville 2003; Romero et al. 2013), which appear to have some level of resistance. Bud rot and FW disease are apparently absent in South East Asia but the fungal species are present. Strict phytosanitary vigilance and control measures against their importation have been implemented as first defense, but other factors may also be at play that have prevented the establishment of the disease in South East Asia.

The phytopathological and host–parasite relationship principles relevant to the breeding for resistance to these major diseases in palm are discussed in Section 5.5.2, while in Section 5.5.3, the applied breeding aspects for these diseases at Cirad are explored. In Section 5.5.4, the perspectives of the BSR disease resistance work undertaken in Malaysia are considered.

5.5.1.2 Pests

Currently, there are no dedicated pest resistance breeding programs being undertaken. Presumably, integrated pest management (IPM) methods are adequate at this point or it would become a priority.

5.5.1.3 Physiological Disorders

Crown Disease (CD) is an important physiological affliction of young (2–3 year old) oil palms in the Far East. CD causes bending and twisting of the young fronds, which can result in losses in early yields, when severe and prolonged. The Deli × AVROS/Yangambi *dura* material appears to be more susceptible. The disorder is caused by a recessive gene, the expression of which is masked by an epistatic gene conferring incomplete penetrance (Blaak 1970). The disorder can be bred out by discarding families with any susceptible progeny.

5.5.2 Aspects of Host–Pathogen Interactions as Applied in Breeding for Disease Resistance against *Ganoderma* Stem Rots, Fusarium Wilt, and Spear/Bud Rots

Richard M. Cooper

5.5.2.1 Background

Oil palm has succumbed to many microbial pathogens at seedling and adult palm stages (Turner 1981; Corley and Tinker 2016) but three major microbial diseases of very different characteristics have emerged on the continents where *E. guineensis* originated and in the current greatest planting areas of South East Asia and Latin America (Figures 5.2 and 7.4). This section will attempt to provide for each disease the basics of the pathogen biology, pathogen variability, the genetics and expression of disease resistance, and screening for disease resistance. An initial overview will consider factors affecting durability of disease resistance. The final section will look to the future considering novel approaches based on the wealth of information currently emerging from many more tractable microbial diseases of crop plants.

FIGURE 5.2 **(See color insert.)** The three main microbial diseases of oil palm on three continents. Fusarium wilt (acute form; Ghana): top left. Ganoderma basal stem rot (Malaysia): top middle; upper stem rot causing stem fracture: top right. Spear or bud rot: lower left; necrosis of developing spear leaves (lower middle); resistance of OxG cross in affected area: lower right (all Colombia).

5.5.2.1.1 *Resistance Durability as Influenced by Host and Pathogen Genetics and Pathogen Biology*

Resistance to specific microbial pathogens can be expressed as extreme immunity controlled by one or few "major" genes (monogenic or oligogenic), through to partial resistance dictated by many genes (polygenic) (Day 1974; Ploetz 2007). The evolution and exploitation of these levels of resistance are partly dictated by the nature of the pathogen, its mode of pathogenesis and epidemiology, as well as crop agronomy.

Obligately *biotrophic* fungal pathogens include rusts, smuts, and powdery and (oomycete) downy mildews (Agrios 2005). *Hemibiotrophic* fungi such as *Colletotrichum* species exhibit a biotrophic phase before inducing necrosis (O'Connell et al. 2012) and some bacteria can invade in this manner. These groups typically coevolve with their hosts in an arms race between their virulence factors (usually protein "effectors") and host resistance (R) genes that recognize (directly or indirectly) these effectors (Zipfel 2014). The resulting evolved resistance is usually monogenic, at least pathogen species-specific and can be race-specific, as new forms of the pathogen arise to overcome R genes, such as by losing a cognate effector (Dangl et al. 2013). The typical expression of resistance is localized host cell death or the hypersensitive reaction (HR), which terminates obligate biotrophs that require living host cells for their nutrition (Agrios 2005). Monogenic resistance also exists against some facultative species, notably *F. oxysporum*, in hosts such as tomato (Rusli et al. 2015) and cabbage (Michielse and Rep 2009).

This type of resistance is notorious in monocultures for being rapidly overcome by airborne pathogens, including rusts and mildews that spread rapidly and diversify, by producing vast numbers of spores. Therefore, typically, resistance is not long lasting or durable, but there are exceptions; knowledge of the biology and genetics of both host and pathogen are essential. Thus, R genes against Fusarium wilts (FWs) of tomato and cabbage have been successful in the long term as the pathogen is soil-borne and slow moving, propagule numbers are low, and the crops are rotated each year. Nevertheless, pathogen races have evolved eventually to overcome resistance genes (e.g., Huang and Lindhout 1997). However, FW of banana is a different situation, where the crop is vegetatively propagated and not rotated, and flooding or irrigation can result in widespread pathogen distribution. Here, R genes can be vulnerable. Therefore, R genes require strategic deployment to create genetic diversity such as by regional distribution, release of multilines, and cultivar mixtures (Day 1974; Newton et al. 2008) or are best avoided. Lack of durability also occurs under tropical conditions on perennial crops, such as for coffee rust and *Microcyclus* on rubber (Ploetz 2007).

Monogenic resistance is rarely, if ever, found against pathogens that from the outset cause host necrosis, the so-called *necrotrophs*, such as *Sclerotinia* and *Sclerotium*. This is because they tend to have wide host ranges and HR can even be to the advantage of pathogens, such as *Botrytis* (Govrin and Levine 2000). *Ganoderma* infecting oil palm shows clear necrotrophic traits, although Rees et al. (2009) revealed ultrastructurally a brief biotrophic phase.

Polygenic resistance typically is not overcome, so it is not race-specific. The advantage of its durability can be offset by incomplete disease control, variation in expression across regions, and plant age (Day 1974; Ploetz 2007). Clearly, breeding is greatly complicated when dealing with unknown gene combinations.

In some cases, multiple gene resistance can be expressed as *tolerance*. This term should be used for when a pathogen colonizes extensively but does not induce symptoms or yield loss. *Resistance* should be reserved for the ability to limit pathogen colonization. There are various examples of the use of tolerance, such as against Verticillium wilt of cotton (Wilhelm 1981), rusts and *Septoria* of cereals and coffee rust (Schafer 1971; Ploetz 2007). However, there is the ongoing danger of pathogen inoculum buildup. The term is often misused, notably in Malaysia, when describing partial resistance to *Ganoderma* stem rots of oil palm, but it might be applicable to FW.

Resistance genes are typically sourced from the center of origin or of diversity where a host or its wild relatives and a pathogen have coevolved, such as *Phytophthora infestans*–potato and *F. oxysporum*–tomato in Latin America. FW of oil palm in Africa is a case in point and will be considered later. New encounters of a host or pathogen can be devastating because coevolution for host resistance will not have had time to take place (Ploetz 2007). The expansion of African oil palm into South East Asia and the rise of *Ganoderma* and so-called spear rot in Latin America are likely examples.

5.5.2.2 Fusarium Wilt

5.5.2.2.1 Biology and Impact

Fusarium vascular wilt has been the most important disease of oil palm and it remains a major threat to the crop. The disease is endemic in western and central Africa (Turner 1981; Oritsejafor 1989). In the acute form, palms die rapidly, and in the slower-progressing, chronic form, yields are severely affected. Some plantations have been abandoned or use has changed to other crops. The pathogen is the soil-borne fungus *F. oxysporum* f. sp. *elaeidis* (*Foe*) that invades intact or wounded roots and then remains constrained within xylem vessels, enabling systemic colonization (Cooper 2011; Cooper and Rusli 2014). The pathogen has not been reported from South East Asia; therefore, quarantine is required for imported palm materials. *F. oxysporum* strains are classified based on their host specificity into formae speciales (ff. spp.) of which there are at least 70 (Lievens et al. 2008). Evolution for virulence to a particular host can occur in different ways as evidenced by the polyphyletic nature of some host-specific forms such as f. sp. *cubense* on banana (Lievens et al. 2008) and f. sp. *lycopersici* on tomato (O'Donnell et al. 1998). It appears that pathogenicity and virulence mechanisms involved in specificity have evolved multiple times, possibly through mutation or transposition or spreading to distantly related strains through parasexuality or horizontal gene or chromosomal transfer (Baayen et al. 2000; Ma et al. 2010).

Our spatial distribution analysis of disease incidence revealed localized and non-random patterns of disease in four Ghanaian plantations (Rusli and Cooper, unpublished) and support the earlier findings of Dumortier et al. (1992). These results imply that tree-to-tree infection via roots plays the main role in the establishment of *Foe* infection, even though the potential for aerial spread by spores has been suggested

(Cooper et al. 1989). Slow movement through the soil is likely to have limited the extent of spread and rate of evolution to new virulence of *Foe*.

5.5.2.2.2 Breeding for Resistance and Resistance Expression

In regions where FW is endemic, the only sustainable method for control is by breeding for disease-resistant oil palms (Cooper 2011; Corley and Tinker 2016). Decades of breeding and selection for resistance took place in the Ivory Coast with selection from the 1960s and nursery testing from the 1970s (Renard et al. 1972; Durand Gasselin et al. 2000; de Franqueville and Louise 2017).

Fusarium resistance in oil palm is claimed to be based on many resistance genes (Meunier et al. 1979), but others have suggested that few genes are involved (de Franqueville and de Greef 1988). The resistance is likely to be polygenically controlled because it appears to be durable, not having been overcome by *Foe* in the 40 years or so of its development and use in Africa according to Cochard et al. (2005). This durability is in spite of the genetic flexibility of *F. oxysporum* species that readily form new races overcoming R genes (Michielse and Rep 2009). Immunity to *Foe* has been reported by Rosenquist et al. (1990) for pure Dumpy Deli *dura* in the nursery (Cameroon) and field (Congo), but elsewhere, progenies have shown only tolerance or partial resistance (Locke and Colhoun 1974; Rajagopalan et al. 1978). In our study on four diverse palm lines, expression of resistance typically resulted in lower levels of pathogen vascular colonization; in one case, however, it included a high degree of tolerance because colonization by *Foe* was systemic in both visually susceptible and "resistant" combinations (Rusli et al. 2015). This level of analysis has rarely been conducted, but is important because while partially resistant or tolerant lines are achieving suppression of FW and acceptable yields, *Foe* inoculum is likely to accumulate in living tolerant palms and could be released back into soils as resistant chlamydospores.

5.5.2.2.3 Pathogen Variability

Currently, little is known about the genetic variability of *Foe*, except that local populations, such as in Zaire, have evolved to be similar according to somatic compatibility evidence (Flood et al. 1992). It is tenable that different levels of pathogen aggressiveness might explain the expression of acute or chronic wilt. Nursery tests using strains from acute and chronic infected palms did not reveal differences in host reaction (Prendergast 1963; de Franqueville 1991). It is possible that host adaptation to water stress, as described by Mepsted et al. (1995a) might explain the different disease expressions.

There remains a risk with a potentially variable pathogen such as *Foe* that material developed in one area might be susceptible elsewhere. In general, this seems unlikely based on findings where inoculation of 14 palm clones with three *Foe* isolates from different parts of Africa showed variation in aggressiveness, but clone–isolate interactions were not significantly different (Mepsted et al. 1994). However, sometimes the ranking of *Foe* isolates by clones varied considerably, which might explain why a few apparently resistant crosses bred for resistance in one area have proved susceptible when planted at distance, such as Ivory Coast progenies in Nigeria, Nigerian progenies in Ivory Coast, and Zaire material to a Brazilian isolate (Flood et al. 1993; Mepsted et al. 1994).

Recently, we found at least one differential interaction between African *Foe* isolates and some palm lines from Malaysia, for example Progeny PK 5463 expressed partial resistance to *Foe* isolate F3 (Congo), but not to isolate 16F (Ivory Coast); both were isolates used in resistance screening in those countries (Rusli et al. 2015). This was unexpected as isolate–cultivar interactions more typify single gene controlled resistance.

The study on fungal isolate–host genotype interactions was linked to a recent study on Malaysian biosecurity (Rusli et al. 2015). There is a risk of inadvertent spread of the disease on *Foe*-contaminated seed or pollen (Flood et al. 1990; Cooper 2011), which are imported from Africa into South East Asia to expand genetic diversity (Soh 2011). Localized disease outbreaks have occurred in Brazil and Ecuador following seed shipments from Ivory Coast or Benin. This likelihood is because the pathogen isolates were genetically very similar (Flood et al. 1992; Dossa and Boisson 1991 and IRHO 1992, cited in Corley and Tinker 2016).

Regular reevaluation of currently grown palm lines toward *Foe* is clearly required. Four Malaysian oil palm progenies, three in current or recent commercial use, proved to be highly susceptible to infection by at least one of two African isolates of *Foe*, representing different countries, aggressiveness, and vegetative compatibility groups (Flood et al. 1992; Mepsted et al. 1994; Rusli et al. 2015). This underlines the vulnerability of Malaysian palm lines to FW and that selection for wilt resistance, which has not been a priority in Malaysia, should be paid attention to by breeders there.

5.5.2.3 *Ganoderma* Stem Rots

5.5.2.3.1 Biology and Impact

Basal stem rot (BSR) incidence caused by the basidiomycete, white rot, fungal pathogen *G. boninense* (several other *Ganoderma* species have also been linked) (Corley and Tinker 2016) continues to increase in oil palm plantings. Typically, there is a decay of the bole leading ultimately to the toppling of the palm. One definitive diagnostic feature is the presence of characteristic basidiocarps on the stems of infected palms, which produce myriad, wind-dispersed basidiospores, such as ≤11,000/m^3 throughout 24 hours recorded by Rees et al. (2012) in Sumatra and also found in high numbers in plantations around Peninsular Malaysia (A. Wahab and R. Cooper, unpublished). Upper stem rot (USR), where decay occurs above 1–1.5 meters, is much less frequent but can be severe in some regions. The two symptoms are usually physically separated and the result of different infection events (Rees et al. 2012). Stem fracture can result in severe cases (Pilotti 2005; Rees et al. 2007). Young palms usually die within 6–24 months after the first appearance of BSR symptoms, whereas mature palms survive for 2–3 years or more. Most severe losses from BSR occur in mature stands in Indonesia and Malaysia with lower incidences recorded in Africa, Papua New Guinea, and Thailand (Idris et al. 2004). In Malaysian coastal areas, an average of 50% yield losses from 80% of 13-year-old plantings were recorded by Lim et al. (1992), but infection of palms as young as 12–24 months after planting occurs (Gurmit 1991). In North Sumatra, at replanting (25 years), 40%–50% of palms have been lost with symptoms on most standing palms. Yield decline was a mean 0.16 t/

ha fresh fruit bunch for every palm lost, and when the stand had declined by 50%, losses were c. 35% (Subagio and Foster 2003).

Our understanding of infection and epidemiology is incomplete, but has important implications for disease control and resistance screening. BSR can result from root infection, presumably following root contact with mycelium in debris or other infected roots. Numerous infection trials using oil palm seedlings and *Ganoderma*-colonized rubber wood blocks (RWBs) have supported this view (Sariah et al. 1994; Hasan and Turner 1998; Lim and Fong 2005; Rees et al. 2007). Rees et al. (2007) showed that colonization by *G. boninense* can occur through unwounded roots, and then progresses mainly through the inner, thin-walled cortex before reaching the bole. In contrast to *Foe*, it does not typically invade xylem vessels. Also, in contrast, it is not a soil inhabitant but a very weak competitor: mycelial growth from an established food base will not occur in nonsterile, diverse soils or in organic debris (Rees et al. 2007; R. Cooper, unpublished). Therefore, it seems that palm roots must contact inoculum rather than mycelium actively contacting host roots via soil; also, it is unlikely that USR arises from colonization of the debris found in the axils of frond bases.

Many have questioned the involvement of root infection (Ariffin et al. 1996; Miller et al. 1999; Pilotti et al. 2003). Genetic studies (mitochondrial DNA markers, randomly amplified polymorphic DNA [RAPDs] mating alleles, somatic [or vegetative] compatibility, randomly amplified microsatellites [RAMs], single sequence repeats [SSRs]) of *G. boninense* from infected palms in Papua New Guinea and Malaysia reveal considerable diversity in oil palm plantations (Miller et al. 1999; Pilotti et al. 2004; Pilotti 2005; Rees et al. 2012; Cooper, Wahab, and Woo, unpublished). Different genotypes are often found within the same tree and between adjacent trees. These must have arisen through recombination involving formation, then dispersal and hybridization of basidiospores of compatible mating types. Also, outbreaks of BSR in new plantations, where *G. boninense* inoculum is presumably not present in debris or soil, combined with initially random appearance of BSR imply the introduction by spores (Sanderson et al. 2000). Outcrossing is favored because *G. boninense* is heterothallic and tetrapolar with multiple alleles at both mating type loci (Pilotti et al. 2002).

Do basidiospores establish an inoculum and then infect indirectly, or do they infect palms directly? In view of the very weak, competitive saprotrophic ability of *G. boninense* (Rees et al. 2007), it is hard to envisage where mycelium might form and flourish to create a sufficient inoculum. One route for direct infection might be the extensive wound sites created by routine harvesting (severing the fruit bunch peduncle) and pruning (of frond base to free the fruit bunch), so the potential for spore infection is very considerable. Rees et al. (2012) showed for the first time that basidiospores can germinate abundantly on cut surfaces (fronds, peduncles, stems) under plantation conditions and on roots *in vitro* (Wahab and Cooper, unpublished). Spores contaminating the surface of cut fronds are withdrawn up to 10 cm into the xylem as a result of negative tension within functional vessels. Here, spores would be relatively protected from microbial competition, dehydration, and solar radiation. The formation of a resulting heterokaryon by anastomosis of spores of opposite mating types is a prerequisite for formation of infective mycelium; monokaryotic hyphae are non-pathogenic (Rees et al. 2007). Sanderson and Pilotti (1997) cut back the rachis of decayed frond bases and followed lesions into the stem base. They suggested that when the palm expands, this initial infection would appear to have originated near the center of the palm base, suggesting root infection, even though it arose from the rachis. Overall, the data show a genetically highly variable pathogen population in North Sumatra, Malaysia, and Papua New Guinea, and the importance of basidiospores in spread and infection. However, infection by artificial spore inoculation remains to be established, although our work is currently investigating this aspect. The high genetic variability could have implications for resistance durability, unless resistance is polygenically controlled.

5.5.2.3.2 Breeding for Resistance and Resistance Expression

Resistant varieties hold the greatest hope for future control of BSR in oil palm in South East Asia as cultural, chemical, and biological control methods are only partially successful or simply delay disease progression (Chung 2011; Cooper et al. 2011). Field observations in North Sumatra revealed that *E. guineensis* of Deli origin from Malaysia and Indonesia was more susceptible than African material (Durand-Gasselin et al. 2005), and other trials have revealed differences in susceptibility, indicating possible genetic resistance within host populations (de Franqueville et al. 2001; Idris et al. 2004; Chung 2011). Socfindo field trials appear to be revealing significant, albeit partial resistance and that the resistance is additive (de Franqueville and Louise 2017).

Coevolution of host and pathogen enabling the development of resistance genes is unlikely to have occurred in the very short period from the 1960s when *E. guineensis* was grown on any scale in South East Asia and then faced relatively recent exposure to intensive infection pressure from indigenous *Ganoderma* sp. Has the oil palm evolved in Africa against *Ganoderma*, albeit at the much lower level of challenge there? Are the *Ganoderma* pathogenic species there the same or similar to those in South East Asia? If not, the search for disease resistance in South East Asia based on imported African breeding materials could be a difficult one. It took decades to select for and implement resistance to *Foe* in Africa, where host and parasite have coexisted long term.

The genetics and nature of partial resistance shown by palms to *Ganoderma* is not known and will await the availability of palm lines with effective resistance. Thus far, "tolerant" lines that have been studied show significant levels of infection.

5.5.2.4 Spear Rot (Lethal Bud Rot; Pudricion de Cogollo [PC]; Fatal Yellowing Amerelecimento Fatal [AF])

Very serious damage has occurred on oil palm plantations in most regions of South and Central America, with some areas completely devastated. Symptoms differ between regions, disease can express as chronic or acute, and many pathogens have been associated with the syndrome. It is possible that more than one disease is being described under the same name (de Franqueville 2003; Corley and Tinker 2016). Typically, symptoms involve a spear rot and yellowing of young fronds (Turner 1981; Gomez et al. 2000). Spread to the growing point can result in death of the palm, but generally, bud rot is not an appropriate term. The disease might have originated in Panama in the 1920s but more recently large-scale destruction has occurred in

Colombia (1960s to present), Suriname (mid-1970 to 1980s), Brazil and Ecuador (1980s to present), and Costa Rica (present). Overall, the disease has destroyed over 150,000 ha and affected another 250,000 ha, although some areas have recovered in Llanos, Colombia and Quepos, Costa Rica (J. Corredor, pers. comm.).

Torres et al. (2010) claimed that *Phytophthora palmivora* is the causal agent, based on pathogen reisolation and infection of immature (white) spear leaves with high zoospore numbers. Poor drainage and high rainfall have also been linked to the disease, as have grasshopper vectors and secondary bacteria causing the main symptoms (Drenth et al. 2012). Other pathogens linked to spear rot have included *Fusarium moniliforme*, *F. oxysporum*, *F. solani*, *Thielaviopsis paradoxica*, *Sclerophoma* sp., and *Botryodiplodia* sp. (Turner 1981; Corley and Tinker 2016). Speculative applications of mixed insecticides and fungicides were ineffective (Gomez et al. 2000). Narvaez (2012) failed to detect *Phytophthora* but obtained *F. oxysporum* and *F. solani*, which reproduced symptoms. Also, *P. palmivora* is widespread in Malaysia but there is no equivalent disease there, even in oil palm plantations adjacent to infected crops such as durian and cacao. An ongoing study funded by MPOB is comparing pathogenicity to oil palm of Malaysian and Colombian isolates.

5.5.2.4.1 Breeding for Resistance and Resistance Expression

Within *E. guineensis*, levels of tolerance or partial resistance have been found when compared with the highly susceptible Deli × AVROS (Turner 1981; Santacruz et al. 2000; de Franqueville 2003). Also see de Franqueville and Louise (2017). Extensive planting of the hybrid *E. oleifera* × *E. guineensis* in Colombia and Ecuador has followed its ability to survive among the death of surrounding African palms (de Franqueville and Louise 2017). The problem of assisted pollination for the hybrid remains. Hybrid crosses do vary in level of resistance, and in West Colombia, 28% infection occurred compared to 95% in *E. guineensis* crosses (Romero et al. 2013). Presumably, resistance traits will eventually be revealed in the hybrid, to be incorporated into *E. guineensis*. Here, interspecific hybridization with the native species that has coevolved with spear rot pathogen(s), largely overcomes the susceptibility of the new introduction of *E. guineensis*. Only once the etiology is firmly established, can the necessary studies on infection, spread, pathogen diversity, screening, and expression of resistance be conducted.

5.5.2.4.2 Screening for Disease Resistance

5.5.2.4.2.1 Fusarium Wilt Screening for lines resistant or tolerant to *Foe* is relatively straightforward and has changed little since first developed by Prendergast (1963). The pathogen produces abundant conidia in submerged culture, a trait of xylem-invading fungi (Cooper 2011), which can be poured at predetermined concentrations onto soil around containerized seedlings. A suitably aggressive isolate needs to be used, ideally from the country where the palms are to be released. Earlier comments on possible differential interactions strongly suggest that more than one isolate should be deployed. In order to avoid changes in aggressiveness, isolate(s) stock cultures should be maintained in a dormant state, such as conidia in 25% (v/v) glycerol at −80°C or as dried chlamydospores in sterilized soil or sand, and then new cultures should be made from these, rather than using frequent transfers, that can select for less pathogenic forms. Inoculum level must be controlled to ensure relatively rapid disease progression but to avoid rapid death.

Seedlings should be of the same age and shaded to prevent excessive soil temperatures under the tropical sun, which will prevent growth of *Foe* and therefore infection (Rees et al. 2007). In Zaire in the 1980s, we brought in changes to all of the above practices and dramatically increased infection rate and reduction in the number of replicates required (J. Flood, R. Mepsted, and R. Cooper, unpublished). The use of a disease index, replication, and statistical analysis of data are reviewed by Corley and Tinker (2016) who highlight de Franqueville's (1984) use of wilt index, angular transformation of percentages for individual plots, and statistical analysis of data; resistance is ascribed if progenies had significantly lower loses than either the trial mean or standard crosses. A reasonable correlation has been reported between nursery testing and wilt incidence in the field by de Franqueville (1984) and Renard et al. (1972, 1980). Replication has varied from 40 (Prendergast 1963) to 160 (Renard et al. 1991; de Franqueville and Louise 2017). Detailed classification of symptoms and controlled inoculum enabled Flood et al. (1989) to reduce replication to only 12 plants, but this involved screening clonal plants. Clearly, defined conditions and inoculum must be practiced to make the scale of such trials more manageable.

The disease still takes several months to develop before evaluation can be made. This led Mepsted et al. (1995b) to develop a rapid test giving results within 8 days. Rachis sections of 2.5 cm length were infiltrated with *Foe* conidia leading to internal browning of susceptible clones but to minimal or no symptoms in disease-resistant clones. This test allows evaluation of individual palms rather than the average performance of a group of seedling palms as in nursery testing. The method needs further testing and refinement as field tests in Zaire were erratic, often reflecting nutrient deficiency or other infections (Buchanan 1999).

Selection has sometimes been made on disease incidence in known *Foe*-infested fields (Corley and Tinker 2016). Clearly, these findings must then be supported by nursery tests as absence of disease might represent noncontact with inoculum or "disease escape." This approach is much used currently for bud/spear rots (see below). Subsequent evaluation of putative resistant or tolerant lines is usually performed by visual surveys. Mepsted et al. (1991) and Cooper and Rusli (2014) showed that this approach for FW is often flawed. Using increment borers to remove cylinders from palm stems reveals infection as discolored xylem vessels, from which *Foe* can be readily isolated. Apparent wilt symptoms often did not result from *Foe* infection. This simple technique should be part of the critical assessment of field performance.

5.5.2.4.2.2 Ganoderma Stem Rots As with FW, a robust (reproducible, statistically valid) and facile screening system for evaluating the performance of oil palm genotypes at nursery stage is crucial. Root infection by mycelium is one means of natural infection; thus large wood blocks infested with *Ganoderma* mycelium comprise the inoculum. Rees et al. (2007) found that rubber wood is superior to palm wood and direct contact of roots with inoculum (attached with parafilm) gives faster and more reliable infection and allows use of smaller inoculum, saving time and space, but there is a minimum size of inoculum to achieve reliable infection. Shading of seedling containers is crucial to keep soil temperatures near the growth optimal range (25–27°C) of *G. boninense*. In full sun, soil temperatures can exceed 40°C, which is inhibitory and even lethal to the pathogen (Rees et al. 2007).

Isolate genetic variation is considerable and, unsurprisingly, aggressiveness can differ. Many companies use an isolate PER71 supplied by MPOB, but some use several isolates and some are claimed to be more aggressive than PER71. Standardization of the method is being sought by several companies. Resistance to isolates of different aggressiveness rank similarly (de Franqueville and Louise 2017). However, SSR comparisons show that isolates from South East Asia, Borneo, and Cameroon do not cluster, so it poses the question as to the validity of data on resistance from screening trials from one country to another (T. Durand-Gasselin, pers. comm.).

Infection is slow, reflecting the weak pathogenicity of *G. boninense*. Rees et al. (2007) found that maximum root invasion by the most aggressive isolate was 3.5 cm per month, so just reaching the bole at this rate would take several months. Evaluation for resistance might be reduced from the present 5–7 months by using a combination of more aggressive isolate, inoculum placed nearer the bole, and/or use of very young seedlings. Germinated seed is currently used for inoculation by Socfindo (T. Durand-Gasselin, pers. comm.). They use 100 replicates per cross, again very costly in terms of time and space. Paradoxically, only when we have a range of resistances can we improve this method. Factors such as excessive inoculum and immature plant morphology could well overwhelm and therefore miss potentially useful resistance or tolerance. There are current investigations, including at University of Bath, to source alternative inocula to rubber wood, in terms of ready availability and cost, ability for more directed inoculum placement, and speed and reduced space required for inoculum preparation. Currently, infested wood remains the standard method.

If it is proven that basidiospore infection via cut petioles/rachis is a common occurrence, then this should be a valid means of resistance screening, especially for palm lines to be released where USR (most likely a result of spore infection) is a major problem. Resistance expression in petioles to spores could be very different to that of root tissues to mycelium. Thus far, spore infection has not been proven and in any case infection might prove to be too slow for routine screening.

5.5.2.4.2.3 Spear Rot(s) As previously stated, until we know the causal pathogen and its biology, controlled inoculation for resistance screening cannot be conducted. It is tenable that several different pathogens are involved; thus this research needs intensive international coordination. All that is available is to plant various crosses in areas with high disease pressure and rely on natural infection. The planting and survival of O × G hybrids in otherwise badly diseased areas of Colombia shows the potential of the method. Ideally, there should be present positive controls (e.g., susceptible Deli × AVROS) and resistant hybrids. Socfindo are using this approach and revealing some lines with partial resistance (de Franqueville and Louise 2017).

5.5.2.5 Innate Immunity Genes as Candidates for Genetic Modification for Disease Resistance?

For many microbial plant diseases, host–pathogen interactions have been dissected to a high degree and novel strategies for disease are at last emerging. These advances are largely based on manipulating components of host innate immunity and the pathogen factors that they recognize. Most, *but not all*, approaches will involve genetic modification, so in the current climate might be deemed inappropriate for use with the oil palm. However, it is important to look ahead and be aware of advances and possible novel approaches for combating microbial diseases.

5.5.2.5.1 Innate Immunity

Plants possess highly sophisticated and complex innate immunity comprising two levels. Basal or level one immunity arguably includes any physical (such as cuticle, cell walls) or chemical (antimicrobial phytoanticipins) barriers to pathogens (Agrios 2005). These components can rarely be altered to more resistant forms, so the current interest is at the level of pathogen recognition.

PRRs or pattern recognition receptors are receptor-like kinases and receptor-like proteins and have evolved to recognize pathogens' conserved, signal molecules known as MAMPs (*m*icrobial *a*ssociated *m*olecular *p*atterns) (Zipfel 2014). From fungi, these signature molecules include chitin and glucan oligosaccharides from the cell wall. From bacteria, MAMPs include flagellin, lipopolysaccharide, peptidoglycan, and elongation factor (Aslam et al. 2008, 2009; Erbs et al. 2008). Other MAMPs are still being discovered. PRRs have been characterized largely from *Arabidopsis* and include receptors against flagellin (FLS2) and elongation factor (EFR) but also in rice against chitin (CEBiP) (Zipfel 2009). In some cases, not all plant families contain a specific PRR, affording an opportunity to transfer recognition across taxa. For example, this has been achieved with the expression of EFR from *Arabidopsis*, which confers response to bacterial elongation factor EF-Tu when introduced into solanaceous plants *Nicotiana benthamiana* and tomato, providing relative resistance to a range of bacteria (Lacombe et al. 2010). Some hope for oil palm comes from the transfer to rice of dicot EFR (Schwessinger et al. 2014). In the light of the vulnerability of most major R genes, some breeders are seeking to combine layers of potentially durable PRRs. The search is on for more MAMPs and their corresponding receptors.

The second tier of immunity concerns genes for disease resistance as described earlier. These have evolved against pathogen virulence effectors (mainly proteins), which typically target and disable basal innate immunity pathways and receptors (Jones and Dangl 2006).

Moving R genes between taxa has so far been largely unsuccessful, leading to the term "restricted taxonomic functionality," for example, *Ve* R gene has been transferred from tomato to potato and *BS2* (and other R genes) from tomato to tobacco, potato, and pepper. However, *BS2* does not function in *Arabidopsis*, nor does *Arabidopsis RPS2* in tomato (Dangl et al. 2013).

Resistance genes show considerable homology; many are based on nucleotide binding, leucine-rich repeat (NBS-LRR) motifs (ca. 125 in *Arabidopsis*), and this allows a genetic search for putative R genes from genomes or DNA. Buchanan (1999), working on FW of oil palm, obtained sequences with characteristics of NBS, but genetic mapping showed no associations with wilt resistance. Cloning R genes then transfer to elite cultivars, is a potentially major advance from conventional breeding, but for some crop species, advances in genomics and transformation protocols are still required. Synteny between relatives, for example, rice and sorghum, has facilitated cloning and identification of R genes. This is useful for crops difficult to breed with long breeding cycles such as coconut and oil palm (Collard et al. 2008).

Durable resistance genes do exist (Yang et al. 2008; Borhan et al. 2010) and have in some cases been identified based on the requirement of a pathogen to retain the cognate effector, such as *Xanthomonas oryzae* and rice (Leach et al. 2001). In other words, overcoming an R gene by dispensing with that effector renders the pathogen nonvirulent or with reduced fitness. An alternative aim is to stack R genes that recognize these so-called "core" effectors (Dangl et al. 2013). In a strange contrast to major R genes, are attempts to manipulate host *susceptibility* genes; these are reprogrammed by pathogens for their survival and proliferation. Loss of function deprives the pathogen of a benefit and recessive R genes are candidate genes being sought for breeding, as reviewed by Dangl et al. (2013).

5.5.2.5.2 Pathogen Effectors and Toxins

Conversely, pathogen effectors are proving valuable for searching and characterizing new R genes, using the so-called effectoromics. High-throughput functional genomics uses effectors for probing plant germplasm to detect R genes. Also, it can map quantitative traits (Vleeshouwers and Oliver 2014). Defining core (essential) effectors facilitates identification of suites of corresponding R genes from wild germplasm by using coexpression assays (Dangl et al. 2013). The approach has mainly been pioneered for *P. infestans* where R gene–effector interactions are being exploited by potato breeders along with marker-assisted selection. Vleeshouwers et al. (2008) employed a repertoire of effector genes predicted from *P. infestans* genome to accelerate the identification, characterization, and cloning of potential broad-spectrum R genes. They used a high-throughput screen with PVX-*Agrobacterium* infection. There remains the challenge of finding functional assays. For Solanaceae, transient assays based on *Agrobacterium* are quite effective, but soybean leaves and some monocots (oil palm?) are not yet amenable (Vleeshouwers and Oliver 2014). As new pathogen genomes are available, other pathosystems are being investigated by effectoromics. *Phytophthora* (and other oomycete) effectors are well documented, thanks to available genomes and characteristic motifs like RXLR (Pais et al. 2013). If *P. palmivora* is shown to be the main causal agent of spear rot in Latin America, its putative effectors could be identified from the genome and possibly used to probe for resistance.

Some effectors, notably from *Xanthomonas*, are facilitating plant genome editing. These transcriptional activator like (TAL) effectors, such as AvrXa7 of *X. oryzae*, contribute to virulence, such as by activating specific rice susceptibility genes (Dangl et al. 2013). TAL effector nucleases (TALENs) are fusion proteins from DNA recognition repeats of native or customized TAL effectors and a DNA cleavage domain that can affect site-specific gene modifications. This approach has been used to edit a susceptibility gene to counteract the virulence strategy of *X. oryzae* (Li et al. 2012). Also, it has created loss-of-function mildew resistance wheat locus *Mlo*, giving broad-spectrum, durable resistance to powdery mildew (Wang et al. 2014). Importantly, no foreign DNA remains, so GM regulation might be by-passed, although a ruling in the United Kingdom is still awaited on whether some forms of gene editing count as genetic modification. It is likely that a number of countries will take their lead from the UK decision, including Malaysia.

Effectors can also assist in pathogen detection. The ultimate means of disease control is to prevent pathogen introduction into areas where it is not found, such as for *Foe*. As stated above, strict quarantine exists for importation of African oil palm material into Malaysia. Quarantine currently can involve confirmation of the presence of *F. oxysporum* on or in seed and pollen by PCR based on the TEF gene (Cooper and Rusli 2014). However, *F. oxysporum* is a common soil-borne organism and therefore a contaminant, but most isolates will not be pathogenic to oil palm; therefore, the resulting destruction of potentially valuable "contaminated" breeding materials may be unnecessary. We recently developed DNA primers that are specific to *Foe* and should therefore be invaluable to quarantine facilities. Briefly, the possibility arose through the understanding of the importance in *Fusarium* virulence of pathogen effectors. Effector proteins were first discovered in tomato xylem infected by *F. oxysporum* f. sp. *lycopersici* (*Fol*). These so-called SIX (*s*ecreted *i*n *x*ylem) effectors were then located from the *Fol* genome (Ma et al. 2010) and homologs found in other ff. spp. (Lievens and Thomma 2008). We investigated effectors in *Foe* and found one (an oxidoreductase) that appears to be unique to *Foe* and allowed the design of specific primers (Cooper and Rusli 2014).

Some microbial plant pathogens are dependent on low-molecular-mass toxins for pathogenicity or as a component of virulence. Toxins have been used with varied success in the screening for disease resistance using seed germination or *in vitro* cultured plant material (Slavov 2014). Alternatively, the identification of a key toxin in disease has led to transgenic approaches to combat the toxin. Examples of the latter include oxalate oxidase from wheat engineered into oil seed rape against oxalic acid of the broad-host-range necrotroph *Sclerotinia sclerotiorum* (Dong et al. 2008) and *albD* gene for an enzyme that detoxifies *X. albilineans* toxin albicidin in sugar cane (Birch 2001). In wheat, sensitivity to *Pyrenophora tritici-repentis* (tan spot) toxin ToxA strongly correlates with disease susceptibility. The current strategy uses ToxA to eliminate from current cultivars the sensitive gene Tsn1, which recognizes ToxA (Vleeshouwers and Oliver 2014). It should be noted that basidiomycete tree pathogens such as *Heterobasidion* (Olson et al. 2012) and related *Ganoderma lucidum* (Chen et al. 2012a) produce a wide range of secondary metabolites and some are phytotoxic. Oxalic acid is also typical of this group. We await any information on potential toxins from *G. boninense*.

5.5.2.5.3 Defense-Related Genes

Following pathogen recognition by levels 1 or 2 of innate immunity, myriad genes linked to defense or stress responses are upregulated. These so-called defense-related genes include those that degrade pathogen cell walls, such as chitinases, antimicrobial peptides like thionins, and biosynthetic enzymes responsible for the synthesis of lignin, cell wall structural proteins, and antimicrobial phytoalexins (van Loon et al. 2006). Defense-related genes have been described from oil palm, such as in response to *Ganoderma* and *Foe*, for example (Ho and Tan 2015; Rusli et al. 2015). Hopes are often expressed of improving BSR resistance by transformation with these genes, but the wider literature based on many crop species and diverse pathogens suggests that chances of success are very low, with no practical disease control to date (Collinge et al. 2010).

5.5.2.6 Conclusions

In conclusion, molecular interactions in other diverse diseases and availability of worldwide germplasm diversity are beginning to provide new strategies. Approaches may be based on understanding pathogen recognition (PRRs and MAMPs) and on using effectors as molecular probes. However, until we have the genome and valid transcriptomes of the major oil palm pathogens, especially *G. boninense*, we cannot search for putative effectors or toxins. Also, we have no knowledge of PRRs in oil palm and which pathogen MAMPs they recognize. For example, lack of a key PRR in oil palm, but which is present in another family, might point to a future strategy for cross-taxa transfer, as for the *elf* example cited earlier. Likewise, what major or minor genes linked to disease resistance exist and what are their functions?

Major R genes have not been clearly identified against any disease in oil palms and seem to be rare in necrotrophic diseases of trees (Olson et al. 2012). Although always of interest to breeders and pathologists, R genes might not be the way to progress with genetically highly diverse and adaptable pathogens, including *Ganoderma* with its sexually produced basidiospores and those pathogens capable of rapid evolution to virulence, notably *Phytophthora* and *F. oxysporum* (one or both as major pathogens of oil palm). While durable R genes might hold the key, discovery probably awaits a better understanding of the genetics and molecular interactions of each disease. One priority should be unraveling the pathogenicity of *G. boninense*. Progress on spear rot(s) still awaits definitive confirmation of its etiology(ies).

Looking even further ahead for oil palm diseases, based on rapid progress in other diseases, we shall have to await much-needed advances with oil palm and its main pathogens to see if benefits will come from predicted breakthroughs predicted in more tractable host–pathogen interactions, that is: creating layers of resistance by stacking R genes (if they exist); transferring and stacking novel PRRs; using pathogen effectors to locate and test likely durability of R genes; and manipulating host susceptibility alleles. We must however be realistic and always bear in mind the additional measures required to deal with diseases of this long-term, perennial, tropical crop, where there is no break in pathogen pressure, unlike diseases of annual or temperate crops.

5.5.3 Breeding for Resistance to Oil Palm Diseases

Hubert de Franqueville and Claude Louise

5.5.3.1 Main Oil Palm Diseases

Oil palm cultivation is subject to major parasitic threats in each of its main growing areas.

5.5.3.1.1 Africa

The main disease is Fusarium vascular wilt, caused by a soil-borne fungus, *F. oxysporum* f. sp. *elaeidis*. Fusarium vascular wilt is the most important disease of oil palm in western and central Africa. It has been observed in several countries such as Ivory Coast, Ghana, Benin, Nigeria, Cameroon, Democratic Republic of Congo (DRC, ex-Zaire), and recently in Liberia. Localized foci have also been observed in Brazil, where the disease appeared in 1983 (Van de Lande 1984) and in Ecuador, where it appeared in 1986 (Renard and de Franqueville 1989). These foci have now disappeared. Vascular wilt has never been described in South East Asia. The disease was first described in DRC (ex-Zaire) by Wardlaw (1946) who isolated *F. oxysporum* f. sp. *elaeidis* from diseased palms. Fraselle (1951) reproduced artificially the disease symptoms on seedlings and thus confirmed the pathogenicity of the isolates.

5.5.3.1.2 South East Asia

This is the current main production zone where BSR caused mainly by *G. boninense* leads to considerable damage in North Sumatra, Indonesia, and in many parts of Peninsular Malaysia. BSR also occurs in Africa and in Latin America, but with lower incidence.

5.5.3.1.3 Latin America

Bud rot type diseases are prevalent here. Several types of symptoms are expressed and it is suspected that they are different diseases or disorders. The role of *P. palmivora* has been investigated in Colombia (Torres et al. 2010). Breeding for resistance to these diseases has been challenging.

5.5.3.2 Fusarium Vascular Wilt

The pathogen penetrates the palm via the root, develops in the xylem, and causes rapid decay. The symptoms (acute or chronic) vary depending on several factors such as the palm age, crop cycle, and susceptibility of the planting material. Palm losses exceeding 50% can be recorded in some plantations. Being a soil inhabitant, the pathogen and its population increase during the first planting, causing damage to the mature palms, but in replanting, FW affects young palms. Some cultural practices can reduce disease incidence (de Franqueville and Renard 1990) if the palms are moderately resistant or moderately susceptible, but have a very poor effect if the palms are highly susceptible. The planting site or cover crop and also fertilization (mineral or organic with empty fruit bunches) favor disease development. As the pathogen spreads mainly via the soil, and given the extent of the contaminated areas, chemical control is not economically feasible. Planting with resistant genetic material is the only feasible method of control.

In the first palm generation, the disease appears around 10 years after planting. In replanting, vascular wilt appears as early as year one and stabilizes 4 years after planting. This is the consequence of the inoculum pressure built up during the first cycle planting. Avoidance of the infected first cycle palm foci in replanting is important. Breeding for resistance is of paramount importance and sources of partial resistance can be found within many genetic origins of *E. guineensis* and there is also a very strong resistance factor available in some *E. oleifera* populations. Selection is achieved through a screening at the prenursery stage, based on the artificial inoculation of the pathogen. This early screening test is closely correlated to the field behavior of the planting material. The best results are obtained when this screening is performed within the framework of organized breeding programs, with strong collaboration between plant pathologists and plant breeders, to combine wilt resistance and high oil yields. Most of the obtained results are based on several decades of screening, without any interruption. Each year, several hundred progenies are screened. This strategy has led to the reduction of FW to negligible levels in all the African plantations planted with the resultant resistant material.

Since the early 1970s, with the IRHO/Cirad/PalmElit approach, artificial inoculation is conducted at the prenursery stage in order to obtain fast results and screen several hundreds of progenies per year. Seedlings are inoculated one and a half months after seed planting and the trials concluded 5 months later. Eventually, the screening tests reached an evaluation capacity of 200 progenies for a single test, with standard crosses to link the different tests. The annual capacity of evaluation is now of 1200 progenies, through two phenotyping units based in Africa. The degree of resistance of a cross is assessed by comparison with the mean of all the crosses in the same test. After the statistical analysis, an index is attributed to each progeny. The base "100" corresponds to the mean of the percentages infection rate observed for all the progenies. The index of a progeny represents the ratio between the percentage of plants of that progeny affected by vascular wilt and the mean of the percentages of affected plants for all the progenies in a given test. An index under 100 indicates greater resistance than the mean of the progenies being assessed in the test. An index over 100 indicates greater susceptibility. Families or parents are characterized by the mean of the crosses in which they are involved, but it must be mentioned that the evaluation in recent years has been based on a range of known testers to assess the Dura families of group A and the parents (*pisifera* or *tenera*) of group B.

The general consistency of the screening test was proven in the ex-Zaire (de Franqueville 1984) and in Ivory Coast (de Franqueville and Renard 1990), where the losses in a 4000-hectare estate were reduced from 35% to 40% in the 1960s, then to 10%–15% in the 1970s, and progressively to less than 3%, with a high amount of recovery, in the 1980s and 1990s. In a recent study, Diabaté et al. (2010) showed that this low level of the disease prevalence is maintained 20 years later, with higher levels only observed in the plots where comparative trials were implemented, with known susceptible progenies detected in prenursery and deliberately planted. Planting material with a mean index below 80 suffers from very few cases of vascular wilt, and a mean index above 110 can lead to losses over 20%–30% without any recovery.

Studies realized in different countries indicate that there is no isolate × host genotype interaction, despite the great differences between strains and between genotypes (de Franqueville 1991; Mepsted et al. 1994).

5.5.3.3 *Ganoderma* Basal Stem Rot

The predominant *Ganoderma* species, *G. boninense*, appears to be the most aggressive (Cooper et al. 2011), even if other species such as *G. ryvardense* have sometimes been mentioned (Kinge and Mih 2011). *G. boninense* can also cause USR, but less frequently than BSR. This disease must be fought using an integrated disease management approach through cultural practices (land preparation, removal of infected palms), biological control (use of antagonistic organisms such as *Trichoderma* sp.), and plant (partial) resistance. The development of the last component is discussed here.

BSR can result in heavy and significant losses in oil palm plantations in South East Asia, especially after replanting (Susanto 2009; Susanto and Huan 2010). BSR impact gets worse as the number of generations of replanting increases; some plantations in North Sumatra have reached their fourth or even fifth generation. BSR is now a major problem throughout Indonesia and not just confined to North Sumatra. Recent estimates show that it is not unusual to have 60%–80% diseased palms by the end of a third or fourth generation of planting oil palm on the same land. Until recently, *Ganoderma* in Africa mostly infected palm trees older than 30 years. It was therefore considered as an opportunistic disease. However, it is now apparent on young replants if they have been planted close to the infected bole tissues of the previous palms.

Akbar et al. (1971) first reported palm differences in susceptibility to BSR. Further evidence was presented in 2001 by de Franqueville et al., and again in 2005 (Durand-Gasselin et al. 2005). Akbar et al. (1971) reported the high susceptibility of pure Deli commercial material (60%–74% of BSR, at year 12–15) when compared to Deli × Yangambi material (below 20%) in two Socfindo estates (Mata Pao and Tanah Gambus). de Franqueville et al. (2001), based on observations on pure Deli, pure La Mé, and pure Yangambi parental material planted in Block 60 in Bangun Bandar plantation (Socfindo), despite the lack of statistical design, revealed that Deli material was highly susceptible to *Ganoderma* while the La Mé and Yangambi materials were partially resistant: Deli origin suffers up to 80% palm losses versus an average 23.5% losses with La Mé and Yangambi materials. High variability of infection within materials was also observed, for example, La Mé (12.4%–40.9%) and also in Deli. Different Deli origins were crossed with the same La Mé origin (LM2T self) and planted in two trials. Highly significant differences were observed. The consistency of these results was shown in other trials indicating that some selected group B (non-Deli) origin can give additional partial resistance (Durand-Gasselin et al. 2005).

5.5.3.3.1 Nursery Screening of Planting Material

5.5.3.3.1.1 Early Screening Tests Turnbull et al. (2014) summarized the different steps developed for this approach. An inoculation method was developed by Cirad in collaboration with two private partners (Socfindo and Lonsum) between 2002 and 2008 in Indonesia (Breton et al. 2009a,b). This method is now used to make an early assessment of the resistance to *Ganoderma* of various oil palm progenies. The research and development of the inoculum and of the environment enabling infection was a fairly complex process, which required several years studying the different growth parameters. The goal was to create a rapid, reproducible, and high-throughput test to screen oil palms. After testing a wide range of growth substrates and inoculation methods, the RWB colonized with *Ganoderma* method described by Idris et al. (2004) was confirmed to be the best procedure. These wood blocks must be as small as possible so they can be placed inside nursery polybags before the seedlings are planted. The output of this test is currently of a 100 progenies tested with 100 plants per progeny per month. Results are available after only a few months. These results are generally expressed as a percentage of palms affected or killed by *Ganoderma*, without considering the scale of symptom severity. The results were used to attribute an index to crosses within each test, similar to the one used for interpreting vascular wilt early resistance tests in Africa. By classifying crosses with respect to each other, and centering them on the mean for each trial, the index enables overall comparisons and makes it possible to rapidly and clearly establish a susceptibility range. In addition, a reference scale using a fixed set of control crosses was developed to be used in every screening. It represents a range of resistance and susceptibility that makes it possible to connect all the tests together. Results are adjusted by minimizing the residual variance observed in the reference set.

5.5.3.3.1.2 Methodology of Testers It has been possible to define a group of reference palms (testers) that would be crossed with the tested parents. The study of the variations of resistance and susceptibility between one cross to another would therefore be possible. To test group A parents, mostly Deli, we selected a group of *pisifera* coming from two progenies: one susceptible and one intermediate resistant to *Ganoderma.* In contrast, group B parents were tested using a group of Deli Dura coming from two progenies susceptible and intermediate resistant. Testers were verified in a factorial design set up to cross group A and group B, in two by two subsets. We had anticipated a mean index within the interval (above 100, i.e., between 105 and 107).

The result of the work using these testers was used as a basis for selecting various sources of resistance within every origin. Overall, we evaluated 5100 crosses; 3066 of them were analyzed by the test cross method. A particular focus was put on group A (2054 tests cross) as there are fewer sources of resistance to be found in this group. Nevertheless, the number of test crosses analyzed for group B is still important (1012 test crosses). The objective was to assess the resistance to *Ganoderma* for every single family currently available in our breeding programs. Ten palms from each family are selected at random to be crossed with both types of testers; every cross done is tested at least twice in order to get 20 data points per family.

5.5.3.3.1.3 Interactions between Ganoderma Isolates and Oil Palm Progenies It is important to check that there is no significant interaction between *Ganoderma* isolates and oil palm progenies tested. Consequently, trials are regularly conducted using a range of isolates and progenies. Results are obtained for every isolate–progeny pair. Despite the efforts made for standardization of the test, the accuracy of each data point remains weak, with a confidence interval at 95% of about ±15 percentage points. Statistical analysis shows that although the interaction isolate × progeny is actually significant; its level is small compared to the progeny or isolate effects. The consequence of this weak interaction is that the classification of the progenies tested is not affected, considering the imprecision of the tests, whatever the isolate is.

5.5.3.3.1.4 Strengthening the Relationship between Field and Nursery Tests Early screening tests results must be validated by a consistent behavior in the field to be used as routine. For this purpose, trials were planted between 2002 and 2005 in areas highly affected by *Ganoderma.* Three sets of trials were planted using a completely randomized design (one plant per replication) covering an area of 310 hectares. Another trial was planted in 2009 to compare the resistance level of the genetic material but also the effect of total randomization as opposed to monocategory blocks in commercial planting that could enhance contamination by *Ganoderma.* These four trials included almost all the genetic diversity available in the joined breeding program of Socfindo and PalmElit in North Sumatra. At the end of 2013, the first trial was the only one showing enough symptoms to be analyzed. Ten years after planting, the sources of resistance identified in the early screen test happened to have the best scores in the field too.

5.5.3.4 Bud Rot Diseases: Pudricion Del Cogollo (PC)

This disease complex (PC) appeared for the first time in 1928 on the Almirante plantation in Panama. A form appeared in the 1960s in Colombia, in Turbo near Panama, and then in the Eastern *Llanos*. Bud rot appeared from the 1970s in all the northern countries of South America with oil palm developments: Brazil in 1974, Surinam in 1976, on the Pacific part of Ecuador, and then in the eastern part in 1979. A spectacular outbreak appeared around 2005–2006 in the north of Ecuador (San Lorenzo region), in the south of Colombia (Tumaco area), and then in the Magdalena region (Puerto Wilches), still in Colombia. Thousands of hectares have been destroyed.

The PC starts, depending of the cases, on the young spears, with a rot or desiccation progressing downward more or less rapidly toward the meristematic zone. These symptoms are frequently associated with the development of a green to orange coloration or with the yellowing of the young central leaves. Important differences in symptomatology exist depending on the case or location: it is therefore highly possible that the palms are facing a complex of various disorders. It is highly difficult to identify a primary pathogen, although Colombian researchers from Cenipalma and their partners work on the possible role of *P. palmivora* (Torres et al. 2010).

Different types of symptoms can be distinguished:

Spear rot (*pudricion de la flecha*—PF), in all the regions. The bunches do not rot and generally, it does not reach the meristem.

Bud rot (*pudricion del cogollo*—PC), mainly in the eastern region of Ecuador. The bunches rot as well as the meristem.

Diffused bud rot (*pudricion difusa de cogollo*—PCD), in the western part of Ecuador as well as in the eastern part. The bunches do not rot but the meristem can be reached by the rot.

La ENI (*enfermedad no identificada*) in the San Lorenzo region and in the Tumaco area and probably the Puerto Wilches.

5.5.3.4.1 Breeding for Resistance to the PC Complex

There is no other economically efficient way of controlling PC other than via the *E. oleifera* species. This species is PC resistant and transmits its resistance to the interspecific hybrid with the *E. guineensis* species. Nevertheless, a source of resistance within *E. guineensis* has also been found and is being exploited following confirmation of the results in Ecuador (Louise et al. 2013). Three programs have been conducted with the Cirad/PalmElit approach in Latin America:

1. A resistance breeding program for the interspecific hybrid *E. oleifera* × *E. guineensis*
2. A resistance breeding program within *E. guineensis*
3. An introgression program of the resistance detected within *E. oleifera* to *E. guineensis*

Interspecific hybrids express a high level of resistance that can appear as complete resistance depending on the locations it is planted. But it shows some disadvantages: the main one is partial sterility, which obliges selection for high fertility and the practice of assisted pollination every two days throughout the economic life of the trees. As a first step, the collections of different populations were assessed in hybrid combinations and the best ones are reproduced for seed production. The results revealed the importance of the *E. guineensis pisifera* progenitor that was used. Among the *E. oleifera* populations, it is important to mention the good results obtained with the accessions from Brazil (Coari, Manicore, Mangenot) as well as the problems encountered with the populations from Central America in which hybridization can lead to lethal chlorophyll deficiency around the third or fourth year of planting. The Taisha population (Ecuador) also shows very good results in terms of bunch production, but the extraction rate is lower because of a low level of oil to mesocarp as compared with the other hybrids. This accession is interesting due to its long stalk and thick mesocarp. Currently, the Taisha population is crossed with the good accessions from Brazil. With a good assisted pollination, the extraction rate of the best crosses can reach 27% or more. The *E. guineensis* population of La Mé origin gives good results in E × O hybridization, while with the Yangambi origin, problems of sterility and yellowing due to chlorophyll deficiency that are not lethal do sometimes occur.

Sources of resistance within *E. guineensis* have been detected and exploited. This material still shows great variability and its improvement is being studied. In order to validate these resistance sources, multilocation trials have been established in Ecuador, Colombia, Peru, and Brazil. In the meantime, other sources of resistance, particularly within the African origins of the "B Group" of the general breeding program, are being investigated. This research is difficult due to the fact that there are no early screening trials as is the case for FW and *Ganoderma* disease. The only trials possible have to reply on field trials on important areas with a high PC incidence and intensity.

For the long term, the aim of the program is based on the introgression of the *E. oleifera* resistance into the *E. guineensis* and to accumulate the resistances from the two species. Other traits are also considered. A sufficient level of resistance must be achieved to be exploited in commercial plantations and without the need for assisted pollination. The production of pollen with high viability and of high affinity for the pollinating insects must be guaranteed. Results show that during the first generation, in the Colombian Llaños, individuals can maintain the resistance. The use of clones has allowed this to be confirmed. However, in the subsequent generation, the level of resistance declined.

5.5.4 Some Aspects of Screening for Resistance to *Ganoderma* sp. in Malaysia

Shenyang Chin

5.5.4.1 Background

In the development of any resistant cultivar or variety toward a pathogen, a technique to perform artificial inoculation must be established to enable the quantification of the degree of resistance or susceptibility. In the case of foliar diseases of annual crops caused by fungal pathogens, assessment of resistance or susceptibility is direct as the symptoms can be appraised directly from the infected leaf. In addition, for the majority of annual crops, specific genotypes can be assessed on multiple occasions to reliably demonstrate the level of resistance or susceptibility. Depending on the type of crop screened, resistant cultivars that are successfully identified can either go directly into commercial production or the resistance be introgressed into elite varieties to improve upon other traits of interest.

The hallmarks of an ideal disease assessment as described by Campbell and Naher (1994) for root rot diseases are that it should be accurate, precise, and reproducible by many observers, applicable over a range of conditions, economical, and simple. However, many of these requirements have not been addressed in the case of *Ganoderma* sp. BSR. Even so, materials that are purportedly partially resistant have been identified in various germplasm sources (Idris et al. 2004; Breton et al. 2009b) and are also commercially available in Indonesia through PT SOCFINDO and in Malaysia through FELDA. These purportedly resistant materials are based on early nursery screening but direct evidence of field resistance of these commercial materials is still awaited. An attempt to summarize the progress of screening efforts toward *Ganoderma* sp. BSR is described below and the differing areas that could require more attention for the development of *Ganoderma* sp.-resistant oil palm varieties are highlighted. Readers are referred to studies carried out by Breton et al. (2009a,b) and Turnbull et al. (2014) as they have attempted to address many of the issues described below. Nevertheless, this section attempts to consolidate and compare studies that have been carried out thus far.

5.5.4.2 Early Nursery Screening Methods

The methods used for screening for *Ganoderma* sp. resistance are primarily based on the work of Khairudin et al. (1991). This publication demonstrated that using RWBs inoculated with *Ganoderma* sp. was a reliable method of infecting nursery seedlings. In a later study, Khairudin (1993) also found that a large inoculum size was important to establish rapid infection to give external foliar symptoms. This and subsequent studies (Sariah et al. 1994) made it possible for different materials to be assessed in the nursery in a period of several months rather than to wait for field results, which would take years. Since then, there have only been a few optimizations of the RWB-based screening method, the most recent ones being Breton et al. (2006) and Rees et al. (2007), where it was found that shading had a significant effect on the rate of infection, which is now a common practice. Both studies also confirmed findings of Khairudin (1993) that the size of RWB inoculum has an effect on infection rate, but Breton et al. (2006) were unsure if this was a direct effect of the inoculum strength or due to the ratio of RWB to contaminated soil.

Although most screening methods used are based on using an RWB as the inoculum, parameters used often differ across studies, for example, size of RWB used, physiological age of seedlings used, and method and period of disease assessment. Other screening methods not using RWBs are also described below (see Table 5.2).

The choice of seedling age and method differs between studies and is often dependent on the method used. For example, 3–4-month-old (MO) seedlings were used

TABLE 5.2
Comparison of *Ganoderma* sp. Resistance Nursery Screening Studies and Parameters Used

Studies	Seedling Age	Method	RWB Incubation Period	Individuals per Replication	Replication
Breton et al. (2009a,b)	Germinated seeds	RWB	9 weeks	20	4
Idris et al. (2004)	12 MO	Root inoculation	NA	40	3
Rahamah et al. (2015)	Germinated seeds, 4 MO, 12 MO	RWB, root inoculation	NA	40	NA
Chong (2012)	12 MO	Spraying suspension of *G. boninense* fragments on roots	Nil	NA	NA

Note: MO—months old, RWB—rubber wood block, NA—not available.

as this coincides with transplanting of seedlings from the prenursery to the main nursery and there are also sufficient roots to attach the seedling to the RWB inoculum. Alternatively, germinated seeds were also used and were planted directly above the RWB inoculum with a spacing of 5 cm (Breton et al. 2006). This allowed for a higher turnover of number of seedlings screened, as seedlings need not be raised to age and smaller footprint required due to the smaller polybags used. Twelve-month-old seedlings were used by Idris et al. (2004) as access to roots at this age is relatively easy and inoculated in a test tube containing substrate (consisting a mixture of paddy and oil palm wood sawdust) supplemented with other nutrients. This method was employed as it possibly mimics how natural infection occurs in the field, that is, through root contact with an inoculum.

5.5.4.3 Physiological Age

Breton et al. (2006) reported that the physiological age of seedlings used (germinated vs. 6-month-old seedlings) did not influence the level of resistance as there was a good correlation of the level of resistance between the two ages used in the seven progenies screened. The results indicated that the 6-month-old seedlings had a higher infection percentage compared to the germinated seeds although the authors did not elaborate if the mean percentage of infection of the two physiological ages differed significantly. One of the drawbacks of using germinated seeds is the difficulty to obtain synchronized germination and inability to weed out runts, though this can be preempted by planting additional seedlings and to perform replacements if necessary (Breton et al. 2006). Although the distance between planted seed and the inoculated RWB is determined, it is not known if root to inoculum contact is important, which could otherwise introduce variation to the experiment.

Rahamah et al. (2015) screened 15 crosses using three different methods, germinated seeds and 4-month-old seedlings using RWBs and 12-month-old seedlings using root inoculation. The authors reported that while no significant differences of mortality rates were observed when using germinated seeds, some crosses were significant when screened using 4-month-old seedlings and 12-month-old seedlings. It was also shown that progenies of the Zaire × Cameroon origin had the lowest mortality using 4-month-old (RWB) and 12-month-old seedlings (root inoculation), which is similar to an earlier study (Idris et al. 2004). However, there was no explanation as to why significant differences between genetic origins were only observed in the 12-month-old seedling (root inoculation) while the germinated seeds (RWB) did not.

Goh et al. (2014c) observed that younger seedlings (3-month-old compared to 5-month-old) would be infected at a faster rate when screened using 11 different isolates. A similar trend was also observed when screening was performed using 5-, 6-, and 7-month-old seedlings (Chin et al., unpublished). This trend however is not observed when germinated seeds were used compared to 6-month-old seedlings (Breton et al. 2006). Could it be possible that the lower infection rate observed when using germinated seeds could be due to other confounding effects? The drawback of using seedlings for screening is that root injury when removing the soil core would affect the outcome of the experiment as Breton et al. (2006) showed that seedlings with injured roots had higher disease incidence. This coupled with transplanting shock of seedlings especially for older seedlings could also possibly increase disease incidence. However, this was not observed in the results shown in both studies (Goh et al. 2014c; Chin et al., unpublished).

5.5.4.4 Disease Assessment

Several methods were used to assess the level of resistance of a family, that is, percentage infected or disease incidence through visual scoring, percentage infection through destructive sampling, percentage mortality, and also disease severity index (DSI) based on categorical scoring (Kok et al. 2013). In addition, assessment of the severity or percentage of the lesion in the bole is also possible through destructive sampling though this would ultimately only yield information at a single time point. Instead of using individual parameters, Tan et al. (2015) used a cumulative ranking system of multiple parameters to assess different progenies. Furthermore, certain biochemical markers such as ergosterol (Chong et al. 2012) allowed assessments to be carried out in a quantitative manner. Often, assessment used for analysis is only based on a single time point determined by the researcher. A cumulative score of the disease progression across periods can be calculated by integrating any of the parameters above to obtain the area under disease progression curve (AUDPC).

Breton et al. (2010) reported that the trial disease incidence around 20–22 weeks, or at 30%, should be used for comparison, as the F-value of the trial was stable during this period. The disease incidence of each progeny was then expressed in relative terms based on a set of controls, which had a baseline of 100. Thus, scores lower than 100 indicate resistance while scores of more than 100 indicate increased susceptibility. There has only been a single study that described the variability expected in a nursery trial using germinated seeds at 20 weeks (Razak Purba et al. 2012). It was found that environment effect was the largest, accounting for 76.28% of the variation while GCA contributed by the male and female parent was only 4.18% and 2.37%, respectively.

5.5.4.5 Alternative Screening Methods

To overcome some of the inherent variability of screening on a progeny level, as individual seedlings are genotypically distinct, several studies (Goh et al. 2014b; Sugiharti et al. 2015) were conducted on clonal plantlets or ramets to enable individual genotypes to be assessed accurately via *in vitro* methods. Nevertheless, factors such as inoculum concentration, physiological age, and disease assessment techniques will need further work and more importantly the reproducibility and nursery-to-field disease trait correlation still need to be considered. Only limited genotypes will be available for screening due to the inherent low cloning amenability of oil palm, which may be a limitation. Even if successful, the production of potentially resistant ramets would be in limited numbers unless large numbers of ortets are included in the production line, although the use of clonal and semiclonal seed might be pursued.

5.5.4.6 Field Resistance and Nursery–Field Screening Relationships

It is generally accepted that the Deli Dura is highly susceptible as reported by de Franqueville et al. (2001) and this may partly be a reflection of limited genetic variability for resistance in this origin. Differences in the level of susceptibility between crosses were observed in the field (Durand-Gasselin et al. 2005) though the authors also cautioned that these trials were set up for varietal improvement programs and not specifically to assess disease resistance.

The epidemiology of the spread of BSR is reported to be random in the initial stages and in subsequent stages becomes clustered, and was also affected by physical barriers such as drainage and roads. It was also reported that the disease severity was higher in acid sulfate soils, thus indicating that soil type could be a factor influencing disease severity in the field (Chen et al. 2012b). Similar results have also been reported by Azahar et al. (2014) where disease distribution was random initially and later clustered, leading to differences in disease pressure. Taken together, the interpretation of field resistance observed in the field would be difficult unless specific trials were set up to be able to account for variation in the field. Current nursery–field relation has not been convincing, with r^2 ranging from 0.27 to 0.33 (Breton et al. 2010) and 0.5 (Razak Purba et al. 2012). Breton et al. (2010) have thus suggested a more holistic approach by also considering genetic origins, families, or categories in the development of a resistant cultivar.

5.5.4.7 Current Molecular Understanding of Oil Palm Defense Response toward *Ganoderma* sp.

A comprehensive review of molecular studies to understand oil palm–*Ganoderma* sp. interaction has been undertaken previously (Ho and Tan 2015) and will not be repeated here. The salient point of the review however suggests that although oil palm is able to mount a defense response toward *Ganoderma* sp. and that many of these defense-related genes have been characterized. However, the authors also noted that the main limitations of these studies are the unavailability of a reference genotype and variability inherent to screening and sampling methods. A recent publication by Ho et al. (2016) provides a better understanding of host response by comparing transcriptomes of oil palm roots in response to *G. boninense* and *Trichoderma harzianum*. This study indicated that *Ganoderma* sp. was able to induce reactive oxygen species (ROS), which could potentially lead to susceptibility as *Ganoderma* sp. is a known hemibiotroph. Furthermore, there was also an indication that salicylic acid, jasmonic acid, and ethylene host responses may be modulated by *Ganoderma* sp. during infection. It was also shown that in *Ganoderma* sp.-infected seedlings, genes related in strengthening of host cell wall and production of pathogenesis-related proteins and secondary metabolites were expressed as part of the defense response.

A different approach was used by several studies to where uninfected or asymptomatic seedlings were analyzed. Tan et al. (2016) compared transcripts of several candidate genes involved in phenylpropanoid and flavonoid pathways of symptomatic, asymptomatic, and control seedlings of Deli dura origin at 6 and 14 months post infection. The study indicated that although expression of PAL (phenylalanine ammonia lyse; key component of secondary metabolite production) was significantly higher for symptomatic seedlings compared to control and asymptomatic seedlings, it did not confer potential resistance. This observation however is in contrast with previous study (Tee et al. 2012; Ho et al. 2016) as PAL was seen to be downregulated during infection, though this could also be attributed to a difference in sampling periods. Furthermore, Tan et al. (2016) also suggested that there could be a change in lignin syringyl/guaiacyl (S/G) composition in asymptomatic seedlings and therefore conferring resistance. Another study (Hama-Ali et al. 2014) was able to identify SSRs that associated with seedlings that were either infected or noninfected 7 months post infection. These studies cast an interesting prospect of identifying potentially resistant seedlings through early screening, rather than to rely on assessment on a progeny level. Nevertheless, more studies are needed to determine the repeatability, heritability, and field correlation of this approach.

5.5.4.8 Conclusions

Though there is no general consensus on the best method to conduct early screening, several putative *Ganoderma* sp.-resistant varieties have already been developed. Extensive screening has been carried out by Turnbull et al. (2014) to show that families of some origins could be potentially resistant. However, field evidence of *Ganoderma* sp. resistance of these commercial materials has yet to fully materialize. Lastly, it has been shown that BSR caused by *Ganoderma* sp. is inoculum dependent (Flood et al. 2005) and the incidence of BSR can be controlled to a large extent with proper phytosanitary measures during replanting (Virdiana et al. 2010). It is therefore necessary for candidate *Ganoderma* sp.-resistant varieties to have good yields comparable with genetic yield improvement of commercial materials (Goh et al. 2014a).

REFERENCES

Agrios, G.N. 2005. *Plant Pathology* 5th edition. Academic Press, New York.

Akbar U., Kusnady M., and Ollagnier M. 1971. Influence de la Nature du Matériel Végétal et de la Nutrition Minérale sur la Pourriture Sèche du Tronc due à *Ganoderma. Oléagineux* 26: 527–34.

Ariffin, D., Idris, A.S., and Azahari, M. 1996. Spread of *Ganoderma boninense* and vegetative compatibility studies of a single field palm isolates. In: Ariffin, D. et al. (eds.) *Proceedings of the PORIM International Palm Oil Congress*, Palm Oil Research Institute of Malaysia, Kuala Lumpur, Malaysia, pp. 317–29.

Aslam, S., Erbs, G., Morrissey, K.L., Newman, M-A., Chinchilla, D., Boller, T., Molinaro, A., Jackson, R.W., and Cooper, R.M. 2009. MAMPs signatures, synergy, size and charge: Their influences on perception or mobility and host defence responses. *Molecular Plant Pathology* 10: 375–87.

Aslam, S. et al. 2008. Bacterial polysaccharides suppress induced innate immunity by calcium chelation. *Current Biology* 18: 1078–83.

Austin, R.B., Ford, M.A., and Morgan, C.L. 1989. Genetic improvement in the yield of winter wheat: A further evaluation. *Journal of Agricultural Sciences* 112: 295–301.

Azahar, T.M., Mustapha, J.C., Mazlihan, S., and Boursier, P. 2014. Temporal analysis of basal stem rot disease in oil palm plantations: An analysis on peat soil. *International Journal of Engineering & Technology IJET-IJENS* 11: 96–101.

Baayen, R.P., O'Donnell, K., Bonants, P.J.M., Cigelnic, E., Kroon, L.P.N.M., Roebroeck, E.J.A., and Waalwijk, C. 2000. Gene genealogies and AFLP in the *Fusarium oxysporum* complex identify monophyletic and non-monophyletic formae speciales causing wilt and rot disease. *Phytopathology* 90: 891–900.

Basri, M.W. 2003. Prioritizing future products for the oil palm industry. *Paper presented at Renaissance Palm Garden Hotel*, Putrajaya, Malaysia. March 31, 2003.

Bah Lias Research Station Annual Report 2010. *PT. London Sumatra Tbk.*, Medan, Sumatra, Indonesia.

Birch, R.G. 2001. *Xanthomonas albilineans* and the antipathogenesis approach to disease control. *Molecular Plant Pathology* 2: 1–11.

Blaak, G. 1970. Epiatasis for crown disease in the oil palm (*Elaeis guineensi* Jacq.). *Euphytica* 19: 22–24.

Borhan, M.H., Holub, E.B., Kindrachuk, C., Omidi, M., Bozorgmanesh-Frad, G., and Rimmer, S.R. 2010. WRR4. A broad spectrum TIR-NB-LRR gene from *Arabidopsis thaliana* that confers white rust resistance in transgenic oilseed brassica crops. *Molecular Plant Pathology* 11: 283–91.

Bourgis, F., Kilaru, A., Cao, X., Ngando-Ebongue, G-F., Drira, N., Ohlrogge, J.B., and Arondel, V. 2011. Comparative transcriptome and metabolite analysis of oil palm and date palm mesocarp that differ dramatically in carbon partitioning. In: *Proceedings of the Academy of Sciences USA* 108: 12526–32.

Breton, F., Hasan, Y., Hariadi, L.Z., and de Franqueville, H. 2006. Characterization of parameters for the development of an early screening test for basal stem rot tolerance in oil palm progenies. *Journal of Oil Palm Research (spec.)* 24–36.

Breton, F., Miranti, R., Lubis, Z., Hayun, Z., Umi, S., Flori, A., Nelson, S.P.C., Durand-Gasselin, T., and de Franqueville, H. 2009a. Early screening test: A routine work to evaluate resistance/susceptibility level of oil palm progenies to basal stem rot disease. In: *PIPOC*, 9–12 November, Kuala Lumpur Convention Centre, Kuala Lumpur, Malaysia.

Breton, F., Miranti, R., Lubis, Z., Hayun, Z., Umi, S., Flori, A., Nelson, S.P.C., Durand-Gasselin, T., Jacquemard, J.C., and de Franqueville, H. 2009b. Implementation of an early artificial inoculation test to screen oil palm progenies for their level of resistance and hypothesis on natural infection: *Ganoderma* disease of the oil palm. In:

16th International Oil Palm Conference and Expopalma. Challenges in Sustainable Oil Palm Development, September 22–25, 2009, Cartegena de Indias, Colombia, 16, 2009-09-22/2009-09-25.

Breton, F., Rahmaningsih, M., Lubis, Z., Syahputra, I., Setiawati, U., Flori, A., and de Franqueville, H. 2010. Evaluation of resistance/susceptibility level of oil palm progenies to basal stem rot disease by the use of an early screening test, relation to field observation. In: *Proceedings of the Second International Seminar Oil Palm Diseases*, Yogyakarta, Indonesia, pp. 33–54.

Breure, C.J. 1985. Relevant factors associated with crown expansion in oil palm (*Elaeis guineensis* Jacq.). *Euphytica* 34: 161–75.

Breure, C.J. 1986. Parent selection for yield and bunch index in the oil palm in West New Britain. *Euphytica* 35: 65–72.

Breure, C.J. 2003. The search for yield in oil palm: Basic principles. In: Fairhurst, T.H. and Hardter, R. (eds.) *The Oil Palm, Management for Large and Sustainable Yields*, Potash & Phosphate Institute (PPI), Potash & Phosphate Institute of Canada (PPIC) and International Potash Institute (IPI), Singapore, pp. 58–99.

Breure, C.J. 2010. Rate of leaf expansion: A criterion for identifying oil palm (*Elaeis guineensis* Jacq.) types suitable for planting at higher densities. *NJAS–Wageningen Journal of Life Sciences* 57: 141–7.

Breure, C.J. and Bos, I. 1992. Development of elite families in oil palm (*Elaeis guineensis* Jacq.). *Euphytica* 64: 99–112.

Breure, C.J. and Corley, R.H.V. 1983. Selection of oil palms for high density planting. *Euphytica* 32: 177–86.

Breure, C.J. and Powell, M.S. 1988. The one-shot method of establishing growth parameters in oil palm. In: Halim Hassan, A., Chew, P.S., Wood, B.J., and Pushparajah, E. (eds.) *Proceedings of the 1987 International Oil Palm Conference: Progress and Prospects*. Palm Oil Research Institute Malaysia, Kuala Lumpur, pp. 203–9.

Breure, C.J. and Soebagyo, F.X. 1991. Factors associated with occurrence of crown disease in oil palm (*Elaeis guineensis* Jacq.) and its effect on growth and yield. *Euphytica* 54: 55–64.

Breure, C.J. and Verdooren, L.R. 1995. Guidelines for testing and selecting parent palms in oil palm. Practical aspects and statistical methods. *ASD Oil Palm Papers,* 9: 1–68.

Buchanan, A.G. 1999. Molecular Genetic Analysis of Fusarium Wilt Resistance in Oil Palm. PhD Thesis, University of Bath, UK.

Caligari, P.D.S. 1993. G × E studies in perennial tree crops: Old familiar friend or awkward unwanted nuisance. In: *Proceedings of 1991 International Society of Oil Palm Breeders Workshop on Genotype–Environment Interaction Studies in Perennial Tree Crops*. Palm Oil Research Institute Malaysia, Kuala Lumpur, pp. 1–11.

Campbell, C.L. and Neher, D.A. 1994. Estimating disease severity and incidence. In: Campbell, C.L. and Buchwalter, D. (eds.) *Epidemiology and Management of Root Diseases*. Springer Berlin, Heidelberg, pp. 117–47.

Chan, K.W., Ong, E.C., Tan, K.S., Lee, C.H., and Law, I.H. 1986. The performance of oil palm genetic laboratory (OPGL) germplasm material. In: *Proceedings of International Workshop Oil Palm Germplasm & Utilisation*. Palm Oil Research Institute Malaysia, Kuala Lumpur. Vol. II. pp. 774–8. Malaysian Palm Oil Board, Kuala Lumpur.

Chen, S. et al. 2012a. Genome sequence of the model medicinal mushroom *Ganoderma lucidum*. *Nature Communications* 3: 369–77.

Chen, Z.Y., Goh, Y.K., Liew, Y.A., Goh, Y.K., Kok, S.Y., and Goh, K.J. 2012b. Dominant drivers associated with the epidemiology of *Ganoderma bonensis* of oil palms on coastal soils in Malaysia. In: *Proceedings of Soil Science Conference on Soil Quality towards Sustainable Agriculture Production*, pp. 535–42.

Chew, P.S. 1998. Prospects for precision plantation practices in oil palm. *The Planter* 74: 661–84.

Chong, K.P., Atong, M., and Rossall, S. 2012. The susceptibility of different varieties of oil palm seedlings to *Ganoderma boninense* infection. *Pakistan Journal of Botany* 44: 2001–4.

Chopart, J.L. 1999. Relations entre état physique du sol, systèmes racinaires et fonctionnement hydrique du peuplement végétal: Outils d'analyse *in situ* et exemples d'études en milieu tropical à risque climatique élevé. *PhD Thesis*, Univ. Joseph Fournier, Grenoble I, France. p. 115.

Chung, G.F. 2011. Management of *Ganoderma* diseases in oil palm plantations. *The Planter* 87: 325–39.

Claeys, H., De Bodt, S., and Inze, D. 2014. Gibberellins and DELLAs: Central nodes in growth regulatory networks. *Trends in Plant Science* 19: 231–9.

Cochard, B., Amblard, P., and Durand-Gasselin, T. 2005. Oil palm genetics and sustainable development. *Oleagineux Corps Gras Lipids* 12: 141–7.

Cochard, B., Noiret, J.M., Baudouin, L., Flori, A., and Amblard, P. 1993. Second cycle reciprocal recurrent selection in oil palm, *Elaeis guineensis* Jacq. Results of Deli × La Mé hybrid tests. *Oléagineux*, 48: 441–51.

Colebrook, E.H., Thomas, S.G., Phillips, A.L., and Hedden, P. 2014. The role of gibberellin signalling in plant responses to abiotic stress. *Journal of Experimental Biology* 217: 67–75.

Collard, B.C.Y., Vera Cruz, C.M., McNally, K.L., Virk, P.S., and Mackill, D.J. 2008. Rice molecular breeding laboratories in the genomics era: Current status and future considerations. *International Journal of Plant Genomics* 2008: 1–25. 524847, doi: 10.1155/2008/524847.

Collinge, D.B., Jorgensen, H.J.L., Lund, O.L., and Lyngkjaer, M.L. 2010. Engineering pathogen resistance in crop plants. Current trends and future prospects. *Annual Review of Phytopathology* 48: 269–91.

Cooper, R.M. 2011. Fusarium wilt of oil palm: A continuing threat to the Malaysian oil palm industry. *The Planter* 87: 409–18.

Cooper, R.M., Flood, J., and Mepsted, R. 1989. Fusarium wilt of oil palm: Transmission, isolate variation, resistance. In: Tjamos, E.C. and Beckman, C.H. (eds.) *Vascular Wilt Diseases of Plants, NATO ASI Series* Vol H-28, 247–58, Springer-Verlag, Berlin.

Cooper, R.M., Flood, J., and Rees, R.W. 2011. *Ganoderma boninense* in oil palm plantations: Current thinking on epidemiology, resistance and pathology. *The Planter* 87: 515–26.

Cooper, R.M. and Rusli, M.H. 2014. Threat from Fusarium wilt disease of oil palm to southeast Asia and suggested control measures. *Journal of Oil Palm Research* 26: 109–19.

Corley, R.H.V. 1977. Oil palm yield components and yields cycles. In: Earp, D.A. and Newall, W. (eds.) *Oil Palm Kuala Lumpur*, Incorp. Soc. of Planters, Malaysia, pp. 116–29.

Corley, R.H.V. 1993. Fifteen years experience with oil palm clones—A review of progress. In: *Proceedings of International Oil Palm Conference—Agriculture*. Palm Oil Research Institute Malaysia, Kuala Lumpur, pp. 69–81.

Corley, R.H.V., Boonrak, T., Donough, C.R., Nelson, S., Dumortier, F., Soebagyio, F.X., and Vallejo, G. 1995. Yield of oil palm clones in different environments. In: Rao, V., Henson I.E., and Rajanaidu, N. (eds.) *Recent Developments in Oil Palm Tissue Culture and Biotechnology*. Palm Oil Research Institute Malaysia, Kuala Lumpur, pp. 145–57.

Corley, R.H.V., Hardon, J.J., and Tan, G.Y. 1971. Analysis of growth of the oil palm (*Elaeis guineensis* Jacq.). Estimation of growth parameters and application in breeding. *Euphytica* 20: 307–15.

Corley, R.H.V. and Lee, C.H. 1992. The physiological basis for genetic improvement of oil palm in Malaysia. *Euphytica* 60: 179–84.

Corley, R.H.V. and Tinker, P.B. 2016. *The Oil Palm*, 5th edition. Wiley Blackwell, Oxford.

Corley, R.H.V., Tan, Y.P., Timti, I.N., and de Greef, W. 1993. Yield of oil palm progenies in Zaire, Cameroun, and Malaysia. In: *Proceedings of the 1991 International Society for Oil Palm Breeder's Workshop—Environment Interaction Studies in Perennial Tree Crops.* Palm Oil Research Institute of Malaysia, Kuala Lumpur, pp. 46–54.

Cornaire, B., Daniel, C., Zuily-Fodil, Y., and Lamade, E. 1994. Le comportement du palmier sous stress hydrique. Données du problème, premiers résultats et voies de recherche. *Oléagineux* 49: 1–12.

Cros, D., Flori, A., Nodichao, L., Omoré, A., and Nouy, B. 2013. Differential response to water balance and bunch load generates diversity of bunch production profiles among oil palm crosses (*Elaeis guineensis*). *Tropical Plant Biology* 6: 26–36.

Da Matta, F.M. 2004. Exploring drought tolerance in coffee: A physiological approach with some insights for plant breeding. *Brazilian Plant Physiology* 16: 1–6.

Dangl, J.L., Horvath, D.M., and Staskawicz, B.J. 2013. Pivoting the plant immune system from dissection to deployment. *Science* 341: 746–51.

Day, P.R. 1974. *Genetics of Host–Parasite Interactions.* Freeman, San Francisco, 238 pp.

de Franqueville, H. 1984. Vascular wilt of the oil palm: Relationship between nursery and field resistance. *Oléagineux* 39: 513–8.

de Franqueville, H. 1991. Former Savannah or former forest: Effect of pathogen isolates on the performance of oil palm families with respect to Fusarium wilt. *Oléagineux* 46: 179–86.

de Franqueville, H. 2003. Oil palm bud rot in Latin America. *Experimental Agriculture* 39: 225–40.

de Franqueville, H., Asmady, H., Jacquemard, J.C., Hayun, Z., and Durand-Gasselin, T. 2001. Indications on sources of oil palm (*Elaeis guineensis* Jacq.) genetic resistance and susceptibility to *Ganoderma* sp., the cause of basal stem rot. In: *Cutting Edge Technologies for Sustained Competitiveness.* Malaysian Palm Oil Board (MPOB), Mutiara Kuala Lumpur, Malaysia, pp. 420–31.

de Franqueville, H. and de Greef, W. 1988. Hereditary transmission of resistance to vascular wilt of the oil palm: Facts and hypotheses. In: Halim Hasan, A. et al. (eds.) *Proceedings of 1987 International Oil Palm Conference. Progress and Prospects.* Palm Oil Research Institute, Kuala Lumpur, Malaysia, pp. 118–29.

de Franqueville, H. and Louise, C. 2017, Breeding for resistance to oil palm diseases. In: Aik Chin Soh, Sean Mayes, and Jeremy Roberts (eds.) *Oil Palm: Breeding, Genetics and Genomics.* CRC Press, Boca Raton, FL.

de Franqueville, H. and Renard, J.L. 1990. Oil palm wilt in replantings: Study methods and determination of certain environment factors on the expresssion of this. *Oléagineux* 43: 149–57.

de Franqueville, H. and Renard, J.L. 1990. Improvement of oil palm vascular wilt tolerance. Results and development of the disease at the R. Michaux plantation. *Oléagineux* 45: 399–405.

Diabaté S., Traoré A., and Boaké K. 2010. Evaluation of the performance of tolerant crosses of oil palm selected in prenursery and replanted on wilt disease area. *Agriculture and Biology Journal of North America* 1: 1273–7.

Dong, X., Guo, J.R., Foster, S.J., Chen, H., Dong, C., Liu, Y., Hu, Q., and Liu, S. 2008. Expressing a gene encoding wheat qxalate oxidase enhances resistance to *Sclerotinia sclerotiorum* in oilseed rape (*Brassica napus*). *Planta* 228: 331–40.

Drenth, A., Torres, G.A., and Martinez, G. 2012. *Phytophthora palmivora*, the cause of bud rot in oil palm. Paper presented at *XVII Conferencia International Sobre Pal de Aceite*, 26–28 September, Cartagena, Colombia.

Dumortier, F., van Amstel, H., and Corley, R.H.V. 1992. *Oil Palm Breeding at Binga, Zaire.* 1970–1990, Unilever Plantations, London.

Durand-Gasselin, T., Asmady, H., Flori, A., Jacquemard, J.C., Hayun, Z., Breton, F., and de Franqueville, H. 2005. Possible sources of genetic resistance in oil palm (*Elaeis guineensis* Jacq.) to basal stem rot caused by *Ganoderma boninense*—Prospects for future breeding. *Mycopathologia* 159: 93–100.

Durand-Gasselin., T., Diabat, S., de Franqueville, H., Cochard, B., and Adon, B. 2000. Assessing and utilizing sources of resistance to Fusarium wilt in oil palm (*Elaeis guineensis* Jacq.). In: *Proceedings of the International Symposium on Oil Palm Genetic Resources and Their Utilization*. Kuala Lumpur, Malaysia, pp. 44–70.

Erbs, G. et al. 2008. Peptidoglycan and muropeptides from pathogens *Agrobacterium* and *Xanthomonas* elicit plant innate immunity: Structure and activity. *Chemistry and Biology* 15: 438–48.

Evans, L.T. and Fischer, R.A. 1999. Yield potential: Its definition, measurement, and significance. *Crop Science* 39: 1544–51.

Falconer, D.S. and Mackay, T.F.C. 1996. *Introduction to Quantitative Genetics*. Longmans Green/John Wiley & Sons, Harlow, Essex.

Flood, J., Cooper, R.M., and Lees, P.E. 1989. An investigation of pathogenicity of four isolates of *Fusarium oxysporum* from South America, Africa and Malaysia to clonal oil palm. *Journal of Phytopathology* 124: 80–8.

Flood, J., Keenan, L., Wayne, S., and Hasan, Y. 2005. Studies on oil palm trunks as sources of infection in the field. *Mycopathologia,* 159: 101–7.

Flood, J., Mepsted, R., and Cooper, R.M. 1990. Potential spread of Fusarium wilt of oil palm on contaminated seed and pollen. *Mycological Research* 94: 708–9.

Flood, J., Mepsted, R., Velez, A., Paul, T., and Cooper, R.M. 1993. Comparison of virulence of isolates of *Fusarium oxysporum* f. sp. *elaeidis* from Africa and South America. *Plant Pathology* 42, 168–71.

Flood, J., Whitehead, D.S., and Cooper, R.M. 1992. Vegetative compatibility and DNA polymorphisms in *Fusarium oxysporum* f. sp. *elaeidis* and their relationship to isolate virulence and origin. *Physiological and Molecular Plant Pathology* 41: 201–15.

Fraselle, J.V. 1951. Experimental evidence of pathogenicity of *Fusarium oxysporum* Schl. F. to the oil palm (*Elaeis guineensis* J.). *Nature* 167: 44.

Gerritsma, W. and Soebagyo, F.X. 1999. Analysis of growth of leaf area of oil palms in Indonesia. *Experimental Agriculture* 35: 293–308.

Goh, K.J., Ng, H.C., Wong, C.K., and Arif, S. 2014a. Yield potential of oil palm and its attainment in Malaysia. *The Planter* 90: 503–20.

Goh, K.M., Ganeson, M., and Supramaniam, C.V. 2014b. Infection potential of vegetative incompatible *Ganoderma boninense* isolates with known ligninolytic enzyme production. *African Journal of Biotechnology* 13: 1056–66.

Goh, Y.K., Ng, F.W., Kok, S.M., Goh, Y.K., and Goh, K.J. 2014c. Aggressiveness of *Ganoderma boninense* isolates on the vegetative growth of oil palm (*Elaeis guineensis*) seedlings at different ages. *Malaysian Applied Biology* 43: 9–16.

Gomez, P.L., Ayala, L., and Munévar, F. 2000. Characteristics and management of bud rot, a disease of oil palm. In: Pushparajah, E. (ed.) *Proceedings of International Planters Conference: Plantation Tree Crops in the New Millenium: The Way Ahead, Incorp. Soc. Planters*, Kuala Lumpur, pp. 545–53.

Govrin, E.M. and Levine, A. 2000. The hypersensitive response facilitates plant infection by the necrotrophic pathogen *Botrytis cinerea*. *Current Biology* 10: 751–7.

Gurmit, 1991. Ganoderma—The scourge of oil palms in the coastal areas. *Planter* 67: 421–444.

Hama-Ali, E.O., Panandam, J.M., Tan, S.G., Sharifah Shahrul Rabiah, S.A., Tan, J.S., Ho, C.L., and Hoh, B.P. 2014. Association between basal stem rot disease and simple sequence repeat markers in oil palm, *Elaeis guineensis* Jacq. *Euphytica* 202: 199–206.

Hardon, J.J., Williams, C.N., and Watson, I. 1969. Leaf area and yield in the oil palm in Malaysia. *Experimental Agriculture* 5: 25–32.

Hasan, Y. and Turner, P.D. 1998. The comparative importance of different oil palm tissues as infection sources for basal stem rot in replantings. *The Planter* 74: 119–35.

Henderson, J. and Osborne, D.J. 1990. Cell separation and anatomy of abscission in the oil palm, *Elaeis guineensis* Jacq. *Journal of Experimental Botany* 41: 203–10.

Henson, I.E. 1993. Assessing frond dry matter production and leaf area development in young oil palms. In: Basiron, Y. et al., *Proceedings of the 1991 PORIM International Palm Oil Conference—Agriculture*, Palm Oil Research Institute Malaysia, Kuala Lumpur, pp. 473–8.

Huang, C.C. and Lindhout, P. 1997. Screening for resistance in wild Lycopersicon species to Fusarium oxysporum f.sp. lycopersici race 1 and race. *Euphytica* 93: 145–153.

Ho, C.L. and Tan, Y.C. 2015. Molecular defense response of oil palm to *Ganoderma* infection. *Phytochemistry* 114: 168–77.

Ho, C.L., Tan, Y.C., Yeoh, K.A., Ghazali, A.K., Yee, W.Y., and Hoh, C.C. 2016. *De novo* transcriptome analyses of host-fungal interactions in oil palm (*Elaeis guineensis* Jacq.). *BMC Genomics* 17: 66.

Houssou, M., Omore, A., and Meunier, J. 1989. Breeding for drought resistance in the oil palm 1. Variability of some crosses for their productivity and their mortality. In: *Paper Presented at International Conference on Palms and Palm Products*, November 21–25, Nigerian Institute for Oil Palm Research, Benin City, Nigeria.

Idris, A.S., Kushairi, D., Ariffin, D., and Basri, M.W. 2004. Technique for inoculation of oil palm germinated seeds with *Ganoderma*. In: *MPOB Information Series*, 321, Malaysian Palm Oil Board, Malaysia.

Idris, A.S., Kushairi, A., Ismail, S., and Ariffin, D. 2004. Selection for partial resistance in oil palm progenies to *Ganoderma* basal stem rot. *Journal of Oil Palm Research*, 16: 12–8.

IPCC. 2014. *Climate Change 2014—Impacts, Adaptation and Vulnerability*. Intergovernmental Panel on Climate Change, Geneva.

Jones, J.D.G. and Dangl, J.L. 2006. The plant immune system. *Nature* 444: 323–9.

Khairudin, H. 1993. Basal stem rot of oil palm caused by *Ganoderma boninense*: An update. In: *Proceedings of the 1993 PORIM International Congress (Agriculture)*, Palm Oil Research Institute of Malaysia, Kuala Lumpur, Malaysia, pp. 739–49.

Khairudin, H., Lim, T.K., and Abdul Razak, A.R. 1991. Pathogenicity of *Ganoderma boninense* Pat. on oil palm seedlings. In: *Proceedings of the 1991 PORIM International Congress (Agriculture)*, Palm Oil Research Institute of Malaysia, Kuala Lumpur, Malaysia, pp. 418–23.

Kim, J. and Patterson, S.E. 2006. Expression divergence and functional redundancy of polygalacturonases in floral organ abscission. *Plant Signalling Behaviour* 1: 281–3.

Kinge, T.R. and Mih, A.M. 2011. *Ganoderma ryvardense* sp. nov. associated with basal stem rot (BSR) disease of oil palm in Cameroon. *Mycosphere* 2: 179–88.

Kok, S.M., Wong, W.C., Tung, H.J., Goh, Y.K., Goh, K.J., and Goh, Y.K. 2013. *In vitro* growth of *Ganoderma* sp. *boninense* isolates on novel palm extract medium and virulence on oil palm (*Elaeis guineensis*) seedlings. *Malaysian Journal of Microbiology* 43: 9–16.

Kushairi, A., Mohd Din, A., and Rajanaidu, N. 2011. Oil palm breeding and seed production. In: Mohd Basri, W., Choo, Y.M., and Chan, K.W. (eds.) *Further Advances in Oil Palm Research (2000–2010)*, Vol 1. Malaysian Palm Oil Board, Kuala Lumpur, pp. 47–93.

Lacombe, S. et al. 2010. Interfamily transfer of a plant pattern-recognition receptor confers broad-spectrum bacterial resistance. *Nature Biotechnology* 28: 365–9.

Leach, J., Vera Cruz, C.M., Bai, J., and Leung, H. 2001. Pathogen fitness penalty as a predictor of durability of disease resistance genes. *Annual Review of Phytopathology* 39: 187–224.

Lee, C.H. and Donough, C.R. 1993. Genotype-environment interaction in oil palm clones. In: *Proceedings of 1991 International Society of Oil Palm Breeders Workshop on Genotype—Environment Interaction Studies in Perennial Tree Crops*. Palm Oil Research Institute Malaysia, Kuala Lumpur, pp. 33–45.

Lee, C.H. and Rajanaidu, N. 1999. Genotype environment interaction in oil palm. In: *Proc. Seminar Science of Oil Palm Breeding (Montpellier, 1992)*. Palm Oil Research Institute Malaysia, Kuala Lumpur, pp. 96–111.

Li, T., Liu, B., Spalding, H., Weeks, D.P., and Yang, B. 2012. High efficiency TALEN-based gene editing produces disease-resistant rice. *Nature Biotechnology* 30: 390–2.

Lievens, B., Rep, M., and Thomma, P.H.J. 2008. Recent developments in the molecular discrimination of formae speciales of *Fusarium oxysporum*. *Pest Management Science* 64: 781–88.

Lim, H.P. and Fong, Y.K. 2005. Research on basal stem rot (BSR) of ornamental palms caused by basidiospores from *Ganoderma boninense*. *Mycopathologia* 159: 171–9.

Lim, T.K., Chung, G.F., and Ko, W.H. 1992. Basal stem rot of oil palm caused by *Ganoderma boninense*. *Plant Pathology Bulletin* 1: 147–52.

Locke, T. and Colhoun, J. 1974. Contributions to a method of testing oil palm seedlings for resistance to *Fusarium oxysporum* Schl. f. sp. *elaeidis* toovey. *Phytopathologische Zeitschrift* 79: 77–92.

Louise, C., Poveda, R., Flori, A., and Amblard, P. 2013. Tolerancia al complejo de Pudrición del cogollo en el material Elaeis guineensis: Caso del material C07 en el oriente ecuatoriano. Tolerance to oil palm bud rot disease in the material *Elaeis guineensis*: Case of the material c07 in eastern Ecuador. *Palmas* 34: 126–34. *Conferencia Internacional Sobre Palma de Aceite*. 17, 2012-09-25/2012-09-28, Cartagena, Colombia.

Ma, L.J. et al. 2010. Comparative genomics reveals mobile pathogenicity chromosomes in *Fusarium*. *Nature* 464: 367–73.

Maillard, G., Daniel, C., and Ochs, R. 1974. Analyse des effets de la sécheresse sur le palmier à huile. *Oléagineux* 29: 395–404.

Mepsted, R., Flood, J., and Cooper, R.M. 1995a. Fusarium wilt of oil palm II. Stunting as a mechanism to reduce water stress. *Physiological Molecular Plant Pathology* 46: 373–87.

Mepsted, R., Flood, J., Paul, T., Airede, C., and Cooper, R.M. 1995b. A model system for rapid selection for resistance and investigation of resistance mechanisms in Fusarium wilt of oil palm. *Plant Pathology* 44: 749–55.

Mepsted, R., Flood, J., Paul, T., and Cooper, R.M. 1994. Virulence and aggressiveness in *Fusarium oxysporum* f. sp. *elaeidis*; implications for screening for disease resistance. *Oleagineux* 49: 209–12.

Mepsted, R., Nyandusa, C., Flood, J., and Cooper, R.M. 1991. A non-destructive, quantitative method for the assessment of infection of oil palm by *Fusarium oxysporum* f. sp. *elaeidis*. *Elaeis*. 3: 329–35.

Meunier, J., Renard, J.L., and Quillic, G. 1979. Heredity of resistance to Fusarium wilt in the oil palm *Elaeis guineensis* Jacq. *Oleagineux* 34: 555–61.

Michielse, C.B. and Rep, M. 2009. Pathogen profile update: *Fusarium oxysporum*. *Molecular Plant Pathology* 10: 311–24.

Miller, R.N.G., Holderness, M., Bridge, P.D., Chung, G.F., and Zakaria, M.H. 1999. Genetic diversity of *Ganoderma* in oil palm plantings. *Plant Pathology* 48: 595–603.

Morcillo, F. et al. 2013. Improving palm oil quality through identification and mapping of the lipase gene causing oil deterioration. *Nature Communications* 4: 2160.

Narvaez, M.P.R. 2012. Etiology of Cogollo rot in oil palm (*Elaeis guineensis* Jacq.) in Ecuador. *Masters thesis*, Universidad de Puerto Rico, Recinto Universitario de Mayaguez.

Newton, A.C., Begg, G.S., and Swanston, J.S. 2008. Deployment of diversity for enhanced crop function. *Annals of Applied Biology* 154: 309–22.

Nodichao, L., Chopart, J-L., Roupsard, O., Vauclin, M., Aké, S., and Jourdan, C. 2011. Genotypic variability of oil palm root system distribution in the field. Consequences for water uptake. *Plant and Soil* 341: 505–20.

Nouy, B., Baudouin, L., Djégui, N., and Omoré, A. 1999. Oil palm under limiting water supply conditions. *Plantation Recherche Developpement* 6: 31–45.

Nouy, B., Omoré, A., and Potier, F. 1996. Oil palm production cycles in different ecologies: Consequences for breeding. In: *1996 PORIM International Palm Oil Congress*, Kuala Lumpur, pp. 62–75.

Obisesan, I.O. and Fatunla, T. 1983. Heritability of fresh fruit bunch yield and its components in the oil-palm (*Elaeis guineensis*, Jacq.). *Theoretical and Applied Genetics* 64: 65–8.

O'Connell, R. et al. 2012. Life style transitions in plant pathogenic *Colletotrichum* fungi deciphered by genome and transcriptome analyses. *Nature Genetics* 44: 1060–5.

O'Donnell, K., Kistler, H.C., Cigelnik, E., and Ploetz, R.C. 1998. Multiple evolutionary origins of the fungus causing panama disease of banana: Concordant evidence from nuclear and mitochondrial gene genealogies. In: *Proceedings of the National Academy of Sciences USA* 95: 2044–9.

Olson, A. et al. 2012. Insight into trade-off between wood decay and parasitism from the genome of a fungal forest pathogen. *New Phytologist* 194: 1001–13.

Ong, E.C., Lee, C.H., Law, I.H., and Ling, A.H. 1986. Genotype—Environment interaction and stability analysis for bunch yield and its components, vegetative growth and bunch characters in the oil palm (*Elaeis guineensis* Jacq.). In: *New Frontiers in Breeding Research—Proceedings of 5th International Congress. SABRAO.* Bangkok, pp. 523–32.

Oritsejafor, J.J. 1989. Status of the oil palm vascular wilt disease in Nigeria. In: *Proceedings of the International Conference of Palms and Palm Products.* Nigerian Institute of Oil Palm Research, Benin City.

Osborne, D.J., Henderson, J., and Corley, R.H.V. 1992. Controlling fruit-shedding in the oil palm. *Endeavour* 16: 173–7.

Pais, M. et al. 2013. From pathogen genomes to host plant processes: The power of plant parasitic oomycetes. *Genome Biology* 14: 211, doi: 10.1186/gb-2013-14-6-211.

Peng, J. et al. 1999. "Green revolution" genes encode mutant gibberellin response modulators. *Nature* 400: 256–61.

Pilotti, C.A. 2005. Stem rots of oil palm caused by *Ganoderma boninense*: Pathogen biology and epidemiology. *Mycopathologia* 159: 129–37.

Pilotti, C.A., Sanderson, F.R., and Aitken, E.A.B. 2002. Sexuality and interactions of monokaryotic and dikaryotic mycelia of *Ganoderma boninense. Mycological Research* 11: 1315–22.

Pilotti, C.A., Sanderson, F.R., and Aitken, E.A.B. 2003. Genetic structure of a population of *Ganoderma boninense* on oil palm. *Plant Pathology* 52: 455–63.

Pilotti, C.A., Sanderson, F.R., Aitken, E.A.B., and Armstrong, W. 2004. Morphological variation and host range of two *Ganoderma* species from Papua New Guinea. *Mycopathologia* 158: 251–6.

Ploetz, R.C. 2007. Diseases of tropical perennial crops: Challenging problems in diverse environments. *Plant Disease* 91: 644–63.

Prendergast, A.G. 1963. A method of testing oil palm progenies at the nursery stage for resistance to vascular wilt disease caused by *Fusarium oxysporum. Journal of West African Institute of Oil Palm Research* 4: 156–75.

Rahamah Bivi, M.S.H., Idris, A.S., Maizatul, S.M., Hohd Din, A., Marhalil, M. 2015. Three methods for screening of oil palm resistance to Ganoderma. In: *Proceedings of the PIPOC 2015 International Oil Palm Congress: Agriculture, Biotechnology, and Sustainability Conference*, Kuala Lumpur Convention Centre, Kuala Lumpur, Malaysia.

Rafii, M.Y., Rajanaidu, N., Kushairi, A., Jalani, B.S., and Din, M.A. 2000. Evaluation of Cameroon and Zaire genetic materials. In Rajanaidu, N. and Ariffin, D. (eds.) *Int. Symp. Oil Palm Genetic Resources and Utilization*, 8–10 June, 2000. Malaysian Palm Oil Board, Kuala Lumpur, pp. 1–34.

Rajagopalan, K., Aderungboye, F.O., and Obasola, C.O. 1978. Evaluation of oil palm progenies for reaction to the vascular wilt disease. *Journal of Nigerian Institute for Oil Palm Research* 5: 87.

Rajanaidu, N. and Jalani, B.S. 1994. Oil palm genetic resources—Collection, evaluation, utilization and conservation. Paper presented at Colloquium "Oil Palm Genetic Resources", Palm Oil Research Institute Malaysia, Bangi, Malaysia.

Rajanaidu, N., Rao, V., Hoong, H.W., Lee, C.H., Tan, Y.P., Tan, S.T., and Jalani, B.S. 1992. Yield potential and genotype × environment (GE) studies in oil palm (*Elaeis guineensis*). In: *Proceedings of Workshop Yield Potential in the Oil Palm*. Int. Soc. Oil Palm Breeders, Kuala Lumpur, pp. 44–57.

Rajanaidu, N., Tan, Y.P., Ong, E.C., and Lee, C.H. 1986. The performance of inter-origin commercial D × P planting material. In: *Proceedings of International Workshop Oil Palm Germplasm & Utilisation*. Palm Oil Research Institute Malaysia, Kuala Lumpur, pp. 155–61.

Rao, V. 1998. Ripening in the virescens oil palm. In: Rajanaidu, N., Henson, I.E., and Jalani, B.S. (eds.) *Oil and Kernel Production in Oil Palm—A Global Perspective. Proceedings of International Conference*. p. 226. Malaysian Palm Oil Board, Kuala Lumpur.

Razak Purba, A., Setiawati, U., Rahmaningsih, M., Yenni, Y., Rahmadi, H.Y., and Nelson, S. 2012. Indonesia's experience of developing *Ganoderma* tolerant/resistant oil palm planting material. In *International Seminar on Breeding for Sustainability in Oil Palm*. Retrieved from http://isopb.mpob.gov.my/index.php?r=site/page&view=publication_2012.

Rees, R.W., Flood, J., Hasan, Y., and Cooper, R.M. 2007. Effects of inoculum potential, shading and soil temperature on root infection of oil palm seedlings by the basal stem rot pathogen *Ganoderma boninense*. *Plant Pathology* 56: 862–70.

Rees, R.W., Flood, J., Hasan, Y., and Cooper, R.M. 2012. *Ganoderma boninense* Basidiospores in oil palm plantations: Evaluation of their possible role in stem rots of oil palm (*Elaeis guineensis*). *Plant Pathology* 61: 567–78.

Rees, R.W., Flood, J., Hasan, Y., Potter, U., and Cooper, R.M. 2009. Basal stem rot of oil palm (*Elaeis guineensis*); mode of root infection and lower stem invasion by *Ganoderma boninense*. *Plant Pathology* 58: 982–9.

Reis de Carvalho, C. 1991. Mécanismes de résistance à la sécheresse chez les plantes jeunes et adultes de palmier à huile. *Thèse*, Université Paris-Sud, Orsay, France.

Renard, J.L. and de Franqueville, H. 1989. Oil palm vascular wilt. *Oléagineux* 44: 342–7.

Renard, J.L., de Franqueville, H., Meunier, J., and Noiret, J.M. 1991. Méthode d'évaluation du comportement du palmier à huile vis-à-vis de la fusariose vasculaire due à *Fusarium oxysporum f sp elaeidis*—Résultats. In: Chalbi, N. and Demarly, Y. (eds.) L'amelioration des plantes pour l'adaptation aux milieux arides John Libbey Eurotext, Paris, pp. 121–134.

Renard, J.L., Gascon, J.P., and Bachy, A. 1972. Research on vascular wilt disease of the oil palm. *Oléagineux* 27: 581–91.

Renard, J.L., Noiret, J.M., and Meunier, J. 1980. Sources et Gammes de Résistance a la Fusariose chez les Palmiers à Huile *Elaeis guineensis* et *Elaeis melanococca*. *Oléagineux* 35: 387–93.

Roberts, J.A. and Zubaidah, R. 2013. Genetic strategies to regulate height in oil palm. In: *PIPOC 2013 International Oil Palm Congress: Agriculture, Biotechnology, and Sustainability Conference*, Kuala Lumpur Convention Centre, Kuala Lumpur, Malaysia.

Romero, H.M., Arias, D., Moreno, L., Rivera, Y., Prada, F., Daza, E., Avila, R., and Forero, D. 2013. The interspecific O × G (*Elaeis oleifera* × *Elaeis guineensis*) is a commercial alternative for oil palm production in the Americas. *PIPOC 2013 International Oil Palm Congress: Agriculture, Biotechnology, and Sustainability Conference*, Kuala Lumpur Convention Centre, Kuala Lumpur, Malaysia.

Roongsattham, P. et al. 2012. Temporal and spatial expression of polygalacturonase gene family members reveals divergent regulation during fleshy fruit ripening and abscission in the monocot species oil palm. *BMC Plant Biology* 12: 150–65.

Rosenquist, E.A. 1982. The genetic base of oil palm breeding populations. In: *Proceedings of the International Workshop Oil Palm Germplasm and Utilisation*. Palm Oil Research Institute Malaysia, Kuala Lumpur, pp. 27–56.

Rosenquist, E.A., Corley, R.H.V., and de Greef, W. 1990. Improvement of tenera populations using germplasm from breeding programmes in Cameroon and Zaire. In: *Proceedings of Workshop on Progress of Oil Palm Breeding Populations*. Palm Oil Research Institute, Malaysia, Kuala Lumpur, pp. 37–69.

Rusli, M., Idris, A., and Cooper, R. 2015. Evaluation of Malaysian oil palm progenies for susceptibility, resistance or tolerance to *Fusarium oxysporum* and defence-related gene expression in roots. *Plant Pathology* 64: 638–47.

Sanderson, F.R. and Pilotti, C.A. 1997. *Ganoderma* basal stem rot: An enigma, or just time to rethink an old problem? *The Planter* 73: 489–93.

Sanderson, F.R., Pilotti, C.A., and Bridge, P. 2000. Basidiospores: Their influence on our thinking regarding a control strategy for basal stem rot of oil palm. In: Flood, J., Bridge, P. and Holderness, M. (eds.) *Ganoderma Diseases of Perennial Crops*. CABI Publishing, Wallingford, UK, pp. 113–9.

Santacruz, L., Zambrano, J., Avila, M., and Calvo, F. 2000. El Complejo Pudrición de Cogollo en la Zona Oriental de Colombia. In: *Paper Presented at Conference, 'Competitividad y Prospectiva de la Palma de aceite'*, 6–8 September, Cartagena, Colombia.

Santner, A. and Estelle, M. 2009. Recent advances and emerging trends in plant hormone signalling. *Nature* 459: 1071–8.

Sariah, M., Hussain, M.Z., Miller, R.N.G., and Holderness, M. 1994. Pathogenicity of *Ganoderma boninense* tested by inoculation of oil palm seedlings. *Plant Pathology* 43: 507–10.

Schafer, J.F. 1971. Tolerance to plant disease. *Annual Review of Phytopathology* 9: 235–52.

Schwessinger, B. et al. 2014. Transgenic expression of the dicotyledonous pattern recognition receptor EFR in rice leads to ligand dependent activation of defense responses. *PLoS Pathogens* 11(4): e1004872, doi: 10.137.

Singh, R. et al. 2013. Oil palm genome sequence reveals divergence of interfertile species in old and new worlds. *Nature* 500: 335–41.

Singh, R. et al. 2014. The oil palm *VIRESCENS* gene controls fruit colour and encodes a R2R3-MYB. *Nature Communications* 5: 4106.

Slavov, S.B. 2014. Phytotoxins and *in vitro* screening for improved disease resistant plants. *Biotechnology & Biotechnology Equipment* 19: 48–55.

Smith, B.G., Donough, C.R., and Corley, R.H.V. 1996. Relationship between oil palm clone phenotype and optimal planting density. In: Ariffin, D. et al. (eds.) *Proceedings of the 1996 PORIM International Palm Oil Congress. Competitiveness for the Twenty-First Century*. Palm Oil Research Institute Malaysia, Kuala Lumpur, pp. 76–86.

Soh, A.C. 1999. Breeding plans and selection methods in oil palm. In: *Proceedings of Seminar Science of Oil Palm Breeding (Montpellier, 1992)*. Palm Oil Research Institute Malaysia, Kuala Lumpur, pp. 65–95.

Soh, A.C. 2011. Genomics and plant breeding. *Journal of Oil Palm Research* 23: 1019–28.

Soh, A.C. and Chow, C.S. 1993. Index selection utilizing plot and family information in oil palm. *Elaeis* 5: 27–37.

Soh, A.C., Wong, G., Hor, T.Y., Tan, C.C., and Chew, P.S. 2003. Oil palm genetic improvement. In: Janick, J. (ed.) *Plant Breeding Reviews*, Vol. 22. John Wiley & Sons, New Jersey, pp. 165–219.

Soh, A.C., Yong, Y.Y., Ho, Y.W., and Rajanaidu, N. 1995. Commercial potential of oil palm clones. In: Rao, V., Henson, I.E., and Rajanaidu, N. (eds.) *Recent Developments in Oil Palm Tissue Culture and Biotechnology*. International Society for Oil Palm Breeders, Bangi, pp. 134–44.

Squire, G.R. 1983. Solar energy and productivity in oil palm. In: *Ann. Res. Rep. 1983*, Palm Oil Research Institute of Malaysia, Kuala Lumpur, pp. 149–156.

Squire, G.R. 1984. *Light Interception, Productivity and Yield of Oil Palm*. Palm Oil Research Institute of Malaysia, Kuala Lumpur.

Subagio, A. and Foster, H.L. 2003. Implications of *Ganoderma* disease on loss in stand and yield production of oil palm in North Sumatra. In: *Proceedings of the MAPPS Conference*, August 2003, Malaysian Plant Protection Society, Kuala Lumpur.

Sugiharti, M., Prihatna, C., Fortunatus, A., Hartali, I., and Suwato, A. 2015. A rapid diagnostic assay of *Ganoderma* disease in oil palm seedlings. In: *Proceedings of the PIPOC 2015 International Oil Palm Congress: Agriculture, Biotechnology, and Sustainability Conference*, Kuala Lumpur Convention Centre, Kuala Lumpur, Malaysia.

Susanto, A. 2009. Basal stem rot in Indonesia: Biology, economic importance, epidemiology, detection, and control. In: *Proceedings of the International Workshop on Awareness, Detection, and Control of Oil Palm Devastating Diseases*, 6 November 2009, Kuala Lumpur Convention Centre (KLCC), Kuala Lumpur, Malaysia, 180 pp.

Susanto, A. and Huan, L.K. 2010. Management of *Ganoderma* in mineral and peat soil in Indonesia. In: *Proceedings of the Second International Seminar Oil Palm Diseases: Advances in Ganoderma Research and Management*. 31 May 2010, Yogyakarta, Indonesia.

Tai, T.H., Dahlbeck, D., Clark, E.T., Gajiwala, P., Pasion, R., Whalen, M.X.C., Stall, R.E., and Staskawicz, B.J. 1999. Expression of Bs2 pepper gene confers resistance to bacterial spot disease in tomato. In: *Proceedings of the National Academy for Sciences USA* 96: 14153–8.

Tan, B.A., Daim, L.D.J., Ithnin, N., Ooi, T.E.K., Md-Noh, N., Mohamed, M., and Kulaveerasingam, H. 2016. Expression of phenylpropanoid and flavonoid pathway genes in oil palm roots during infection by *Ganoderma boninense*. *Plant Gene* 7: 11–20.

Tan, C.C., Ng, W.J., Choo, C.N., Kumar, K., Aida, N., Chin, S., Melody, M., and Wong, C.K. 2015. Modern planting materials and their characters. *Paper Presented at MOSTA Agronomy Best Practices Workshop 2015*, 17–19 August 2015, Miri, Sarawak, Malaysia.

Tan, J.S. 2015. Breeding for oil palm resistant (*Ganoderma boninense*) varieties. *Presented at the MOSTA Oil Palm Best Practices Workshop 2015*. Miri, Sarawak, Malaysia.

Tee, S.S., Tan, Y.C., Abdullah, F., Ong-Abdullah, M., and Ho, C.L. 2012. Transcriptome of oil palm (*Elaeis guineensis* Jacq.) roots treated with *Ganoderma boninense*. *Tree Genetics & Genomes* 9: 377–86.

Torres, G.A., Sarria, G.A., Varon, F., Coffey, M.D., Elliott, M.L., and Martinez, G. 2010. First report of bud rot caused by *Phytophthora palmivora* on African oil palm in Colombia. *Plant Disease* 94: 1163.

Tranbarger, T.J. et al. 2011. Regulatory mechanisms underlying oil palm fruit mesocarp maturation, ripening, and functional specialization in lipid and carotenoid metabolism. *Plant Physiology* 156: 564–84.

Turnbull, N., de Franqueville, H., Breton, F., Jeyen, S., Syahputra, I., Cochard, B., Poulain, V., and Durand-Gasselin, T. 2014. Breeding methodology to select oil palm planting material partially resistant to *Ganoderma boninense*. Presented at the *5th Quadrennial International Oil Palm Conference*, Nusa Dua, Indonesia.

Turner, P.D. 1981. *Oil Palm Diseases and Disorders*. Oxford University Press, Kuala Lumpur.

Van de Lande, H.L. 1984. Vascular wilt disease of oil palm (*Elaeis guineensis* Jacq.) in Para, Brazil. *Oil Palm News*, 28: 6–10.

Van Loon, L.C., Rep, M., and Pieterse, C.M.J. 2006. Significance of inducible defense-related proteins in infected plants. *Annual Review of Phytopathology* 44: 135–62.

Virdiana, I., Hasan, Y., Aditya, R., and Flood, J. 2010. Testing the effects of oil palm replanting practices (windrowing, fallowing and poisoning) on incidence of *Ganoderma*. In: *Proceedings International Oil Palm Conference*. Yogyakarta, Indonesia. AGR. p. 2.8.

Vleeshouwers, V.G.A.A. et al. 2008. Effector genomics accelerates discovery and functional profiling of potato disease resistance and *Phytophthora infestans (Pi)* avirulence genes. *PLoS One* 3, e2875.

Vleeshouwers, V.G.A.A. and Oliver, R.P. 2014. Effectors as tools in disease resistance breeding against biotrophic, hemibiotrophic and necrotrophic plant pathogens. *Molecular Plant–Microbe Interactions* 27: 196–206.

Wang, Y., Cheng, X., Shan, Q., Zhang, Y., Liu, J., Gao, C., and Qui, J.L. 2014. Simultaneous editing of three homoalleles in hexaploid bread wheat confers heritable resistance to powdery mildew. *Nature Biotechnology* 32: 947–52.

Wardlaw, C.W. 1946. *Fusarium oxysporum* on the oil palm. *Nature* 158: 712.

White, T.L. and Hodge, G.R. 1989. *Predicting Breeding Values with Applications in Forest Tree Improvement.* Kluwer Academic Publishers, Boston.

Wilhelm, S. 1981. Sources and genetics of host resistance in field and fruit crops. In: Mace, M.E., Bell, A.A., and Beckman, C.H. (eds.) *Fungal Wilt Diseases of Plants.* Academic Press, New York, pp. 299–376.

Yamada, Y. 1962. Genotype by environment interaction and genetic correlation of the same trait under different environments. *Japanese Journal of Genetics* 37: 498–509.

Yamada, Y. 1993. Interaction: Its analysis and interpretations. In: *Proceedings of 1991 International Society of Oil Palm Breeders Workshop on Genotype—Environment Interaction Studies in Perennial Tree Crops.* Palm Oil Research Institute Malaysia, Kuala Lumpur, pp. 131–9.

Yamaguchi, S. 2008. Gibberellin metabolism and its regulation. *Annual Review of Plant Biology* 59: 225–51.

Yang, S., Gao, M., Xu, C., Gao, J., Deshpande, S., Lin, S., Roe, B.A., and Zhu, H. 2008. *Alfalfa* benefits from *Medicago sativa*: The *RTC1* gene confers broad-spectrum resistance to anthracnose in *Alfalfa.* In: *Proceedings of the National Academy of Sciences USA* 105: 12164–9.

Zeven, A.C. 1967. *The Semi-Wild Oil Palm and Its Industry in Africa.* Agric. Res. Rep. 689. Wageningen University, Wageningen.

Zipfel, C. 2009. Early molecular events in PAMP-triggered immunity. *Current Opinion in Plant Biology* 12: 414–420.

Zipfel, C. 2014. Plant pattern recognition receptors. *Trends in Immunology* 35: 345–51.

6 Breeding Plans and Selection Methods

Aik Chin Soh, Sean Mayes, Jeremy Roberts, David Cros, and Razak Purba

CONTENTS

6.1 Introductory Overview (*Aik Chin Soh, Sean Mayes, and Jeremy Roberts*) 143
6.2 Breeding and Selection Methods and Variety Types in Oil Palm (*Aik Chin Soh, Sean Mayes, and Jeremy Roberts*) 144
6.2.1 Historical Perspective 144
6.2.1.1 Mass Selection 144
6.2.1.2 Recurrent Selection 145
6.2.2 Current Perspective 145
6.2.2.1 Top-Cross Test Breeding 145
6.2.2.2 Inbred-Hybrid Variety Breeding 145
6.2.2.3 Recombinant Inbred Variety/Line Breeding (RIV/L) 146
6.2.2.4 Single Seed Descent and Doubled Haploid Breeding 146
6.2.2.5 Backcross Breeding 146
6.2.2.6 Isogenic and Near Isogenic Lines 147
6.2.2.7 Breeding for Clonal Propagation 147
6.2.2.8 Clonal Hybrid Seeds 148
6.2.2.9 Clonal Variety 148
6.2.3 Parent Palm Selection Methods 148
6.2.3.1 Single Trait Selection 148
6.2.3.2 Multiple Trait Selection 149
6.3 Conclusions 150
Appendix 6A.1 SI Incorporating Full-Sib, Half-Sib Family, and Plot Information (*Aik Chin Soh*) 150
Appendix 6B.1 SIs for Multiple Traits in Oil Palm (*Aik Chin Soh*) 151
Appendix 6C.1 Best Linear Unbiased Prediction (*Aik Chin Soh and Razak Purba*) 153
Appendix 6D.1 Genomic Selection (*David Cros*) 156
References 160

6.1 INTRODUCTORY OVERVIEW

Aik Chin Soh, Sean Mayes, and Jeremy Roberts

It is often said that plant breeding is both a science and an art. The science part is derived from a sound understanding of not only the principles of plant genetics (qualitative, quantitative, molecular), statistics, and field experimentation but also plant and crop physiology, agronomy, and host–parasite relationships as the other chapters and their inclusion in the book attest. The art part is in decision making in terms of the desirable variety traits and type for future market needs; the breeding populations and parents to deploy; the most appropriate breeding method; the trait combination and relative emphases; and the number of families and family size carried forward to each subsequent breeding cycle. These issues could be investigated scientifically. However, the results of such studies are usually unavailable at the time of decision making, hence the art or intuitive approach comes in. Nevertheless, it is still founded on good science and experience. A breeding program comprises the three stages of creation of genetic variability, the selection of the desirable variants within the variable population, and stabilization of the desirable variant cum multiplication for commercial variety or cultivar release. Before embarking on a breeding program, a breeding plan must be in place. The breeding plan, comprising the varietal type to produce and the breeding process or scheme to achieve it, is largely directed by the breeding or propagation system of the crop and the target clientele, for example, small farmers or large corporate farms. More genetically variable varieties, for example, composites and synthetics which are likely to be more resilient to biotic and abiotic environmental stresses and lower agricultural inputs are more appropriate for small farmers and are cheaper to produce and sell. Large corporate farms are more inclined toward highly homogeneous, high input responsive, high yielding F_1 or single-cross hybrids or clones and are prepared to pay higher seed prices. These considerations hold true for an oil palm breeder whether they start a new program or inherit one.

6.2 BREEDING AND SELECTION METHODS AND VARIETY TYPES IN OIL PALM

Aik Chin Soh, Sean Mayes, and Jeremy Roberts

6.2.1 Historical Perspective

6.2.1.1 Mass Selection

Since its inception, oil palm breeding was guided by the need to produce sufficient improved planting materials to meet the requirements of a rapidly expanding plantation industry. Early oil palm breeding presumably began with collecting open-pollinated seeds from mass or phenotypically selected palms and was soon followed by organized intercrosses among them. These would be African T × T crosses in W. Africa (WA) and Deli D × D in S.E. Asia. Inbreeding incurred advertently or inadvertently.

With the revelation of the monogenic inheritance of the shell trait and the heterotic D × P or T hybrid expressed not only in terms of higher oil content due to the thicker mesocarp but also higher bunch yield and better growth vigor, major oil palm breeding programs have gravitated to the production of inter-population D × P hybrids with the highly selected Deli D population (uniform high oil yielding and bigger bunches) as the maternal parent and the highly selected WA (AVROS, Yangambi, La Mé, Ekona, NIFOR) P/T (uniform high oil yielding and high bunch number) as the paternal parent. This would then involve D × P/T progeny-testing and recurrent selection, that is, selected (mass, progeny-tested) parents are intercrossed to form the parents for the next generation or cycle of selection, adapting the selection methodologies developed from maize breeding (Allard 1960; Simmonds and Smartt 1999).

6.2.1.2 Recurrent Selection

This is essentially a population improvement method. In the modified recurrent selection (MRS) scheme (Soh 1986, 1999) sometimes also referred to as family and individual palm selection (FIPS) (Rosenquist 1990), the opposing Deli D and WA T parents are mass selected (by FIPS) for recurrent cycles of breeding from the Deli D × D and WA T × T crosses. The Ps based on their T sib performance in the T × T progeny trials (i.e., sib selection) are then progeny-tested against a sample of mass selected Deli D mother palms in a top-cross test or nested/North Carolina Model 1 (NCM1) mating design. The selected progeny-tested Ps are then used as male parents to cross with all the mass selected Deli D mother palms in commercial D × P hybrid production. This scheme is based on exploitation of general combining ability (GCA). Depending on the genetic relationships and inbreeding/selection status of the parents used, the commercial hybrid seeds are mixed hybrids similar to synthetics (Allard 1960) with varying levels of genetic variability.

In the modified reciprocal recurrent selection (MRRS), the Deli D and WA P/T parents for both recurrent breeding and hybrid seed production are from the self-pollination (selfs) of the corresponding D and T parents of selected progeny-test D × T cross. Owing to severe inbreeding depression affecting seed and pollen production, reciprocal recurrent selection (RRS) with each hybrid combination is usually limited to two cycles. There is usually also a concurrent recombinant phase or cycle involving crosses within the component parental populations for longer term population and hybrid improvement (Meunier and Gascon 1972). This scheme exploits both GCA and specific combining ability (SCA) and the commercial D × P hybrids are genetically more uniform, close to near true F1 or SX (single cross) hybrids.

Refer to Chapter 7 of this book and Soh (1999) for further discussions.

6.2.2 Current Perspective

Many oil palm breeding programs have moved or added on to the traditional MRS and MRRS breeding schemes or methods.

6.2.2.1 Top-Cross Test Breeding

This is a "hybrid" between MRS and MRRS. A set of proven D parents or lines used in current commercial seed production is used to top-cross test with a new set of Ps and vice versa before acceptance as parents into the main breeding and seed production program. The new Ds and Ps need not be the traditional Deli's and WA Ps.

6.2.2.2 Inbred-Hybrid Variety Breeding

Inbred lines up to F_5 or more through conventional selfing have been obtained for Deli Ds and WA Ts without inbreeding depression and have been used in hybrid seed production in some programs. Conventional inbred development is tedious and time consuming. Innovative approaches have been made to circumvent this as discussed below.

6.2.2.3 Recombinant Inbred Variety/Line Breeding (RIV/L)

The premise of the Recombinant Inbred Variety (RIV) approach is that hybrid vigor is not a consequence of heterosis but of dominance or pseudo-dominance. As such it is possible to recapture the superior performance of a hybrid through continuous inbreeding to produce a recombinant inbred variety. This was proposed for oil palm but it did not take off (Pooni et al. 1989). However, Recombinant Inbred Line (RIL) breeding for D and T parents is currently being practiced. RILs representing the spectrum of random inbred homozygous lines arising from a cross are currently in vogue in oil palms. RILs are useful in linkage, mapping, and genomic studies and for development of F_1 hybrid seed production ready parents, potentially allowing rapid multiplication of the best parental lines.

6.2.2.4 Single Seed Descent and Doubled Haploid Breeding

These are essentially rapid methods of developing RIL. In single seed descent (SSD), starting from a segregating population (usually F_2), random single seed per random palm line are propagated rapidly (once the palm starts fruiting) over a number of generations of self or sib pollination. When sufficient uniformity is achieved (about F_6), the lines are progeny-tested and the best hybrids reproduced from crosses using selfs of the tested parents (Simmonds and Smart 1999). In doubled haploid (DH) breeding, fully homozygous or inbred parents are achieved in a single generation. Haploids occur naturally and can be screened for or induced, for example, pollen irradiation, microspore culture, wide-crossing, inducer lines (Touraev et al. 2009). DHs can also occur naturally or be induced (tissue culture, colchicine treatment). In both SSD and DH breeding, selection is postponed to the last cycle as compared to conventional selfing methods which carry out selection each generation. The SSD and DH approaches thus save time, space, and effort. However, to be able to capture the best lines as inbreds requires large numbers of homozygous lines, as selection has not been used to remove the poorer combinations of genes and traits as the generations have progressed. Some major breeding programs are actively pursuing these approaches (Iswandar et al. 2010; Nelson et al. 2009).

6.2.2.5 Backcross Breeding

Backcross (BC) breeding is usually performed with the objective of transferring a desirable gene or trait from a donor parent of less desirable genetic background by recurrent crossings to a parent with an otherwise superior genetic background, that is, recurrent parent. At least five BCs are necessary to incorporate the donor trait in near pure recurrent parent genotype, although it is fairly common in wheat breeding programs to develop the BC_3 and to follow this with a self-pollinated generation, to evaluate the trait and progress of the introgression, before deciding on whether to continue or to use a marker-based approach to cross complementary BC_3S_1 lines. For a recessive gene, cycle(s) of self-pollination may be interjected to confirm and fix the trait. BC breeding is often an adjunct to standard breeding schemes either at the start to broaden the genetic base and introduce additional traits or at the end to further improve what is otherwise a good established variety.

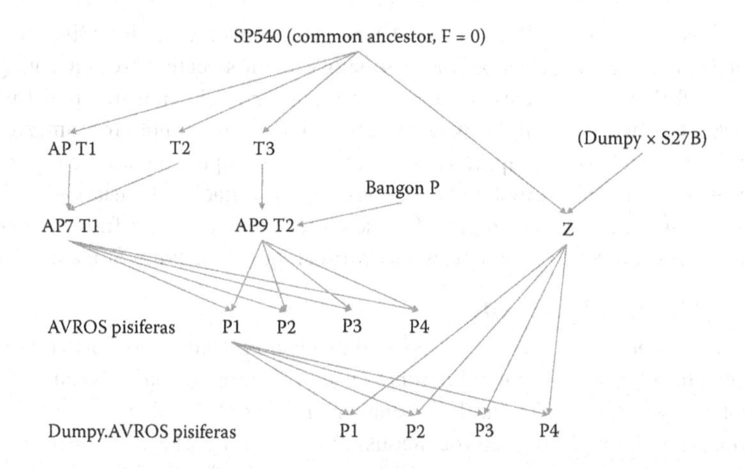

FIGURE 6.1 Lineage of AVROS and Dumpy.AVROS pisifera male parents. (Adapted from Soh, A.C. 1999. Breeding plans and selection methods in oil palm. In: Rajanaidu, N. and Jalani, B.S. (eds.), *Proc. Symp. The Science of Oil Palm Breeding*. Palm Oil Research Institute of Malaysia, Kuala Lumpur, pp. 65–95.)

6.2.2.6 Isogenic and Near Isogenic Lines

These are developed from backcrossing with only one differing gene or trait among the inbred lines. Isogenic (IL) and near isogenic (NIL) lines are also useful for genetic studies and for F_1 hybrid seed production ready parent development. True backcrossing with a constant recurrent palm is difficult in oil palm as the recurrent palm may not be available in subsequent cycles and selfs/sibs or other relatives are commonly used as surrogates. This approach has been used in the development of the Dumpy. AVROS and Ulu Remis tenera (URT) breeding parents of Applied Agricultural Resources (AAR) and Chemara, respectively (see Figure 6.1). BC breeding will feature prominently in current and future oil palm breeding programs with the need to introgress novel (specialty products) and sustainability (resistance, adaptability, efficiency) traits from wild progenitors especially *Elaeis oleifera* into the high yielding but narrow genetically based *Elaeis guineensis* advanced breeding populations, with each breeding program also differing from the others. The availability of rapid generation turnover (SSD, DH) techniques and genomic tools (trans/cis-genics, marker-assisted selection [MAS]) will encourage further BC breeding efforts. The need to introduce speciality traits into oil palm elite lines is perhaps one of the strongest arguments in favor of genetic modification.

6.2.2.7 Breeding for Clonal Propagation

Many crops are vegetatively or clonally propagated due to sexual or fertility barriers (polyploidy, incompatibility), production of few or recalcitrant or poor viability seeds, for example, rubber, forest tree crops, potato, cassava, sugarcane, banana or because commercial clonal propagation of improved genotypes is more expedient than via seed breeding, especially for perennial tree crops which are mostly highly heterozygous, making seed breeding a tedious process. The last rationale was the basis for venturing into (and succeeding in) tissue culture clonal propagation in oil palms (Jones 1974; Wong et al. 1996). The conventional breeding approach for clonally propagated crops begins with single plant selection from a segregating population followed by cycles of cloning, testing, and selection in small to large-scale trials (Brown et al. 1988; Kawano et al. 1998; Simmonds and Smartt 1999; Tan 1987). Such a breeding approach has yet to be attempted due to the inefficiency of cloning and the major problem encountered with somaclonal variation leading to abnormal flowering but may change if these issues are resolved in future. Currently, there are two approaches to exploiting oil palm clones (Soh 1986; Soh et al. 2003).

6.2.2.8 Clonal Hybrid Seeds

These are obtained from crosses reproduced using clonal parents of superior crosses; mono or semi-clonal if one clonal parent is used in a cross (usually D) and bi-clonal with both clonal parents (D and P). Clonal hybrid variety tends to be more genetically uniform (although still heterogeneous) than mixed hybrids as many more seeds can be generated in repeated crosses between multiple clones of the same parental genotype and also tends to be higher yielding as they are reproduced from the very best cross combinations (Prasetyo et al. 2011; Wong et al. 2010). The seed production process follows that for conventional hybrid seeds and hence is cheaper than mass propagation of clones through tissue culture.

6.2.2.9 Clonal Variety

The variety, in the form of *in vitro* plantlets or "ramets," is derived from clonal propagation of selected superior individual plants or "ortets." Again the members of a clone are phenotypically uniform but genetically heterozygous and frequently exhibit higher yields than clonal hybrids. The cost of clonal plants is much higher due to the significant investment in clean laboratory infrastructure, skilled labor, and the inherent process inefficiency (Soh 2012; Soh et al. 2011). (Refer to Clonal Propagation chapter for further discussion.)

6.2.3 Parent Palm Selection Methods

Individual palms (genotypes) are usually selected as parents to start a crossing cum breeding program, especially for a perennial tree crop such as the oil palm which is highly heterozygous.

6.2.3.1 Single Trait Selection

Mass, family, and sib selection which are based on the phenotype (visual, measurement data) have been alluded to earlier.

Individual/mass selection: This is based on the phenotype of an individual. This method is most efficient with highly heritable traits and in relatively unimproved populations.

Family selection: Selection is based on a family mean basis and is most favored with traits of low heritability and from large families with large family differences between average trait values.

Sib selection: This is the form of family selection practiced when selection is limited by sex (exemplified by selection for yielding ability in female sterile

P palms) or when the original palms are no longer available, particularly due to Ganoderma BSR in Malaysia and Indonesia.

Within family selection: This method is advisable when differences among families are small or have common environment complication, for example, abiotic (moisture temperature) or biotic (disease, pest) stress effects when all families are susceptible.

Combined selection: This is when the best individuals are selected from the best families, as in FIPS in oil palm. This is commonly practiced and is usually most efficient.

This is sometimes applied in the form of a *selection index (SI)*, or *best linear unbiased prediction (BLUP)* where information from various relatives (parents, full-sibs, half-sibs) and other sources are included in the selection decision. These could be experimental (plot, block, treatment means), commercial field data even from highly unbalanced mating and experimental design and without standard check variety linkage (Bernardo 2002; Falconer and Mackay 1996; White and Hodge 1989).

Interested readers to refer to Appendixes 6A.1 and 6C.1 for further discussions on these concepts and applications to oil palm breeding.

Progeny-testing: In this method, the parent is only selected based on its progeny-test results.

Correlated trait/MAS: This approach is sought when the highly associated trait or marker can be more efficiently selected than the target trait. MAS for highly heritable traits are discussed in more detail in Chapter 9 while genomics (GS) or genome-wide (GWS) selection incorporating genome-wide markers in BLUP selection for quantitative traits (much in vogue currently) are dealt with later in this chapter (Appendix 6D.1).

6.2.3.2 Multiple Trait Selection

Oil palm breeders usually select for at least 5–10 traits at any one time. The traits include primary (objective) traits, for example, palm oil yield, kernel oil yield, slow height growth, high oil unsaturation, and secondary (component) traits, for example, thick mesocarp and high oil in mesocarp, kernel size.

6.2.3.2.1 Tandem and Independent Culling-Level Selection

Selecting one trait at a time (tandem) is seldom practiced as it is a protracted process while selecting for a number of traits simultaneously (independent culling level) and would result in too few or no individuals being selected or a reduction in the selection intensity per trait.

SI. Using an SI is also a simultaneous multiple trait selection in which selection is based on total merit or genetic worth, in the form of an index derived from statistical integration of the heritabilities, genetic correlations, and relative economic values of the desirable traits (Baker 1986). There are many variants of SI, for example: Smith-Hazel (Optimum) SI, Base SI, Desired Gains SI. SIs can be constructed on just the primary traits, or a combination of objective/primary and secondary/component traits, for example, percent mesocarp to fruit, percent kernel to fruit, and percent oil to mesocarp. Applications of multiple trait selection in the form of optimum, desired gains, and base SIs in oil palm are illustrated in Appendix 6B.1.

6.3 CONCLUSIONS

This chapter has evaluated the application and applicability (sometimes with modifications due to the biology of the crop) of the classical breeding and selection methods and techniques in oil palm. It has also illustrated that oil palm breeding has been keeping abreast with the latest quantitative genetic techniques and tools which incorporate molecular markers and information and rapid generation turnover methods used in many plants (including forest trees) and animal breeding. However, the feasibility of the different options also varies in practice due to the long generation cycles of oil palm. Potentially supplementing conventional breeding with genetic modification (whether conventional or through genome editing) becomes an attractive way to add specific important traits into already elite backgrounds. Public acceptability and continuing concerns over abnormal flowering in tissue culture are at issue, although the latter at least, is beginning to be understood at the genetic level.

APPENDIX 6A.1 SI INCORPORATING FULL-SIB, HALF-SIB FAMILY, AND PLOT INFORMATION

Aik Chin Soh

Selection index (I) or genetic worth of individual (W) $= b_1P_1 + b_2P_2 + \cdots\cdots b_nP_n$

where,

P's = phenotypic values (measurements) of individual, relative group, plot
b's = weightings (partial regressions) for each measurement

Simultaneous equations can be set up to solve for the b's

$$b_1P_{11} + b_2P_{12} + \cdots b_kP_{1k} = g_{11}$$
$$b_1P_{21} + b_2P_{22} + \cdots b_kP_{2k} = g_{21}$$
$$b_1P_{k1} + b_2P_{k2} + \cdots b_kP_{kk} = g_{k1}$$

where P_{11}, P_{22}, and P_{kk} are the phenotypic variances of the various measurements; P_{12}, P_{1k}, and P_{2k} are the phenotypic covariances of the measurements; g_{12}, g_{1k}, and g_{2k} are the genetic or breeding value covariances of the measurements; and g_{11} is the breeding value variances of the individual measurements.

The solutions are easily done using matrices:

y = vector of phenotypic observations or measurements of the individual
b = vector of index coefficients to be estimated
g = vector of unobservable breeding values

In matrix algebra terms,

$$g = a + b'y$$

where a = 1 (number of traits)

$$b = V^{-1}Ca$$

Genetic worth, $w = a'C'V^{-1}(y-\alpha)$

where,

V = matrix of variances and covariances among the observations

C = Cov (y,g′) matrix of covariance between observations and the predicted breeding value

In the oil palm example (Soh and Chow 1993):

Observed data

$$y = \begin{pmatrix} y_{ijAElm} \\ y_{ijAEl} \\ y_{ijAE} \\ y_{ijAF} \\ y_{ijAG} \end{pmatrix} \begin{matrix} \text{-Individual observation} \\ \text{-Plot mean} \\ \text{-Full-sib family mean} \\ \text{-Half-sib family mean 1} \\ \text{-Half-sib family mean 2} \end{matrix}$$

Model

$$y_{ijkm} = U + E_i + B_{ij} + f_k + m_l + p_{ijkl} + w_{ijklm}$$

where,

y_{ijkm}—individual observation

U—general mean

E_i—fixed effect of the ith test environment

B_{ij}—random effect of the jth block in the ith test

f_k—random effect of kth female the ith test, $E(f_k) = 0$, $Var(f_k) = \sigma_f^2$

m_l—random effect of the ith male $E(m_l) = 0$, $Var(m_l) = \sigma_m^2$

p_{ijkl}—random plot error of klth family in jth block of the ith test, $E(p_{ijkl}) = 0$, $var(p_{ijkl}) = \sigma_p^2$

w_{ijklm}—random tree error of wth tree in ijklth plot, $E(w_{ijklm}) = 0$, $var(w_{ijklm}) = \sigma_w^2$

The results from this study showed that inclusion of family information would be most useful in palm selection.

APPENDIX 6B.1 SIS FOR MULTIPLE TRAITS IN OIL PALM

Aik Chin Soh

The construction of the optimum SI (OSI) (I) takes the form

$$\text{Selection index, } I = b_1P_1 + b_2P_2 + \cdots b_nP_n$$

where P's = phenotypic values of traits and b's = weightings (partial regressions) for each trait

$$\text{Aggregate genotypic value/worth, } H = a_1G_1 + a_2G_2 + \cdots a_kG_k,$$

where a's = relative economic values of the traits and G's = genotypic values of the traits.

Normal simultaneous equations can be set up to solve for the b's

$$b_1P_{11} + b_2P_{12} + \cdots b_kP_{1k} = a_1G_{11} + a_2G_{12} + a_kG_{1k}$$
$$b_1P_{21} + b_2P_{22} + \cdots b_kP_{2k} = a_1G_{21} + a_2G_{22} + a_kG_{2k}$$
$$b_1P_{k1} + b_2P_{k2} + \cdots b_kP_{kk} = a_1G_{1k} + a_2G_{2k} + a_kG_{kk}$$

where P_{11}, P_{22}, and P_{kk} are the phenotypic variances of the traits; P_{12}, P_{k1}, and P_{k2} are the phenotypic covariances of the traits; G_{11}, G_{22}, and G_{kk} are the genetic variances of the traits; and G_{12}, G_{1k}, and G_{2k} are the genetic covariances of the traits.

The solution of the normal equations for multiple trait OSI based on matrix algebra is similar to earlier form.

Again

$$Pb = Ga$$

where,

P is the matrix of phenotypic variance and covariances.

G the matrix of estimated genotypic variances and covariances, a the vector of relative economic values, and b the vector of index coefficients obtained by solving

$$b = P^{-1}Ga$$

where P^{-1} is the inverse of the phenotypic variance–covariance matrix.

Soh and Chow (1989) computed OSIs to select palms (ortets) for cloning using a combination of primary traits (OY, KOY, and economic traits), secondary traits (BNo, M/F%, K/F%; highly heritable traits and genetically correlated to yield), and physiological efficiency traits (bunch index [BI], leaf area ratio [LAR], and height [HT]). As the objective was to select palms for cloning, genotypic variance and covariance and broad-sense heritabilities were appropriately computed and used in constructing the indexes. For a single primary trait, the relative economic values (a) for the primary trait are 1 and 0 for the secondary traits. With two primary traits (OY and KOY), they were based on the ratio of their 10 years' average prices and on various relative perceived emphases.

The results from this exercise showed that in terms of expected relative efficiency (compared to selection on oil yield alone) the yield component traits contributed most while the contributions from the physiological traits were not significant. It was argued that the real advantage of the physiological traits could only be demonstrated in the clonal trials derived.

Desired gains index: Sometimes oil palm breeders may be more interested in the desired improvements in particular traits rather than their relative economic weights or that the latter are difficult to establish (Soh et al. 1994). The desired gains index is obtained by solving

$$b = (G'R)^{-1}Q$$

where Q is the vector of genetic changes for the desired traits and R is the matrix of Wright's coefficient of relationship **r** (Yamada et al. 1975).

The number of generations (q) needed to achieve the desired gain can also be computed

$$q = \sqrt{b'Pb/i_t}$$

where i_t is the selection intensity.

Base SI: If reliable estimates of genotypic and phenotypic variances and covariances are not available, a facile index method is the base SI based just on the relative economic values (Baker 1986; Williams 1962)

$$I = a_1P_1 + a_2P_2 + \cdots + a_kP_k$$

Some oil palm breeders have used this approach.

APPENDIX 6C.1 BEST LINEAR UNBIASED PREDICTION

Aik Chin Soh and Razak Purba

SI assumes the means, variances, and covariance of the joint distribution of g (genetic values) and y (observed values) are known although in practice they are not and are usually estimated. Estimates of the fixed effects (block, trial, etc.) have been obtained by taking simple averages (ordinary least squares) prior to predicting the breeding values. The estimates are assumed to be both accurate and precise and are treated as known constants throughout the whole process. For many "messy" or very unbalanced data sets and experiments (large livestock, tree crops), ordinary least squares estimates of fixed effects are unsatisfactory. The BLUP procedure developed by Henderson (1984) uses best linear unbiased estimates (BLUE) to remove the messy fixed effects through generalized least squares before computing the BLUP of the random genetic (breeding value) effects:

Linear mixed model (matrix):

$$y = X\beta + Z\mu + e$$

where,

y = vector of observed data records
β = vector of fixed effects (blocks, environment, genetic groups)
X = incidence matrix (0's and 1's) of fixed effects relating to β
μ = vector of random genetic effects
Z = incidence matrix (0's and 1's) relating μ to y

e = random error effects (plot, within plot)
BLUP breeding value, $g = C'V^{-1}(y - X\beta)$

where,

$\beta = (X'V^{-1}X)^{-} (X'V^{-1}y)$ and $(X'V^{-1}X)^{-}$ = generalized inverse of $X'V^{-1}X$

The BLUP methodology enables the estimation of genetic variances from the data that are generated from routine breeding programs even without the need for mating designs and are particularly useful in inter-population hybrid crosses (Bernardo 2002). The random breeding values allow for the estimation of the performance of missing hybrids in the mating program and of future hybrids.

Two BLUP examples in oil palm are illustrated here: the sire model or hierarchical/nested design and the *single-crosses model* or factorial design.

6C.1.1 Sire Model (Soh 1994)

Three D × P trials: Trial 1 progeny-tested 2 AVROS Ps with 10 and 7 Deli Ds, respectively, planted in a randomized block design with 5 replicates of 10 palm plot size; Trial 2 with 4 AVROS Ps (2 Ps in common with Trial 1) tested with 1, 1, 10, 1 Deli Ds in randomized complete block (RCB) design with 4 replicates of 12 palm plots; and Trial 3 with 1 common AVROS P tested in Trial 1 and Trial 2 and 5 Dumpy. AVROS Ps derived from the common AVROS P (Figure 6.1) crossed with 4, 3, 3, 2, 3 Deli Ds in RCB design with 3 replicates of 20 palm plots. The traits studied were bunch yield, bunch number, mean bunch weight, oil to bunch, and height increment.

6C.1.1.1 Analysis

The linear mixed model which takes into account the genetic (pisifera or P) grouping:

$$y = X\beta + Qg + Z\mu + e$$

where,

y = vector of observed data
β = vector of fixed effects for experiments
X = incidence matrix relating β's to y's
g = vector of P group fixed effects
Q = incidence matrix relations record to each P effect
μ = random breeding value effects
Z = incidence matrix relating μ's to y's
e = random error effects

The mixed model equation:

$$\begin{pmatrix} X'X & X'Q & X'Z \\ Q'X & Q'Q & Q'Z \\ Z'X & Z'Q & Z'Z+\sigma_e^2G^{-1} \end{pmatrix} \begin{pmatrix} \beta \\ g \\ \mu \end{pmatrix} = \begin{pmatrix} X'y \\ Q'y \\ Z'y \end{pmatrix}$$

where,

$\sigma_e^2 G^{-1} = A^{-1}\zeta$, and

A = additive relationship matrix among the P's computed from the pedigree (Figure 6.1) following the procedure of Van Fleck (1988) and Kempthorne (1957).

$$\zeta = \sigma_e^2/\sigma_s^2 = 4 - h^2/h^2$$

where σ_s^2 = P group variance, h^2 = heritability.

Heritability values used were average values obtained from published and unpublished estimates of similar genetic populations.

Breeding value, BV = **2** $(\mathbf{g_i} + \mu_i)$, with g_i obtained from the solution of the above matrix equation.

The aggregate genetic worth or merit of each P was also computed using the function:

$$W = a'g \quad \text{or} \quad W = a_1g_1 + a_2g_2 + a_3g_3,$$

as in base SI and the **a**'s = relative economic values of the traits.

Owing to the highly unbalanced nature of the data, the BLUP prediction errors were large. Consequently, selection of the P parents based on their rankings in the predicted BV would be advisable than on their actual predicted values.

6C.1.2 Single-Crosses Model (Purba et al. 2001)

The data were obtained from 401 hybrids planted in 26 progeny-test trials in two locations in North Sumatra, Indonesia. The hybrids were derived from crossing 154 Deli D with 135 WA parents of different sources (Zaire 79, Cameroon 39, Ivory Coast 14, Nigeria 3). The trials were planted in a randomized block design with 5–6 replicates and 12–16 palm plots. The unbalanced agronomic data (variable number or parents and their cross combinations, no check crosses, replicate and plot sizes), and pedigree information were analyzed using restricted maximum likelihood (REML) and BLUP.

The linear mixed model:

$$y_{AB} = X\beta + Z_1\alpha_A + Z_2\alpha_B + Z_D + e$$

where,

$Z_1\alpha_A$ = vector of random additive effects of Deli parents (different from earlier)
$Z_2\alpha_B$ = vector of random additive effects of African parents
Z_D = vector of random dominance effects

The pedigree information was obtained from Lubis (1985), Rosenquist (1986), and Centre de cooperation internationale en recherché agronomique pour le developpement (Cirad) archives.

The coefficient of parentage or additive relationship was obtained using PROC INBREED of SAS software (SAS Institute Inc. 1996a).

To test the precision of the BLUP estimates, the observed performance of some tested hybrids was compared (correlated) with their estimations when using the other hybrids as predictors. The correlations between predicted and observed hybrid performances were good at about 0.4–0.7. With a coefficient of 0.6, the probability for the best single cross to be included in the 20% selected hybrids was greater than 80% (Bernardo 1996).

APPENDIX 6D.1 GENOMIC SELECTION

David Cros

6D.1.1 Principles of Genomic Selection

Genomic selection (GS) is a MAS method efficient to improve quantitative traits (Meuwissen et al. 2001). It relies on dense marker coverage of the whole genome and on statistical methods able to take advantage of the information of all the markers jointly. It is usually used to estimate the additive genetic value (genomic estimated breeding value [GEBV]) of selection candidates, but it can also estimate the genotypic (sum additive and nonadditive genetic) values. All the markers are used for selection which differs from previous MAS approaches where only a subset of significant markers was kept. The model of GS is calibrated on a pool of individuals called "training set" using their data records (usually phenotypes) and their genotypes; and applied to a pool of individuals (selection candidates) with genotypes for the same markers but no data records. The standard molecular markers in GS are single-nucleotide polymorphisms (SNPs), although other types of markers could be used. The challenge for GS is to increase the rate of genetic gain compared to conventional phenotypic breeding, that is, to improve the ratio $r \times i \times \sigma_a/L$, with r the accuracy of selection (correlation between true and estimated breeding values), i the intensity of selection, σ_a the additive standard deviation, and L the generation interval in years (Falconer and Mackay 1996; Lynch and Walsh 1998).

As the number of markers (predictors) is usually larger than the number of individuals, the estimation of marker effects requires the use of the BLUP methodology and Bayesian approaches or nonparametric methods. With n_{ind} training individuals and n_{SNP} SNPs, the general form of the model that predict the GEBVs is

$$y = X\beta + Zm + e \tag{6D.1}$$

where y is the vector of data records ($n_{ind} \times 1$), β is the vector of fixed effects (overall mean, trial, block, etc.) associated to a design matrix X, m is the vector of the random effects of SNPs ($n_{SNP} \times 1$) with design matrix Z ($n_{ind} \times n_{SNP}$) containing the number of copies (0, 1, or 2) of the most frequent alleles, and e is the vector of residuals ($n_{ind} \times 1$). This gives an estimate of m, an estimate of the additive effect associated to each marker. The GEBVs of the selection candidates are obtained by summing SNP effects over the whole genome, for an individual i: $GEBV_i = \Sigma_{j=1}^{n_{SNP}} Z_{ij}\hat{m}_j$, with $\hat{m}$ the vector of estimated marker effects. Some of the statistical methods that are used to solve model (6D.1) consider a common genetic variance (σ_m^2) for all the markers, which is appropriate for genetic architectures strictly following the infinitesimal model. This is the case of the random regression BLUP (RR-BLUP) (Meuwissen et al. 2001) and the Bayesian random regression (BRR) (Pérez et al. 2010). With RR-BLUP, the marker effects are obtained by estimating the variances and solving the mixed model equations of Henderson (1984). Other methods like BayesA (Meuwissen et al. 2001) and the Bayesian least absolute shrinkage selection operator (LASSO) regression (de los Campos et al. 2009) consider a genetic variance specific to each marker. This allows giving more weight to a few markers, thus matching with quantitative traits whose determinism also includes some major genes. The model (6D.1) is similar to

$$y = X\beta + g + e \tag{6D.2}$$

with g the vector ($n_{ind} \times 1$) of random additive values of individuals (GEBV) following $N(0, G\sigma_g^2)$, with σ_g^2 the additive variance and G the genomic relationship matrix ($n_{ind} \times n_{ind}$). The G matrix contains the realized additive relationships between the genotyped individuals, taking into account the random sampling of alleles that occurred at meiosis (Mendelian sampling). Consequently, using the G matrix can lead to estimates of genetic values more accurate than the pedigree-based relationship matrix A used in traditional BLUP analysis, which ignores Mendelian sampling and contains expected relationships. By contrast with model (6D.1), model (6D.2) directly estimates the GEBV without estimating marker effects. The genomic best linear unbiased prediction (GBLUP) (Habier et al. 2007; VanRaden 2007) is the standard statistical method to solve the model (6D.2), and also the most common method used in GS (Heslot et al. 2015). The G matrix is computed from the genotypes, and for SNPs the formula is usually

$$G = \frac{X^t(X)}{2\sum_{l=1}^{n_{SNP}} p_l(1-p_l)}$$

with X = Z–P, Z containing the genotypes coded in 0, 1, and 2, and P a matrix with elements $2p_l$, with p_l the minor allele frequency at locus l.

For more details about the statistical aspects of GEBV prediction, an abundant literature is available (de los Campos et al. 2013; Desta and Ortiz 2014; Heslot et al. 2012; Jannink et al. 2010; Lorenz et al. 2011; Neves et al. 2012).

6D.1.2 Accuracy of Genomic Selection

The accuracy r of GS is the correlation between the true (unknown) genetic value and the estimate given by the GS model. It is a major parameter for breeding, as genetic gain is directly affected by r. The GS accuracy is altered by several parameters, most of them being interrelated. This includes the relationship between the training set and the selection candidates, the number of individuals in the training set, the linkage disequilibrium (LD) between markers and genes, the trait heritability, the marker density, the genetic architecture of the trait (number of genes, distribution of their effects, …), and the statistical method used to predict the genetic value (Grattapaglia 2014; Jannink et al. 2010; Lorenz et al. 2011). The accuracy of GS increases with the size of the training set (Desta and Ortiz 2014; Grattapaglia 2014; Jannink et al. 2010; Lorenz et al. 2011). The training set must correspond to an operational phase of the breeding program, to allow a direct application in the selection candidates, generally progenies or sibs. Indeed, the relationship between the training set and the selection candidates is a key factor of the accuracy of GS (see e.g., Daetwyler et al. 2013; Gowda et al. 2014; Pszczola et al. 2012). The accuracy of GS increases with the number of markers before reaching a plateau (Calus and Veerkamp 2007; Calus et al. 2008; de Roos et al. 2009; Desta and Ortiz 2014; Meuwissen 2009; Solberg et al. 2008). The marker density required to reach a given accuracy depends on the size of the training set and the LD of the population (Grattapaglia 2014; Heffner et al. 2009; Jannink et al. 2010; Lorenz et al. 2011). Implementing GS in a population with a high LD requires fewer markers than in a population with low LD. According to Calus and Veerkamp (2007), a marker density giving a mean r^2 of 0.15 between adjacent markers would be sufficient for a trait with a heritability of 0.5, and an r^2 of 0.2 for a heritability of 0.1. The level of LD in a population is often represented by the effective size (N_e) of the population, the two parameters being negatively correlated. Marker density and training population size have to scale with N_e to maintain GS accuracy. The practical application of GS for breeding implies large-scale genotyping capabilities, at reasonable cost. With the development of next generation sequencing (van Dijk et al. 2014), a high density SNP coverage of the entire genome is possible for all species. So far, most GS studies used SNP arrays, but genotyping by sequencing (GBS) (Elshire et al. 2011) is a suitable alternative, as indicated for instance by El-Basyoni et al. (2013) and Zhang et al. (2015).

6D.1.3 GS for Hybrid Performances

Oil palm genetic improvement relies on hybrid breeding. As a consequence, the genetic values to estimate are the GCA of the parents of the hybrid crosses and, although of secondary importance compared to GCA, the SCA of the hybrid crosses. The GCA of an individual is defined as half its breeding value in hybrid crosses. It results from the additive value of the gene alleles that the individual transmits on average to its hybrid progenies, expressed as a difference with the mean value μ_{AB} of the hybrid population. The SCA of a cross $A_i \times B_i$ is the difference between the genetic value expected from the parental GCAs (i.e., $\mu_{AB} + GCA_{A_i} + GCA_{B_i}$) and the actual genetic value. For hybrids between unrelated populations A and B, the estimation of the GCA of the parents and the SCA of the crosses is made with the model (Lynch and Walsh 1998; Stuber and Cockerham 1966)

$$y_{AB} = X\beta + GCA_A + GCA_B + SCA_{AB} + e_{AB} \quad (6D.3)$$

with $GCA_A \sim N(0, A_A\sigma^2_{aA})$ and $GCA_B \sim N(0, A_B\sigma^2_{aB})$, and A_A and A_B matrices of additive coancestry among A and B individuals computed from pedigrees, respectively. This means, for example, for individuals x and y in population A that $A_{Axy} = \{f_{xy}\}$, with f the Malécot's coefficient of coancestry (Malécot 1948), computed from the pedigrees. σ^2_{aA} and σ^2_{aB} are the additive variances of the A and B parents in $A \times B$ hybrid crosses, respectively. $SCA_{AB} \sim N(0, D\sigma^2_d)$, with D the dominance coancestry matrix between hybrid crosses, with elements $D_{AB,A'B'} = f_{AA'}f_{BB'}$ (i.e., $D = A_A \otimes A_B$), and σ^2_d the variance of the dominance effects in the $A \times B$ population.

The traditional analysis of this model only used pedigree data, as published for maize (Bernardo 1996) and oil palm (Purba et al. 2001) hybrids.

In the GS framework, this model for hybrid breeding was revisited to take into account molecular information. Thus, a genomic version of model (6D.3) is

$$y_{AB} = X\beta + Z_A m_A + Z_B m_B + Z_d d_{AB} + e_{AB}$$

where y_{AB} is the vector of hybrid data records ($n_{ind(A \times B)} \times 1$), m_A and m_B are the vectors of the additive effect associated to each SNP in parental populations A and B ($n_{SNP} \times 1$) with Z_A and Z_B incidence matrices ($n_{ind(A \times B)} \times n_{SNP}$), d_{AB} the vector of dominance effect associated to each SNP ($n_{SNP} \times 1$) with incidence matrices Z_d reflecting the genotype of hybrids ($n_{ind(A \times B)} \times n_{SNP}$), with elements $Z_{dij} = 1$ if hybrid individual i is heterozygote at SNP j and 0 if it is homozygote. This is a population-specific allele model, which can be solved using for instance RR-BLUP. Alternatively, model (6D.3) can be directly analyzed with a GBLUP approach, with $GCA_A \sim N(0, G_A\sigma^2_{gA})$, $GCA_B \sim N(0, G_B\sigma^2_{gB})$, and $SCA_{AB} \sim N(0, D\sigma^2_{gAB})$, with $D = G_A \otimes G_B$, G_A and G_B the genomic coancestry matrices and σ^2_{gA} and σ^2_{gB} the additive variances for populations A and B in hybrid crosses, respectively. These genomic models for hybrids between unrelated populations were used, for example, in maize (Massman et al. 2013; Technow et al. 2012) and eucalyptus (Bouvet et al. 2015).

6D.1.4 GS for Perennial Crops

GS is particularly interesting for perennial crops (Grattapaglia 2014; Isik 2014; van Nocker and Gardiner 2014). In these species, the individual evaluation can be very long, leading to a selection phase that takes place years after reproductive maturity. In this case, major reductions in the generation interval are possible. In addition, the difficulty and costs associated with such long-term evaluations limit the number of evaluated individuals, resulting in low selection intensity. The potential of GS is therefore obvious here. However, depending on the characteristics of the species, traits, and conventional breeding methodology, the accuracy of GS can be lower than the accuracy of phenotypic selection. Therefore, the challenge is to conceive a genomic alternative to the conventional breeding scheme, where the resulting rate of genetic gain would increase. Empirical results have already been obtained for a few perennial crops of major economic importance: apple (Kumar et al. 2012), eucalyptus (Resende et al. 2012a), loblolly pine (Resende et al. 2012b), spruce (Beaulieu et al. 2014), oil palm (Cros et al. 2015b), and maritime pine (Isik et al. 2015).

6D.1.4.1 GS for Oil Palm

In oil palm, conventional breeding usually relies on RRS between two parental groups A (Deli and Angola populations) and B (other African populations). The genetic trials give reliable estimates of parental GCAs with accuracy around 0.9 for yield components (Cros et al. 2015b) when analyzed with traditional pedigree-based mixed models. As with other perennials, the major drawbacks of oil palm breeding are the long generation interval, currently around 20 years, and the low selection intensity, with usually less than 200 individuals progeny-tested per generation and parental group. Two different approaches of GS have been studied so far to increase oil palm yield: a single-crosses approach (Wong and Bernardo 2008) and a population-level approach (Cros et al. 2015a,b). These studies are summarized and discussed in Section 9.6.

REFERENCES

Allard, R.W. 1960. *Principles of Plant Breeding*. John Wiley & Sons, New York and London.

Baker, R.J. 1986. *Selection Indices in Plant Breeding*. CRC Press.

Beaulieu, J., Doerksen, T., Clement, S. et al. 2014. Accuracy of genomic selection models in a large population of open-pollinated families in white spruce. *Heredity* 113: 343–352.

Bernardo, R. 1996. Best linear unbiased prediction of maize single-cross performance. *Crop Science* 36: 50–56.

Bernardo, R. 2002. *Breeding for Quantitative Traits in Plants*. Stemma Press, Woodbury, Minnesota.

Bouvet, J.-M., Makouanzi, G., Cros, D., and Vigneron, P. 2015. Modeling additive and non-additive effects in a hybrid population using genome-wide genotyping: Prediction accuracy implications. *Heredity* 116: 146–157.

Brown, J., Caligari, P.D.S., Dale, M.F.B., Swan, G.E.L., and Mackay, G.R.. 1988. The use of cross prediction methods in a practical potato breeding programme. *Theoritical and Applied Genetics* 76: 33–38.

Calus, M.P.L., Meuwissen, T.H.E., de Roos, A.P.W., and Veerkamp, R.F. 2008. Accuracy of genomic selection using different methods to define haplotypes. *Genetics* 178: 553–561.

Calus, M.P.L. and Veerkamp, R.F. 2007. Accuracy of breeding values when using and ignoring the polygenic effect in genomic breeding value estimation with a marker density of one SNP per cM. *Journal of Animal Breeding Genetics* 124: 362–368.

Cros, D., Denis, M., Bouvet, J.-M., and Sanchez, L. 2015a. Long-term genomic selection for heterosis without dominance in multiplicative traits: Case study of bunch production in oil palm. *BMC Genomics* 16: 651.

Cros, D., Denis, M., Sánchez, L. et al. 2015b. Genomic selection prediction accuracy in a perennial crop: Case study of oil palm (*Elaeis guineensis* Jacq.). *Theoretical and Applied Genetics* 128: 397–410.

Daetwyler, H.D., Calus, M.P.L., Pong-Wong, R. et al. 2013. Genomic prediction in animals and plants: Simulation of data, validation, reporting, and benchmarking. *Genetics* 193: 347–365.

de los Campos, G., Naya, H., Gianola, D. et al. 2009. Predicting quantitative traits with regression models for dense molecular markers and pedigree. *Genetics* 182: 375–385.

de los Campos, G., Pérez, P., Vazquez, A., and Crossa, J. 2013. Genome-enabled prediction using the BLR (Bayesian linear regression) R-package. In: Gondro, C., van der Werf, J., Hayes, B. (eds.), *Genome-Wide Association Studies and Genomic Prediction*. Humana Press, New York, pp. 299–320.

Desta, Z.A. and Ortiz, R. 2014. Genomic selection: Genome-wide prediction in plant improvement. *Trends in Plant Science* 19: 592–601.

El-Basyoni, I., Lorenz, A.J., Akhunov, E. et al. 2013. A comparison between genotyping-by-sequencing and array-based scoring of SNPs for genomic prediction accuracy in winter wheat. *ASA, CSSA, and SSSA International Annual*, November 3–6, 2013, Tampa, FL, USA.

Elshire, R.J., Glaubitz, J.C., Sun, Q. et al. 2011. A robust, simple genotyping-by-sequencing (GBS) approach for high diversity species. *PLoS ONE* 6: e19379.

Falconer, D. and Mackay, T. 1996. *Introduction to Quantitative Genetics*. 4th edn. Longman: Harlow, UK.

Gowda, M., Zhao, Y., Wurschum, T. et al. 2014. Relatedness severely impacts accuracy of marker-assisted selection for disease resistance in hybrid wheat. *Heredity* 112: 552–561.

Grattapaglia, D. 2014. Breeding forest trees by genomic selection: Current progress and the way Forward. In: Tuberosa, R., Graner, A., Frison, E. (eds.), *Genomics of Plant Genetic Resources*. Springer, Berlin/Heidelberg, pp. 651–682.

Habier, D., Fernando, R.L., and Dekkers, J.C.M. 2007. The impact of genetic relationship information on genome-assisted breeding values. *Genetics* 177: 2389–2397.

Heffner, E.L., Sorrells, M.E., and Jannink, J.-L. 2009. Genomic selection for crop improvement. *Crop Science* 49: 1–12.

Henderson, C.R. 1984. *Application of Linear Models in Animal Breeding*. University of Guelph: Guelph, Canada.

Heslot, N., Jannink, J.-L., and Sorrells, M.E. 2015. Perspectives for genomic selection applications and research in plants. *Crop Science* 55: 1–12.

Heslot, N., Yang, H.-P., Sorrells, M.E., and Jannink, J.-L. 2012. Genomic selection in plant breeding: A comparison of models. *Crop Science* 52: 146–160.

Isik, F. 2014. Genomic selection in forest tree breeding: The concept and an outlook to the future. *New Forests* 45: 379–401.

Isik, F., Bartholomé, J., Farjat, A. et al. 2015. Genomic selection in maritime pine. *Plant Science* 242: 108–119.

Iswandar, H.E., Dunwell, J.M., Forster, B.P., Nelson, S.P.C., and Caligari, P.D.S. 2010. Doubled haploid ramets *via* embryogenesis of haploid tissue cultures. *Paper presented at ISOPB 2010 Seminar*, Bali, Indonesia.

Jones, L.H. 1974. Propagation of clonal oil palms by tissue culture. *Oil Palm News* 17: 1–8.

Jannink, J.-L., Lorenz, A.J., and Iwata, H. 2010. Genomic selection in plant breeding: From theory to practice. *Briefings in Functional Genomics* 9: 166–177.

Kawano, K., Narintaraporn, K., Nariataraporn, P., Sarakarn, S., Limsila, A., Limsila, J., Suparhan, D., Sarawat, V., and Watananonta, W. 1998. Yield improvement multistage breeding program for cassava. *Crop Science* 38: 325–332.

Kempthorne, O. 1957. *An Introduction to Genetic Statistics*. John Wiley and Sons, New York.

Kumar, S., Chagné, D., Bink, M.C.A.M. et al. 2012. Genomic selection for fruit quality traits in apple (*Malus* × *domestica* Borkh.). *PLoS ONE* 7: e36674. doi: 10.1371/journal.pone.0036674.

Lorenz, A.J., Chao, S., Asoro, F.G. et al. 2011. Genomic selection in plant breeding: Knowledge and prospects. In: Donald, L. Sparks (ed.), *Advances in Agronomy*. Academic Press, Elsevier, Amsterdam, pp. 77–123.

Lubis, A.U. 1985. *Riwayat dan Sumber Asa Bahan Tanaman Yang Dipergunakan Pada Permuliaan Kelapa Sawit Di Indonesia*. Pusat Penelitian Marihat.

Lynch, M. and Walsh, B. 1998. *Genetics and Analysis of Quantitative Traits*. Sinauer Associates, Inc., Sunderland, Massachusetts.

Malécot, G. 1948. *Les Mathématiques de l'Hérédité*. Masson and Cie, Paris.

Massman, J., Gordillo, A., Lorenzana, R., and Bernardo, R. 2013. Genomewide predictions from maize single-cross data. *Theoretical and Applied Genetics* 126: 13–22.

Meunier, J. and Gascon, J.P. 1972. General scheme for oil palm improvement at IRHO. *Oleagineux* 27: 1–12.

Meuwissen, T. 2009. Accuracy of breeding values of "Unrelated" individuals predicted by dense SNP genotyping. *Genetics Selection Evolution* 41: 35.

Meuwissen, T., Hayes, B.J., and Goddard, M.E. 2001. Prediction of total genetic value using genome-wide dense marker maps. *Genetics* 157: 1819–1829.

Nelson, S.P.C., Wilkinson, M.J., Dunwell, J.M., Forster, B.P., Wening, S., Sitorus, A., Croxford A., Ford, C., and Caligari, P.D.S. 2009. Breeding for high productivity lines via haploid technology. In: (Unedit.). *Proc. of Agriculture, Biotechnology, and Sustainability Conference* Vol. 1, p 203–225. PIPOC 2009. Malaysian Palm Oil Board, Kuala Lumpur.

Neves, H.H., Carvalheiro, R., and Queiroz, S. 2012. A comparison of statistical methods for genomic selection in a mice population. *BMC Genetics* 13: 100.

Pérez, P., de los Campos, G., Crossa, J., and Gianola, D. 2010. Genomic-enabled prediction based on molecular markers and pedigree using the Bayesian linear regression package in R. *Plant Genetics* 3: 106–116.

Pooni, H.S., Cornish, M.A., Kearsey, M.J., and Lawrence, M.J. 1989. The production of superior lines and second cycle hybrids by inbreeding and selection. *Elaeis* 1: 17–30.

Prasetyo, J.H.H., Sitepu, B., Iswandar, H.E., Djuhjana, J., and Stephen, P.C., Nelson, S.P.C. 2011. Performance of SUMBIO semi-clonal progenies. *Paper presented at PIPOC 2011.* Malaysian Palm Oil Board, Kuala Lumpur.

Pszczola, M., Strabel, T., Mulder, H.A., and Calus, M.P.L. 2012. Reliability of direct genomic values for animals with different relationships within and to the reference population. *Journal of Dairy Science* 95: 389–400.

Purba, A.R., Flori, A., Baudouin, L., and Hamon, S. 2001. Prediction of oil palm (*Elaeis guineensis* Jacq.) Agronomic performances using the best linear unbiased predictor (BLUP). *Theoretical and Applied Genetics* 102: 787–792.

Resende, M.D.V., Resende, M.F.R., Sansaloni, C.P. et al. 2012a. Genomic selection for growth and wood quality in eucalyptus: Capturing the missing heritability and accelerating breeding for complex traits in forest trees. *New Phytologist* 194: 116–128.

Resende, M.F.R., Muñoz, P., Resende, M.D.V. et al. 2012b. Accuracy of genomic selection methods in a standard data set of loblolly pine (*Pinus taeda* L.). *Genetics* 190: 1503–1510.

Rosenquist, E.A. 1986. The genetic base of oil palm breeding Populations. In: Soh, A.C., Rajanaidu, N., and Nasir, M. (eds.) *Proc. Int. Workshop on Oil Palm Germplasm and Utilization.* Palm Oil Research Institute of Malaysia, Kuala Lumpur, pp. 16–27.

Rosenquist, E.A. 1990. An overview of breeding technology and selection in *Elaeis guineensis.* In: *Proc. 1989 Int. Palm Oil Development Conference—Agriculture.* Palm Oil Research Institute of Malaysia, Kuala Lumpur, pp. 5–26.

Simmonds, N.W. and Smart, J. 1999. Principles of crop improvement, 2nd edn, Blackwell Science. Oxford, UK, 412pp.

Soh, A.C. 1986. Expected yield increase with selected oil palm clones from current DxP seedling materials and its implications on clonal propagation, breeding and ortet selection. *Oleagineux* 41: 51–56.

Soh, A.C. 1994. Ranking parents by best linear unbiased prediction (BLUP) breeding values in oil palm. *Euphytica* 76: 13–21.

Soh, A.C. 1999. Breeding plans and selection methods in oil palm. In: Rajanaidu, N. and Jalani, B.S. (eds.), *Proc. Symp. The Science of Oil Palm Breeding.* Palm Oil Research Institute of Malaysia, Kuala Lumpur, pp. 65–95.

Soh, A.C. 2012. The future of oil palm clones in Malaysia. *Journal of Oil Palm and the Environment* 3: 93–97.

Soh, A.C. and Chow, C.S. 1989. Index selection in oil palm for cloning. In: Iyama, S. and Takeda, G. (eds.), *Proc. 6th Int. Congr. SABRAO.* Tsukuba, Japan, pp. 713–716.

Soh, A.C. and Chow, C.S. 1993. Index selection utilizing plot and family information in oil palm. *Elaeis* 5: 27–37.

Soh, A.C., Chow, S.C., Iyama, S., and Yamada, Y. 1994. Candidate traits for index selection in choice of oil palm ortets for clonal propagation. *Euphytica* 76: 23–32.

Soh, A.C., Wong, G., Hor, T.Y., Tan, C.C., and Chew, P.S. 2003. Oil palm genetic improvement. In: Janick, J. (ed.), *Plant Breeding Reviews*. John Wiley and Sons, New Jersey, Vol. 22, pp. 165–219.

Soh, A.C., Wong, G., Tan, C.C., Chew, P.S., Chong, S.P., Ho, Y.W., Wong, C.K., Choo, C.N., Nor Azura, H., and Kumar, K. 2011. Commercial-scale propagation and planting of elite oil palm clones: Research and development towards realization. *Journal of Oil Palm Research* 23: 935–952.

Solberg, T.R., Sonesson, A.K., Woolliams, J.A., and Meuwissen, T.H.E. 2008. Genomic selection using different marker types and densities. *Journal of Animal Science* 86: 2447–2454.

Stuber, C.W. and Cockerham, C.C. 1966. Gene effects and variances in hybrid populations. *Genetics* 54: 1279–1286.

Tan, H. 1987. Strategies in rubber tree breeding. In: Abbot, A.J. and Atkin, R.K. (eds.), *Improvement of Vegetatively Propagated Plants*. Academic Press, Elsevier, Amsterdam, pp. 28–62.

Technow, F., Riedelsheimer, C., Schrag, T., and Melchinger, A. 2012. Genomic prediction of hybrid performance in maize with models incorporating dominance and population specific marker effects. *Theoretical and Applied Genetics* 125: 1181–1194.

Touraev, A., Forster, B.P., and Jain, S.M. (eds.). 2009. *Advances in Haploid Production in Higher Plants*. Springer, Berlin/Heidelberg.

van Dijk, E.L., Auger, H., Jaszczyszyn, Y., and Thermes, C. 2014. Ten years of next-generation sequencing technology. *Trends in Genetics* 30: 418–426.

Van Fleck, D. 1988. *Notes on the Theory and Application of Selection Principles for the Genetic Improvement of Animals*. Cornell University, Ithaca, New York.

van Nocker, S. and Gardiner, S.E. 2014. Breeding better cultivars, faster: Applications of new technologies for the rapid deployment of superior horticultural tree crops *Horticulture Research* 1:14022.

VanRaden, P.M. 2007. Genomic measures of relationship and inbreeding. *Interbull Bulletin* 37: 33–36.

White, T. L. and G. R. Hodge. 1989. *Predicting Breeding Values with Applications in Forest Tree Improvement*. Kluwer Academic Publishers, Dordrecht.

Williams, J.S. 1962. The evaluation of a selection index. *Biometrics* 18: 375–393.

Wong, C.K. and Bernardo, R. 2008. Genomewide selection in oil palm: Increasing selection gain per unit time and cost with small populations. *Theoretical and Applied Genetics* 116: 815–824.

Wong, C.K., Choo, C.N., Liew, Y.R., Ng, W.J., Krishnan, K., and Tan, C.C. 2010. Benchmarking best performing semi-clonal seeds against tenera clones: Avenue for superior clonal ortet identification. *Paper presented in ISOPB 2010 Seminar*, Bali.

Wong, G., Tan, C.C., and Soh, A.C. 1996. Large-scale propagation of oil palm clones—Experiences to date. *Acta Horticulturae* 447: 649–658.

Yamada, Y., Yokouchi, K., and Nishida, A. 1975. Selection index when genetic gains of individual traits are of primary concern. *Japan Journal of Genetics* 50: 33–41.

Zhang, X., Perez-Rodriguez, P., Semagn, K. et al. 2015. Genomic prediction in biparental tropical maize populations in water-stressed and well-watered environments using low-density and GBS SNPs. *Heredity* 114: 291–299.

7 Breeding Programs and Genetic Progress

Aik Chin Soh, Sean Mayes, Jeremy Roberts, Nicolas Philippe Daniel Turnbull, Tristan Durand-Gasselin, Benoît Cochard, Choo Kien Wong, Tatang Kusnadi, and Yayan Juhyana

CONTENTS

7.1 Introductory Overview (*Aik Chin Soh, Sean Mayes, and Jeremy Roberts*) 166
7.2 PalmElit/CIRAD's Breeding Program and Genetic Progress (*Nicolas Philippe Daniel Turnbull, Tristan Durand-Gasselin, and Benoît Cochard*)....168
7.2.1 History of the Breeding Program 168
7.2.1.1 Côte d'Ivoire 168
7.2.1.2 Bénin 168
7.2.1.3 Socfin Indonesia 168
7.2.2 Pre-Reciprocal Recurrent Selection Strategy 169
7.2.2.1 La Mé F_0 and F_1 Breeding Populations 169
7.2.2.2 The International Experiment 169
7.2.3 Breeding Strategy 169
7.2.3.1 Selection Criteria 169
7.2.3.2 RRS Scheme 170
7.2.3.3 Diversity and Improvement of Breeding Populations 171
7.2.4 Current Program 171
7.2.4.1 Bénin—Presco—Colé 172
7.2.4.2 Socfin Indonesia 172
7.2.5 Breeding Tools 172
7.2.5.1 Identity Checking 172
7.2.5.2 *In Vitro* Culture 172
7.2.5.3 Linkage Mapping: Genomic Breeding 173
7.2.6 Genetic Progress 173
7.2.6.1 Oil Yield Progress 173
7.2.6.2 Specific Progress of a True Second Cycle of Selection 174
7.2.6.3 Height Increment 174
7.2.6.4 Disease Resistance 175
7.3 Breeding Programs in AAR, SAIN, and SUMBIO (*Aik Chin Soh, Choo Kien Wong, Tatang Kusnadi, and Yayan Juhyana*) 175
7.3.1 Background 175
7.3.2 AAR 175

7.3.3 SAIN 179
7.3.4 SUMBIO 179
7.4 Conclusions (*Aik Chin Soh, Sean Mayes, and Jeremy Roberts*) 183
References 185

7.1 INTRODUCTORY OVERVIEW

Aik Chin Soh, Sean Mayes, and Jeremy Roberts

As discussed in the previous chapter, systematic oil palm breeding has gravitated to two basic schemes: the modified recurrent selection (MRS) scheme (Rosenquist, 1990; Soh, 1999) and modified reciprocal recurrent selection (MRRS) scheme (Spanaaij, 1969; Meunier and Gascon, 1972). The MRS scheme (Figure 7.1) promulgated by the Oil Palm Genetics Laboratory, a consortium comprising H&C, Guthrie, Pamol, and Dunlop, has been adopted by the breeding programs in Zaire (UniPamol), Thailand (Univanich), Colombia (Cenipalma), Malaysia (Sime Darby, Kulim, IOI, MPOB), Indonesia (SUMBIO, SMART), and Papua New Guinea (Dami). The breeding populations of the latter were also based mainly on Deli, AVROS, Yangambi, Ekona, and Binga. The MRS scheme perhaps arose from the early need to produce a large amount of seeds to meet the rapidly expanding plantation demand and the paucity of the available West African (WA) *pisifera* (P) pollen, which had to be developed via introgression of WA P/*tenera* (T) into the Deli to give

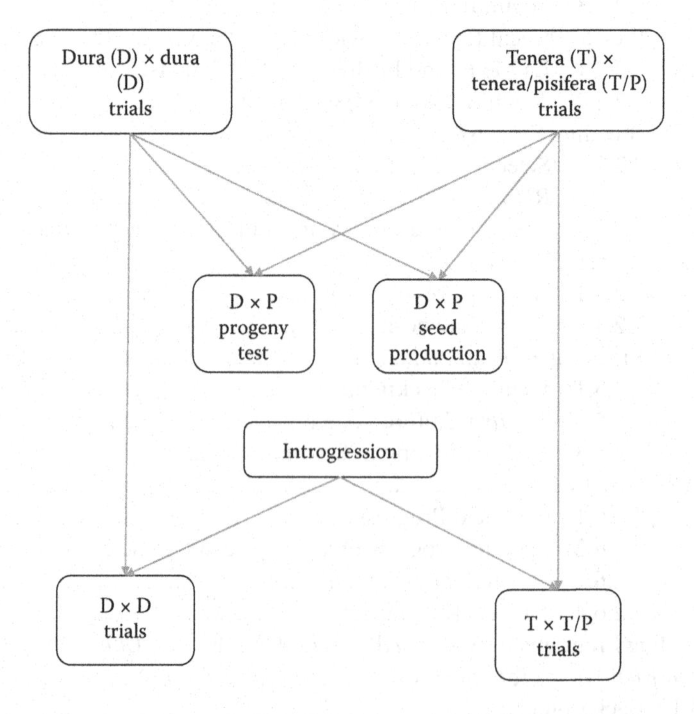

FIGURE 7.1 Generic MRS scheme in oil palm. (Adapted from Soh, A.C. 1999. *Proc. Symp. The Science of Oil Palm Breeding*, Palm Oil Res. Inst., Kuala Lumpur, Malaysia, pp. 65–95.)

rise to the Ulu Remis T/Ps. This went against the principles of clear separation of the heterotic parental populations to maximize hybrid vigor or "inter-origin effect" as espoused by IRHO's MRRS scheme.

The MRRS scheme (Figure 7.2) has been adopted by the breeding programs in West Africa: NIFOR in Nigeria, GOPRI in Ghana, and Cirad coordinated programs in Cote d'Ivoire, Cameroun, Benin, Niger, and Guinea-Bissau; and in the SE Asia by IOPRI and SOCFINDO in Indonesia based on breeding populations derived from Cirad in Cote d'Ivoire (Deli, Angola, Yangambi, La Mé, Yocobue). Other Indonesian programs, for example, Sampoerna Agro, SAIN, Tania Selatan, and Bakrie, also adopted this approach strategically, to quick start their seed production and breeding programs utilizing breeding populations derived from ASD de Costa Rica (via seed purchase or franchising). MRRS has also crept into programs in Malaysia, for example, AAR and MPOB.

This chapter describes in more detail the historical development of PalmElit/Cirad/IRHO's comprehensive and extensive MRRS-based breeding programs and illustrates the salient features of AAR's and SUMBIO's breeding programs, which originally practiced the MRS scheme but have since moved or added on to modified versions of MRRS, MRS, and other breeding methods, and finally SAIN's breeding program to exemplify how some Indonesian companies found entry into the oil palm breeding business. MRS-based programs have been well discussed by Rosenquist (1986, 1990), Hardon et al. (1976), and Soh (1999) previously.

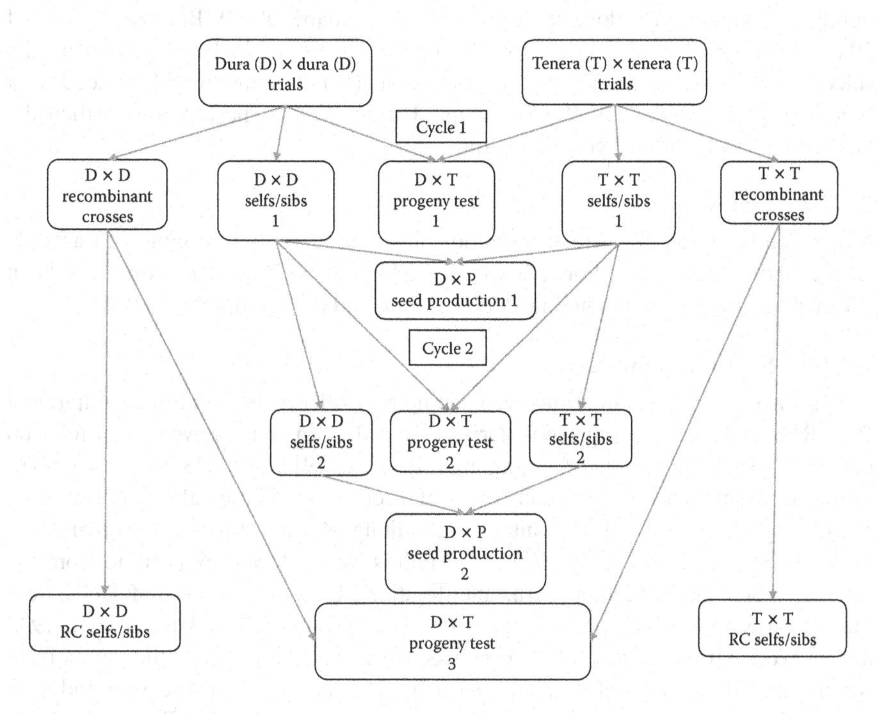

FIGURE 7.2 Generic MRRS scheme in oil palm. (Adapted from Meunier, J. and Gascon, J.P. 1972. *Oléagineux* 27: 1–12.)

7.2 PALMELIT/CIRAD'S BREEDING PROGRAM AND GENETIC PROGRESS

Nicolas Philippe Daniel Turnbull, Tristan Durand-Gasselin, and Benoît Cochard

7.2.1 History of the Breeding Program

Prior to the World War II, French agricultural services were in charge of the selection work in all of the French West African colonies. After the war, all this work was transferred from the French agricultural services to the newly formed Institut de Recherches pour les Huiles et Oléagineux (IRHO). Spanning more than 70 years, IRHO (then Cirad, now PalmElit) has implemented breeding programs with a number of public centers and private companies.

7.2.1.1 Côte d'Ivoire

The first selection work started in the early 1920s and comprised observation of natural oil palm groves in the area of Bingerville, which led to a first selection of 133 palms. Out of this batch, 50 individuals were further studied and finally 29 palms were selected for breeding (Gascon and de Berchoux, 1964). Some of these palms formed the base of the "La Mé" genetic origin or breeding population of restricted origin (BPRO). Between 1925 and 1934, 1727 ha palms were planted in Dabou by L'Union Tropicale de Plantation (UTP) using Deli seeds derived from 17 palms selected by Ferrand in different plantations in Semadam, Medang Ara, and Tendjong Genteng in Indonesia (Boyer, 1962; Cochard, 2008). Between 1946 and 1948, IRHO conducted a survey of 3000 palms in these blocks, which led to the selection of 200–250 parent palms for the production of commercial Deli seeds and breeding work (Brédas, 1969; Corley and Tinker, 2003). These palms formed the base of the "Deli Dabou" genetic origin.

7.2.1.2 Bénin

Concurrently to Côte d'Ivoire, prospection began in Benin in the regions of Banigbé-Lagbé, Adja-Oukré, and Porto Novo. By 1927, 38 wild *tenera* palms had been selected and selfs were planted in Pobè (Gascon and de Berchoux, 1964).

7.2.1.3 Socfin Indonesia

Socfin's first oil palm plantations were set up by Hallet in 1911 in Sungei Liput and Pulu Raja in Sumatra using seeds derived from the ornamental avenues of tobacco companies based in the Deli Serdang area. Between 1913 and 1918, three new plantations were set up using seeds coming from three palms. These palms were selected out of larger group of palms planted in Tandjung Morawa Kiri tobacco plantation by Hallet (Jagoe, 1952). In 1923/1924, 17 palms were selected by Ferrand from the estates of Semadam, Medang Ara, and Tendjong Genteng. The selfed seeds were planted in Sumatra in Mopoli Estate in 1927 and Bangun Bandar Estate in 1933 and in West Africa in Dabou Estate between 1924 and 1930. Later, the palms from Mopoli and Bangun Bandar underwent a second generation of selection and selfing or sib-crossing to produce seeds that were planted in Bangun Bandar between 1936 and 1943 (Cochard, 2008). These selected palms yielded fresh fruit bunches

(FFB) of about 200 kg/palm/year with a F/B ratio of 60%–61% and a M/F ratio of 68%–69% (Corley and Tinker, 2003).

7.2.2 Pre-Reciprocal Recurrent Selection Strategy

7.2.2.1 La Mé F_0 and F_1 Breeding Populations

The first breeding populations developed by IRHO were based on palms selected in Bingerville botanical garden between 1920 and 1924 by General Inspector Houard. These palms were selected because of their good productivity and because they followed the "ideal type" standard comprising palms bearing fruits with more than 60% M/F, 20% S/F, and 20% K/F (Houard, 1926, 1927). The selected palms were selfed, and this first generation (F_0) was planted in La Mé between 1924 and 1930 (Cochard, 2008). The best palms from this generation were selfed again to produce the F_1 generation, which was planted between 1938 and 1942. This second generation of self-pollination attracted little further breeding interest because of strong inbreeding depression.

7.2.2.2 The International Experiment

After the publication of the inheritance of shell thickness by Beirnaert and Vanderweyen (1941) and with the available knowledge on allogamous plant breeding, IRHO organized between 1947 and 1955 an international exchange program between five different plantations (INEAC in Yangambi, IRHO in La Mé, Pobè and Sibiti and Socfin Malaysia in Johore Labis). The first Deli D × WA T crosses were made and compared to pure Deli and pure WA crosses. The following major conclusions were drawn from these comparisons: (1) genetic materials can be differentiated into three groups (A, B, and AB) based on their component traits for bunch production with low variability within groups and high variability between groups; (2) bunch component traits are quantitative with high additive effects; and (3) Deli D (A) × WA T (B) crosses are superior to the pure crosses because of complementarity of the traits between Deli D and WA T and the heterosis effect between dissimilar origins (Gascon and de Berchoux, 1964; Bénard, 1965; Noiret et al., 1966).

7.2.3 Breeding Strategy

The conclusions drawn from the International Experiment led to the implementation of a reciprocal recurrent selection (RRS) scheme by 1957 at IRHO (Meunier and Gascon, 1972). This selection process that enables selection of low heritable traits through progeny testing is followed in combination with a family and individual palm selection (FIPS) method based on pedigree and more heritable traits in order to select the best parents for the next cycle on the most heritable traits and to limit inbreeding between the A and B parents (Hardon et al., 1976; Gascon et al., 1988; Durand-Gasselin et al., 2009).

7.2.3.1 Selection Criteria

Palms selected for breeding are based on their progeny-test performance for oil yield. In order to improve the selection efficiency, oil yield is subdivided into a series of component traits: FFB, BN, ABW, F/B, M/F, and O/M. Expected mill oil extraction rate (OER) is calculated using the formula F/B × M/F × O/M, which is then adjusted with an efficiency correction factor of 0.855. These traits are assessed until the palm is 10 years old on 60–72 palms per cross (Gascon et al., 1988); for specific experiments, up to 96 palms may be assessed.

Selection for resistance to diseases is also carried out routinely using specific nursery plant screening units for both Fusarium wilt and Ganoderma basal stem rot. This started in 1972 in Dabou (Côte d'Ivoire) and in 2005 in Semé Podji (Benin) for Fusarium wilt and in 2009 in Socfindo (Indonesia) for Ganoderma basal stem rot.

Vegetative criteria are also recorded routinely in the selection process. The main interests are in the measurements of height (HT), leaf area index (LAI), and leaf area (LA) in order to select for optimal palm architecture.

7.2.3.2 RRS Scheme

7.2.3.2.1 Côte d'Ivoire: Genetic Setup "Block 500"

The first-cycle RRS of IRHO was planted between 1959 and 1966 in La Mé under the name "Block 500." Parents derived from both the International Experiment and new introductions were tested with a total of 529 crosses planted (Durand-Gasselin et al., 2009). The progeny-test trials as well as recombinant crosses and selfs of the parents were planted concurrently. This setup enabled the inbreeding effect on yield to be highlighted (Gascon et al., 1969) (see Figure 7.3) and the heritability estimates of all major selection criteria determined (Meunier et al., 1970). This cycle fully impacted on the seed production by 1976 with 15 crosses selected.

7.2.3.2.2 Indonesia: Aek Kwasan I and La Mé

Between 1975 and 1986, a series of second-cycle trials testing 288 Deli D parents and 93 WA T/P were planted. In both groups, the best breeding parents were selected based on their performance in the first-cycle selection; selfs or intra-group

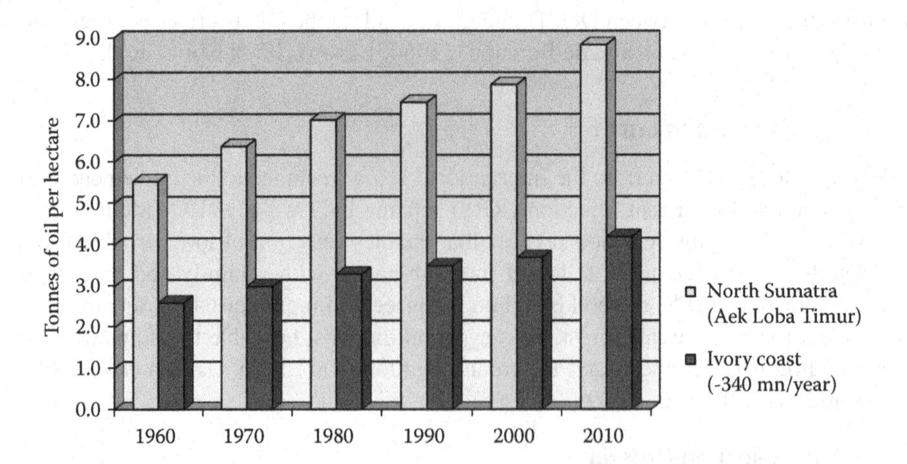

FIGURE 7.3 Genetic progress made during RRS cycles. The *y*-axis shows the genetic value in terms of oil yield (tons of oil/ha/year) of the commercial hybrids in Indonesia (no water deficit) and in Côte d'Ivoire (340 mm of water deficit).

recombinant crosses were then done. From these selfs and recombinant crosses, the parents for the second cycle were selected based on their phenotypic value for the most heritable traits (HT and M/F). After 7 years of assessment of the A × B crosses, it appeared that recombination between intra-group parents with complementary characteristics led to significant breeding progress as compared to selfs for Group A but not for Group B. This is because the Deli (Group A) was more genetically differentiated into subpopulations through prior history of selfings and sibbings and thus gave more scope for recombinations in the intra-group crosses. This was not the case with the La Mé population (Group B). The breeding strategy could therefore be adapted to each group by promoting recombination within the Deli group and selfs within the La Mé. Overall, by selecting the best crosses, this second-cycle selection gave genetic progress of around 18% compared to the standard cross from the first cycle (Gascon et al., 1988).

7.2.3.3 Diversity and Improvement of Breeding Populations

7.2.3.3.1 Population Improvement in the RRS Scheme

IRHO/Cirad/PalmElit's RRS program started with a very narrow genetic base, but continuous work toward introducing a new material into both RRS groups was undertaken (Meunier and Gascon, 1972; Gascon et al., 1988). Several populations resulting from prospections (Yocoboué, Nigeria, Widikum) and exchange programs (Angola, Lobé, Aba, Calabar, Sibiti, Yangambi, and various Deli BPROs) were introduced in La Mé and Pobé between the 1960s and 1980s. These populations were assessed on both their individual trait data and breeding values in order to introduce the best parents into the breeding program. The whole process of introgression, testing, and recombination was carried out for two cycles in the Angola population (Adon et al., 1995) but only partially applied in other populations (Cochard et al., 2006).

7.2.3.3.2 Diversity of Origins Used

Most of the breeding work done by PalmElit and its partners is based on families from Deli Dabou, Deli Socfin, and Deli Socfindo origins for Group A. A few progenies coming from Deli Ulu Remis, Deli Nifor, and Angola have also been included for testing. For Group B, most of the work is based on families from La Mé and Yangambi origins, but some work is also done with Socfindo Yangambi, Nifor, and Ekona material.

Through exchange agreements, Socfindo has recently been able to acquire breeding material from the Pamol breeding program. In parallel, multiparty prospection missions organized in mid-2000s in Angola and Cameroun collected more than 200 accessions. These new accessions are currently being assessed for subsequent introduction into the breeding programs.

7.2.4 Current Program

The second-cycle trials were planted between 1975 and 2001 in Côte d'Ivoire and Indonesia, and results have already impacted on commercial seed production. Since 2005, third-cycle trials have started in Indonesia, Ecuador, and Nigeria in collaboration with three different partners (Socfindo, Murrin, and Siat groups). Specific methodological trials have also been planted in parallel in the past few years with special genetic and experimental designs in order to study new breeding methods and develop innovative molecular tools.

7.2.4.1 Bénin—Presco—Colé

Third-cycle trials are being set up with the aim of testing recombinants and selfs of the best parents from the second cycle showing good resistance to Fusarium wilt. A total of 255 Group A parents and 145 Group B parents are to be tested in Nigeria and Ecuador on 800 ha. Planting started from 2010 and should end by 2016.

7.2.4.2 Socfin Indonesia

The genetic block of the second cycle planted in Socfindo is coming to an end, and most results have already been incorporated into seed production since mid-2000s. In parallel, a new genetic block has recently been set up with the aim of testing 149 new Group A parents and 123 new Group B parents. The uniqueness of this block is the introduction of some new progenitors never used before in PalmElit–Socfindo joint breeding program. These trials were planted between 2005 and 2013, and the first results have already been obtained.

7.2.5 Breeding Tools

7.2.5.1 Identity Checking

PalmElit currently has an extensive program of verification of legitimacy for all the families used in seed production and breeding as well as for off-type families from all the joint research programs. For this purpose, a set of 12 microsatellite markers have been identified out of 400 and have been validated on a pool of 421 parental palms. These markers are highly polymorphic and present no problems during amplification or scoring. To date, 967 progenies have been verified with some of them planted in the 1960s, and as Corley (2005) mentioned, contamination was negligible. Around 7% of the progenies had illegitimacy although traditional yearly controls showed less than 0.5%. Since 2005, all seed gardens are routinely checked and illegitimate families are discarded.

7.2.5.2 *In Vitro* Culture

The IRHO developed its first protocol for vegetative propagation in the mid-1970s (Rabéchault and Martin, 1976) with some improvements added in the early 1980s (Pannetier et al., 1981). The first clones coming from fast growing callus, planted in 1976, all showed somaclonal variation, producing mantled flowers and fruits and bunch failure, which led to sterile palms (Corley et al., 1986). From 1982 to early 2000s, clones came from somatic embryos proliferating on primary callus (Durand-Gasselin et al., 1986; Duval et al., 1988). These clones were assessed for their agronomic value in various trials (Cochard et al., 1999; Nouy et al., 2006). Currently, PalmElit uses a new methodology where proliferation is done by means of an embryo suspension in liquid media (Durand-Gasselin et al., 2010). Although abnormalities can still be observed in the field, these represent less than 5% of the palms and are distributed randomly among the various clones. Various hypothesis were made regarding the cause of floral abnormalities in oil palm (Rival et al., 2009) such as the disturbance of the expression of the genes linked to cytokines metabolism (Besse et al., 1992) or disruptions linked to the methylation of DNA (Matthes et al., 2001). Most recent work on this topic is aimed at studying the perturbation of the regulation of gene expression, whether it be linked to methylation of other regulatory mechanisms (Jaligot et al., 2011; Rival and Jaligot, 2011; Rival et al., 2013). The existence of potential markers for these abnormalities has been investigated, but no global marker has yet been found. The recent publication from Ong-Abdullah et al. (2015) identifies the likely causative mutation and holds promise to help the development of a diagnostic test for application during the tissue culture process.

7.2.5.3 Linkage Mapping: Genomic Breeding

A first saturated genetic map was obtained in 2005 based on the control cross LM2T × DA10D, which was planted in all genetic trials until early 2000. This map was made of 255 SSR (short sequence repeats) markers, 688 AFLP (amplified fragment length polymorphism) markers, and the Sh gene marker (Billotte et al., 2005). These 944 markers are scattered over the 16 linkage groups and cover 1743 cM. As the experimental designs used for the evaluation of parent palms in the breeding programs do not allow for QTL (quantitative trait loci) detection, a specific strategy was developed based on multiparental populations. A consensus map was built using 251 microsatellite markers, the Sh gene marker, and an AFLP marker and was used on a population of 299 individuals. A set of 76 QTLs involved in 24 different phenotypic traits were thus detected (Billotte et al., 2010). Another linkage mapping method was recently used, which compensated for the low number of palms per progeny. It is a pedigree-based approach (Cochard et al., 2015). Results were similar to the consensus linkage map developed by Billotte et al. (2010). Since then, QTL studies have focused on the detection of traits linked to the fatty acid composition of *Elaeis guineensis* (Montoya et al., 2014) and of the interspecific back-crosses (*E. guineensis* × *E. oleifera*) × *E. guineensis* (Montoya et al., 2013). New methods based on genomic selection have been developed more recently (see Chapter 9).

7.2.6 Genetic Progress

7.2.6.1 Oil Yield Progress (Figure 7.3)

The first inter-origin crosses planted during the International Experiment were found to produce 40% more oil than the pure origin crosses. Oil production increased from an average of 2.3 t/ha/year for the pure crosses to 3.3 t/ha/year for the Deli × WA crosses in La Mé (Gascon and de Berchoux, 1964; Gascon et al., 1988). The best cross was LM2T × DA10D with 3.6 t/ha/year of oil, which represented an improvement of 9% over the mean of the inter-origin crosses (Gascon et al., 1981, 1988). This improvement was mainly due to the complementarity of Deli and African origins in terms of bunch production (Gascon and de Berchoux, 1964). After the completion of the first-cycle MRRS scheme, planted in La Mé in the mid-1970s, 15 crosses were selected for their outstanding performance with potential production of 3.9 t/ha/year of oil in La Mé (Gascon et al., 1981), 12%–15% above the standard LM2T × DA10D cross at 3.45 t/ha/year (Gascon et al., 1988). This was mostly due to an increase in FFB yield (Nouy et al., 1991). The second cycle was planted in a network of research centers in Africa, Asia, and Latin America and was completed mid-1990s. In Aek Kwasan, the mean oil production of the second cycle was of 6.71 and 4.52 t/ha/year in La Mé (Gascon et al., 1988). By selecting the best crosses of the second cycle, the improvement was around 30% compared to the LM2T × DA10D standard cross, hence an improvement of 15%–18% compared to the best first-cycle crosses of Gascon et al.(1988) and Nouy et al. (1991).

With the implementation of the RRS scheme in 1957, an overall improvement of 1% a year has been achieved in the IRHO/Cirad/PalmElit program, mostly through the progress in bunch production (Cochard et al., 1993; Corley and Tinker, 2003; Durand-Gasselin et al., 2009).

7.2.6.2 Specific Progress of a True Second Cycle of Selection

The first-cycle and early second-cycle trials were exclusively designed to exploit SCA that had been highlighted in the "Block 500" planted in the 1960s (Gascon et al., 1976). In order to represent a cross (to exploit GCA and SCA) in commercial seed production, at least 10 *pisifera* parents are needed (Jacquemard, 1979). It became apparent that to limit reproduction of the original crosses in seed production would be difficult and require much space. To maximize seed production, it was subsequently decided to refocus on estimating GCA more precisely in the choice of mating design in the breeding experiments. With this, the best GCA parents and their selfs can be used in more crosses for seed production. Since 1995, all the trials planted enabled more precise estimation of the GCAs of all the parents tested.

However, these designs were unable to properly estimate the global combining values (GCA + SCA) commonly exploited in RRS schemes. Since then, new experimental designs that allow good estimation of GCA have been used. All trials planted since 1995 follow this design, which enables calculation of the GCAs of all the parents used.

The first project following this objective was planted between 1995 and 2000 in partnership with Socfin Indonesia (Socfindo). A total of 142 parents from Group A and 153 parents from Group B were tested in 561 crosses. The standard cross LM2T × DA10D was represented in some of the trials; its oil yield varied between 6.8 and 7.4 t/ha/year, which was similar to its level in the first-cycle experiments (Cochard et al., 1993). The mean oil yield of all the trials was 7.4 t/ha/year for the 6- to 9-year-old period compared to 6.7 t/ha/year for the 6- to 7-year-old period in first-cycle trials. For the 9% best Group A parents, mean GCA for oil yield was +8.3% and for the 16% best Group B parents, it was +7.8% (Jacquemard et al., 2010). Improvements were made on FFB (+5.5% for Group A and +4% for Group B) and on OER (+2.8% for Group A and +3.8% for Group B).

7.2.6.3 Height Increment

Height increment (HINC) is one of the major objective traits in IRHO/Cirad/PalmElit's breeding programs (Jacquemard, 1979). Although average growth rate in the second-cycle crosses is the same as for the standard cross LM2T × DA10D (57 cm/year), many parents showed growth rates at least 10% slower. For Group A as for Group B, the smallest 10% of parents gave a 12% lower HINC (Jacquemard et al., 2010). It is therefore possible to offer commercial hybrids with growth rates of 46–50 cm/year in areas with no water deficit and 44–48 cm/year in areas with 200 mm water deficit. Furthermore, a special program aimed at the introgression of slow growing palm genotypes into the breeding scheme was launched in the early 1980s in La Mé. After a first generation of selection, average growth rates of the standard cross was 47 cm/year and the best crosses had a growth rate lower than 30 cm/year in an environment with 200 mm water deficit (Adon et al., 2001).

7.2.6.4 Disease Resistance

Oil palm is a crop susceptible to a wide range of pests and diseases, some of them with dramatic results in several parts of the world. The most destructive diseases are Fusarium wilt in Africa and Ganoderma basal stem rot in South East Asia, two soil-borne fungal diseases, and the bud rot complex of unknown origin and with a broad range of symptoms, which is widespread in Latin America. Major programs aimed at finding genetic resistance for breeding were implemented as early as 1972 for Fusarium wilt (Renard et al., 1972) and in the early 2000s for Ganoderma basal stem rot (Breton et al., 2009a,b) and bud rot diseases (Louise, 2013, 2014) (see Chapter 4).

Fusarium wilt-resistant varieties developed by PalmElit have been available earlier and field proven since (Renard and de Franqueville, 1989; Durand-Gasselin et al. 2006). Socfindo in collaboration with PalmElit has commercially released its moderately tolerant variety for Ganoderma recently (Turnbull et al. 2014) (see Figure 7.4).

7.3 BREEDING PROGRAMS IN AAR, SAIN, AND SUMBIO

Aik Chin Soh, Choo Kien Wong, Tatang Kusnadi, and Yayan Juhyana

7.3.1 Background

The main breeding programs of AAR (Applied Agricultural Resources, Malaysia), SAIN (Sarana Inti Pratama, Indonesia), and SUMBIO (Sumatra Bioscience, Indonesia), which practice variant or fusion/hybrid MRS and MRRS breeding schemes and breeding populations, are illustrated here.

7.3.2 AAR

The original base breeding populations were the Ulu Remis Deli D (Figure 7.5) and the introgressed Dumpy.AVROS and Yangambi.AVROS T/P populations (Figure 7.6). Subsequent expansion included the introgression of Dabou Deli Ds and Binga, Ekona, Calabar, La Mé Ts, and MPOB's Nigerian genetic materials through exchanges. The original main breeding scheme was MRS but gradually evolved into MRRS. It is now in the first recombinant cycle. Top-cross progeny testing with elite tester D and P/T lines is being practiced instead of the conventional progeny testing of contemporaneous recurrent parents. Auxiliary or secondary breeding programs to incorporate more heritable traits, for example, Ganoderma disease resistance, oil quality, long stalk, large and virescens fruits, OxG hybrids, into the advanced breeding parents would involve some form of backcrossing. Near true single cross hybrids

FIGURE 7.4 **(See color insert.)** Breeding for resistance to Ganoderma basal stem rot: (a) diseased young palm, (b) diseased old palm, (c) Ganoderma fruiting bodies, (d) field susceptible and resistant progenies, (e) nursery resistance screening results, and (f) relationship between nursery and field screening results. (Courtesies of Idris A.S., Chung G.F., and Yayan J.)

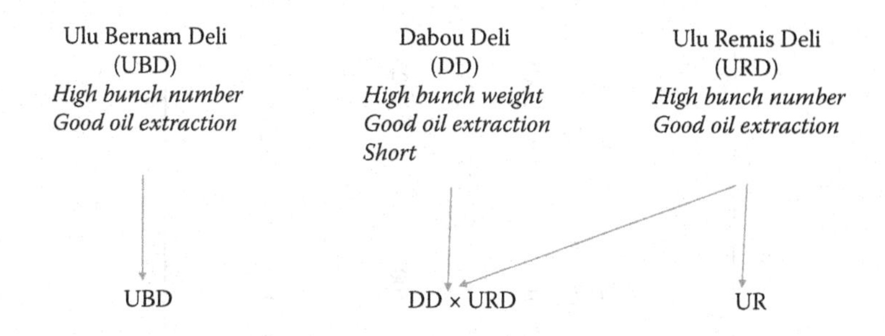

FIGURE 7.5 AAR's AA Hybrida Deli *Dura* lineage.

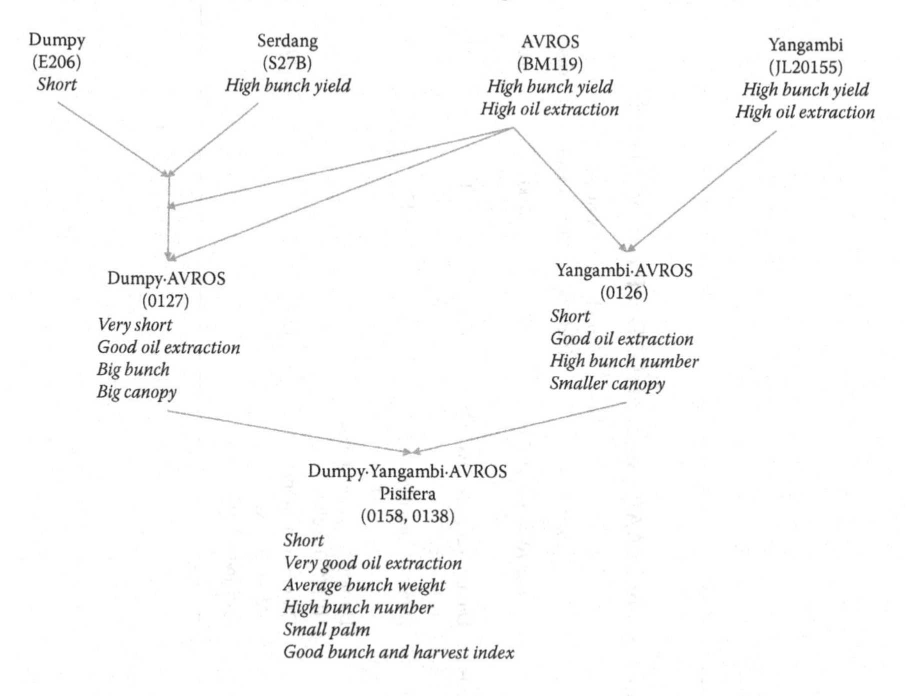

FIGURE 7.6 AAR's Hybrida I Pisifera lineage.

and the elite top semi-clonal (one clonal parent) hybrids and clones are currently the commercial products.

The primary objective and characteristic features of AAR's main breeding program and commercial planting materials are high oil yielding short and high harvest index palms that can be planted at higher density. The realized and expected breeding progress in AAR's D × P materials is given in Table 7.1. Genome-wide selection is currently being investigated within the AAR breeding program.

The AAR breeding program is backed by an advanced joint collaborative biotechnology research laboratory with the University of Nottingham Malaysia Campus, which provides the genomics and molecular breeding expertise and support and is backed by an in-house strong clonal oil palm R&D program and large commercial clone production laboratory.

TABLE 7.1
Realized and Expected Breeding Improvement in AAR Commercial D × P Hybrid Materials

Variety	*Dura*	*Pisifera*	Bunch Yield (%)	Oil to Bunch (%)	Oil Yield (%)	Palm Height (%)	Seeds Available (Year)
AA D × P (Dumpy.AVR)	Deli	Dumpy.AVROS (0127)	100	100	100	100	1992–2003
AA Hybrida I	Deli, mainly 016/01, 003/31 selfs and clones	Dumpy.Yangambi.AVROS (0158 family)	111	111	122	100	2004–2014
AA Hybrida II	Deli, 0738 family (selfs or clones) Deli, 0744 (family selfs or clones)	Dumpy.Yangambi.AVROS (0138/04 selfs) Dumpy.Yangambi.Lame (0150/07 selfs)	125	126	157	100	2015–2025
AA Hybrida III	Deli, 016/01 × 003/31	Dumpy.Yangambi. AVROS × Nifor (0138/04 × 0228/04)	130	133	173	100	2026–2036

7.3.3 SAIN

SAIN's main breeding program exemplifies those of the group of plantation companies particularly from Indonesia (e.g., SampoernaAgro, Asian Agri, Tania Selatan, Bakrie) that leap-frogged into the oil palm breeding and seed production business by purchasing a wide range of advanced breeding materials from ASD de Costa Rica in the late 1990s and early 2000s. Typically, selected D × P hybrid progeny test seeds as well as their parental selfed/sib-crossed (sibbed) seeds and clones from ASD were planted in the client's plantation. Once the progeny-test results were available, "reproduction" of the selected hybrids using their selfed/sibbed parents or clones could be effected for commercial seed production, essentially following the MRRS breeding scheme. The Deli Ds were derived from Banting, Chemara, MARDI, and Socfin origins while the African Ts/Ps were from Ekona, Calabar, Cameroun, Dami (AVROS) Ghana, La Mé, and Yangambi origins; advanced *E. guineensis* genetic materials were acquired by ASD through exchanges with other breeding stations using their *E. oleifera* accessions.

Owing to the large number of D and P parents selected and consequently very large complete factorial mating combinations possible, the Alpha (incomplete factorial) mating design was adopted by SAIN (Table 7.2). Even so, this resulted in 96 hybrid progenies achieved (out of 136 progenies programmed), an unwieldy treatment size for efficient field experimentation. The Alpha incomplete block experimental design was used to circumvent this (see Chapter 12). Owing to the wide range of genotypes from various genetic origins to be tested in various agroecological zones, genotypes with general and specific adaptability could be selected. The achieved and expected genetic improvements in SAIN's main breeding program are given in Table 7.3. SAIN is a sister company of SUMBIO, and collaborative breeding efforts between them can be expected.

7.3.4 SUMBIO

SUMBIO's main breeding program has been based on the MRS and UR Deli × AVROS-based breeding populations as its parent company LONSUM was originally owned by H&C. Under the Combined Breeding Programme consortium, introduction and introgressions of Binga, Zaire, Calabar, and Yangambi and from local accessions, for example, Gunung Malayu, were made. There were no clear distinctions as to whether the introgressions were to be made on the Deli or AVROS side (although lately this has been guided through molecular genetic diversity analysis). Reciprocal top-cross progeny testing of new or introgressed parents with selected advanced seed production Deli D and AVROS P testers are made (Figure 7.7).

In recent years, SUMBIO has pioneered the applications of single seed descent, molecular screening for homozygosity and ploidy levels and the production of doubled haploids in its quest for the production of F_1 or single cross hybrids in oil palm. It also leads in the breeding program to incorporate the long stalk, virescens and delayed abscission fruit ease of harvesting traits into high yielding varieties (Figure 7.8). The progress made in the different generations of SUMBIO's D × P materials is given in Table 7.4 while Table 7.5 illustrates the comparative expected time-frames to achieve true F_1 hybrid seed production using conventional inbreeding,

TABLE 7.2
Alpha Mating Design Crosses (Part of) in SAIN's RRS II D × T/P Progeny Test

Parents	P1	P2	P3	P4	P5	P6	P7	P8	P9	P10	P11	T12	T13	T14	T15	T16	T17	T18	T19	T20	T21	T22	T23	T24	No. Crosses
D1	x												x												2
D2		x												x											2
D3			x												x										2
D4				x												x									2
D5					x												x								2
D6						x												x							2
D7							x												x						2
D8								x												x					2
D9									x												x				2
D10										x												x			2
D11											x												x		2
D12												x												x	2
D13	x												x												2
D14		x												x											2
▼																									2
D68								x														x			2
No. Crosses	6	6	6	6	6	6	6	6	5	5	5	5	5	5	6	6	6	6	6	6	6	6	5	5	136

TABLE 7.3
Achieved and Expected Genetic Improvements in SAIN's Main Breeding Program

Improved Hybrids	Total Oil Yield, t/ha (%)	Palm Oil Yield, t/ha (%)	Kernel Oil Yield, t/ha (%)	Fresh Bunch Yield, t/ha (%)	Oil Extraction Rate (%)	Height Increment, cm/Year	Harvest Index	Potential Seed Production, Million/Year
RRS II (LD1–3 Trials)	7.0 (100)	6.5 (100)	0.6 (100)	27.4 (100)	23.6 (100)			
H1 RRS II (selected hybrids)	8.0 (114)	7.4 (114)	0.6 (100)	29.7 (108)	24.8 (106)	ca. 60	ca. 2.0	10 (2011)
cH1 (creamed hybrids)	8.2 (117)					ca. 60	ca. 2.0	10 (2025)
H2 RRS III	9.0 (128)					ca. 55	ca. 2.2	10 (2026)

Note: Based on the first 7 years' mean yield data: oil extraction rate = 0.855 × oil to bunch.

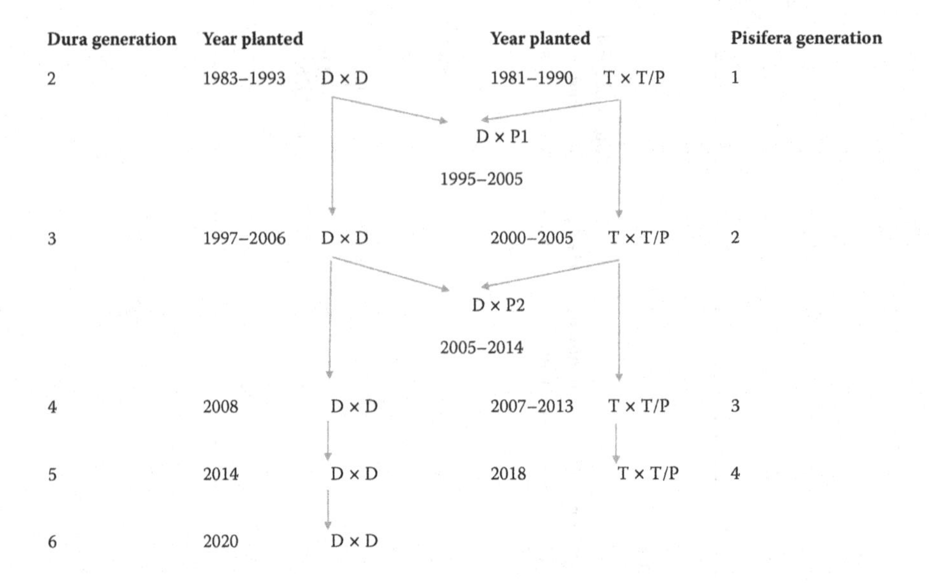

FIGURE 7.7 Outline of SUMBIO's main breeding program.

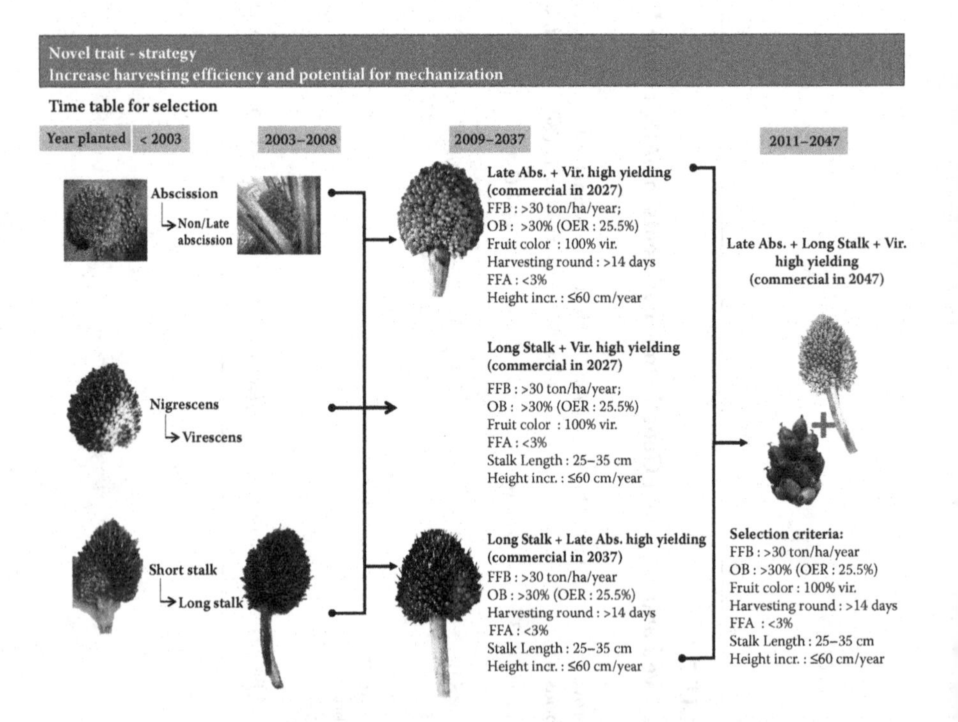

FIGURE 7.8 **(See color insert.)** Breeding for ease of harvesting traits.

TABLE 7.4
Achieved and Expected Genetic Improvements in SUMBIO's Main Breeding Program

D × P Generation Crosses	Fresh Fruit Bunch Yield, t/ha/Year	Total Oil Yield t/ha/Year	Oil to Bunch, %	Palm Height (Year 5), cm
D1 × P1	28.1	9.9	30.8	98
D1 × P2	32.2	10.5	32.9	112
D2 × P2	32.9	10.9	31.0	111
D2 × P3	33.3	11.3	32.1	124
D3 × P3[a]	34.0[a]	11.5[a]	33.0[a]	<100[a]
D4 × P3[a]	35.0[a]	12.0[a]	33.5[a]	<90[a]
D5 × P4[a]	36.0[a]	12.5[a]	36.0[a]	<80[a]

[a] Expected crosses and values.

single seed descent, single seed descent cum homozygosity screening and doubled haploid breeding methods. SUMBIO has also a very active Ganoderma basal stem rot resistance breeding program (Figure 7.4).

7.4 CONCLUSIONS

Aik Chin Soh, Sean Mayes, and Jeremy Roberts

Davidson (1993) estimated that breeding improvement has contributed about 50% to the oil palm plantation yield increase in Malaysia, the rest to agronomy and management. Cirad (above) indicated that it has achieved an average 1% annual improvement (10%–15% per 10-year cycle) in oil yield with its adoption of RRS in 1957. This estimate should be reasonably accurate as it has had a constant standard cross (D10D × L2T) linking across all the RRS trials. With the MRS trials without a constant standard cross-linking across all the trials, the computation of breeding progress is less direct. Nevertheless, both Hardon et al. (1987) and Soh et al. (2003) with slightly different computation methods also arrived at the 1%–1.5% figure. Similar achievements were also demonstrated in the breeding programs of AAR, SAIN, and SUMBIO. The breeding progress in yield made in the oil palm thus has been comparable to those made in annual crops despite its much longer breeding cycle. This could perhaps be attributed to the relatively lower starting point, with oil palm only having been through 6 or so generations of selection from wild material as it is a relatively newly bred crop as compared to that of the other traditional annual crops, for example, wheat and rice, which have undergone a large number breeding cycles or generations. Nevertheless, yields are expected to plateau with continued use of the current advanced breeding pure Deli, AVROS, Yangambi, and La Mé populations, and there is a need to infuse genes from other genetic sources and this has been actively pursued by some groups. To breach the yield plateau, the palm architecture needs to be modified to be more efficient in resource (light, water, nutrients) use in terms of their capture and conversion to biomass for increased yield potential (refer Chapters 2, 5, and 13).

TABLE 7.5
Comparative Expected Time-Frames to Produce F_1 Hybrids from Conventional Inbreeding, Single Seed Descent, Single Seed Descent + Homozygosity Screening, and Double Haploid Breeding Programs

Generation	Cycle	Conventional Inbreeding Program	Single Seed Descent Program (SSD)	SSD + Homozygosity Screening Program (SSD + Ho)	Doubled Haploid Program (DH)	D × P Progeny Testing and Seed Production
F_2	0	Year 0			DH palms →	Year 10
				SSD + Ho palms (Year 5) →	→	Year 15
S_1	1	Year 10	SSD2	→	→	Year 20
S_3		Year 25	SSD5	→	→	Year 35
S_5	5	Year 50	→	→	→	Year 60

It is important to differentiate between the yield improvements achieved from breeding trials and those reported in national crop statistics. The former is a reflection of the genetic variability available and selection efficiency achieved while the latter depends on the age and corresponding yield profiles of the national crop plantings. An ideal or optimum profile from agro-management considerations (economic, cash flow, labor management) is to have about 5% immature (<5 years old) and 5% aged (>20 years old) plantings with low yield and 90% in the prime yielding ages (6–19 years old). Delayed replanting resulting in higher proportion of low yielding palms and concomitant labor and management constraints have been attributed to be the cause of the Malaysian national yield stagnation or plateau at 5 t/ha oil for the past 20 years. The best plantation yields achieved are around 6–8 t/ha while for trials and trial plots they are 10–14 t/ha (see Chapters 2, 5, and 13).

So far, the breeding progress tracking exercise has understandably focused only on yield. A similar exercise should also be done for other traits, for example, quality, disease, pest and stress resistance, and vegetative and physiological efficiency traits, although the last was attempted on the Deli D (Corley and Lee, 1992) retrospectively and by Cirad for height in the D × P.

Lastly, although breeding progress for yield in terms of improvement per unit time (year) in oil palm is similar to that of annual crops, the lag time in replacement with new improved varieties is about 20–25 years and only for less than 5% of the existing plantings per year, as compared to annual crops that can be done *in toto* in one season. Also, the oil palm varieties have not changed much with each subsequent replanting, hence, the need for new much improved oil palm varieties to be available and in shorter time, which is now made possible with the latest breeding and biotechnological techniques and tools.

REFERENCES

Adon, B., Baudouin, L., Durand-Gasselin, T., and Kouamé, B. 1995. Utilisation de Matériel non Amélioré pour la Sélection du almier à Huile: l'Origine Angola. *Plantation Recherche Développement* 5: 201–207.

Adon, B., Cochard, B., Flori, A., Potier, F., Quencez, P., and Durand-Gasselin, T. 2001. Introgression of slow vertical growth in improved oil palm (*E. guineensis* Jacq.) populations. *Proceedings of the 2001 PIPOC International Palm Oil Congress, Agriculture Conference*, Kuala Lumpur, Malaysia, August 20–22, 2001: pp. 210–217.

Beirnaert, A. and Vanderweyen, R. 1941. *Contribution à i'Etude Génétique et Biométrique des Variétés d'Elaeis guineensis Jacq.* Publications de l'Institut National pour l'Etude Agronomique du Congo Belge. Série scientifique no. 27.

Bénard, G. 1965. Caractéristiques Qualitatives du Régime d'Elaeis guineensis Jacq.—Teneur en Huile de la Pulpe des Diverses Origines et des Croisements Interorigines. *Oléagineux* 20: 163–168.

Besse, I., Verdeil, J.L., Duval, Y., Sotta, B., Maddiney, R., and Miginiac, E. 1992. Oil palm (*Elaeis guineensis* Jacq.) clonal fidelity: Endogenous cytokinins and indoleacetic acid in embryogenic callus cultures. *Journal of Experimental Botany* 43: 983–989.

Billotte, N., Jourjon, M.F., Marseillac, N. et al. 2010. QTL detection by multi-parent linkage mapping in oil palm (*Elaeis guineensis* Jacq.). *Theoretical and Applied Genetics* 120: 1673–1687.

Billotte, N., Marseillac, N., Risterucci, A.M. et al. 2005. Microsatellite-based high density linkage map in oil palm (*Elaeis guineensis* Jacq.). *Theoretical and Applied Genetics* 110: 754–765.

Boyer, P. 1962. La Plantation Expérimentale de Dabou. *Oléagineux* 17: 297–306.

Brédas, J. 1969. La Selection du Palmier à Huile à la Plantation Robert Michaux de Dabou (Côte d'Ivoire). La Recherche de Géniteurs. *Oléagineux* 24: 63–68.

Breton, F., Miranti, R., Lubis, Z., Hayun, Z., Umi, S., Flori, A., Nelson, S.P.C., Durand-Gasselin, T., and de Franqueville, H. 2009a. Early screening test: A routine work to evaluate resistance/susceptibility level of oil palm progenies to basal stem rot disease. *PIPOC 2009*, November 9–12, 2009, Kuala Lumpur Convention Centre, Malaysia.

Breton, F., Miranti, R., Lubis, Z., Hayun, Z., Umi, S., Flori, A., Nelson, S.P.C., Durand-Gasselin, T., Jacquemard, J.C., and de Franqueville, H. 2009b. Implementation of an early artificial inoculation test to screen oil palm progenies for their level of resistance and hypothesis on natural infection: Ganoderma disease of the oil palm. *16th International Oil Palm Conference and Expopalma—Challenges in Sustainable Oil Palm Development*, September 22–25, 2009, Cartegena de Indias, Colombia.

Cochard, B. 2008. Etude de la Diversité Génétique et du Déséquilibre de Liaison au sein de Populations Améliorées de Palmier à Huile (*Elaeis guineensis* Jacq.). *Thesis*. Supagro, Montpellier.

Cochard, B., Carrasco-lacombe, C., Pomiès, V., Dufayard, J.F., Suryana, E., Omoré, A., Durand-Gasselin, T., and Tisné, S. 2015. Pedigree-based linkage map in two genetic groups of oil palm. *Tree Genetics and Genomes* 11: 68.

Cochard, B., Durand-Gasselin, T., and Adon, B. 2006. Oil palm genetic resources in the Côte d'Ivoire. Composition, Assessment and Use. In: N. Rajanaidu, I.E. Henson, A. Darus (Eds.), *International Symposium on Oil Palm Genetic Resources and Their Utilization*, June 8–10, 2000, Malaysian Palm Oil Board, Kuala Lumpur: pp. 81–100.

Cochard, B., Durand-Gasselin, T., Amblard, P., Konan, E., and Gogor, S. 1999. Performance of adult oil palm clones. *1999 PORIM Int. P.O. Congr. "Emerging Technologies and Opportunities in the Next Millenium"*, 2000, Palm Oil Research Institute, Malaysia: pp. 53–64.

Cochard, B., Noiret, J.M., Baudouin, L., Flori, A., and Amblard, P. 1993. Second cycle reciprocal recurrent selection in oil palm, *Elaeis guineensis* Jacq. Results of Deli × La Mé hybrid tests. *Oléagineux* 48: 441–451.

Corley, R.H.V. 2005. Illegitimacy in oil palm breeding—A review. *Journal of Oil Palm Research* 17: 64–69.

Corley, R.H.V. and Lee, C.H. 1992. The physiological basis for genetic improvement of oil palm in Malaysia. *Euphytica* 60: 179–184.

Corley, R.H.V., Lee, C.H., Law, I.H., and Wong, C.Y. 1986. Abnormal flower development in oil palm clones. *Planter* 62: 233–240.

Corley, R.H.V. and Tinker, P.B. 2003. *The Oil Palm*. Wiley Blackwell, Oxford, UK: 639 pp.

Davidson, L. 1993. Management for efficient cost-effective and productive oil palm plantations. *Proc. 1991 PORIM Int. Palm Oil Conference Agriculture Module*, Palm Oil Res. Inst., Kuala Lumpur, Malaysia: pp. 153–167.

Durand-Gasselin, T., Cochard, B., Amblard, P., and Nouy, B. 2009. Exploitation de l'Hétérosis dans l'Amélioration Génétique du Palmier à Huile (*Elaeis guineensis*). Hétérosis et Variétés Hybrides en Amélioration des Plantes, 100 ans après Shull. *Association des Sélectionneurs Français*, 5 février, INRA de Versailles – Versailles, France.

Durand-Gasselin, T., de Franqueville, H., Diabaté, S., Cochard, B., and Adon, B. 2006. Assessing and utilizing sources of resistance for Fusarium wilt in oil palm (*Elaeis guineensis* Jacq.) genetic resources. In: N. Rajanaidu, I.E. Henson, A. Darus (Eds.),

International Symposium on Oil Palm Genetic Resources and Their Utilization, June 8–10, 2000, Malaysian Palm Oil Board, Kuala Lumpur: pp. 446–470. ISBN 967-961-122-1.

Durand-Gasselin, T., Konan, K., Duval, Y., and Pannetier, C. 1986. Development of oil palm micropropagation through somatic embryogenesis. *Poster in Sixth International Congress of Plant Tissue Culture*, August 3–8, 1986, University of Minnesota, Minneapolis USA.

Durand-Gasselin, T., Labeyrie, A., Amblard, P., Potier, F., Cochard, B., de Franqueville, H., and Nouy, B. 2010. Strategic considerations on clonal propagation of oil palm. *The International Society for Oil Palm Breeders (ISOPB) International Seminar on Advances in Oil Palm Tissue Culture*, May 29, 2010, Yogyakarta, Indonesia.

Duval, Y., Durand-Gasselin, T., Konan, K., and Pannetier, C. 1988. In vitro vegetative micropropagation of oil palm (*Elaeis guineensis* Jacq.): Strategy and results. In: A. Halim Hassan et al. (Eds.), *Proc. 1987 Int. Oil Palm Conf. "Progress and Prospects,"* Palm Oil Res. Inst., Kuala Lumpur, Malaysia: pp. 191–196.

Gascon, J.P. and de Berchoux, C. 1964. Caractéristiques de la Production d'Elaeis guineensis (Jacq.) de Diverses Origines et de Leurs Croisements. Application à la Sélection du Palmier à Huile. *Oléagineux* 19: 75–84.

Gascon, J.P., Jacquemard, J.C., Houssou, M., Boutin, D., Chaillard, H., and Kanga-fondjo, F. 1981. La Production de Semences Sélectionnées de Palmier à Huile *Elaeis guineensis. Oléagineux* 36: 476–486.

Gascon, J.P., Le Guen, V., Nouy, B., and Asmadyand Kamga, F. 1988. Résultats d'Essais de Second Cycle de Sélection Récurrente Réciproque chez le Palmier à Huile *Elaeis guineensis* Jacq. *Oléagineux* 43: 1–5.

Gascon, J.P., Noiret, J.M., and Meunier, J. 1969. Effet de la Consanguinité chez *Elaeis guineensis* Jacq. *Oléagineux* 24: 603–611.

Gascon, J.P., Noiret, J.M., and Meunier, J. 1976. The IRHO Programme. In: R.H.V. Corley, J.J. Hardon, B.J. Wood (Eds.), *Oil Palm Research*, Elsevier, Amsterdam: pp. 110–118.

Hardon, J.J., Corley, R.H.V., and Lee, C.H. 1987. Breeding and selecting the oil palm. In: A.J. Abbot and R.K. Atkin (Eds.), *Improving Vegetatively Propagated Crops*, Academic Press, London: pp. 63–81.

Hardon, J.J., Gascon, J.P., Noiret, J.M., Meunier, J., Tan, G.Y., and Tam, T.K. 1976. Major oil palm breeding programmes. In: R.H.V. Corley, J.J. Hardon and B.J. Wood (Eds.), *Oil Palm Research*, Elsevier, Amsterdam: pp. 109–125.

Houard, A. 1926. *La Sélection du Palmier à Huile à la Station Expérimentale de La Mé.* Rapport technique de 1925. Ed. Larose, Paris: 183 pp.

Houard, A. 1927. *Contribution à l'Etude du Palmier à Huile en Afrique Occidentale Française*. Ed. Emile Larose, Paris: 98 pp.

Jacquemard, J.C. 1979. Contribution to the study of the height growth of the stems of *Elaeis guineensis* Jacq. study of the L2T × D10D cross. *Oléagineux* 34: 492–497.

Jacquemard, J.C., Suryana, E., Cochard, B., de Franqueville, H., Breton, F., Syaputra, I., Dermawan, E., and Permadi, P. 2010. Intensification of oil palm (*Elaeis guineensis* Jacq.) plantation efficiency through planting material: New results and developments. *IOPRI International Oil Palm Conference (IOPC 2010): Transforming Oil Palm Industry*, June 1–3, 2010, Yogyakarta, Indonesia: 36 p.

Jagoe, R.B. 1952. Deli oil palm and early introduction of *Elaeis guineensis* to Malaya. *Malaysian Agricultural Journal* 35: 3.

Jaligot, E., Adler, S., Debladis, É., Beulé, T., Richaud, F., Ilbert, P., Finnegan, E.J., and Rival, A. 2011. Epigenetic imbalance and the floral developmental abnormality of the in vitro-regenerated oil palm *Elaeis guineensis* Jacq. *Annals of Botany* 108: 1453–1462.

Louise, C. 2014. Alternativas Frente al Complejo PC en Ecuador. Ecupalm. *Congresso Internacional de Palma Aceitera. Santo Domingo de los Tsáchitas*, 8–11 de abril 2014, Ecuador.

Louise, C., Poveda, R., Flori, A., and Amblard, P. 2013. Tolerancia al Complejo de Pudrición del Cogollo en el Material *Elaeis guineensis*: Caso del Material C07 en el Oriente Ecuatoriano = Tolerance to Oil Palm Bud Rot Disease in the Material *Elaeis guineensis*: Case of the Material C07 in Eastern Ecuador. *Palmas* 34(1) (spec): 126–134. *Conferencia Internacional sobre Palma de Aceite*. 17, 2012-09-25/2012-09-28, Cartagena, Colombia.

Matthes, M., Singh, R., Cheah, S.C., and Karp, A. 2001. Variation in oil palm (*Elaeis guineensis* Jacq.) tissue culture-derived regenerants revealed by AFLPs with methylation-sensitive enzymes. *Theoretical and Applied Genetics* 102: 971–979.

Meunier, J. and Gascon, J.P. 1972. Le Schéma Général d'Amélioration du Palmier à Huile à l'IRHO. *Oléagineux* 27: 1–12.

Meunier, J., Gascon, J.P., and Noiret, J.M. 1970. Hérédité des Caractéristiques du Régime d'*Elaeis guineensis* Jacq. En Côte d'Ivoire. *Oléagineux* 25: 377–382.

Montoya, C., Cochard, B., Flori, A. et al. 2014. Genetic architecture of palm oil fatty acid composition in cultivated oil palm (*Elaeis guineensis* Jacq.) compared to its wild relative *E. oleifera* (H.B.K) Corte's. *PLoS One* 9, e95412. doi: 10.1371/journal.pone.0095412.

Montoya, C., Lopes, R., Flori, A. et al. 2013. Quantitative trait loci (QTLs) analysis of palm oil fatty acid composition in an interspecific pseudo-backcross from *Elaeis oleifera* (HBK) Cortés and oil palm (*Elaeis guineensis* Jacq.). *Tree Genetics and Genomes* 9: 1207–1225.

Noiret, J.M., Gascon, J.-P., and Bénard, G. 1966. Contribution de l'Hérédité des Caractéristiques de la Qualité du Régime et du Fruit d'Elaeis Guineensis Jacq. Application à la Sélection du Palmier à Huile. *Oléagineux* 21: 343–349.

Nouy, B., Jacquemard, J.C., Suryana, E., Potier, F., Konan, K.E., and Durand-Gasselin T. 2006. The expected and observed characteristics of several oil palm (*Elaeis guineensis* Jacq.) clones. *IOPRI. International Oil Palm Conference (IOPC)*, June 19–23, 2006, Bali, Indonesia: 7 pp.

Nouy, B., Lubis, R.A., Kusnadi, T.T., and Akiyatand Samritaan, G. 1991. Oil palm (*Elaeis guineensis*) production potential results for Deli × La Mé hybrids in North Sumatra. *Oleagineux* 46: 91–99.

Ong-Abdullah, M., Ordway, J.M., Jiang, N. et al. 2015. Loss of Karma transposon methylation underlies the mantled somaclonal variant of oil palm. *Nature* 525: 533–537.

Pannetier, C., Arthuis, P., and Liévoux, C. 1981. Néoformation de Jeunes Plants de *Elaeis guineensis* à Partir de Cals Primaires Obtenus sur Fragments Foliaires Cultivés *in vitro*. *Oléagineux* 36: 119–122.

Rabéchault, H. and Martin, J.P. 1976. Multiplication Végétative du Palmier à Huile (*E. guineensis* Jacq.) à l'Aide de Tissus Foliaire. *Comptes Rendus de l'Académie des Sciences Paris, Série D*, 238: 1735–1737.

Renard, J.L. and de Franqueville, H. 1989. Oil palm vascular wilt. *Oleagineux*, 44: 341–349.

Renard, J.L., Gascon, J.P., and Bachy, A. 1972. Research on vascular wilt disease on the oil palm. *Oléagineux* 27: 581–591.

Rival, A., Ilbert P., Labeyrie, A., Torres, E., Doulbeau, S., Personne, A., Dussert, S., Beulé, T., Durand-Gasselin, T., Tregear, J., and Jaligot, E. 2013. Variations in genomic DNA methylation during the long-term in vitro proliferation of oil palm embryogenic suspension cultures. *Plant Cell Reports* 32: 359–368.

Rival, A. and Jaligot, E. 2011. Epigenetics and plant breeding. *CAB Reviews: Perspectives in Agriculture, Veterinary Science, Nutrition and Natural Resources* 6: 048.

Rival, A., Jaligot, E., Beulé, T., Tregear, J., and Finneganand, E.J. 2009. The oil palm "Mantled" Somaclonal variation: A model for epigenetic studies in higher plants. *Acta Horticulturae* 829: 177–182.

Rosenquist, E.A. 1986. The genetic base of oil palm breeding populations. In: A.C. Soh, N. Rajanaidu and M. Nasir (Eds.), *Proc. Int. Workshop on Oil Palm Germplasm and Utilization*, Palm Oil Res. Inst., Kuala Lumpur, Malaysia: pp. 16–27.

Rosenquist, E.A. 1990. An overview of breeding technology and selection in *Elaeis guineensis. Proc. 1989 Int. Palm Oil Development Conference—Agriculture*, Palm Oil Res Inst. Inst., Kuala Lumpur, Malaysia: pp. 5–26.

Soh, A.C. 1999. Breeding plans and selection methods in oil palm. In: N. Rajanaidu and B.S. Jalani (Eds.), *Proc. Symp. The Science of Oil Palm Breeding*, Palm Oil Res. Inst., Kuala Lumpur, Malaysia, pp. 65–95.

Soh, A.C.,Wong, G., Hor, T.Y., Tan, C.C., and Chew, P.S. 2003. Oil palm genetic improvement. In: Janick F. (Ed.), *Plant Breeding Reviews Vol. 22*, John Wiley & Sons, New York, pp. 165–219.

Sparnaaij, L.D. 1969. Oil palm (*Elaeis guineensis* Jacq.). In: F.P. Ferwerda and F. Wit (eds.), *Outlines of Perennial Crop Breeding in the Tropics*, Misc.Papers, paper number 4, pp. 339–387. Wageningen University, Wageningen.

Turnbull, N., de Franqueville, H., Breton, F., Jeyen, S., Syahputra, I., Cochard, B., Poulain, V., and Durand-Gasselin, T. 2014. Breeding methodology to select oil palm planting material partially resistant to *Ganoderma boninense. Presentation at the 5th Quadrennial International Oil Palm Conference*, June 17–19, 2014, Bali Nusa Dua Convention Center, Indonesia.

8 Clonal Propagation

Aik Chin Soh, Sean Mayes, Jeremy Roberts, Ong-Abdullah Meilina, Siew Eng Ooi, Ahmad Tarmizi Hashim, Zamzuri Ishak, Samsul Kamal Rosli, Wei Chee Wong, Chin Nee Choo, Sau Yee Kok, Nuraziyan Azimi, and Norashikin Sarpan

CONTENTS

8.1 Introductory Overview (*Aik Chin Soh, Sean Mayes, and Jeremy Roberts*) 192
8.2 Historical Development, Clonal Improvement, Breeding, and Cloning Strategies (*Aik Chin Soh, Sean Mayes, and Jeremy Roberts*) 193
8.2.1 Historical Development 193
8.2.2 Genetic Improvement with Clones 194
8.2.3 Breeding Strategies for Clonal Propagation 195
8.2.4 Ortet Selection and Cloning Strategies 195
8.3 Revolutionizing OPTC Research (*Ong-Abdullah Meilina, Siew Eng Ooi, Ahmad Tarmizi Hashim, Zamzuri Ishak, Samsul Kamal Rosli, Wei Chee Wong, Chin Nee Choo, Sau Yee Kok, Nuraziyan Azimi, and Norashikin Sarpan*) 197
8.3.1 Background 197
8.3.2 Understanding the Embryogenesis Process 199
8.3.2.1 Molecular Aspects of SE 199
8.3.2.2 Influence of Hormones on SE 200
8.3.3 Refining the *In Vitro* Protocols 201
8.3.3.1 Manipulating the Explants 201
8.3.4 Liquid Culture System 201
8.3.5 Somaclonal Variation Associated with Micro-Propagation 203
8.3.5.1 Understanding Flower Development 204
8.3.5.2 Epigenome: Moving above and beyond the Genome 205
8.3.6 Potential of Organelle (Chloroplast and Mitochondria) Research in Tissue Culture 208
8.3.7 Post-Genomics (and a Systems Biology Approach) for *In Vitro* Cultures 209
8.4 Conclusions 211
Appendix 8A.1 Oil Palm Liquid Culture: MPOB Protocol 212
Appendix 8B.1 Other Innovations 214
References 216

8.1 INTRODUCTORY OVERVIEW

Aik Chin Soh, Sean Mayes, and Jeremy Roberts

Perennial tree crops are notorious for their outbreeding behavior and consequent high heterozygosity within individuals and broad genetic variability among hybrid seedling populations (Simmonds and Smartt 1999). They have a long breeding cycle (commonly more than 5 years) and other compelling reasons that make generation of commercial material by seedling production nonideal, for example, few, recalcitrant or nonviable seeds; sterility due to polyploidy; and genetic incompatibility. Vegetative or clonal propagation provides the means to circumvent some of these issues either in the form of commercial ramet (clonal offspring or propagule) production from selected ortets (clonal parent) or in the form of clonal hybrid seeds from cloned parents: biclonal (both clonal parents, e.g., cocoa); mono/semiclonal (one clonal parent), polycross (from open-pollination of isolated clonal parents in a seed-garden, e.g., rubber). Clonal hybrid seeds are easier and cheaper to produce but have a higher level of genetic variability and consequently lower expected yield potential as compared to the best clonal lines.

The practice of vegetative propagation using grafting, cuttings, suckers, rhizomes, and corms dates back to ancient if not prehistoric times. Growers in China have been using grafting techniques on fruit trees for over 3000 years, since preclassical times. The plantation industry, derived mostly from semi-perennial herbs, for example, sugarcane, banana, cassava, pepper, and especially perennial trees, for example, tea, coffee, fruit trees, has been based on initial improvement via breeding programs, followed by multiplication of the best lines by clonal propagation. The successful rubber industry in Malaysia and Indonesia, before its supersession by the seed-based oil palm, has been based on clonal propagation (initially polycross seeds, rapidly superseded by clones). The desire for uniformity in rubber was adopted for aesthetic, commercial, and management considerations in addition to the relative ease of clonal propagation by conventional vegetative means.

Plant tissue culture originated from the ideas of Hildebrandt at the start of the twentieth century (Thorpe 2007). The intervening years until 1990 saw further development in its science and applications, for example, studies on nutrition, morphogenesis, embryogenesis, and their application toward disease-free plants, germplasm conservation, genetic manipulation, and clonal propagation. The 1990s saw the expansion of its use for mass uniform crop propagation, especially for horticultural species. It is based on the principle of *totipotency*: every plant cell contains all the genetic information needed to regenerate into a whole plant (when given the right conditions). Nevertheless, most of the tissues used are the meristematic cells in the apex, adventitious shoot or bud. This is sometimes called meristem culture. In other cases, the tissue has to go through a callus phase before it can be induced, via the manipulation of media and physical growth conditions, to differentiate into an embryogenic phase to produce plantlets. The process is called somatic embryogenesis (SE) and is used to describe the oil palm tissue culture (OPTC) process (Rabechault and Martin 1976) although others contend that it is largely organogenesis with shoot formation occurring first and root formation induced subsequently. True embryoids with shoot and root primordia occur less frequently.

The oil palm has a long breeding cycle (ca. 10 years) and protracted inbreeding of the parents is needed to produce reasonably uniform heterozygote hybrids. It is thus not surprising that the *in vitro* clonal propagation of oil palm was sought, especially with claims that more than 30% yield improvement could be achieved (Jones 1974; Hardon et al. 1987).

The R&D aspects of oil palm clonal propagation are discussed in this chapter, while the commercial production and field planting aspects are considered in Chapter 11.

8.2 HISTORICAL DEVELOPMENT, CLONAL IMPROVEMENT, BREEDING, AND CLONING STRATEGIES

Aik Chin Soh, Sean Mayes, and Jeremy Roberts

8.2.1 Historical Development

The apical meristem oil palm has been considered to be recalcitrant to tissue culture. The announcements of the first plantlet regeneration from the *in vitro* culture of palm in the 1970s heralded a new technological breakthrough for the genetic improvement of the oil palm (Jones 1974; Rabechault and Martin 1976) at Unilever and IRHO (now Cirad). With their successful first field plantings, both groups set up their commercial laboratories (Unifield/Bakasawit, Tropiclone) and began selling the clonal ramets. Other breeding groups (MPOB, AAR, UP, Guthrie, Agrocom) soon followed.

In 1986, somaclonal variation in the form of flowering and fruiting (mantling) abnormality leading to sterility was observed in the Unilever oil palm clones (Corley et al. 1986). Similar "mantling" abnormality (the growth of the rudimentary androecium in the fruitlets into additional carpels) was subsequently observed in the clones of all other groups. Unilever and Cirad halted their commercial clone production and reverted to continuing R&D. Some smaller labs, for example, AAR, FELDA, AGROCOM, however, persisted. In 1997, AAR announced to an international audience and published in a peer-reviewed journal the feasibility of large-scale propagation of oil palm clones by gel culture (Wong et al. 1997). The same group subsequently further announced the feasibilities of recloning and the liquid suspension techniques that would facilitate large-scale production of proven clones (Wong et al. 1999; Soh et al. 2003a,b,c). In 2005, AAR commissioned its new OPTC lab with a production capacity of 1.5 million plantlets. Other labs operated by companies such as FELDA, TSH, and UP later followed.

Despite the "commercial breakthrough," annual ramet production in Malaysia and the world still remains at less than 10 million compared to about 150 and 4500 million seeds, respectively, produced each year, with 1.5 million ramets as the maximum production of any lab. OPTC clonal propagation is still an inefficient process with less than 5% of the palms sampled producing actively proliferating cultures. However, the risk of mantling has been the lingering issue which is still stifling attempts to expand plantlet production from reclones and liquid suspension cultures. The prudent approach adopted is to reduce the risk of mantling in commercial field material and achieve commercial production through production from a "basket" of clones (derived from a large number of new ortets) and through limiting the production per culture, with stringent quality control procedures in the production process. The inability to exploit the production of very large clonal numbers from the most elite clones represents a serious limitation to the increased uptake of clonal material in commercial fields.

Recently, MPOB, from research in collaboration with AAR, Felda, UP, and Orion Genomics, have elucidated the likely epigenomic basis of the mantling syndrome (Ong-Abdullah et al. 2015). A molecular marker for mantling is under development which, when launched, could be a "game changer" in terms of providing confidence to the industry to scale up production using reclones and suspension cultures. Low efficiency of embryogenesis in terms of obtaining proliferating cultures (especially from suspension cultures) from a wider range of genotypes still remains an issue, limiting the pool of potential genotypes which can be successfully cloned and proliferated, although research is ongoing.

Oil palm ramets are costly to produce and accordingly higher priced; currently about 5–10 times more than seedlings. The price should come down with improvements in cloning efficiency. Nevertheless, it would still be more expensive than seeds and clients would only buy the more expensive ramets if they have been demonstrated to produce consistently more or higher quality oil than the seed-derived palms.

8.2.2 Genetic Improvement with Clones

The impetus to venture into clonal propagation of oil palm was purported more than 30% yield improvement achievable with clones (Jones 1974; Hardon et al. 1987). This was based on the frequency distribution in oil yield (2 years' data) of individual palms in a D × P progeny-test trial. Through a series of related quantitative analyses based on theoretical principles (Hallauer and Miranda 1981; Falconer and Mackay 1996) and experimental evidence, the expected mean yield improvement of clones derived from the first round ortet selection from progeny-test trials of advanced breeding parents would be about 10%–15% (Soh 1986; Soh et al. 2003b, 2006a,b). More than a 30% improvement would be possible by recloning the best clones. But by then the next breeding cycle of D × P hybrids with an improvement of 10%–15% would be available, reducing the advantage of the clones back to 10%–15%. Potier et al. (2006) from their trial results combined data from 42 clone-ortet sets and their 17 parental crosses from Cirad's second reciprocal recurrent cycle (RRS) and found that the mean oil yield of the clones was 7% higher than the D × P control (representative of the previous generation's commercial D × P) but was 8.6% below the mean of the crosses from which the clones were derived. The best clones were only on a par with the best crosses. One important caveat with all clone analysis (and trial work in general in oil palm) is that in most cases, the genetic identity of the material has not been confirmed by genetic fingerprinting (Corley 2005). Incorrect identity is a quality control issue throughout plantations with growing evidence of higher than expected levels of incorrect identity, compared to field plans and expected pedigree.

This also brings up the need for clonal tests with appropriate check varieties and multiple locations to ascertain objectively the yield improvement advantage of clones over contemporary D × P hybrid cultivars. In field crops, such tests (adaptability) are made prior to new variety release, evaluating environment/location effects and agronomic (spacing, fertilizer, irrigation) treatments, through the use of an appropriate set of check crosses/genotypes. Such standard check material should link to previous generations of trial and commercial material, including the current best germplasm with general adaptability across environments and the current best locally (specific) adapted germplasm. Many, if not all, oil palm clonal test trials that have been carried out have lacked these, preventing a clear assessment of clonal advantage over breeding improvement, but as importantly, not taking into account the effects of G × E adequately in the assessment of clonal advantage. Commonly, the test was done in one location and the check treatment was either a non-constant standard cross used in previous trials or a sample of the current commercial D × P, which will involve a sampling bias and a comparison to mixed hybrids. Claims of clonal superiority over D × P hybrids in yield have often been based on such trials.

Finally, the above arguments pertain only to selection for yield which is a poorly heritable trait. Oil quality, dwarf stature, and stress resistance (biotic, abiotic) are more heritable traits and would likely to be readily and demonstrably captured by cloning.

8.2.3 Breeding Strategies for Clonal Propagation

The breeding method for clonally propagated crops, for example, cassava and rubber, usually adopts the following process:

- Hybridization is made between two or more genetically divergent parents but with complementary desirable traits.
- Selection begins in the widely segregating F_1 or F_2 of the cross of two or more parents. The final desirable recombinant genotype is predicted to be present in this widely segregating population (Simmonds and Smartt 1999).
- Repeated cycles of selection, cloning, and replicated field tests of increasing rigor until at least the fifth cycle (C_5) are used to identify the selected clones, before being considered for release as cultivars or varieties.

This method presumes cloning and recloning is feasible. This cannot be said with confidence for the oil palm, due to the risk of abnormal flowering, although a molecular marker for mantling is being developed. Even then, as hybrid seed breeding is such an established and advanced industry, oil palm clonal propagation is likely to remain as an added value adjunct to commercial oil palm hybrid seed production and as a invaluable breeding tool.

8.2.4 Ortet Selection and Cloning Strategies

As many oil palm breeding programs are producing near true F_1 hybrid seeds, ortet selection within such genetic materials will not be efficient due to the low heritability for oil yield as confirmed by experiences to date (Soh 1986; 1998; Soh et al. 2003b, 2006a,b; Potier et al. 2006). Ortet selection would be more efficient in the earlier stages of a new breeding program or the recombinant phase of a mature program where genetic variability and heritability are higher. Ortets are usually sought from (but not limited to) the D × T/P hybrid progeny-test trials in these programs. Ortets can also be sought from the recombinant (T × T/P) parent crosses, which usually manifest greater genetic variability for the desirable traits. However, as transgressive segregants for high yield are less frequent, perhaps due to reduced heterosis or because the selected palms are also the important parents for further breeding, few ortets have been selected from such materials. Ortet selection should not be based on oil yield alone but should be combined with other more heritable traits such as oil to bunch, fruit weight, oil quality, disease resistance, and palm height. A selection index approach combining family information and relative economic weights in multiple trait selection, for example, oil yield and kernel oil yield, would be useful (Soh and Chow 1989, 1993; Soh et al. 1994).

Cloning from less advanced populations would also be advantageous in terms of early commercial exploitation of new genetic material and for broadening the genetic base of commercial field plantings to reduce the risk of genetic vulnerability to pest and disease epiphytotics and environmental stresses because of mono-genotype culture. This would be particularly true with wide crosses and introgression programs, for example, new germplasm × BPRO (breeding parents of restricted origin), *E. oleifera* × *E. guineensis* (*OG*) programs (Escobar and Alvarado 2003), where superior segregants in the intercross or backcross populations may be expeditiously exploited as commercial materials. Presumably, the superiority of these segregants arose from linkage disequilibrium, chance recombination, epistasis, or heterosis which may be difficult to recapture by a conventional breeding approach.

The above arguments would also apply to hybrid seed production from clonal parents, which are alternative cloning strategies, to exploit both general and specific combining abilities in large scale hybrid production. A more mundane (but important) use would be to expand the seed parent base of the top crosses, to overcome limited seed production from the best seed parents, and avoid inbreeding effects in the next generation of selfed and intercrossed parents.

Production of semiclonal and biclonal seeds is an established technology with a number of seed producers offering such commercial seeds, for example, AAR (Wong et al. 2010), SUMBIO (Prasetyo et al. 2011), and UP (Musa et al. 2011).

Owing to the inefficiency of cloning, the availability of sufficient good ortets from breeding trials as a starting candidate pool for cloning has always been a limitation in expanding production in most oil palm tissue culture labs (OPTCLs). Ortet selection from commercial fields has been shown not to be a worthwhile exercise but would be the only option for OPTCLs without breeding trials, other than outsourcing from those plantations who have sufficient breeding trials. The use of embryos, seedlings, or field plantings of reproduced best crosses or crosses from the best combining parents to serve as "ortet gardens" has been suggested to circumvent this issue (Soh et al. 2003c). However, as with recloning, the resultant clones would be one generation behind the concurrent new hybrid in terms of genetic advancement. Hence, this approach has not taken off. A modified version of this approach which aims to ensure that cloning improvement runs ahead of breeding improvement has been suggested and is being investigated (Wong Choo Kien, Durand du Gasselin personal communication). The approach involves cloning a sample of the seed embryos/seedlings of all D × P crosses undergoing progeny testing. The seedling-derived clones and the corresponding D × P families are field tested at the same time. Only the best clones which exceed the best family in the clone cum progeny test are recloned to produce commercial clones which are better than the best reproduced commercial hybrid seeds. This approach may be used for both the advanced and less advanced hybrid populations. Another approach is to clone the nursery progeny seedlings of the best cross. The recovered nursery seedlings are planted in ortet gardens. Once the field test results of the resultant clones are known, the clones can be produced from the original ortet garden palms (SUMBIO Annual Report 2010). As molecular genetics becomes more advanced, there is the potential to use marker-assisted selection on seedling populations to identify those close to a breeding objective (e.g., during trait introgression into elite material) or the predicted highest yielding palms (based on either oil yield itself or the components in a selection index) and to clone the most promising genotypes, according to the predicted performance, allowing proliferation of those genotypes through cloning while awaiting confirmation from the field. There are a number of ways that marker-assisted selection (see Chapter 9) can be coupled with clonal multiplication and these are being actively investigated.

Finally, adaptability trial tests are an integral part of a breeding and cultivar development program. The issues of commercial field plantings of clones in terms of their composition, planting configuration, pollination efficiency, pest and disease control, and the agro-management practices to fully exploit the advantage of clones and the field trials needed have been highlighted earlier (Soh et al. 2003a, 2006a,b) and developing clone-specific conditions (whether soil, planting spacing, integrated pest management, etc.) has the potential to recover some of the yield gap that is evident between breeders' trials and farmers' fields.

In general, however, there is a paucity of such research that has been undertaken or described in the literature.

8.3 REVOLUTIONIZING OPTC RESEARCH

Ong-Abdullah Meilina, Siew Eng Ooi, Ahmad Tarmizi Hashim, Zamzuri Ishak, Samsul Kamal Rosli, Wei Chee Wong, Chin Nee Choo, Sau Yee Kok, Nuraziyan Azimi, and Norashikin Sarpan

8.3.1 Background

To date a little over 2% of the total oil palm planted area in Malaysia is occupied by OPTC-derived materials. Despite claims that clonal oil palm is able to boost yields by 20%–30% the technology, although having been widely embraced by the industry, has yet to make any significant impact. The common reasons cited are clonal plantlets or ramets are costlier (prices range from RM25 to RM35) than seed-derived planting materials (currently ranging at RM2.70 to RM12 per seedling depending on the development stage of the plantlet at point of purchase) and the main challenge remains, clonal abnormality.

Nevertheless, the clonal industry has continued to grow albeit at a slower pace than expected, during which time conventional breeding has persisted in producing improved quality planting materials. As pointed out by Soh et al. (2006a,b), conventional hybrid breeding requiring inbred parents would take at least three generations (or more than 20 years) to achieve superior yields of individual hybrid palms. Thus, to fast track the introduction of high yielding materials into the field mass propagation via tissue culture seemed the right choice.

Oil palm clonal propagation has become an established technology and business enterprise despite having some important limitations still to be resolved. The OPTC clonal propagation process for both the gel and liquid suspension pathways has been summarized in Figure 8.1.

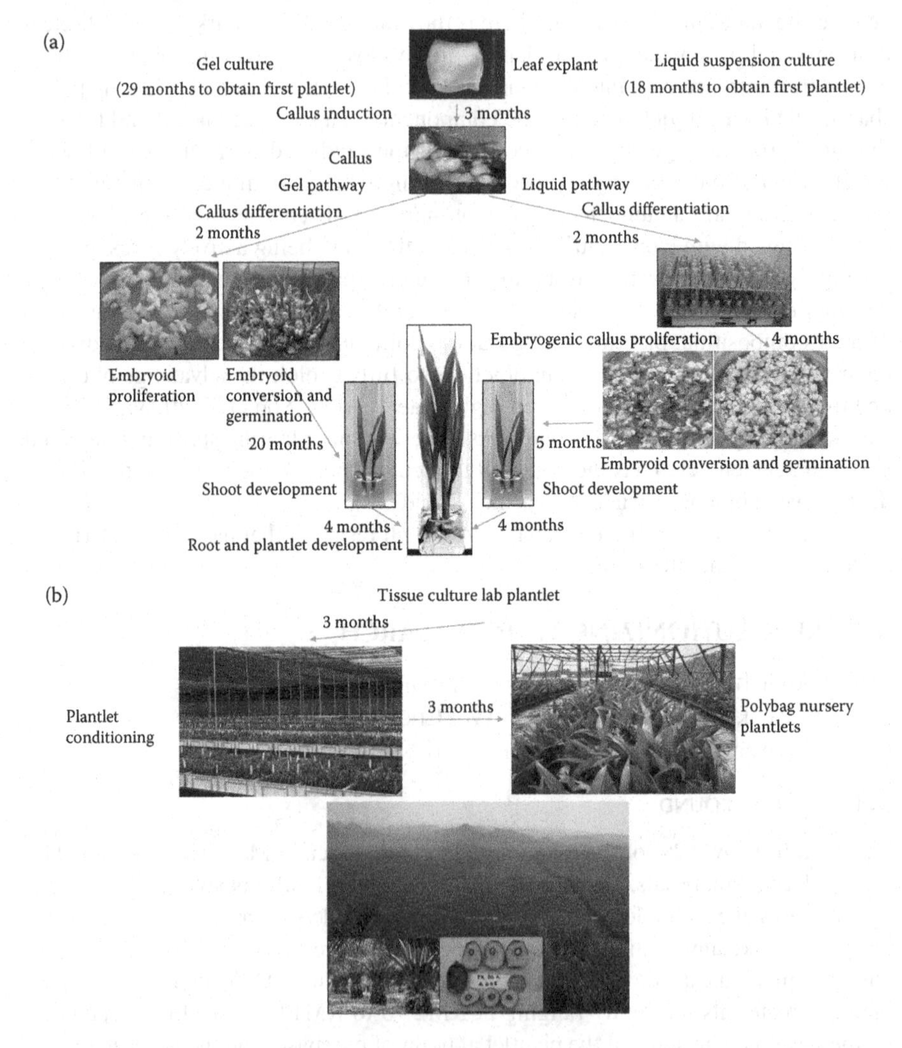

FIGURE 8.1 **(See color insert.)** (a) Culture stages and durations to generate first batch of lab plantlets using gel and liquid suspension pathways. (b) Plantlet conditioning, nursery planting, and commercial clonal plantings. (Adapted from Wong, G., Tan, C.C. and Soh, A.C. 1997. *Acta Horticulturae* 447: 649–658; Wong, G. et al. 1999. Liquid suspension culture—A potential technique for mass production of oil palm clones. In: *Proc. 1999 International Palm Oil Conference on Emerging Technologies and Opportunities in the Next Millennium.* PORIM, Bangi, 3–11; Soh A.C. et al. 2011. *Journal of Oil Palm Research* 23: 935–952.)

In Chapter 11, the practical aspects of commercializing the process at AAR are discussed including the protocols developed and results obtained to improve cloning efficiency without the attendant increase in risks of somaclonal variation and the relative merits and demerits of cloning versus recloning and gel versus liquid suspension process.

Over the years, OPTC processes have undergone improvements and innovative ideas that are translated into patent-worthy technologies have been developed. Cutting-edge molecular tools have further enhanced the feasibility of clonally propagating high yielding materials for a more sustainable industry. In the next section, this approach is discussed further.

8.3.2 Understanding the Embryogenesis Process

Research undertaken by AAR, during the 1970s and 1980s, identified that improvements to the OPTC process could be delivered by manipulating the phytohormones in the growth medium to improve clonability (Jones 1974; Ahée et al. 1981; Pannetier et al. 1981). Schwendiman et al. (1988) further enhanced the results by histologically describing the SE process detailing the formation of callus from cut tissues (explants) and their subsequent transformation into somatic embryos with shoot and root apices. Even then it had already been shown by Turnham and Northcote (1982) that acetyl-CoA carboxylase as a biochemical indicator was able to predict the degree of embryogenesis in oil palm cultures. The turn of the following decade, saw the advent of molecular biology, which aided further the understanding of SE by adding information on the molecular triggers which switch somatic cells into an embryogenic state (Dudits et al. 1995).

8.3.2.1 Molecular Aspects of SE

The reason for critically addressing the embryogenesis issue in the oil palm is because despite the fact that most of the oil palms selected for cloning and recloning have 72%–88% regeneration success rates, their callogenesis and embryogenesis rates are much lower being, 14%–19% and 3%–7%, respectively (Kushairi et al. 2010). This has led to an intensified investigation to understand the SE process at the molecular level with the hope of either developing a trigger mechanism to induce the process or a predictive test to gauge the clonability of the selected palms.

In this aspect, molecular indicators or biomarkers, highly predictive of biological processes, would be an invaluable tool to develop. With the recent surge in applications of high-throughput "omics" technologies especially in agriculture, plant research is witnessing a trend toward using big data to focus on important and meaningful biomarkers at various levels of regulation (Deusch et al. 2015; Kumar et al. 2015).

Amongst some of these biomarkers particularly related to the oil palm, Ong-Abdullah and Ooi (2007) demonstrated the potential of a number of these markers to provide information on the regulation of parts of the OPTC process. This earlier work initiated more intensified efforts toward developing embryogenic specific markers. Some of the leads followed were based on knowledge of the effect of hormones on cultures (Roowi et al. 2010; Ooi et al. 2012, 2013). However, hormone regulation (whether internally or externally controlled) is not the sole pathway involved in SE.

It has been argued that, stress factors, namely osmotic shock, water stress, changes in the pH of the culture medium, etc., are equally inductive in promoting cells to dedifferentiate and then differentiate into the desired somatic embryoids (Zavattieri et al. 2010).

Based on what we now know, induction of SE is not a simple one-gene trigger, therefore, subsequent studies have tended to employ a more global approach. Before sequencing became common place, building massive libraries of extended sequence tags (ESTs) from various tissues and screening these tags was a means to gain a holistic view of the molecular state of the organism at the point of analysis based on gene expression, thus allowing a better understanding of the process at hand (Low et al. 2008). These ESTs were also instrumental in the development of customized microarrays for the oil palm, which are essentially immobilized ESTs on a glass slide, that further facilitated the discovery of more biomarkers related to embryogenesis (Low et al. 2006; Chan et al. 2010; Ong-Abdullah et al. 2012).

Potential biomarkers such as *EgPER1*, *EgHOX1*, *EgPK1*, and *OPSC10* were found to target processes in early SE. Functional studies were carried out to elucidate their involvement in the process and the data are currently being analyzed (Ong-Abdullah and Ooi 2007). The tapetum development1-like gene *Eg707*, is a strong candidate as an embryogenic marker as it was found to be localized in embryos and embryogenic tissues. However, using rice as a model system, it seemed to affect both plant architecture and fertility (Thuc et al. 2011). *EgSERK* (an SE receptor kinase; in press), *EgHAD* (a hydrolase; unpublished), and *EgSAPK* (a serine threonine protein kinase; unpublished) are some of the other genes that have also made it onto the list of potential molecular markers for SE.

8.3.2.2 Influence of Hormones on SE

As mentioned earlier, hormone regulation is an important aspect of the tissue culture process. A combination of 2,4-D with napthaleneacetic acid (NAA) was found to be effective for promoting the growth of embryogenic suspensions in liquid culture (Tarmizi 2002). This led to a more concerted effort to investigate the response of selected plant hormone-responsive genes in explant cultures toward exogenously applied auxins, namely NAA and 2,4-D, using real-time quantitative polymerase chain reaction (PCR) (Ooi et al. 2009). Putative members of the Aux/IAA gene family, *EgIAA9* and *EgIAA13*, were found to be potentially viable as indicators for ortet response toward auxin and tissue culture amenity. In addition, Chan et al. (2010) and Low et al. (2006) reported more genes related to the embryogenesis process as well as some highly expressed in non-embryogenic conditions, respectively.

Due to our lack of a mechanistic understanding of the process as well as the complexity of the biological system, attempts have been made to try to link the theoretical and experimental components to establish a predictive tool for the biological process at hand. Ooi et al. (2013) established a predicted R^2 value of 68% for callogenesis potential in newly established tissue culture lines. This model was calculated based on the changes in expression of genes involved in hormone signaling or metabolism such as a putative *brassinosteroid leucine-rich repeat (LRR) receptor kinase (EgBrRK)*, a putative *cytokinin dehydrogenase (EgCKX)*, and a putative *response regulator type A gene (EgRR1)*. Prior to this, another model was established to predict embryogenesis potential on the basis of the expression of selected genes, in this case expression of the putative Aux/IAA gene, *EgIAA9,* was correlated with SE. The regression model for embryogenesis indicated a predictive R^2 value of 64%. Although both predictive values are not very tight, they are still viable as a simple tool for predicting callogenesis and embryogenesis, respectively.

8.3.3 Refining the *In Vitro* Protocols

The general OPTC protocol is discussed in Chapter 11, and this section focuses on some of the improvements that have been introduced to expedite and facilitate the process.

Generally, OPTC is costly and labor intensive, requiring specific equipment, laboratory space, and skilled workers. Other related activities such as media preparation; culturing of explants; data recording; washing of seedlings; and environment and sanitary control also require a large number of workers. High turnover of staff and the continuous training of new workers all add to the high cost of production and time. Therefore, increasing the output of ramets would render the tissue culture process more cost effective.

Changes at any of the cloning stages that reduce the contamination rate, increase work output, shorten the time in culture, and overcome "bottlenecks" in the process would be useful. These changes can range from improvements or modifications of the culture medium to modifications to the cloning protocol, or even incorporation of new or modified equipment that can deliver better results, through to the screening of palms with desired biomarkers for a more efficient selection of tissue culture-derived palms.

8.3.3.1 Manipulating the Explants

Conventionally, oil palm leaf explants are cut into 1 cm^2 leaf segments before placing them on nutrient media. The callus develops along the cut edges, at the vascular regions. By slicing the 1 cm^2 explants into four or five smaller strips with widths ranging from 0.10 to 0.25 cm, more vascular meristems are exposed to direct contact with the media. Studies have shown that by adopting this strategy, the callusing rate increased significantly as shown in Table 8.1 (Rohani et al. 2003) with callus forming along the cut edges as in Figure 8.2. It is important to create as much callus as possible since the formation of primary embryoids depends on it. Callus bulk also plays an important role in transformation studies, which require the friable embryogenic callus type as starting material.

Conventionally, shoots were rooted individually in glass tubes. This was time consuming and laborious. Zamzuri (1998, 2001a) showed that by overlaying the solid shoot development medium with liquid rooting medium (Figure 8.3), the work output could be increased by 18-fold. This method is now routinely used for rooting shoots in the laboratory.

8.3.4 Liquid Culture System

The gelled culture system is regarded as the "first generation" technique, which is characterized by high manual labor input and low degree of automation. For mass

TABLE 8.1
Callusing Rates in 1 cm² Leaf Segments versus 1 cm² Segments That Were Cut into 4–5 Strips

		Percent Callusing (Number of Sterile Explants per Treatment)	
Expt.	**Ortet**	**Treatment A 1 cm² Wide Leaf Segment**	**Treatment B 1 cm² Leaf Segment Cut into Strips[a]**
X11-0	0.179/342	0 (90)	85.7 (84)
X18-0[b]	0.189/3446	0 (68)	73.9 (69)
X20-0	0.189/3263	5.3 (95)	19.2 (78)
X22-0	0.189/3337	0 (57)	64.2 (53)

[a] For treatment B, any one or all the strips from a 1 cm segment, which produced callus, was considered as a callused explant.

[b] All experiments used leaf explants from the top region from base of fronds number −4, −5, −6, and −7 except for Experiment X18-0, in which the top region of fronds number −6, −7, −8, and −9 was used. The initial number of explants was 96 per treatment for all the experiments.

propagation, the liquid culture system is one option to address the high labor input. At MPOB, the liquid culture system was developed to overcome the low proliferation rate of friable embryogenic calli. The system basically manipulates shaking or stirring activity so that the cultures are exposed more efficiently to oxygen and nutrients as compared to the gelled system with only partial exposure. Liquid suspension culture technology offers reproducibility as well as the amenability for scaling up

FIGURE 8.2 **(See color insert.)** A 6-month-old plate of leaf explant strips with increased cut regions showing callus development.

FIGURE 8.3 **(See color insert.)** A double-layer rooting culture.

the production of oil palm cloning. The liquid system is also of a shorter duration than the gelled system (see Figure 8.1). Meanwhile, MPOB has employed the bioreactor technology with semi or full automation of the process for mass oil palm clonal production (Tarmizi et al. 2003a,b). This and the further innovations needed and achieved leading to the prospective production of artificial seeds are illustrated in Appendix 8A.1.

8.3.5 Somaclonal Variation Associated with Micro-Propagation

Micro-propagation allows the rapid multiplication of the source planting material and in theory these exact copies of the original have the advantage of being true-to-type. Following this line of research and its success in establishing a suitable cloning procedure, the oil palm industry has invested in the large-scale production of clonal planting materials since the early 1980s (Jones 1974; Rabechault and Martin 1976). As we now know, true-to-typeness does not quite hold in this case. The process of changing from a differentiated state (explant) to a dedifferentiated state (callus) and back again can cause incomplete resetting of the DNA of the somatic cells. These changes can cause chemical tags to be added (genes get turned off) or removed (genes get turned on) to the DNA. This is known as an epigenetic mechanism as opposed to a genetic one, as no change in the primary sequence of the DNA is detected. This mechanism will be elaborated further in the subsequent sections.

The scale up of ramet production in the early 1980s resulted in an increased frequency of abnormal clones following field transplantation. The observed abnormalities were androgynous inflorescences, mantled and parthenocarpic fruits. These abnormalities frequently result in bunch rotting with subsequently no harvestable fruit bunches. The frequency of abnormalities among clones increased from 30% in 1981 to 90% in 1983 (Meins et al. 1991). In 1986, information about abnormal clones was made public which resulted in diminished interest in clonal propagation of oil palm and commercial ramet production was halted (Corley et al. 1986). From then on, there has been an international race to decipher oil palm clonal abnormality.

8.3.5.1 Understanding Flower Development

The most commonly observed abnormality in the plantations is the production of mantled fruit (Corley et al. 1986). As the mantled fruit form affects oil yield, interests and efforts are particularly focused on the female flower. Being monoecious, oil palm produces both male and female inflorescences on the same tree. The female inflorescence is a compound rachis comprising about 150 rachillae (Adam et al. 2005). There are 5–30 floral triads on the rachilla of a female inflorescence. Five key developmental stages of floral development have been determined (Adam et al. 2005). The first stage is defined by the presence of the developing floral meristem, while the second stage involves the developing floral triad which consists of the floral meristem flanked by two perianth organs giving rise, at a later stage, to accompanying staminate flowers. Subsequently, the third stage in floral development is where the perianth organs and initiation of the reproductive organs is observed. The staminodium can be seen developing next to the gynoecium. The accompanying staminate flowers are usually no longer seen next to the pistillate flower as their peduncles have elongated and they have become located above the pistillate flower. Floral anatomy at stages 1–3 is closely similar between the normal and mantled inflorescences (Adam et al. 2005; Sarpan et al. 2015). It is only from stage 4 that the elongating carpel in normal inflorescences is flanked by developmentally halted staminodes whereas the formation of carpel-like structures in place of the staminodes occurs in mantled inflorescences. Finally, the pistillate flower will contain a fully formed ovule at the fifth stage of development (Figure 8.4).

Floral organ identity is tightly regulated by a combination of organ-identity genes mainly from the MADS-box gene family (Coen and Meyerowitz 1991; Dornelas et al. 2011). Analysis of floral organ mutations led to the formulation of the classic ABC model for specification of floral organ identity (Bowman et al. 1991; Coen and Meyerowitz 1991; Weigel and Meyerowitz 1994), which was later expanded to the ABCE or ABCDE model (Rijpkema et al. 2010; Murai 2013; Wellmer et al. 2014).

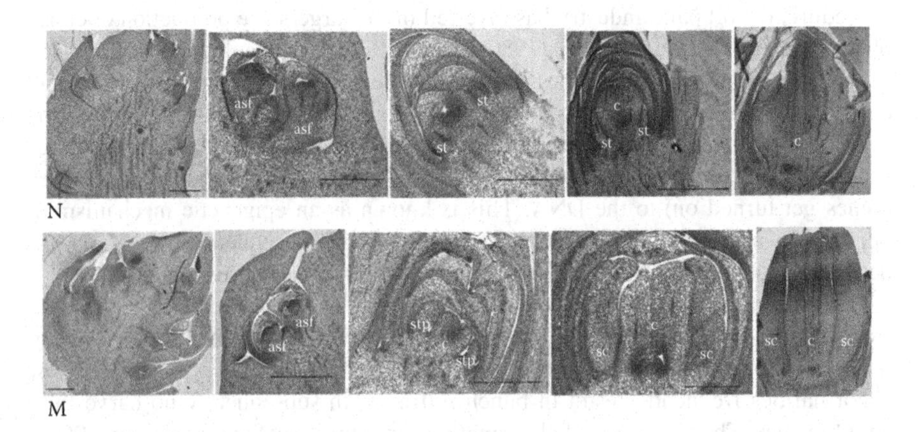

FIGURE 8.4 **(See color insert.)** The five key stages of oil palm flower development in Normal (N) and Mantled (M) female inflorescences. (From Sarpan, N. et al. 2015. *Journal of Oil Palm Research* 27: 315–325.) Bar = 400 μm. asf: accompanying staminate flower, st: staminodia, c: carpel, stp: staminodiaprimordia, sc: supernumerary carpel.

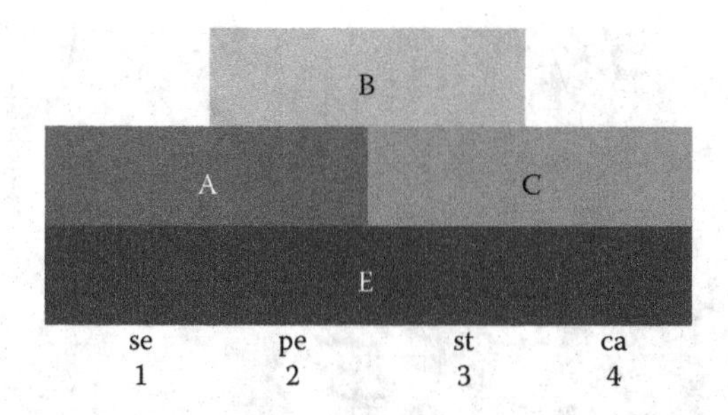

FIGURE 8.5 Schematic diagram of the ABCE model of floral development. se: sepal (1st whorl), pe: petal (2nd whorl), st: stamen (3rd whorl), ca: carpel (4th whorl).

The ABCE model (Figure 8.5) specifies the four whorls of developing floral organs, while the ABCDE model includes specification of the ovule as well. The A function genes specify sepal identity from the first whorl. A- and B-type genes specify petal identity from the second whorl. B- and C-activity specifies stamens at the third whorl, while C-type gene activities specify carpel identity in the fourth whorl (reviewed by Irish 2010). All A-, B-, and C-type genes work in concert with the E-type genes to specify floral organ identities. In Arabidopsis, class A–E genes encode MADS-box transcription factors except for class A gene APETALA2 (Riechmann and Meyerowitz 1998).

The mantled abnormality in oil palm is hypothesized to be similar to the B-type mutant phenotypes in other plants, whereby organ differentiation at the third whorl leads to carpel-like structures instead of stamens (Nagasawa et al. 2003; Yao et al. 2008). The molecular programs that specify stamen and carpel identities seemed to be well conserved across angiosperms (Litt and Kramer 2010). In Arabidopsis, the B-type genes are the *AP3* and *PI* genes (Jack et al. 1992; Goto and Meyerowitz 1994). The oil palm *PISTILLATA* (*PI*) orthologue, *EgMADS16* (or *FEG1*) was able to complement the Arabidopsis *pi* mutant, converting the carpelloid structures back into stamenoid structures in the third whorl (Syed Alwee et al. 2006). This also shows that the oil palm B-type gene is able to function in a dicot system, even though their flower morphologies are clearly different.

The expression of the B-type genes is also dependent on the activity of other genes such as *AG*, *AP1*, *LFY*, and *UFO* (reviewed by Airoldi 2010; Irish 2010). UFO binds LFY on the AP3 regulatory region and promotes transcriptional activity of LFY (Chae et al. 2008). After AP1 activation, AP3 and PI interact to repress AP1 in a feedback loop (Sundstrom et al. 2006). In addition, AP3 and PI proteins maintain their own expression, possibly with SEP3 in a positive feedback loop (Castillejo et al. 2005; Gómez-Mena et al. 2005). This B-function mechanism is conserved in other plants such as Antirrhinum and Petunia (Halfter et al. 1994; Zachgo et al. 1995; Vandenbussche et al. 2004).

8.3.5.2 Epigenome: Moving above and beyond the Genome

Tissue culture propagation can induce genetic and epigenetic changes in regenerated plants. In oil palm ramets, these changes affect the morphology of the flower, where formation of supernumerary carpels lead to loss of flower bunches (Figure 8.6).

FIGURE 8.6 **(See color insert.)** Some clones, due to an altered epigenetic mark, bear mantled bunches. These bunches consist of fruitlets that could be fully mantled with supernumerary carpels (inset) or mixed with normal fruits on the same bunch depending on the severity of the abnormality.

This blocks the wide-scale application of tissue culture propagation of superior trees. The high frequency of the abnormality, when present, the graded phenotypes and high reversal rate suggest that this is due to an epigenetic change. Epigenetic research uses a wide range of molecular biologic techniques to further our understanding of the epigenetic phenomena, including chromatin immunoprecipitation (together with its large-scale variants ChIP-on-chip and ChIP-seq), fluorescent *in situ* hybridization, methylation-sensitive restriction enzymes, DNA adenine methyltransferase identification (DamID), and bisulfite sequencing.

8.3.5.2.1 DNA Methylation

Previous studies have found an overall decrease in DNA methylation in mantled palms relative to ortets and normal ramets (Jaligot et al. 2000, 2002, 2004; Matthes et al. 2001). These results are similar to observations in Arabidopsis and other plant cell cultures, in which transposable elements (TEs) become hypomethylated and expressed (Castilho et al. 2000; Kubis et al. 2003; Tanurdzic et al. 2008; Miguel and Marum 2011). In addition to TEs, somaclonal regenerants in rice and maize undergo extensive gene hypomethylation (Stroud et al. 2013; Stelpflug et al. 2014), which might also contribute to somaclonal variation in oil palm and other crops. The homeotic transformations observed in mantled palms resemble defects in B-function MADS-box genes, suggesting that retroelements within one or more MADS-box genes, or the MADS-box genes themselves are candidates for epigenetic modification (Adam et al. 2005). However, decades of research into DNA methylation changes in candidate retroelements (Castilho et al. 2000; Kubis et al. 2003; Jaligot et al. 2014) and candidate homeotic genes (Syed Alwee et al. 2006; Adam et al. 2007;

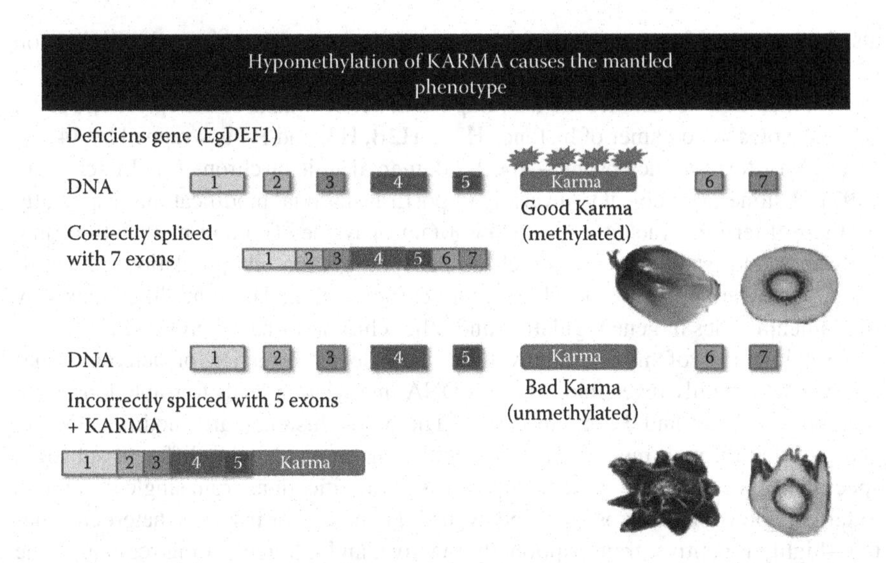

FIGURE 8.7 A mechanism for mantling. (Adapted from Ong-Abdullah, M. et al. 2015. *Nature* 525: 533–537.)

Jaligot et al. 2014) were not successful in uncovering epigenetic changes that were consistently found in somaclonal mantled palms. Furthermore, recent studies of rice and Arabidopsis plants regenerated from tissue culture have implicated genetic rather than epigenetic mechanisms as being responsible for somaclonal variation (Jiang et al. 2011; Miyao et al. 2012).

In 2015, Ong-Abdullah et al. finally reported the successful identification of the *Karma* retrotransposon and its involvement in defining the mantled phenotype. When this epigenetic mark is methylated, also known as *Good Karma*, normal fruits are produced whilst *Bad Karma* denotes a situation where methylation is lost resulting in mantled fruits. The mechanism is depicted in Figure 8.7. These findings were made possible through the combination of a high-throughput and high-resolution technology with a robust genome and powerful statistics. A detailed recording and upkeep of field data of the clonal materials used in the study were also instrumental in the discovery. This breakthrough not only bears scientific importance but also has a major economic consequence as the development of a biomarker for the mantled phenotype would enable the industry to reliably market high yielding clonal materials as alternative planting materials. The confidence in planting clonal palms comes from the ability to screen these materials prior to field planting. This will further encourage larger areas to be planted with these high yielders, and thus boost national palm oil production whilst ensuring sustainable agriculture is being practiced.

8.3.5.2.2 Histone Modification

Recent research has shown that DNA methylation and histone modifications have a close relationship to each other. Histone modifications might be involved in establishing the patterns of DNA methylation and vice versa, and DNA methylation may also be important for maintaining histone modification patterns through cell division (Cedar and Bergman 2009).

Modifications of histones are also important determinants of the epigenetic state. In eukaryotes, an octamer of histones H2A, H2B, H3, and H4 is wrapped by 147 bp of DNA to form a nucleosome—the fundamental unit of chromatin (Luger et al. 1997). Histones are subject to a variety of posttranslational modifications, especially on their N-termini, including acetylation (ac) of lysines (K) and methylation (me) of lysines and arginines (R) as well as phosphorylation, ubiquitylation, glycosylation, sumoylation, ADP-ribosylation, and carbonylation. These modifications play fundamental roles in gene regulation and other chromatin-based processes.

Certain forms of histone methylation cause local formation of heterochromatin, which is readily reversible, whereas DNA methylation leads to stable long-term repression (Cedar and Bergman 2009). Therefore, histones are modified locally and globally through multiple histone-modifying enzymes with different substrate specificities, generating hierarchical patterns of modifications from single promoters to large regions of chromosomes and even single cells. For instance, heterochromatin—highly repetitive, transcriptionally inactive, and late-replicating regions of the genome—is generally deacetylated but could be enriched for H3K9me2/3 (Horn and Peterson 2006). This is in contrast to euchromatin—gene rich and transcriptionally active regions of the genome—which is associated with H3K4me2 and increased acetylation (Strahl et al. 1999). Therefore, genomic distribution of specific histone modifications may reflect the underlying organization of the genome. In addition, transition between the heterochromatin and euchromatin state is gradual and may require multiple cell division cycles (Katan-Khaykovich and Struhl 2005).

It has been hypothesized that during cell-organism differentiation and maturation, genomic DNA becomes more methylated and histones suffer specific posttranslational modifications, resulting in different local states of the chromatin structure (Valledor et al. 2010). Restrictive heterochromatin is associated with hypermethylation of the DNA with histone modifications such as $H3K9^{Me}$, with presence of low percentages of acetylated histone H4 (AcH4), while open euchromatin is associated with hypomethylation and increased presence of AcH4 and methylation of histone H3 at lysine 4 ($H3K4^{Me}$) (Tariq and Paszkwoski 2004). Immature needles of pine showed a global DNA methylation level of 15.7% which rises to 17.6% after 12 months of development to maturity (Valledor et al. 2010). In Arabidopsis, a chemically induced 3% decrease in DNA methylation in cell suspensions led to changes in expression of 1794 sequences (Berdasco et al. 2008). It has been suggested that H3K4 methylation primes certain genes for an increase in H3K9 and H4K16 acetylation (Wang et al. 2009). However, this did not lead to an increase in transcription of these genes. In addition, H3K4me2 has been shown to be stable for several generations in *Caenorhabditis elegans* (Katz et al. 2009).

8.3.6 Potential of Organelle (Chloroplast and Mitochondria) Research in Tissue Culture

Mitochondria and chloroplast are the major sites of energy conversion in the plant cell. Mitochondria are semiautonomous (acting independently, to some degree) organelles, originated from a single endosymbiotic event involving an α-proteobacterial (Gray et al. 2001; Martin and Mentel 2010; Gray 2012) progenitor, whilst chloroplasts are photosynthetic intracellular organelles, derived from cyanobacteria that have undergone at least two independent endosymbiotic events (Martin et al. 2002; Wicke et al. 2011). Mitochondria and chloroplasts have their own genome, which is distinct from the nucleus. Both of these genomes contain sources of genetic diversity. Although the organellar genomes code for only a limited number of proteins, their genes are essential for energy metabolism and photosynthesis (Gunning and Steer 1996).

In many higher plants, structural changes in the nuclear and mitochondrial genomes have shown to be induced by cell culture *in vitro* (Cloutier and Landry 1994; Sadoch et al. 2000) and the stability of the mitochondrial genome is controlled by nuclear loci (Abdelnoor et al. 2003). Research has been conducted in coffee to access the genetic integrity of the nuclear, mitochondrial, and chloroplast genomes in somatic embryo-derived plants. Multiple DNA markers were tested and it was discovered that both nuclear and mitochondrial genomes changed in different but characteristic ways that produced genetic variability and novel genome organization in somatic embryo-derived plants (Rani et al. 2000).

Plant mitochondrial genomes are characterized by an unusually high level of recombination that results in a great structural complexity of the genome (Tada and Souza 2006; Shedge et al. 2010). Mitochondrial genome recombination involves large- and intermediate-sized repeats. The later recombination process produces asymmetric DNA exchange (Shedge et al. 2007), which mediates rapid stoichiometric changes in the genome (Arrieta-Montiel et al. 2009). Studies conducted by Sun et al. 2012 on tobacco and Arabidopsis showed that mitochondrial DNA genome rearrangement was enhanced under tissue culture conditions, where asymmetric DNA exchange occurs at particular repeated sequence within the genome. It was postulated that, reduction in the expression of nuclear genes of MSH1 and RECA3 caused such recombination events to occur. Nevertheless, these changes are reversible in tobacco upon plant regeneration from callus (Kanazawa et al. 1994). Therefore, interactions between the nuclear gene, MSH1 and mitochondria can be further explored with regard to abnormalities seen in OPTC.

On the other hand, the chloroplast genome appears to be more stable during somatic tissue culture (Henry et al. 1998). However, prolonged tissue culture may result in the partial deletion of the chloroplast genome, which can lead to changes in plastid morphology (Kawata et al. 1995). Furthermore, a large deletion of rice chloroplast DNA is often associated with albino plants regenerated from tissue culture (Harada et al. 1991; Shimron-Abarbanell and Breiman 1991).

8.3.7 Post-Genomics (and a Systems Biology Approach) for *In Vitro* Cultures

Over decades of efforts in the OPTC technique, cloning efficiency has been improved through selection of high-quality ortets, reduced number of subcultures, stringent culling procedures, and development of innovative technologies to simplify the process (Kushairi et al. 2010). Despite this, the rate of embryogenesis still remains low and highly genotype dependent. As yet, little is known about the mechanism underlying the induction and development of oil palm during *in vitro* propagation.

The completion of the genome sequencing in Arabidopsis (The Arabidopsis Genome Initiative 2000), rice (International Rice Genome Sequencing Project 2005), soy bean, and oil palm (Singh et al. 2013) has stimulated the application of functional genomics in most fields of plant research. There is a transformation from the genomic era to modern post-genomic times. By knowing the gene-function data, this enables more value to be added to the nucleotide sequence collection, which should lead to the improvement of crop plants (Holtorg et al. 2002). Hence, efforts to describe embryogenesis have been investigated at the molecular level by using "-omics" methodologies, in particular transcriptomic, proteomic, and metabolomic approaches which has led to better understanding of the *in vitro* culture system.

Proteomics analysis is the most direct method to define function of their associated genes (Yin et al. 2007). Recently, proteomic approaches have been employed to study SE in different plant species such as *Oryza sativa* L. (Yin et al. 2007), *Vitis vinifera* (Marsoni et al. 2008), *Medicago truncatula* (Imin et al. 2004; Almeida et al. 2012), *Cyclamen persicum* (Mwangi et al. 2013), *Picea glauca* (Lippert et al. 2005), *Elaeis guinneesis* (Silva et al. 2014), *Citrus sinensis* Osbeck (Pan et al. 2009), *Quercus suber* (Gomez-Garay et al. 2013), and *Acca sellowiana* (Cangahuala-Inocente et al. 2009). Two-dimensional electrophoresis (2-DE) coupled with mass spectroscopy (MS) and iTRAQ are the most widely used approaches for protein classification and identification in different proteome states and environments.

Basically, SE is an adaptive process of plant cells to stress (Fehér et al. 2003). Stress can reprogram somatic cells toward embryogenesis but the relationship between stress and SE is still not well understood (Liu et al. 2015). Proteome studies found that a high abundance of stress-related proteins are present during SE. Heat-shock proteins (HSP) are a very common family found in embryogenic cultures whose family members have been detected at the start of the SE process in *Cyclamen persicum*, alfafa, tobacco, and carrot (Györgey et al. 1991; Zarsky et al. 1995; Kitamya et al. 2000; Winkelmann et al. 2006). In addition, several studies found that pathogenesis-related (PR) proteins, for example, thaumatin, chitinase, osmotin-like proteins, and β-1,3 glucanase were potentially involved in embryogenic competence (Silva et al. 2014; Liu et al. 2015).

Meanwhile, metabolomics is a comprehensive analysis technology in which all the metabolites of an organism or tissue are identified and quantified. A metabolome denotes the ultimate phenotype of cells deduced by the perturbation of gene expression and the modulation of protein function which are triggered by environmental factors or the occurrence of mutations (Saito and Matsuda 2010). Metabolomic approaches have been applied to study spruce SE (Dowlatabadi et al. 2009) and vanilla callus differentiation (Palama et al. 2010).

Compared to genomics and proteomics, metabolomics deals with a range of small molecules with different characteristics. Over several decades, methods with high sensitivity and accuracy have been established for highly complex mixtures of small molecules. These methods included gas chromatography mass spectrometry (GC-MS), liquid chromatography mass spectrometry (LC-MS), capillary electrophoresis mass spectrometry (CE-MS), and Fourier transform ion cyclotron resonance mass spectrometry (FT-ICR-MS). Recently, nuclear magnetic resonance (NMR) spectroscopy is gaining popularity as it offers a wide spectrum of chemical analysis techniques which are nondestructive, reproducible, and offer a nontargeted, quantitative identification of metabolites.

8.4 CONCLUSIONS

The method of *in vitro* propagating the oil palm has generally remained quite traditional since its inception. Conducting tissue culture used to be regarded as an art as it merely involved the mass production of identical copies of the plant of interest. However, as its applications become more sophisticated, so did the understanding of the cloning process. With the advent of technologies, namely molecular biology, genome technology, and more recently post-genome strategies, we now approach tissue culture with renewed interest and one of the best examples is the discovery of *Karma*. The option of being able to now conduct risk assessment on clones has regained industry's confidence in producing and utilizing clones on a large-scale basis as planting materials and the economies of scale would potentially bring down the cost of production thus making clonal oil palms more affordable and competitive. The anticipated increase in yield and lower cost make this improvement a boost to the oil palm industry as a whole.

To recapitulate, it was 40 years ago when the first plant regeneration from OPTC was announced leading confidently to the establishment of commercial ramet production labs with industrial confidentiality considerations and attempted patenting of the unique protocols actively pursued. Ten years down the road the spectre of somaclonal variation forced the closure of the pioneer commercial OPTCLs and their reversion to R&D. A few smaller OPTCLs persisted and eventually around 2000, confidence returned with the experience of much reduced mantling levels. Scale-up production and new OPTCs resulted but production still remains modest today constrained by somaclonal variation and process inefficiency. Scale-up production is achieved through empirical protocol manipulation: cloning a basket of ortets each time, reduced phytohormone treatments, reduced culture multiplication and time, stringent culture selection and good general "housekeeping" in the OPTCL.

The industry has earlier recognized that fundamental research (physiological, molecular) and perhaps the development of biomarkers were needed to resolve the mantling and inefficient embryogenesis issues and urged MPOB's newly formed biotechnology unit to seek collaboration from international centers of excellence in biotechnology and somaclonal variation research *cum* building up its own expertise. This has indeed "paid-off" with the recent revelation of the genomic basis of the shell, *virescens* traits, and the epigenomic basis of mantling.

A biomarker when developed will increase confidence in scaling-up efforts in ramet production. Nevertheless, even with the mantling biomarker developed, to make a significant in-road into the large 450 million oil palm seed market the feasibility of recloning and liquid culture systems must be in place: significant yield advantage of commercial clones over contemporary hybrid progeny palms and reduction in high ramet cost are needed. The low frequencies of embryogenesis in the recloning and liquid culture systems remain an impediment, especially the latter, and are largely genotype dependent, and may require time to sort out even with the latest genomics tools. Also with the prevailing climate change and sustainability concerns and negative sentiments regarding large-scale monoculture, plantings of a few selected clones would not be prudent. That being said cloning allows early commercial exploitation of diverse genotypes and recombinant crosses in new plantings and replantings, potentially contributing to the plantation's agrobiodiversity and sustainability. Other advantages even if the yield increase obtained is not substantial would be improved oil quality, abiotic and biotic stress resistance, adaptability (both broad and to specific environments), and resource use efficiency traits.

The previous subchapter has highlighted the usefulness of clones in other molecular studies. Clones like isogenic, recombinant inbred lines with uniform genetic constitution (genetic replication within clone and genetic segregation between clones) are invaluable in a host of physiological, phytopathological, molecular genetic, genetic, and quantitative genetic studies besides being a useful tool in plant breeding to expedite the production of new cultivars.

APPENDIX 8A.1 OIL PALM LIQUID CULTURE: MPOB PROTOCOL

As depicted in Figure 8.1, MPOB further developed its own basic protocol for the liquid culture system (Tarmizi 2002), loosely based on the shake flask method by Wong et al. (1999). It generally follows the flow shown in Figure 8.1 but subsequently moves into the many inventions developed to enhance the liquid culture system.

Briefly, these innovations are tailored towards increasing the fresh weight of the cultures while simplifying the transfer process. For example, MPOB Fast Transfer Technique (MoFaTT) (Tarmizi and Zaiton 2005) and the Two -in -One MPOB Simple Impeller (2-in-1 MoSLIM) (Tarmizi and Zaiton 2006a) systems are able to minimize the steps required in culture and provide a more convenient alternative as on-site replenishment of medium can be executed. In addition, the combination of these two systems, conveniently called SLIM-FaTT, simplifies a conventional 8-step protocol into a single media replenishment step (Tarmizi and Zaiton 2006b).

The transition from the simple shake flask method to the bioreactor system appears to be conceivably straightforward. The present findings in the bioreactor system show the possibility of large-scale production of embryogenic suspension cultures in a single run. Earlier studies at MPOB have shown fresh weight of cultures from selected clones could multiply up to 4–20-fold after 50–80 days in a conventional B.Braun Bench-top 2-L bioreactor and fivefold after about 60 days in a Biotron bioreactor. Figure 8A.1 shows the distinctive growth increase of cultures in a bioreactor as compared to the shake flask system.

The progress made in the use of this system would pave the way for future semi- or fully automated processes for oil palm clonal production.

Subsequent innovations, such as the MPOB modified vessel (MoVess) and its improvement the MPOB motorized vessel (MPOB-Motovess), were designed to overcome the high cost of commercial bioreactors with a simpler inoculation procedure (Tarmizi et al. 2007). Fresh weight increments of up to 35-fold were obtained for cultures of selected oil palm clones after 30–60 days growth in MoVess.

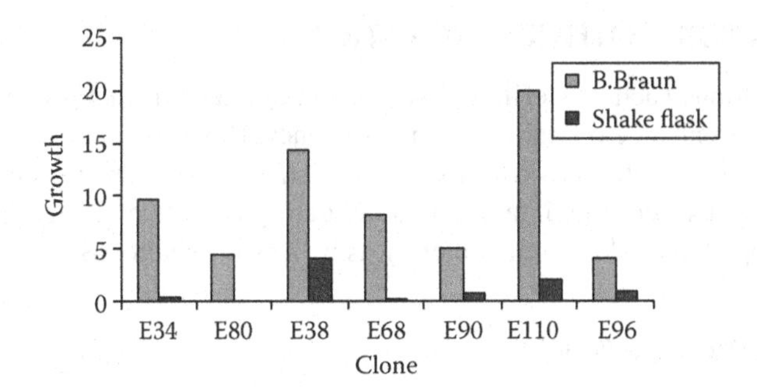

FIGURE 8A.1 Growth comparison of oil palm cultures maintained in a B.Braun bioreactor or shake flasks after 50 days of incubation.

More recently, Tarmizi et al. (2009) designed a 5–10 L capacity motorized vessel equipped with a fast media replenishment attachment known as the Motorized vessel with fast media transfer (Movefast) (see Figure 8A.1). This apparatus is economical, practical, and is now routinely used. It reduces the risk of contamination, it is convenient and can be easily adopted for other fluidic culture systems. A fresh weight increment of about five-fold for cultures of selected oil palm clones was identified after about 40 days in the MoVeFast system (Tarmizi et al. 2012). Figure 8A.2 briefly illustrates the series of innovations.

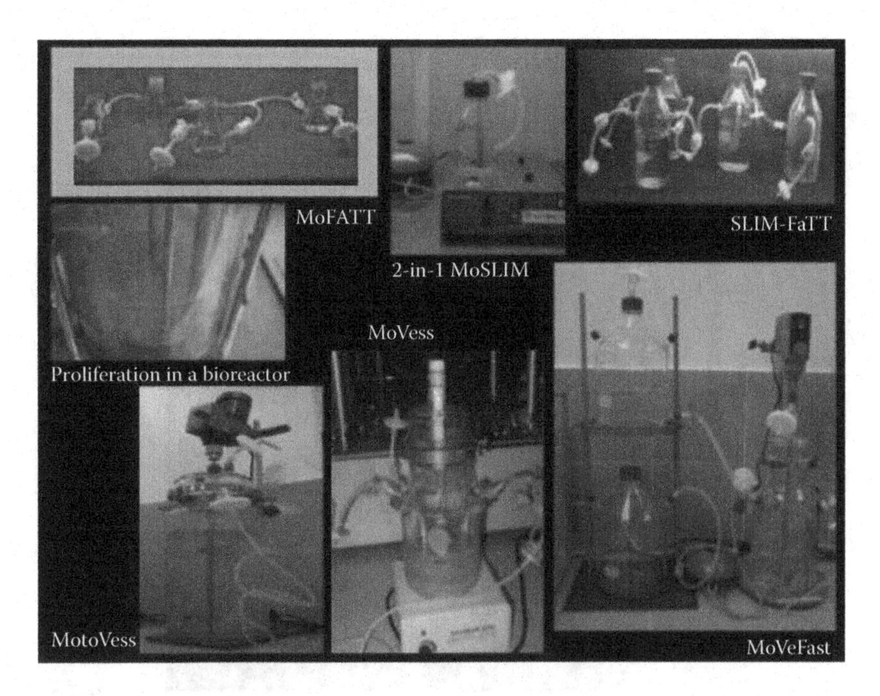

FIGURE 8A.2 (See color insert.) A suite of innovations for Liquid culture propagation systems.

APPENDIX 8B.1 OTHER INNOVATIONS

Other innovations such as sterilizing equipment and a data tracking system would help to further enhance oil palm cloning efficiency. Development of artificial seeds is an approach to enable growth of organized structures in *ex vivo* condition thus saving time and space if maintained in a laboratory condition. Biomarkers are also being developed to predict clonal amenity as well as for quality assurance.

8B.1.1 Flameless Sterilizer

The flameless sterilizer is an electrical device designed for heat sterilization of tissue culture glass containers (for the mouth-rim region) and surgical apparatus (Zamzuri 2002). It is an improvement over flaming that is commonly done using a Bunsen burner or alcohol lamp. The sterilizer (Figure 8B.1) has a circular heating coil in the center whereby the culture vessels can be inserted from below, in an upright position, for heating of their mouthparts. The platform is adjustable to cater for vessels of different heights, such as flasks, jars, and test tubes. In addition, at the top of the device is a chamber that holds the glass beads for sterilization of surgical apparatus (such as forceps and scalpels). The hot coil emanates heat to as high as 300°C. This single heating coil sterilizes the vessel mouthparts below it and the surgical apparatus above, hence maximizing use of the heat generated.

FIGURE 8B.1 Flameless sterilizer.

8B.1.2 OPTRACKS (Oil Palm Tissue Culture Tracking System)

Previously, all the data in OPTC were recorded manually on paper and filed for future reference. The data for each ortet, from ortet selection to field testing, required proper recording for tracking purposes. The volume of data was often substantial and the recordings became increasingly complex. Furthermore, manual recording is prone to human error. Tracking of data was time consuming and tedious and hence, inefficient. MPOB developed a tissue database system using a relational database management software for computerized audit trail (Zamzuri 2001b) and further enhanced with bar coding for monitoring and recording purposes in OPTRACKS (Tarmizi et al. 2003a,b). This system has been licensed to two oil palm agencies.

8B.1.3 Development of Oil Palm Artificial Seeds

Synthetic seed or artificial seed can be defined as the artificial encapsulation of somatic embryos, shoot buds, embryogenic aggregates, or any tissue with the ability to form a plant in *in vitro* or *ex vivo* conditions. At MPOB, attempts were made to encapsulate various tissues of oil palm *in vitro* cultures in sodium alginate for the production of synthetic seeds. Those selected tissues were zygotic embryos, embryogenic aggregates, and shoot apices. All encapsulated tissues were successfully germinated on MS media under sterile conditions (Figures 8B.2, 8B.3, and 8B.4).

Comparatively, the shoots apices were the best tissue for synthetic seed production of oil palm *in vitro* cultures. They could be easily germinated and further developed into single plantlets. Rooting of the germinated synthetic seeds was further improved by using MoFaTT system (Figure 8B.5). Hence, the development of oil palm synthetic seeds offers a convenient and practical means for the long distance delivery of oil palm *in vitro* cultures.

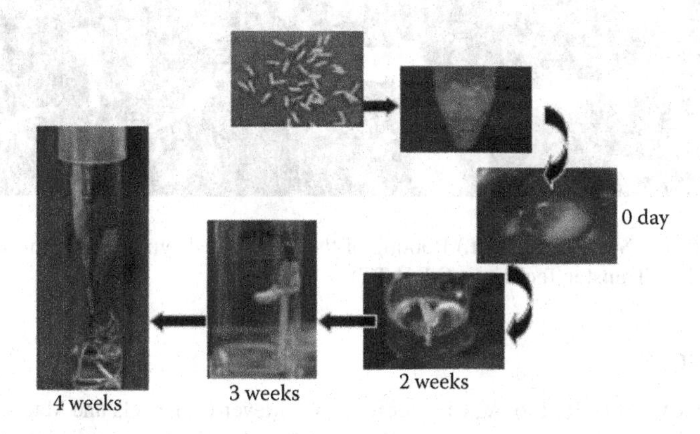

FIGURE 8B.2 **(See color insert.)** Synthetic seed production. As an example, zygotic embryos extracted from seeds were encapsulated in alginate beads and allowed to germinate into a plantlet.

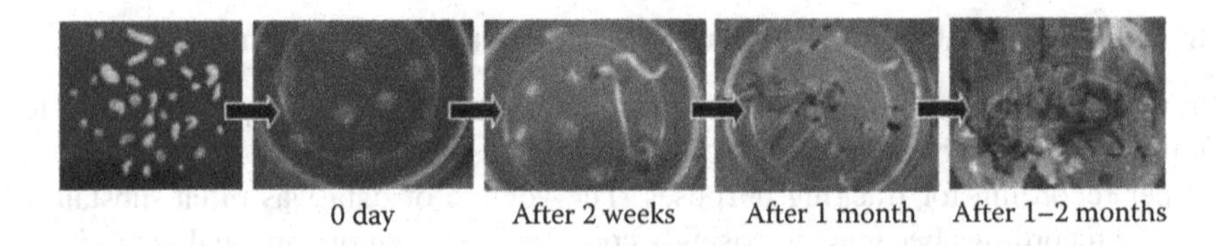

FIGURE 8B.3 **(See color insert.)** Encapsulation can also be done on embryogenic aggregates. This illustrates the progression of germination of the encapsulated aggregates.

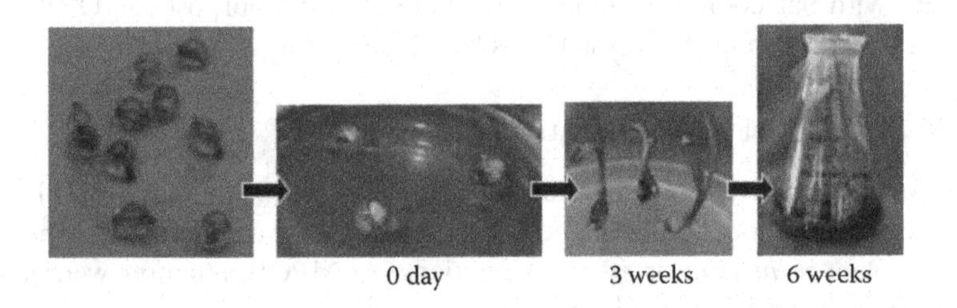

FIGURE 8B.4 **(See color insert.)** Alternatively synthetic seeds can also be established through encapsulation of shoot apices as shown here.

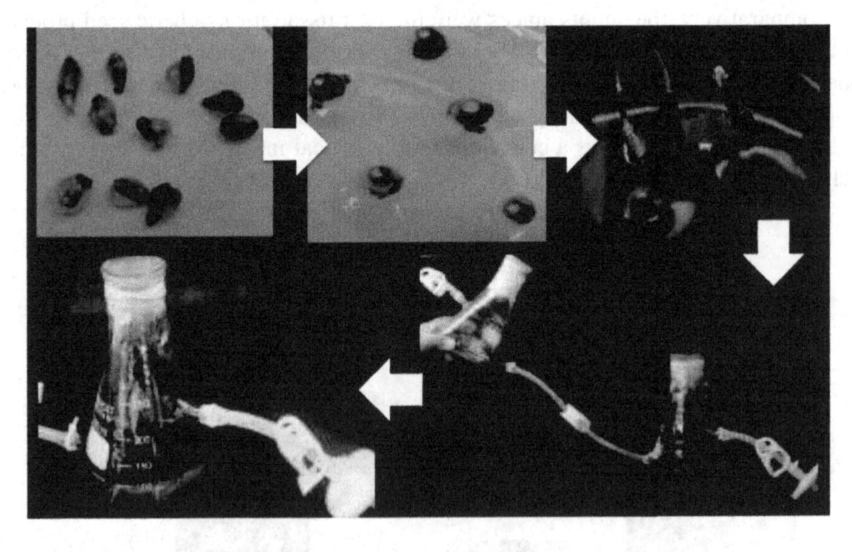

FIGURE 8B.5 **(See color insert.)** Rooting of the germinated synthetic seed is assisted by the MPOB Fast Transfer Technique (MoFaTT).

REFERENCES

Abdelnoor, R.V., Yule, R., Elo, A., Christensen, A.C., Meyer-Gauen, G., and Mackenzie, S.A. 2003. Substoichiometric shifting in the plant mitochondrial genome is influenced by a gene homologous to MutS. In: *Proceedings of the National Academy of Sciences, USA* 100: 5968–5973.

Adam, H., Jouannic, S., Escoute, J., Duval, Y., Verdeil, J.L., and Tregear, J.W. 2005. Reproductive developmental complexity in the African oil palm (*Elaeis guineensis*, Arecaceae). *American Journal of Botany* 92: 1836–1852.

Adam, H., Jouannic, S., Morcillo, F., Verdeil, J.L., Duval, Y., and Tregear, J.W. 2007. Determination of flower structure in *Elaeis guineensis*: Do palms use the same homeotic genes as other species? *Annals of Botany* 100: 1–12.

Ahée, J. et al. 1981. La multiplication vegetative in vitro de Palmier à Huile par Embryogenèse Somatique. *Oléagineux*, 36: 113–118.

Airoldi, C.A. 2010. Determination of sexual organ development. *Sexual Plant Reproduction* 23: 53–62.

Almeida, A.M., Parreira, J.R., Santos, R., Duque, A.S., Franciso, R., Tomé, D.F., Ricardo, C.P., Coelho, A.V., and Fevereiro, P. 2012. A proteomic study of induction somatic embryogenesis in *Medicago truncatula* using 2DE and MALDI-TOF/TOF. *Physiologia Plantarum* 146: 236–249.

Arrieta-Montiel, M.P., Shedge, V., Davila, J., Christensen, A.C., and Mackenzie, S.A. 2009. Diversity of the arabidopsis mitochondrial genome occurs via nuclear-controlled recombination activity. *Genetics* 183: 1261–1268.

Berdasco, M. et al. 2008. Promoter DNA hypermethylation and gene repression in undifferentiated arabidopsis cells. *PLoS ONE* 3: e3306. doi: 10.1371/journal.pone.0003306.

Bowman, J.L., Smyth, D.R., and Meyerowitz, E.M. 1991. Genetic interactions among floral homeotic genes of arabidopsis. *Development* 112: 1–20.

Cangahuala-Inocente, G., Villarino, A., Seixas, D., Dumas-Gaudot, E., Terenzi, H., and Guerra, M. 2009. Differential proteomic analysis of developmental stages of *Acca sellowiana* somatic embryos. *Acta Physiologia Plantarum* 31: 501–514.

Castilho, A., Vershinin, A., and Heslop-Harrison, J.S. 2000. Repetetive DNA and the chromosomes in the genome of oil palm (*Elaeis guineensis*). *Annals of Botany* 85: 837–844.

Castillejo, C., Romera-Branchat, M., and Pelaz, S. 2005. A new role of the arabidopsis SEPALLATA3 gene revealed by its constitutive expression. *Plant Journal* 43: 586–596.

Cedar, H. and Bergman, Y. 2009. Linking DNA methylation and histone modification: Patterns and paradigms. *Nature Reviews Genetics* 10: 295–304.

Chae, E., Tan, Q.K., Hill, T.A., and Irish, V.F. 2008. An arabidopsis F-box protein acts as a transcriptional co-factor to regulate floral development. *Development* 135: 1235–1245.

Chan, P.L., Ma, L.S., Low, E.T.L., Elyana, M.S., Ooi, L.C.L., Cheah, S.C., and Singh, R. 2010. Identification of somatic embryogenesis-related genes from normalized embryoid cDNA library of oil palm. *Electronic Journal of Biotechnology* 13: 14.

Cloutier, S. and Landry, B.S. 1994. Molecular markers applied to plant tissue culture. *In Vitro Cell Developmental Biology* 30: 32–39.

Coen, E.S. and Meyerowitz, E.M. 1991. The war of the whorls: Genetic interactions controlling flower development. *Nature* 353: 31–37.

Corley, R.H.V. 2005. Illegitimacy in oil palm breeding—A review. *Journal of Oil Palm Research* 17: 64–69.

Corley, R.H.V., Lee, C.H., Law, I.H., and Wong, C.Y. 1986. Abnormal flower development in oil palm clones. *Planter* 62: 233–240.

Deusch, S., Tilocca, B., Camarinha-Silva, A., and Seifert, J. 2015. News in livestock research—Use of Omics-technologies to study the microbiota in the gastrointestinal tract of farm animals. *Computational and Structural Biotechnology Journal* 13: 55–63.

Dornelas, M.C., Patreze, C.M., Angenent, G.C., and Immink, R.G.H. 2011. MADS: The missing link between identity and growth? *Trends in Plant Science* 16: 89–97.

Dowlatabadi, R., Weljie, A.M., Thorpe, T.A., Yeung, E.C., and Vogel, H.J. 2009. Metabolic footprinting study of white spruce somatic embryogenesis using NMR spectroscopy. *Plant Physiology and Biochemistry* 47: 343–350.

Dudits, D., Gyorgyey, J., Bogre, L., and Bako, L. 1995. Molecular biology of somatic embryogenesis. In: Thorpe, T.A. (Ed.), *In Vitro Embryogenesis in Plants. Current Plant Science and Biotechnology in Agriculture.* Kluwer Academic Publishers, Dordrecht, pp. 267–308.

Escobar, R. and Alvarado, A. 2003. Strategies in the production of oil palm compact clones and seeds. In: *Proceedings of the MPOB International Palm Oil Congress (PIPOC) 2003.* Malaysian Palm Oil Board, Kuala Lumpur, pp. 75–90.

Falconer, D.S. and Mackay, T.F.C. 1996 *Introduction to Quantitative Genetics.* Longman, Harlow.

Fehér, A., Pasternak, T.P., and Dudits, D. 2003. Transition of somatic plant cells to an embryogenic state. *Plant Cell Tissue and Organ Culture* 74: 201–228.

Gomez-Garay, A., Lopez, J.A., Camafeita, E., Bueno, M.A., and Pintos, B. 2013. Proteomic perspective of *Quercus suber* somatic embryogenesis. *Journal of Proteomics* 93: 314–325.

Gómez-Mena, C., de Folter, S., Costa, M.M.R., Angenent, G.C., and Sablowski, R. 2005. Transcriptional program controlled by the floral homeotic gene AGAMOUS during early organogenesis. *Development* 132: 429–438.

Goto, K. and Meyerowitz, E.M. 1994. Function and regulation of the arabidopsis floral homeotic gene PISTILLATA. *Genes and Development* 8: 1548–1560.

Gray, M.W. 2012. Mitochondrial evolution. *Cold Spring Harbor Perspectives in Biology* 4: a011403.

Gray, M.W., Burger, G., Lang, B.F. 2001. The origin and early evolution of mitochondria. *Genome Biology* 2: 1–5.

Gunning, B.E.S. and Steer, M.W. 1996. *Plant Cell Biology: Structure and Function.* 1st Edition. Jones and Bartlett Publishers, Sudbury, Massachusetts.

Györgey, J. et al. 1991. Alfafa heat shock genes are differentially expressed during somatic embryogenesis. *Plant Molecular Biology* 16: 999–1007.

Halfter, U., Ali, N., Stockhaus, J., Ren, L., and Chua, N.H. 1994. Ectopic expression of a single homeotic gene, the petunia gene GREEN PETAL, is sufficient to convert sepals to petaloid organs. *EMBO Journal* 13:1443–1449.

Hallauer, H.R. and Miranda, J.B. 1981. *Quantitative Genetics in Maize Breeding.* Iowa State Press, Ames, Iowa.

Harada, T., Sato, T., Asaka, D., and Matsukawa, I. 1991. Large-scale deletions of rice plastid DNA in anther culture. *Theoretical and Applied Genetics* 81:157–161.

Hardon, J.J., Corley, R.H.V., and Lee, C.H. 1987. Breeding and selecting the oil palm. In: Abbot, A.J. and Atkin, R.K. (Eds.), *Improving Vegetatively Propagated Crops.* Academic Press, London, pp. 63–81.

Henry, Y., Nato, A., and de Buyser, J. 1998. Genetic fidelity of plants regenerated from somatic embryos of cereals. In: Jain, S.M., Brar, D.S., and Ahloowalia, B.S. (Eds.), *Somaclonal Variation and Induced Mutations in Crop Improvement.* Springer Science+Business Media, Dordrecht, Netherlands, pp. 65–80.

Holtorg, H., Guitton, M.-C.C., and Reski, R. 2002. Plant functional genomics. *Naturwissenschaften* 89: 235–249.

Horn, P.J. and Peterson, C.L. 2006. Heterochromatin assembly: A new twist on an old model. *Chromosome Research* 14: 83–94.

Imin, N., De Jong, F., Mathesius, U., Van Noorden, G., Saeed, N.A., Wang, X.D., Rose, R.J., and Rolfe, B.G. 2004. Proteome reference maps of *Medicago truncatula* embryogenic cell cultures generated from single protoplasts. *Proteomics* 4: 1883–1896.

International Rice Genome Sequencing Project. 2005. The map-based sequence of the rice genome. *Nature* 436: 793–800.

Irish, V.F. 2010. The flowering of Arabidopsis flower development. *Plant Journal* 61: 1014–1028.

Jack, T., Brockman, L.L., and Meyerowitz, E.M. 1992. The homeotic gene APETALA3 of *Arabidopsis thaliana* encodes a MADS box and is expressed in petals and stamens. *Cell* 68: 683–697.

Jaligot, E., Beulé, T., and Rival, A. 2002. Methylation-sensitive RFLPs: Characterisation of two oil palm markers showing somaclonal variation-associated polymorphism. *Theoretical and Applied Genetics* 104: 1263–1269.

Jaligot, E., Beulé, T., Baurens, F.C., Billotte, N., and Rival, A. 2004. Search for methylation-sensitive amplification polymorphisms associated with the mantled variant phenotype in oil palm (*Elaeis guineensis* Jacq.). *Genome* 47: 224–228.

Jaligot, E, Rival, A., Beulé, T., Dussert, S., and Verdeil, J.L. 2000. Somaclonal variation in oil palm (*Elaeis guineensis* Jacq.): The DNA methylation hypothesis. *Plant Cell Reports* 19: 684–690.

Jaligot, E. et al. 2014. DNA methylation and expression of the EgDEF1 gene and neighboring retrotransposons in mantled somaclonal variants of oil palm. *PLoS ONE* 9: e91896.

Jiang, C. et al. 2011. Regenerant arabidopsis lineages display a distinct genome-wide spectrum of mutations conferring variant phenotypes. *Current Biology* 21: 1385–1390.

Jones, L.H. 1974. Propagation of clonal oil palm by tissue culture. *Oil Palm News* 17: 1–8.

Kanazawa, A., Tsutsumi, N., and Hirai, A. 1994. Reversible changes in the composition of the population of mtDNAs during dedifferentiation and regeneration in tobacco. *Genetics* 138: 865–870.

Katan-Khaykovich, Y. and Struhl, K. 2005. Heterochromatin formation involves changes in histone modifications over multiple cell generations. *EMBO Journal* 24: 2138–2149.

Katz, D.J., Edwards, T.M., Reinke, V., and Kelly, W.G. 2009. A *C. elegans* LSD1 demethylase contributes to germline immortality by reprogramming epigenetic memory. *Cell* 137: 308–320.

Kawata, M., Ohmiya, A., Shimamoto, Y., Oono, K., and Takaiwa, F. 1995. Structural changes in the plastid DNA of rice (*Oryza sativa* L.) during tissue culture. *Theoretical and Applied Genetics* 90: 364–371.

Kitamya, E., Suzuki, S., Sano, T., and Nagata, N. 2000. Isolation of two genes that were induced upon the initiation of somatic embryogenesis on carrot hypocotyls by high concentration of 2,4-D. *Plant Cell Reports* 19: 551–557.

Kubis, S.E., Castilho, A.M.M.F., Vershinin, A.V., and Heslop-Harrison, J.S. 2003. Retroelements, transposons and methylation status in the genome of oil palm (*Elaeis guineensis*) and the relationship to somaclonal variation. *Plant Molecular Biology* 52: 69–79.

Kumar, A., Taj, G., Pandey, D., Pathak, R.K., Tiwari, A., and Avashthi, H. 2015. High-throughput omics data for mining of important genes/traits linked to agricultural productivity: A national bioinformatics workshop report. *International Journal of Computational Bioinformatics and In Silico Modeling* 4: 749–752.

Kushairi, A., Tarmizi, A.H., Zamzuri, I., Ong-Abdullah, M., Samsul, K.R., Ooi, S.E., and Rajanaidu, N. 2010. Production, performance and advances in oil palm tissue culture. In: International Seminar on Advanced in Oil Palm Tissue Culture, Organized by the International Society for Oil Palm Breeders (ISOPB), Yogyakarta, Indonesia, May 29, 2010.

Lippert, D., Zhuang, J., Ralph, S., Ellis, D.E., Gilbert, M., Olafson, R., Ritland, K., Ellis, B., Douglas, C.J., and Bohlmann, J. 2005. Proteome analysis of early somatic embryogenesis in *Picea glauca*. *Proteomics* 5: 461–473.

Litt, A. and Kramer, E.M. 2010. The ABC model and the diversification of floral organ identity. *Seminars in Cell and Developmental Biology* 21: 129–137.

Liu, C.P., Yang, L., and Shen, H.L. 2015. Proteomic analysis of immature *Fraxinus mandshurica* cotyledon tissues during somatic embryogenesis: Effects of explant browning on somatic embryogenesis. *International Journal of Molecular Sciences* 16: 13692–13713.

Low, E.T.L., Alias, H., Boon, S.H., Shariff, E.M., Tan, C.Y.A., Ooi, L.C.L., Cheah, S.C., Raha, A.R., Wan, K.L., and Singh, R. 2008. Oil palm (*Elaeis guineensis* Jacq.) tissue culture ESTs: Identifying genes associated with callogenesis and embryogenesis. *BMC Plant Biology* 8: 62.

Low, E.T.L. et al. 2006. Developments towards the application of DNA chip technology in oil palm tissue culture. *Journal of Oil Palm Research* (Special Issue): 87–98.

Luger, K., Mader, A.W., Richmond, R.K., Sargent, D.F., and Richmond, T.J. 1997. Crystal structure of the nucleosome core particle at 2.8 A resolution. *Nature Reviews Genetics* 389: 251–260.

Marsoni, M., Bracale, M., Espen, L., Prinsi, B., Negri, A.S., and Vannini, C. 2008. Proteomic analysis of somatic embryogenesis in *Vitis vinifera*. *Plant Cell Reports* 27: 347–356.

Martin, W. and Mentel, M. 2010. The origin of mitochondria. *Nature Education* 3: 58.

Martin, W., Rujan, T., Richly, E., Hansen, A., Cornelsen, S, Lins, T., Leister, D., Stoebe, B., Hasegawa, M., and Penny, D. 2002. Evolutionary analysis of arabidopsis, cyanobacterial, and chloroplast genomes reveals plastid phylogeny and thousands of cyanobacterial genes in the nucleus. In: *Proceedings of the National Academy of Sciences, USA* 99: 12246–12251.

Matthes, M., Singh, R., Cheah, S.C., and Kar, A. 2001. Variation in oil palm (*Elaeis guineensis* Jacq.) tissue culture-derived regenerants revealed by AFLPs with methylation-sensitive enzymes. *Theoretical and Applied Genetics* 102: 971–979.

Meins, F., Beversdorf, W.D., Oono, K., and Hamzah, I. 1991. Final Report Oil Palm Tissue Culture Consultant Panel.

Miguel, C. and Marum, L. 2011. An epigenetic view of plant cells cultured in vitro: Somaclonal variation and beyond. *Journal of Experimental Botany* 62: 3713–3725.

Miyao, A. et al. 2012. Molecular spectrum of somaclonal variation in regenerated rice revealed by whole-genome sequencing. *Plant Cell Physiology* 53: 256–264.

Murai, K. 2013. Homeotic genes and the ABCDE model for floral organ formation in wheat. *Plants* 2: 379–395.

Musa, B, Kandha, S., and Xaviar, A. 2011. Performance of bi-clonal D × P planting material at United Plantations Berhad. In: *International Society of Oil Palm Breeders*. Kuala Lumpur, Indonesia, pp. 88–109.

Mwangi, J.W., Rode, C., Colditz, F., Haase, C., Braun, H.-P., and Winkelmann, T. 2013. Proteomic and histological analyses of endosperm development in *Cyclamen persicum* as a basis for optimization of somatic embryogenesis. *Plant Science* 201–202: 52–65.

Nagasawa, N., Miyoshi, M., Sano, Y., Satoh, H., Hirano, H.-Y., Sakai, H., and Nagato, Y. 2003. SUPERWOMAN1 and DROOPING LEAF genes control floral organ identity in rice. *Development* 130: 705–718.

Ong-Abdullah, M., Maria, M., Low, E.T.L., Maizura, I., and Rajinder, S. 2012. Oil palm genomics in palm oil: Production, processing, characterization and uses. In: Lai, O.-M., Tan, C.-P., and Akoh, C.C. (Eds.), *AOCS Monograph Series*. AOCS Press, Urbana, Illinois, pp. 59–86.

Ong-Abdullah, M. and Ooi, S.E. 2007. Biomarkers: Finding a niche in oil palm tissue culture. Part 2—Targeting the transcriptome. *Oil Palm Bulletin* 54: 68–88.

Ong-Abdullah, M. et al. 2015. Loss of karma transposon methylation underlies the mantled somaclonal variant of oil palm. *Nature* 525: 533–537.

Ooi, S.-E., Choo, C.-N., Ishak, Z., and Ong-Abdullah, M. 2012. A candidate auxin responsive expression marker gene, EgIAA9, for somatic embryogenesis in oil palm (*Elaeis guineensis* Jacq.). *Plant Cell, Tissue and Organ Culture* 110: 201–212.

Ooi, S.-E., Choo, C.-N., Zamzuri, I., and Ong-Abdullah, M. 2013. Candidate hormone-responsive markers for callogenesis of oil palm (*Elaeis guineensis* Jacq.). *Journal of Oil Palm Research* 25: 9–21.

Ooi, S.E., Zamzuri, I., Ang, C.L., Mohd Azmir, M., and Ong-Abdullah, M. 2009. Response of hormone-responsive genes to 2,4-D vs. NAA treatment of explant cultures via gene expression analysis. In: *Proceedings of the Agriculture, Biotechnology and Sustainability Conference*, PIPOC 2009 International Palm Oil Congress, Vol. III. KLCC, Kuala Lumpur, pp. 1339.

Palama, T.L., Menard, P., Fock, I., Choi, Y.H., Bourdon, E., Govinden-Soulange, J., Bahut, M., Payet, B., Verpoorte, R., and Kodja, H. 2010. Shoot differentiation from protocorm callus cultures of *Vanilla planifolia* (Orchidaceae): Proteomic and metabolic responses at early stage. *BMC Plant Biology* 10: 82.

Pan, Z., Guan, R., Zhu, S., and Deng, X. 2009. Proteomic analysis of somatic embryogenesis in Valencia sweet orange (*Citrus sinensis* Osbeck). *Physiology and Biochemistry* 28: 281–289.

Pannetier, C., Arthuis, P., and Lievoux, D. 1981. Neoformation of young *Elaeis guineensis* plants from primary calluses obtained from leaf fragments cultured in vitro. *Oleagineux* 36: 119–122.

Potier, F., Nouy, B., Flori, A., Jacquarmard, J.C., Edyana Suryna, H., and Durand-Gasselin, T. 2006. Yield potential of oil palm (*Elaeis guineensis* Jacq.) clones: Preliminary results observed in the Aek Loba genetic block in Indonesia. In: ISOPB Seminar on Yield Potential in Oil Palm, Phuket, Thailand, December 2006.

Prasetyo, J.H.H., Setipu, B., Ishwandar, H.E., Djuhjana, J., and Nelson, S.P.C. 2011. Performance of Sumbio semi-clonal progenies. In Int. Palm Oil Congr. 'Palm Oil Fortifying and Energising the World', Malaysian Palm Oil Board, Kuala Lumpur, November 15–17.

Rabechault, H. and Martin, J.P. 1976. Multiplication vegetative du Palmier a Huile (*Elaeis guineensis* Jacq.) a l'Aide de cultures de tissues foliares. *Comptes Rendus de l'Académie des Sciences Paris, Serie D* 283: 1735–1737.

Rani, V., Singh, K.P., Shiran, B., Nandy, S., Goel, S., Devarumath, R.M., Sreenath, H.L., and Raina, S.N. 2000. Evidence for new nuclear and mitochondrial genome organizations among high-frequency somatic embryogenesis-derived plants of allotetraploid *Coffea arabica* L. (Rubiaceae). *Plant Cell Reports* 19: 1013–1020.

Riechmann, J.L. and Meyerowitz, E.M. 1998. The AP2/EREBP family of plant transcription factors. *Biological Chemistry* 379: 633–646.

Rijpkema, A.S., Vandenbussche, M., Koes, R., Heijmans, K., and Gerats, T. 2010. Variations on a theme: Changes in the floral ABCs in angiosperms. *Seminars in Cell and Developmental Biology* 21: 100–107.

Rohani, O., Zamzuri, I., and Tarmizi, A.H. 2003. Oil palm cloning: MPOB protocol. *MPOB Technology No. 26.*

Roowi, S.H., Ho, C.-L., Syed Alwee, S.S.R., Abdullah, M.O., and Napis, S. 2010. Isolation and characterization of differentially expressed transcripts from the suspension cells of oil palm (*Elaeis guineensis* Jacq.) in response to different concentration of auxins. *Molecular Biotechnology* 46: 1–19.

Sadoch, Z., Majewska-Sawka, A., Jazdzewska, E., and Niklas, A. 2000. Changes in sugar beet mitochondrial DNA induced during callus stage. *Plant Breeding* 119: 107–110.

Saito, K. and Matsuda, F. 2010. Metabolomics for functional genomics, systems biology, and biotechnology. *Annual Review of Plant Biology* 61: 463–489.

Sarpan, N., Kok, S.-Y., Chai, S.-K., Fitrianto, A., Nuraziyan, A., Zamzuri, I., Ong-Abdullah, M., and Ooi, S.-E. 2015. A model for predicting flower development in *Elaeis guineensis* Jacq. *Journal of Oil Palm Research* 27: 315–325.

Schwendiman, J., Pannetier, C., and Michaux-Ferrière, N. 1988. Histology of somatic embryogenesis from leaf explants of the oil palm *Elaeis guineensis*. *Annals of Botany* 62: 43–52.

Shedge, V., Arrieta-Montiel, M., Christensen, A.C., and Mackenzie, S.A. 2007. Plant mitochondrial recombination surveillance requires unusual RecA and MutS homologs. *Plant Cell* 19: 1251–1264.

Shedge, V., Davila, J., Arrieta-Montiel, M.P., Mohammed, S., and Mackenzie, S.A. 2010. Extensive rearrangement of the arabidopsis mitochondrial genome elicits cellular conditions for thermotolerance. *Plant Physiology* 152: 1960–1970.

Shimron-Abarbanell, D. and Breiman, A. 1991. Comprehensive molecular characterization of tissue culture-derived *Hordeum marinum* plants. *Theoretical and Applied Genetics* 83: 71–80.

Silva, R.D.C., Carmo, L.S.T., Luis, Z.G., Silva, L.P., Scherwinski-Pereira, J.E., and Mehta, A. 2014. Proteomic identification of differentially expressed proteins during the acquisition of somatic embryogenesis in oil palm (*Elaeis guineensis* Jacq.). *Journal of Proteomics* 104: 112–127.

Simmonds, N.W. and Smartt J. 1999. *Principles of Crop Improvement*. Blackwell Science, Oxford.

Singh, R. et al. 2013. Oil palm genome sequence reveals divergence of interfertile species in old and new worlds. *Nature* 500: 335–339.

Soh, A.C. 1986. Expected yield increase with selected oil palm clones from current D × P seedling materials and its implications on clonal propagation, breeding and ortet selection. *Oleagineux* 41: 51–56.

Soh, A.C. 1998. Review of ortet selection in oil palm. *The Planter* 74: 217–226.

Soh, A.C. and Chow, C.S. 1989. Index selection in oil palm for cloning. In: Iyama, S. and Takeda, G. (Eds.), *Proceedings of the 6th International Congress of SABRAO*. Tsukuba, Japan, pp. 713–716.

Soh, A.C. and Chow, C.S. 1993. Index selection utilizing plot and family information in oil palm. *Elaeis* 5: 27–37.

Soh, A.C., Chow, S.C., Iyama, S., and Yamada, Y. 1994. Candidate traits for index selection in choice of oil palm ortets for clonal propagation. *Euphytica* 76: 23–32.

Soh, A.C., Gan, H.H., Wong, G., Hor, T.Y., and Tan, C.C. 2003b. Estimates of within family genetic variability for clonal selection in oil palm. *Euphytica* 133: 147–163.

Soh, A.C., Wong, G., Hor, T.Y., Tan, C.C., and Chew, P.S. 2003a. Oil palm genetic improvement. In: Janick, J. (Ed.), *Plant Breeding Reviews* Vol. 22. John Wiley and Sons, Hoboken, New Jersey, pp. 165–219.

Soh, A.C., Wong, G., Tan, C.C., Chew, P.S., Chong, S.P., Ho, Y.W., Wong, C.K., Choo, C.N., Nor Azura, H., and Kumar, K. 2011. Commercial-scale propagation of and planting of elite oil palm clones: Research and development towards realization. *Journal of Oil Palm Research* 23: 935–952.

Soh, A.C., Wong, G., Tan, C.C., Chong, S.P., Choo, C.N., Nor Azura, A., and Ho, Y.W. 2006a. Progress and challenges in commercial mass propagation of clonal oil palm. In: Paper presented at the IOPC 2006 International Oil Palm Conference, Bali, June 19–23, 2006a.

Soh, A.C., Wong, G., Tan, C.C., Chong, S.P., Choo, C.N., Nor Azura, A., and Ho, Y.W. 2006b. Advances and issues in commercial propagation of oil palm clones. In Kushairi, A., Ravigadevi, S., Ong-Abdullah, M., and Chang, K.C. (Eds.), In: *Proceedings of the Clonal and Quality Replanting Materials Workshop: Towards Increasing the Annual National Productivitiy by One Tonne FFB/Ha/Year.* MPOB, Selangor, Malaysia. August 10, 2006. pp. 35–55.

Soh, A.C., Wong, G., Tan, C.C., Hor, T.Y., and Wong, C.K. 2003c. Revisited: Cloning seedlings of reproduced best D × P cross strategy. In: *Proceedings of the ISOPB Seminar 2003*, Medan, Indonesia, October 6–7, 2003.

Stelpflug, S.C., Eichten, S.R., Hermanson, P.J., Springer, N.M., and Kaeppler, S.M. 2014. Consistent and heritable alterations of DNA methylation are induced by tissue culture in maize. *Genetics* 198: 209–218.

Strahl, B.D., Ohba, R., Cook, R.G., and Allis, C.D. 1999. Methylation of histone H3 at lysine 4 is highly conserved and correlates with transcriptionally active nuclei in tetrahymena. In: *Proceedings of the National Academy of Sciences, USA* 96: 14967–14972.

Stroud, H., Ding, B., Simon, S.A., Feng, S., Bellizzi, M., Pellegrini, M., Wang, G.-L., Meyers, B.C., and Jacobsen, S.E. 2013. Plants regenerated from tissue culture contain stable epigenome changes in rice. *Elife* 2: e00354.

Sun, P., Arrieta-Montiel, M.P., and Mackenzie, S.A. 2012. Utility of in vitro culture to the study of plant mitochondrial genome configuration and its dynamic features. *Theoretical and Applied Genetics* 125: 449–454.

Sundstrom, J.F., Nakayama, N., Glimelius, K., and Irish, V.F. 2006. Direct regulation of the floral homeotic APETALA1 gene by APETALA3 and PISTILLATA in arabidopsis. *Plant Journal* 46: 593–600.

Syed Alwee, S., van der Linden, C.G., van der Schoot, J., de Folter, S., Angenent, G.C., Cheah, S.-C., Smulders, M.J.M. 2006. Characterization of oil palm MADS box genes in relation to the mantled flower abnormality. *Plant Cell Tissue Organ Culture* 85: 331–344.

Tada, S.F.S. and Souza, A.P. 2006. A recombination point is conserved in the mitochondrial genome of higher plant species and located downstream from the *cox2* pseudogene in *Solanum tuberosum* L. *Genetics and Molecular Biology* 29: 83–89.

Tanurdzic, M., Vaughn, M.W., Jiang, H., Lee, T.-J., Slotkin, R.K., Sosinski, B, Thompson, W.F., Doerge, R.W., and Marteinssen, R.A. 2008. Epigenomic consequences of immortalized plant cell suspension culture. *PLoS Biology* 6: 2880–2895.

Tariq, M. and Paszkwoski, J. 2004. DNA and histone methylation in plants. *Trends in Genetics* 20: 244–251.

Tarmizi, A.H. 2002. Oil Palm Liquid Culture—MPOB Protocol. *MPOB Information Series* TT No. 138.

Tarmizi, A.H., Norjihan, M.A., Samsul Kamal, R., Zaiton, R., and Cheah, S.C. 2003a. Mass propagation of oil palm planting materials using liquid culture and bioreactor technology. In: *Proceedings of the 2003 PIPOC* August 24–28, 2003, Putrajaya, pp. 130–144.

Tarmizi, A.H. and Zaiton, R. 2005. MPOB Fast Transfer Technique (MoFATT) in Liquid Culture System. *MPOB Information Series* TT No. 261.

Tarmizi, A.H. and Zaiton, R. 2006a. Two-in-One MPOB Simple Impeller (2-in-1 MoSLIM) in Liquid System. *MPOB Information Series* TT No. 303.

Tarmizi, A.H. and Zaiton, R. 2006b. Simple Impeller with Fast Transfer Techniques (SLIMFaTT) in Liquid Culture System. *MPOB Information Series* TT No. 304.

Tarmizi, A.H., Zaiton, R., and Rosli, M.Y. 2007. MPOB Modified Vessel (MoVess) for Liquid Tissue Culture System. *MPOB Information Series* TT No. 355.

Tarmizi, A.H., Zaiton, R., and Rosli, M.Y. 2009. MPOB Motorized Vessel (MPOB-MotoVess) for Liquid Tissue Culture System. *MPOB Information Series* TT No. 413.

Tarmizi, A.H., Zaiton, R., and Rosli, M.Y. 2012. Motorized Vessel with Fast Media Transfer (MoVeFast). *MPOB Information Series* TT No. 511.

Tarmizi, A.H., Zamzuri, I., and Hashim, H. 2003b. Oil Palm Tissue Culture Tracking System (OPTRACKS): Version 1. *MPOB Information Series* TT No. 185.

The Arabidopsis Genome Initiative. 2000. Analysis of the genome sequence of flowering plant *Arabidopsis thaliana*. *Nature* 408: 796–815.

Thorpe, T.A. 2007. History of plant tissue culture. *Molecular Biotechnology* 37: 169–180.

Thuc, L.V. et al. 2011. A novel transcript of oil palm (*Elaeis guineensis* Jacq.), Eg707, is specifically upregulated in tissues related to totipotency. *Molecular Biotechnology* 48: 156–164.

Turnham, E. and Northcote, D.H. 1982. The use of acetyl-CoA carboxylase activity and changes in wall composition as measures of embryogenesis in tissue cultures of oil palm (*Elaeis guineensis*). *Biochemical Journal* 208: 323–332.

Valledor, L., Meijon, M., Hasbun, R., Canal, M.J., and Rodrıguez, R. 2010. Variations in DNA methylation, acetylated histone H4, and methylated histone H3 during *Pinus radiata* needle maturation in relation to the loss of in vitro organogenic capability. *Journal of Plant Physiology* 167: 351–357.

Vandenbussche, M., Zethof, J., Royaert, S., Weterings, K., and Gerats, T. 2004. The duplicated B-class heterodimer model: Whorl-specific effects and complex genetic interactions in *Petunia hybrida* flower development. *Plant Cell* 16: 741–754.

Wang, Z., Zang, C., Cui, K., Schones, D.E., Barski, A., Peng, W., and Zhao, K. 2009. Genome-wide mapping of HATs and HDACs reveals distinct functions in active and inactive genes. *Cell* 138: 1019–1031.

Weigel, D. and Meyerowitz, E.M. 1994. The ABCs of floral homeotic genes. *Cell* 78: 203–209.

Wellmer, F., Graciet, E., and Riechmann, J.L. 2014. Specification of floral organs in arabidopsis. *Journal of Experimental Botany* 65: 1–9.

Wicke, S., Schneeweiss, G.M., de Pamphilis, C.W., Muller, K.F., and Quandt, D. 2011. The evolution of the plastid chromosome in land plants: Gene content, gene order, gene function. *Plant Mol Biology* 76: 273–297.

Winkelmann, T., Heintz, D., Van Dorsseleaer, A., Serek, M., and Braun, H.P. 2006. Proteomic analysis of somatic and zygotic embryo of *Cyclamen persicum* Mill. reveal new insights into seed germination physiology. *Planta* 224: 508–519.

Wong, C.K., Choo, C.N., Liew, Y.R., Ng, W.J., Krishnan, K., and Tan, C.C. 2010. Benchmarking best performing semi-clonal seeds against tenera clones: Avenue for superior clonal ortet identification. In: *Proceedings of the ISOPB Seminar 2010*, Bali, Indonesia, pp. 67–70.

Wong, G., Chong, S.P., Tan, C.C., and Soh, A.C. 1999. Liquid suspension culture—A potential technique for mass production of oil palm clones. In: *Proceedings of the 1999 International Palm Oil Conference on Emerging Technologies and Opportunities in the Next Millennium.* PORIM, Bangi, 3–11.

Wong, G., Tan, C.C., and Soh, A.C. 1997. Large-scale propagation of oil palm clones—Experiences to date. *Acta Horticulturae* 447: 649–658.

Yao, S.-G., Ohmori, S., Kimizu, M., and Yoshida, H. 2008. Unequal genetic redundancy of rice PISTILLATA orthologs, OsMADS2 and OsMADS4, in lodicule and stamen development. *Plant Cell Physiology* 49: 853–857.

Yin, L., Tao, Y., Zhao, K., Shao, J., Li, X., Liu, G., Liu, S., and Zhu, L. 2007. Proteomic and transcriptomic analysis of rice mature seed-derived callus differentiation. *Proteomics* 7: 755–768.

Zachgo, S., Silva Ede, A., Motte, P., Trobner, W., Saedler, H., and Schwarz-Sommer, Z. 1995. Functional analysis of the antirrhinum floral homeotic DEFICIENS gene in vivo and in vitro by using a temperature-sensitive mutant. *Development* 121: 2861–2875.

Zamzuri, I. 1998. Efficient rooting of oil palm in vitro plantlets using double-layer technique. *PORIM Bulletin* 36: 23–36.

Zamzuri, I. 2001a. Double-Layer Technique in Rooting of Oil Palm In Vitro Plantlets. *MPOB Information Series* No. 99.

Zamzuri, I. 2001b. Tissue Culture Database System. MPOB Viva No. 184/2001(51).

Zamzuri, I. 2002. Flameless Sterilizer. *MPOB Information Series* TT No. 139.

Zarsky, V. et al. 1995. The expression of small heat shock gene is activated during induction of tobacco pollen embryogenesis by starvation. *Plant Cell Environment* 18: 139–147.

Zavattieri, M., Frederico, A., Lima, M., Sabino, R., and Arnholdt-Schmitt, B. 2010. Induction of somatic embryogenesis as an example of stress-related plant reactions. *Electronic Journal of Biotechnology,* North America, 13: 1–9.

9 Molecular Genetics and Breeding

Aik Chin Soh, Sean Mayes, Jeremy Roberts, Noorhariza Mohd Zaki, Maria Madon, Trude Schwarzacher, Pat Heslop-Harrison, Maizura Binti Ithnin, Amiruddin Mohd Din, Umi Salamah Ramli, Abrizah Othman, Yaakub Zulkifli, Leslie Eng Ti Low, Meilina Ong-Abdullah, Rajinder Singh, Chee Keng Teh, Qi Bin Kwong, Ai Ling Ong, Mohaimi Mohamed, Sukganah Apparow, Fook Tim Chew, David Ross Appleton, Harikrishna Kulaveerasingam, Huey Fang Teh, Katharina Mebus, Bee Keat Neoh, Tony Eng Keong Ooi, Yick Ching Wong, and David Cros

CONTENTS

9.1 Introductory Overview (*Sean Mayes, Aik Chin Soh, and Jeremy Roberts*) 227
9.2 Chromosomes, Cytology, and Molecular Cytogenetics of Oil Palm (*Noorhariza Mohd Zaki, Maria Madon, Trude Schwarzacher, and J.S. (Pat) Heslop-Harrison*) 229
9.2.1 Chromosome Analysis 229
9.2.2 Molecular Cytogenetics of Oil Palm: Oil Palm Chromosomes 232
9.2.3 Epigenetic Modulation of Genomes and DNA Methylation 235
9.2.4 Conclusions 236
9.3 Conventional to Molecular Guided Breeding: A New Era in Improving Oil Palm Productivity (*Maizura Binti Ithnin, Amiruddin Mohd Din, Umi Salamah Ramli, Abrizah Othman, Yaakub Zulkifli, Suzana Mustafa, Leslie Eng Ti Low, Meilina Ong-Abdullah, and Rajinder Singh*) 236
9.3.1 Background 236
9.3.2 Limitations of Conventional Oil Palm Breeding 237
9.3.3 Germplasm Management, Characterization, and Conservation Using Molecular-Based Assays 237
9.3.4 Unraveling the Genome and the Discovery of Genes Influencing Agronomic Traits 238

9.3.4.1 Sequencing the Hypomethylated Genome of Oil Palm 240
9.3.4.2 Whole-Genome Sequencing 241
9.3.5 Realization of Marker-Assisted Selection to Improve Breeding Efficiency 241
9.3.6 Moving to the Post-Genomics Era: Proteomics and Metabolomics Applications in Oil Palm Improvement 243
9.3.7 Conclusions 246
9.4 Application of Genomic Tools in Oil Palm Breeding (*Chee Keng Teh, Qi Bin Kwong, Ai Ling Ong, Mohaimi Mohamed, Sukganah Apparow, Fook Tim Chew, Sean Mayes, David Ross Appleton, and Harikrishna Kulaveerasingam*) 246
9.4.1 Conventional Breeding Programs 246
9.4.1.1 Breeding Selection 246
9.4.1.2 Constraints 247
9.4.2 Boom in DNA Sequencing 248
9.4.2.1 Whole-Genome Sequence 248
9.4.2.2 Resequencing of Various Palm Populations 249
9.4.3 Development of DNA Markers 250
9.4.4 Genetic Dissection 250
9.4.4.1 Linkage Mapping 250
9.4.4.2 GWAS and GS 251
9.4.5 Marker Application to Oil Palm Breeding 252
9.4.5.1 Marker-Assisted Selection 252
9.4.5.2 "Genome Select" 254
9.4.6 Conclusions 254
9.5 Omics Technologies toward Improvement of Oil Palm Yield (*Huey Fang Teh, Katharina Mebus, Bee Keat Neoh, Tony Eng Keong Ooi, Yick Ching Wong, Harikrishna Kulaveerasingam, and David Ross Appleton*) 256
9.5.1 Twenty-First Century Omics Platforms in Oil Crop Studies 256
9.5.2 Omics Platforms in Oil Palm Yield Studies 258
9.5.3 Pitfalls and Challenges in Omics Studies 262
9.5.4 Conclusions and Future Perspectives 263
9.6 Genomic Selection for Oil Palm (*David Cros*) 263
9.6.1 GS Studies Based on Simulations 263
9.6.2 GS Studies Based on Empirical Data 265
9.6.3 Conclusion and Perspectives 265
9.6.3.1 Using GS to Increase the Rate of Genetic Gain in Oil Palm Yield 265
9.6.3.2 Increasing Clonal Values Using GS 268
9.6.3.3 Extending the GS Approach to Other Traits 268
9.6.3.4 Integrating Multiple Omics and Genetic Data in GS 269
9.6.3.5 Interactions with the Environment 269
References 269

9.1 INTRODUCTORY OVERVIEW

Sean Mayes, Aik Chin Soh, and Jeremy Roberts

Molecular genetics is essentially the study of the mechanisms and effects of trait inheritance at the molecular level. The subject area can range from the ability to follow a targeted crop introgression program through fluorescently painting the chromosomes, through to a detailed understanding of a single gene and the consequences of inheriting different versions of that gene. The analysis of the inheritance of allelic forms of unknown functional DNA or even "nonfunctional" DNA across the entire genome of a species is a key subcomponent of molecular genetics. Ultimately the aim is to improve our understanding of trait inheritance and, preferably, to be able to predict and manipulate it. For crop molecular genetics, this is specifically in crop plants and is often integrated into breeding programs, through marker-assisted selection.

Understanding the individual genes involved in a trait of economic importance may allow the development of "perfect markers," which specifically test for the presence of the causal mutation in a gene that specifies a trait. In wheat, some of the gene tests that are used routinely by breeders include the adaptive genes for photoperiod and for vernalization requirement (Li et al. 2013a,b; Langer et al. 2014).

In oil palm, the cloning of the *shell* gene by Malaysian Palm Oil Board (MPOB) and partners has allowed the development of a specific test for the different allelic versions of the causative gene using either the *SureSawit*™Shell or through the use of the underlying allelic information. There is no question that the shell-type gene is of great importance in oil palm, providing a 30% oil yield advantage for thin-shelled *tenera* palms over the thick-shelled *dura* palms. Whether the *SureSawit*™ Shell kit itself is commercially viable depends on the value attached to determining the trait in commercial or breeding material or, perhaps more specifically, whether there are other (nonpatented) approaches that essentially yield the same data (such as genetic fingerprinting of commercial *tenera* to confirm that they are hybrids). Whether or not the kit is a commercial success itself, the fact that it exists illustrates the potential for gene tests for specific oil palm traits where the causative gene has been cloned, with the aim of being able to test for the presence of a particular allele of the gene, before expression of the associated phenotype in subsequent years. A kit designed to test for *virescens* fruit and a kit for diagnostic testing of abnormal flowering potential in tissue culture material are also in development.

In oil palm, there are, however, relatively few known major genes which have a significant influence phenotypically (shell-type, *virescens*, *albescens*, crown disease, long-stalk, low fruit lipase, potentially some trunk height increment genes and oil quality), although the difficulty of accurate phenotyping and the influence of the environment can make it hard to define categorical differences between the phenotypes observed, even if the locus responsible makes a major contribution to the genetic variation for the trait. As such, it is likely that not only are many traits influenced by multiple genes—and the more complex and "developmentally downstream" the trait, the more genes can potentially effect it, as in the case of yield—but there may also be many hidden genes of major effect that are only present in certain germplasm or are widely present but are heavily influenced by the environment. Hence, the potential value of molecular genetics and, specifically, marker analysis to dissect such trait inheritance at the genetic level and to develop diagnostic tests for plant breeding is clear.

An important point to emphasize is that while knowing the causative mutation for a trait gene is very useful (and can allow a deeper understanding of how the trait works and how it might be manipulated further), it is not necessary to know the causative gene to be able to use the trait locus in breeding programs. The development of anonymous markers across the genome allows an understanding of gene–trait relationships in breeding germplasm to be determined, alongside elucidating the genetic structure of breeding material and germplasm (Chapter 3) and also the identification of markers that are genetically linked to alleles of breeding interest, which could be used to imply the presence of the associated allele in the offspring. The accuracy of linked markers is primarily dependent upon how close (in recombinational terms, rather than physical terms) the marker is to the causative gene.

The development of methods to follow the inheritance of the chromosomes (and to know how many chromosomes crops have and how they behave in mitosis and meiosis) underpins much of the current study of inheritance. New tools, such as fluorescent *in situ* hybridization (FISH) and genomic *in situ* hybridization (GISH), have renewed the use and interest in direct chromosome studies (Section 9.2), particularly in terms of introgression work and the development of defined cytogenetic lines. This work is at a very different scale to much of the marker work that has developed over the last 30 years and ultimately both strands come together in the development of the genome sequence of a species. Some of the first marker types developed in oil palm were based on the scoring of the isoforms of enzymes (isozymes), with the first report published in 1984 (Ghesquiere 1984) looking at the genetic control of nine isoenzymes. A wide range of marker systems have been developed in the subsequent years, often in the human research field before they are adopted into plant and crop research. The technology is now approaching the stage at which it will become feasible to sequence the entire genome for the group of individual oil palm of interest. This can be done technically, but is still too expensive to be routine for large numbers for the moment. Evaluating trait variation at the epigenetic level (methylation of DNA/acetylation of histones) is also becoming more feasible across the oil palm genome. In oil palm, the adoption of marker systems developed first in other species led to the construction of the first genetic map (Mayes et al. 1997), numerous population/germplasm structure analyses in both species (see Chapter 3), quantitative trait loci (QTL) analyses in numerous populations (Rance et al. 2001; Billotte et al. 2010 and others), and the identification of markers for potential marker-assisted selection. As technology has improved, the marker density has increased and the cost per data point decreased significantly. This made possible association genetics approaches (Sections 9.3 and 9.4 and theory in Appendix 4A.5), which exploit historical recombination events in a population, potentially increasing the accuracy of location of candidate genes for traits. As marker densities reach the level where there is at least one marker per linkage disequilibrium (LD) block, the potential for genomic selection (GS) and variants of the approach increases, which holds promise to reduce the generation times of the breeding program.

Today, single nucleotide polymorphism (SNP) markers are set to predominate in oil palm genetics, particularly given the release of the oil palm genome sequence from MPOB (Section 9.3). Simple sequence repeats (SSRs) continue, for now, to make a contribution, as microsatellite repeat sequences are multiallelic, whereas SNPs are generally biallelic. SSRs also evolve at a faster rate than SNP loci. For population studies, this can be useful, as the two marker types are sampling different rates of evolution within a population.

Alongside the development of genomics technology, facilitated by next-generation sequencing (NGS), other "omics" technology—transcriptomics, proteomics, metabolomics, ionomics, etc.—have advanced significantly (Sections 9.3 and 9.5). The ability to measure the changes of large numbers of transcripts, proteins, metabolites, and other components of the cell potentially allows us to bridge the different scales from sequence, to expressed genes, to translated genes, to metabolic products of gene action, to protein turnover and degradation. Experiments that explicitly couple more than one level of analysis hold great potential (e.g., eQTL analysis, which uses differences in transcript abundance within a genetically structured population to locate the factors controlling the differences onto a genetic map constructed from a subset of the same data).

The revolutions in omics technology and particularly NGS for genomics have removed a long-standing roadblock from research. It is now possible to generate, relatively easily and relatively cheaply, the genetic and cellular trait data needed for an analysis. Until the generation of the oil palm genome sequence(s), the number of available markers was always limited and a major effort and component of any research program was generating the markers and the subsequent data. The development of oil palm SNP chips (Sections 9.3 and 9.4) has removed this obstacle. In a similar way, the development of other omics is rapidly removing limitations imposed by other levels of data.

Inevitably, attention must now turn to phenotyping methods. While "phenomics" is a widely used term for the very high-throughput phenotyping of individual plants, it is arguably the least developed (and most difficult to develop) "omics" platform, with current offerings (e.g., LemnaTec phenomics platforms) working reasonably for some traits and for some species, but inadequately for others. For oil palm, given that it is a long-lived perennial that can produce a bunch of fruit per month for 25 years, it is accurate phenotyping approaches that reflect the underlying genetic values (let alone "phenomics") that are the immediate requirement, to be able to exploit the power of the other "omics." If custom phenomics approaches can be developed in oil palm, it will have great potential for the traits that can be addressed and could remove the current major bottleneck to trait analysis.

Some suggested areas of interest that are being actively investigated in other species that may have relevance in oil palm are given in Chapter 13.

9.2 CHROMOSOMES, CYTOLOGY, AND MOLECULAR CYTOGENETICS OF OIL PALM

Noorhariza Mohd Zaki, Maria Madon, Trude Schwarzacher, and J.S. (Pat) Heslop-Harrison

9.2.1 Chromosome Analysis

Cytogenetic analysis—the study of the numbers, structure, and organization of the chromosomes packaging the DNA within the cell nucleus—is important for understanding evolution, genetics, epigenetics, genetic recombination, and nuclear stability in all species. Cytogenetic manipulation of the chromosome complement of crop plants is one of the most valuable methods available to plant breeders for introducing entirely new variation into crop varieties through wide hybridization, using crosses beyond current agricultural varieties. In some crops, there is the potential to manipulate ploidy to alter organ size or bring together different genes. The basic aims and methods of the cytogenetic analysis of oil palm chromosomes (both for *Elaeis guineensis* and *Elaeis oleifera*) are similar to those for any other plant species.

Chromosome number needs to be determined to identify haploid plants (Dunwell et al. 2010) or aneuploids. The identification of individual chromosomes in a species provides a reference for demarcating structural differences—both inter- and intraspecies—and as a platform for developing high-resolution cytogenetic maps. Cytogenetic maps show the physical length of individual chromosomes in micrometers as measured through the microscope and the position of genetically mapped markers relative to cytological landmarks such as centromeres, telomeres, heterochromatin, and nucleolar organizer regions (NORs). Genetic linkage maps show the linear order of sequences and markers along the chromosomes, and the amount of recombination between linked markers. With the advancement of DNA sequencing and genomics research, cytogenetic maps are not only valuable for integrating and organizing genetic, molecular, and cytological information, but also provide a unique insight into genome organization in the context of chromosomes (Heslop-Harrison and Schwarzacher 2011). The different kinds of genomic maps differ greatly in the method of production and the ways they are viewed; the integration of the maps is important to gain a comprehensive view of genome structure and behavior. In many cases, the study of the physical chromosomes is the most efficient approach to discover translocations between chromosomes, introgression of chromosomes or chromosome segments from other species, and aneuploidy. Knowledge of the structures and organization of the chromosomes of oil palm, as well as in many other crop species, has been valuable for the development of new lines, making hybrids, understanding the causes of some abnormalities or infertility, and characterizing differences between related species or even breeding lines.

Chromosomes are most commonly studied by light microscopy using the highest magnification available with oil-immersion optics (objective lens magnification of ×64 or ×100) in metaphase preparations. Most typically, living root tips with many dividing cells from seedlings, plants growing in pots or in soil, are pretreated for up to 24 h to arrest the cell cycle and accumulate cells at metaphase when the chromosomes are condensed onto the metaphase plate (see Schwarzacher 2016 for more detailed methods). The roots are then fixed, softened with enzyme, acid, or alkali treatments, and squashed to spread metaphase chromosomes on a glass microscope slide before staining and examination. Meiotic chromosomes are also studied using preparations made at different stages of meiosis. Apical meristems or floral tissues are also convenient sources of metaphases in dividing tissues. It is possible but challenging to obtain metaphases from tissue cultures, where there are few divisions, and cell walls often make spreading difficult. Nuclei and chromosomes may be stained before spreading of the tissue (typically with Feulgen, which stains DNA bright red), during spread preparation (typically with aceto-orcein, also a red stain), or with fluorescent stains for DNA such as DAPI (4′,6-diamidino-2-phenylindole); chromosomes can also be seen without staining by phase contrast microscopy. Pollen mother cells can also provide dividing cells that are suitable for cytological analysis (Madon et al. 2005) although it may be challenging to obtain them in the field in oil palm. Interphase nuclei from root tips, callus, or leaf tissue may also be analyzed for some purposes. Various alternatives and refinements of these basic methods may be used.

The morphological study of metaphase chromosomes, in particular measuring the absolute or relative sizes (normally taken as the length) of each chromosome and the relative sizes of the two arms divided by the centromere, allows many chromosomes in a species to be identified individually. Relative arm sizes are measured as the "arm ratio" (the size of the larger chromosome arm divided by the size of the smaller arm) or "centromeric index" (the size of the shorter arm divided by the size of the whole chromosome, which may be expressed as a percentage; may also be called "centromere index"). Chromosomes may vary in their arm ratio from being telocentric (where the centromere is seen in the microscope to lie at the end of the chromosome), through acrocentric and subacrocentric, to submetacentric and metacentric, where the centromere divides the chromosome into two equal arms. In practice, measurement inaccuracies and unequal condensation of arms during prophase of mitosis can make classification of the long and short arm difficult where the arms are similar sizes. Many species have groups, or even all, chromosomes that are too similar in size and arm ratio, or show a continuous size distribution from larger to smaller, which makes individual identification of all chromosomes impossible. The presence of secondary constrictions at the NORs (the sites of ribosomal RNA genes) at characteristic locations are often seen on one or more pairs of chromosomes and enables additional chromosomes to be identified in metaphase preparations.

DNA *in situ* hybridization is able to show the presence and locations of labeled DNA sequences along chromosomes; the method can be used at on metaphases and interphases (see Schwarzacher and Heslop-Harrison 2000 for detailed methods). Using different probes, a molecular cytogenetic map enables the classification and identification of each chromosome arm by direct localization of DNA sequences (clones or polymerase chain reaction [PCR] products) on mitotic metaphase or pachytene chromosomes. Now, probe hybridization is normally detected by fluorescent labels, and the technique is known as FISH. Where probes are suitable, FISH gives reliable results and the technique has been widely used for cytogenetic and genome research, sometimes allowing visualization of the physical positions of the DNA sequences associated with molecular markers along a given chromosome. Such maps combining chromosome structure with recombination rate and physical distance integrate the molecular and physical measurements of genome organization. High-resolution cytogenetic maps are available for individual chromosomes in a number of crop species, including, for example, maize (Figueroa and Bass 2012), rice (Kao et al. 2006), tomato (Szinay et al. 2008), soybean (Walling et al. 2006), and papaya (Wai et al. 2012), and for whole genomes and chromosomes in *Sorghum* (Kim et al. 2005), potato (Tang et al. 2008), sugar beet (Paesold et al. 2012), and common bean (Pedrosa-Harand et al. 2009, Fonseca et al. 2010).

9.2.2 Molecular Cytogenetics of Oil Palm: Oil Palm Chromosomes

Oil palm (*E. guineensis* L.) has 2n = 32 chromosomes, like the other oil-bearing species *E. oleifera*. Within the Palmeae family, the Kew genome size database (http://data.kew.org/cvalues/) lists chromosome numbers for 80 diverse palm species: most species have between 26 and 36 chromosomes, although counts as high as 2n = 596 have been reported. The 16 pairs of chromosomes of oil palm are all submetacentric and show a continuous range of size, ranging from 3 to 8 µm long at metaphase in a typical preparation. Metaphase chromosomes of oil palm are shown in Figure 9.1 (Castilho et al. 2000). Apart from the smallest chromosome with the secondary constriction at the NOR, most individual chromosomes are difficult to distinguish cytogenetically and in most cases there are two or three pairs with similar morphology. Along with the stained chromosomes, Figure 9.1 also shows the locations of two highly conserved DNA sequences by *in situ* hybridization, with the ends of each chromosome marked by the telomeric DNA sequence.

Several authors have analyzed the karyotype and aimed to identify or group the oil palm chromosomes (e.g., Sato 1949; Sharma and Sarkar 1956; Madon et al. 1995) based on their morphology—chromosome length, arm ratio, and presence of NORs. Using *in situ* hybridization of repetitive sequences and the 45S rDNA sequence,

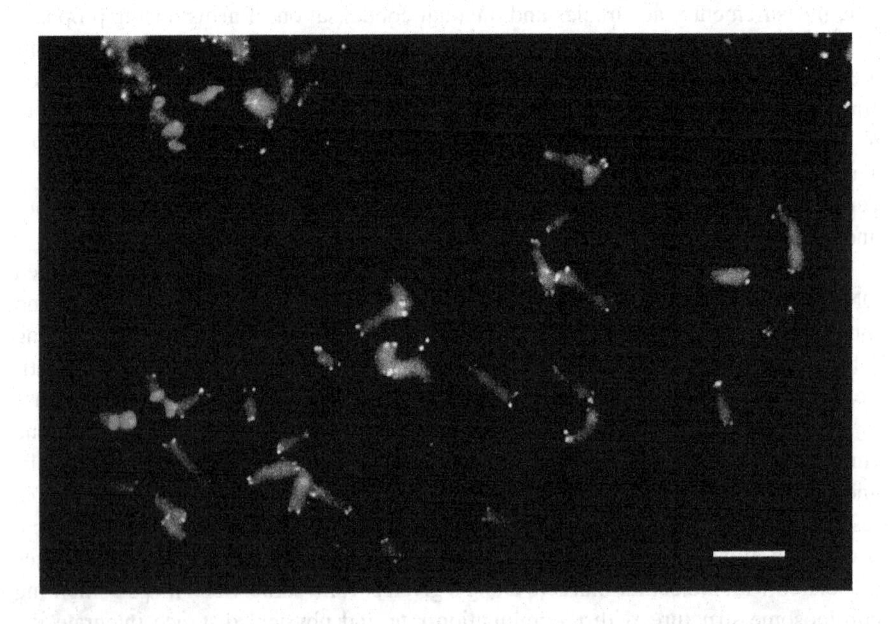

FIGURE 9.1 **(See color insert.)** Metaphase chromosomes of oil palm, *Elaeis guineensis* (2n = 32), seen by fluorescence microscopy after *in situ* hybridization. There are parts of prophase and interphase nuclei seen at the top of the image. The DNA and chromosomes are fluorescing blue with the DNA stain DAPI, while the site of the 45S rRNA genes are labeled red on one pair of chromosomes at the two sides of the secondary constriction on chromosomes with the nucleolar organizing region (NOR). The telomeres at the ends of the chromosomes are labeled green with a synthetic probe of the sequence TTTAGGG. Bar 5 µm. (After Castilho, A. et al. 2000. *Annals of Botany* 856: 837–844.)

Castilho et al. (2000) placed the chromosomes into four groups, comprising the largest chromosome (which hybridizes to 5S rRNA genes), a group of eight medium chromosomes; a group with six smaller chromosomes; and the smallest chromosome carrying the 18S–25S rRNA genes and a secondary constriction at the NOR (Madon et al. 1995, 2012; Castilho et al. 2000). As in most species, the genome of oil palm has abundant repetitive DNA sequences, DNA motifs that are repeated hundreds or thousands of times in the genome. The locations of three different types of highly abundant repetitive sequences are shown in Figure 9.2: a retrotransposon fragment that amplifies through an RNA intermediate, a clone that was selected to be abundant in the genome, and a microsatellite with the sequence motif (GATA). The three sequences, like their equivalents in other crop species (Heslop-Harrison and Schwarzacher 2011), have characteristic positions and abundances in the genome.

The whole-genome sequence of oil palm was published by Singh et al. (2013b). Comparison of the sequence assembly to the genetic linkage maps resulted in 16 genetic scaffolds representing 16 pseudochromosomes of oil palm. The total oil palm genome size has been estimated as 1.8 Gb (Rival et al. 1997), and the length of the assembled oil palm genome was 1.535 Gb, representing some 85% of total genome size although only 43% of this was assembled and assigned to chromosomes. The

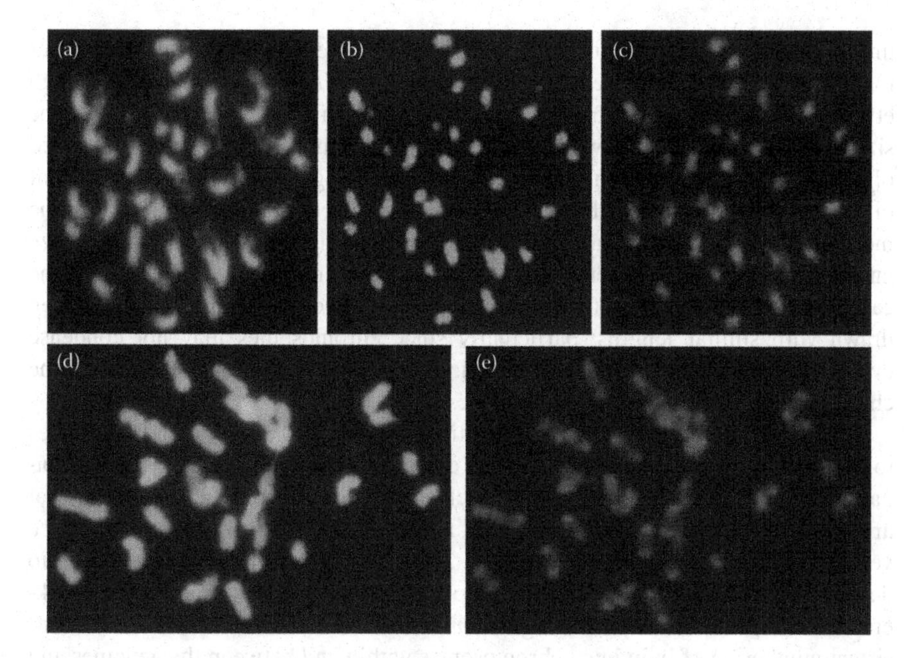

FIGURE 9.2 **(See color insert.)** Locations of some abundant repetitive DNA sequences on oil palm chromosomes shown by *in situ* hybridization. (a) DAPI stained metaphase chromosomes hybridized with (b) pEgKB17 (red) and (c) a fragment of a copia retroelement (green), showing that these sequences are abundant near the centromeres of all chromosomes. (d) DAPI stained metaphase chromosomes hybridized with (e) the synthetic oligonucleotide $(GATA)_4$, showing that it is more abundant in terminal regions of all chromosomes. (After Castilho, A. et al. 2000. *Annals of Botany* 856: 837–844.)

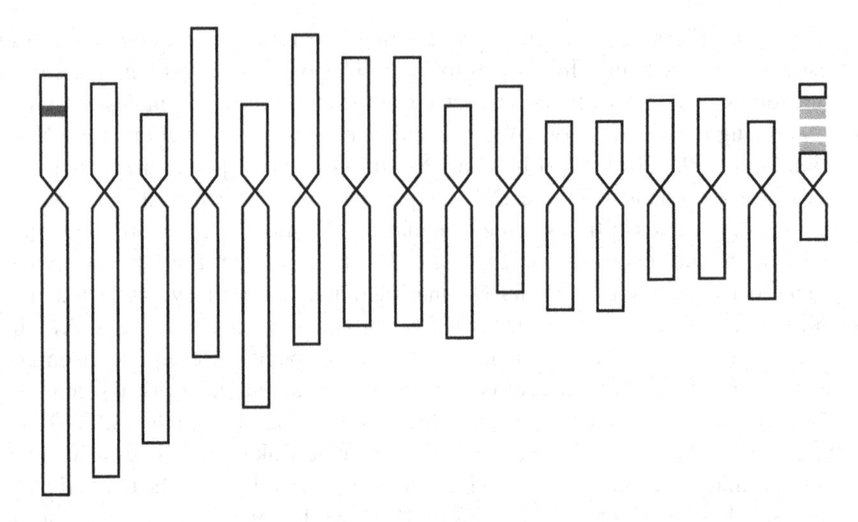

FIGURE 9.3 Ideogram showing the relative sizes and morphology of the 16 pairs of chromosomes in oil palm and location of 45S rRNA genes at the NOR (vertical constriction on the smallest chromosome).

unassembled sequences are likely to include mainly repetitive DNA sequences rather than low- and single-copy sequence. Based on the DNA sequence assembly, Singh et al. (2013b) has assigned numbers to the individual chromosomes according to the size of sequence scaffolds, which correspond to the linkage group in the selected oil palm mapping population. Figure 9.3 shows an ideogram of the chromosomes of oil palm with the arm sizes based on the lengths and centromere position from the most likely location based on chromosome measurements; the numbering is by size and not yet correlated by genetic or sequence data to the linkage groups. In some cases, groups of two pairs of chromosomes cannot be distinguished and have been drawn with similar lengths; particularly since chromosomes may not condense evenly and uniformly through mitosis, it is not possible to distinguish reliably some chromosomes based on morphology alone.

It is important to link the physical position of DNA sequences for whole genomes to individual chromosomes. In particular, crossovers giving rise to meiotic recombination are not uniformly distributed over chromosome arms and as a result, loci that are physically apart on the chromosomes may be linked on linkage maps and vice versa (Heslop-Harrison 1991; Wang et al. 2006; Sun et al. 2013). There is a need to develop a high-resolution cytogenetic map based on chromosomes and DNA markers for both species of oil palm (*E. guineensis* and *E. oleifera*). This will enable consistent numbering of individual chromosomes within and between the genomes and allow characterization of any translocations or recombinants in hybrids, integrating knowledge of the physical chromosomes with sequence data. The availability of the genomic DNA sequence and analysis to find low-copy regions will be able to identify probes to characterize the physical chromosomes. Preliminary work using random clones of low-copy sequences for *in situ* hybridization was able to show that many sequences were present on two pairs of chromosomes, confirmed by the sequence assembly data showing the evidence for an ancestral whole-genome duplication event (Singh et al. 2013b).

9.2.3 Epigenetic Modulation of Genomes and DNA Methylation

Molecular cytogenetics approaches are also able to examine the epigenetic modification of both DNA by cytosine methylation, and the histone proteins associated with the double-stranded DNA within the nucleus (Jaligot et al. 2011; Heslop-Harrison and Schwarzacher 2011, 2013). DNA methylation along chromosomes can be studied using antibodies to methyl-cytosine, showing where there are changes in bulk methylation at the chromosomal level. Figure 9.4 shows changes in overall levels of DNA methylation in oil palm nuclei. Complementing results from HPLC and restriction enzyme analysis of repetitive DNA sequences (Kubis et al. 2003), reduction in methylation following tissue culture can be seen, and it is clear that some methylation changes are related to abnormalities seen following tissue culture (see Ong-Abdullah

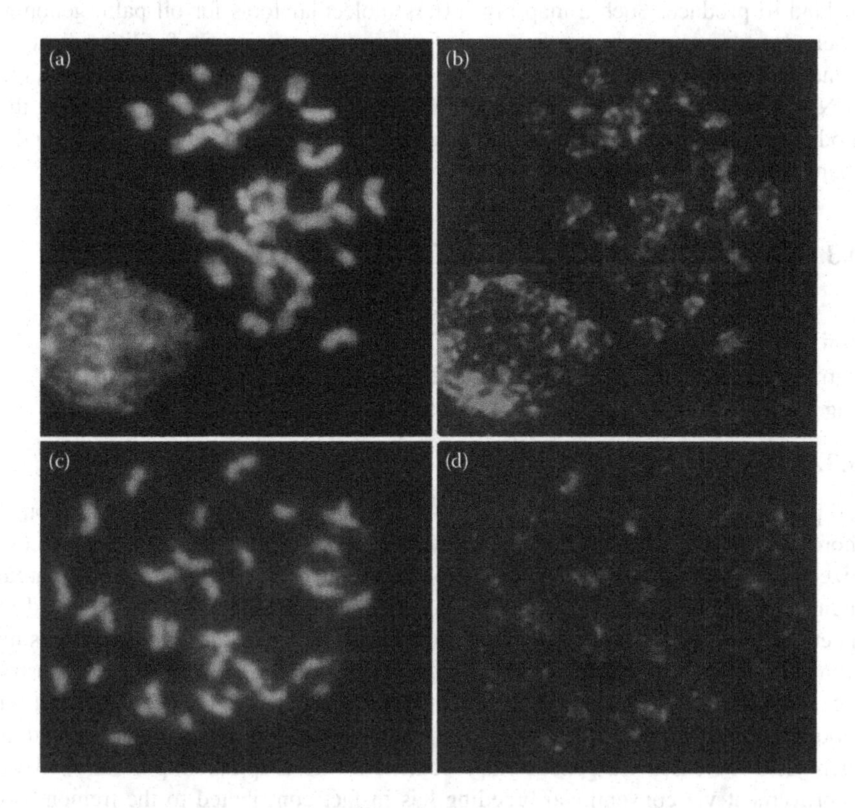

FIGURE 9.4 (See color insert.) Changes in overall levels of DNA methylation in oil palm nuclei and chromosomes after tissue culture, as detected using an anti-methyl-cytosine (b, d) to metaphase chromosomes stained blue with DAPI (a, c). Preliminary results indicate that (a, b) the parental chromosomes (ortet) show more methylation compared to the mantled regenerants (left, less methylated). (After Kubis, S.E. et al. 2003. *Plant Molecular Biology* 52: 69–79.)

et al. 2015). There are also prospects for more detailed immunocytochemistry to examine modifications of chromosomal histone proteins associated with gene expression or with meiotic pairing and recombination.

9.2.4 Conclusions

Study of the chromosomes of oil palm and other crops gives information about their structure, evolution, organization, and inheritance that is valuable for both fundamental research and application in breeding and hybridization. For application of the genomic sequence data, the incorporation of cytogenetic data (physical chromosomes) with genetic linkage data will contribute significantly to the improvement of sequence assembly by confirming the physical positions of markers on the linkage groups, and evaluating the size of the putative remaining gaps. This ability is particularly important in crops such as oil palm where there are only limited segregating populations for genetic analysis since these take decades and tens of hectares of land to produce. Such a map also offers molecular tools for oil palm genomic research, comparative genomics, and evolutionary studies. It facilitates understanding the inheritance of specific traits in oil palm. Further work with chromatin (DNA modifications and associated proteins) will be valuable for understanding the modulation of genomes during development and in epigenetic studies, including the changes occurring during tissue culture.

9.3 CONVENTIONAL TO MOLECULAR GUIDED BREEDING: A NEW ERA IN IMPROVING OIL PALM PRODUCTIVITY

Maizura Binti Ithnin, Amiruddin Mohd Din, Umi Salamah Ramli, Abrizah Othman, Yaakub Zulkifli, Susana Mustafa, Leslie Eng Ti Low, Meilina Ong-Abdullah, and Rajinder Singh

9.3.1 Background

Oil palm is a perennial species mostly grown as a source of vegetable oil, although more recently it is also gaining popularity as a biofuel feedstock. Oil palm is recognized as the highest oil-yielding crop, producing fourfold higher (per unit land area) than its competitors like soybean and rapeseed oils. The crop is ideally suited to meet the growing demand for oil, fats, and biofuel. However, meeting the increasing demand sustainably has to be the way forward, considering the scrutiny the industry receives from environmentalist groups worldwide. Satisfying the growing need for food and fuel is thus best achieved by producing more from existing areas, and as such yield increases via genetic improvement of the crop are important. Genetic improvement via conventional breeding has in fact contributed to the tremendous increase in oil palm yield. Yield increases over threefold from the early 1920s to the 1990s have been attributed to progress in oil palm breeding and agronomy (Basiron 2000). Breeding identifies and utilizes genetic variation for improvement. The long-term goal is to maintain or increase the levels of genetic diversity in breeding material, to allow improvements in productivity and resistance to pests and disease, in addition to incorporating novel features like a change in the fatty acid profile. In this context, the Malaysian oil palm industry has been fortunate, with the large germplasm collection amassed since the 1970s playing a pivotal role in moving the oil palm traits within breeding programs in several directions, such as producing more liquid oil in higher-yielding palms with low height increment. Conventional breeding will be the way forward for genetic improvement although the efficiency has to be improved using modern genomic tools.

9.3.2 Limitations of Conventional Oil Palm Breeding

The efficiency of a breeding and selection program for a crop can be assessed in many ways, including the frequency at which new varieties are released. The reality for oil palm is that improved varieties are slow to develop via traditional breeding as one breeding cycle can take up to 10 years (Rajanaidu et al. 2000). This, coupled with the fact that many rounds of evaluation and selection are required to develop well-tested progenies, makes conventional breeding a challenging endeavor. Another major logistic and cost issue is the fact that only 136–160 palms can be evaluated in one hectare of land, limiting the number of lines that can be carried forward in the evaluation and selection program. However, reasonably large numbers of recombinant progeny are required as desirable genes can be in genetic linkage with undesirable traits. As such, careful observation of large number of palms in the breeding trial is important to get the desirable mix, a process that contributes to the high cost of field trials. Despite these constraints, oil palm breeders have made tremendous progress in developing newer planting materials with improved yields (Corley and Lee 1992).

However, to move forward, considering the scarcity of arable land, the competition from other vegetable oil crops and the desire to improve productivity in a sustainable manner, the application of modern genomic tools in oil palm breeding has become a necessity. This is especially so when there is a need to widen the genetic base and allow the introgression of desirable alleles from selected germplasm into the advanced breeding lines.

The efforts carried out in characterizing the germplasm using molecular markers for effective selection of palms for conservation and for associating genes or markers to specific phenotypes for early selection with an aim to enhance breeding efficiency are described below.

9.3.3 Germplasm Management, Characterization, and Conservation Using Molecular-Based Assays

MPOB holds a comprehensive collection of oil palm genetic materials. More than 2000 accessions have been accumulated for *E. guineensis* as well as *E. oleifera* from their current centers of distribution, respectively, in Africa and South and Central America. The genetic materials are maintained in *ex situ* living field plots and at a planting density of 148 palms/hectare, with approximately 600 hectares of land allocated to accommodate more than 100,000 germplasm palms for conservation. A comprehensive collection of the wild genetic pool offers a wide selection of genes and traits for crop improvement. However, conservation efforts for perennial species like oil palm are expensive and laborious, due to the high long-term maintenance cost, huge land areas involved, and large number of workers employed. Characterization of the MPOB germplasm collection and the potential development of a core collection based on marker analysis has been described in Chapter 3. Access to diverse germplasm is the fundamental requirement for oil palm breeders.

9.3.4 Unraveling the Genome and the Discovery of Genes Influencing Agronomic Traits

In order to implement marker-assisted selection (MAS) or a genomics guided breeding program, there is a necessity to link marker(s) and/or gene(s) to traits of interest in oil palm. This information is critical to select palms early, even before the characteristic in question is manifested. At the very least, a reasonable number of markers well distributed across the entire genome are needed for GS, where it is not necessary to explicitly link markers to a trait and is an endeavor that is now being explored in oil palm (Wong and Bernardo 2008; Cros et al. 2015a,b).

Although markers can be generated in relatively large number without sequence information (e.g., amplified fragment length polymorphism [AFLP], random amplified polymorphic DNA [RAPD]), these are usually dominant in nature and may have limited utility in large-scale application for crop improvement. The more robust markers like SNPs usually require some form of sequence information. The lack of in-depth sequence information for oil palm up to the late 1990s only allowed the exploitation of a limited number of restriction fragment length polymorphism (RFLP) markers (Jack et al. 1995) or dominant markers like RAPD (Shah et al. 1994) and AFLP (Singh et al. 1998) for genome analysis. In order to circumvent the lack of sequence information, Billotte et al. (2001) successfully constructed an SSR-enriched library. This proved to be an important source of markers for genetic map construction (Billotte et al. 2005) and the mapping of the shell gene locus, an important monogenic trait described in greater detail below. Extending on this work, Billotte et al. (2010), constructed a multiparental linkage map for identifying QTLs linked to yield components and vegetative parameters. Although the increased population size brought about by the use of multiple linked families identified a large number of QTLs—76 altogether—linked to the 24 traits analyzed, only a small number were consistently detected across the different families, indicating their limited applicability in a large-scale MAS program.

Starting in early 2000s, there was a concerted effort by several groups to start generating sequence information for oil palm. As in most crops, the generation of expressed sequence tags (ESTs) was initially used to understand the complexity of gene expression and interaction in oil palm (Jouannic et al. 2005; Ho et al. 2007; Low et al. 2014; Shariff et al. 2008). The ESTs approach was especially useful for oil palm as only a handful of genes had been cloned in the early days and most of these were associated with fatty acid synthesis (Shah and Rashid 1996). The initial work on ESTs led to cataloguing of genes associated with male and female inflorescences, shoot apices, zygotic embryos, and roots (Jouannic et al. 2005; Ho et al. 2007). Subsequently, Low et al. (2014) obtained an overview of genes expressed during oil palm tissue culture, in order to try to understand the molecular mechanisms associated with clonal propagation in oil palm. More importantly, the work catalogued over 200 putative genes with full-length open reading frames, as well as identified lipid transfer proteins that were highly expressed in embryogenic tissues—providing a possible lead in differentiating embryogenic and nonembryogenic tissues.

However, in ESTs projects, intermediate and highly expressed genes tend to be catalogued more effectively. Genes that are expressed in lower abundance and whose expression is restricted to certain tissue types are more difficult to uncover. These genes need to be identified as they play an important role in plant development (Shary and Guha-Mukherjee 2004). It is for this reason that MPOB also embarked on a program to identify all possible classes of genes by constructing normalized cDNA libraries for oil palm (Chan et al. 2010), which theoretically should contain equal representation of genes in a particular tissue. In this endeavor, close to 59% of the ESTs identified in the normalized libraries were not observed in standard cDNA libraries constructed previously (Chan et al. 2010). However, constructing normalized libraries was a technically challenging endeavor, and not employed widely.

The ESTs generated for oil palm were actively used as RFLP molecular probes (Cheah et al. 2000; Singh et al. 2008a) and also proved to be an excellent source for mining of SSRs (Singh et al. 2008b) and SNPs (Riju et al. 2007). These markers were incorporated into the ongoing genome mapping program at MPOB (Singh et al. 2011). They proved to be an excellent source for probes of known identity, and as such, the molecular maps generated were constructed from known genes. The RFLP and SSR probes generated were eventually used to identify genomic regions associated with fatty acid composition (Singh et al. 2009) and tissue culture amenity (Ting et al. 2013). However, at this stage, the marker–trait linkage was not close enough for routine application in breeding.

Nevertheless, the RFLP and SSR markers developed from the EST collection proved to be an excellent source of probes for DNA fingerprinting, both for tissue culture and breeding. The RFLP probes were shown to be useful for monitoring uniformity of tissue culture lines and for distinguishing different clones (Cheah et al. 2000). Such a method for distinguishing clones has important implications when the clones are registered as a new variety under the Plant Variety Act (Chapter 11). Although the DNA fingerprint profile itself is not a precondition for the registration of a new variety, it is useful in case of dispute.

The generation of ESTs (via conventional and normalized libraries) had enabled a tissue-specific profile of expressed genes to be created for oil palm. At the same time, they also proved to be a valuable source of molecular markers. Nevertheless, the EST approach does have its limitations, as it only samples expressed genes, which represent a small percentage of the oil palm genome for any particular tissue. ESTs also do not provide any clues on the regulatory regions that control the expression of these genes. In fact, in maize, it was recently demonstrated that 70% of the variance associated with complex traits can actually be mapped back to regulatory regions (Wallace 2014). By focusing only on ESTs, which in the early 2000s were the only cost-effective method for a complex genome like oil palm, the ability to understand the functional components of the nearly 1.8 billion bases in oil palm was severely limited.

9.3.4.1 Sequencing the Hypomethylated Genome of Oil Palm

In order to move away from just sequencing the expressed genes, MPOB in 2004 embarked on a program to exploit the *GeneThresher*™ technology developed at Cold Spring Harbor Laboratory (Rabinowicz et al. 1999). This technique allows the preferential selection of hypomethylated regions of the genome, which can then be sequenced using Sanger technology. As only the hypomethylated region is sequenced, which represents a small section of the genome, the cost for sequencing a complex genome like oil palm is thus dramatically reduced. In the effort carried out, close to 300,000 and 150,000 sequences from the hypomethylated regions of *E. guineensis* and *E. oleifera* genomes, respectively, were generated (Low et al. 2014; Budiman et al. 2005). The estimated hypomethylated genome size of *E. guineensis* was 705 MB (Figure 9.5) (Low et al. 2014). In addition to mining of interesting genes such as those associated with diseases resistance, transcription factors and 14 microRNA (miRNA) were also uncovered, providing an important resource to start deciphering the molecular mechanism of important agronomic traits in oil palm.

In addition to rapid gene discovery, the *GeneThresher* technology was used to uncover tens of thousands of genetic markers in several lines. Optimal discovery of markers for molecular breeding guided the final selection of palms sequenced in the gene and marker discovery stage of the oil palm *GeneThresher* project. More than 33,000 SSR and over 40,000 high-quality SNP markers were identified, representing an impressive collection of genetic markers for oil palm (Low et al. 2014). This would prove to be an important resource for genetic mapping and association studies. In fact, the SNP discovery led to the development of the first oil palm SNP array, a 4500 customized SNP panel that allowed the construction of high-density

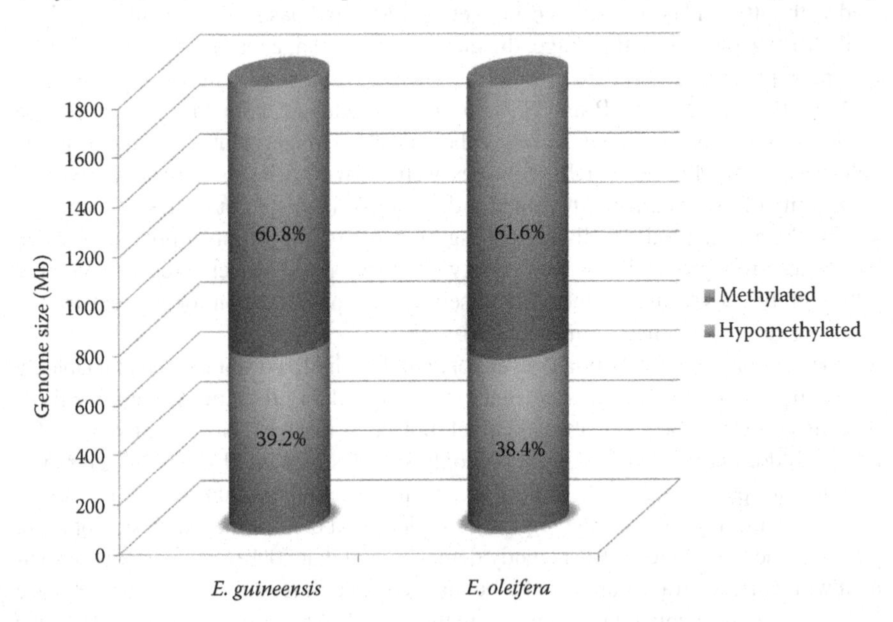

FIGURE 9.5 Hypomethylated regions of the oil palm genome determined using *GeneThresher* technology.

genetic maps of two independent oil palm hybrids (Ting et al. 2014). The genetic maps lay the foundation for determining the chromosomal location of genes affecting simple and complex traits. This information is critical in implementing MAS or molecular breeding, an approach that is well suited to a perennial crop like oil palm.

The discovery of a large number of SNPs is encouraging as SNP marker technology will undoubtedly form the basis of a comprehensive yet cost-effective platform for the development of new varieties of oil palm via marker-assisted breeding.

9.3.4.2 Whole-Genome Sequencing

Although deciphering the ESTs and the hypomethylated regions of oil palm are the way forward for oil palm improvement, it still falls short of decoding the whole genome. The need to further understand the genome led a private plantation company, Asiatic Centre for Genome Technology (ACGT), to announce the sequencing of the oil palm genome in 2008. About a year later, another private plantation company, Sime Darby, announced that they had also sequenced the oil palm genome using the NGS technology. Although both efforts were commendable, the data were not available for public access. Also exploiting the NGS technology, MPOB also embarked on sequencing the genomes of both species of oil palm in 2009. Using 454/Roche technology, MPOB sequenced the genome of *E. guineensis* and *E. oleifera* palms and made the data publicly available (Singh et al. 2013a). Additionally, from a 12× Bacterial Artificial Chromosome (BAC) library, a high information content physical map and associated BAC end sequences were produced. Genome assembly of *E. guineensis* was enhanced by inclusion of BAC end sequences. BAC fingerprint contigs were merged with sequence scaffolds and with an oil palm genetic map so as to provide an integrated genome-wide sequence map of oil palm. The reference genomes were annotated with 454-based transcriptome sequences of more than 30 tissues. The 1.535 Gb of assembled sequence in 40,360 scaffolds (N50 = 1.045 Mb) predicted at least 34,802 genes (Singh et al. 2013a). Comparison with the draft sequence of *E. oleifera* suggested a paleotetraploid origin for palm trees.

It was also interesting to note that when the sequences generated from the hypomethylated region were mapped back to the genome build, they mapped perfectly to the gene-rich regions in each chromosome (Singh et al. 2013a). This provided clear evidence that sequencing of the hypomethylated region earlier had uncovered the gene space for oil palm. This is consistent with the expectation that in plants, genic regions are usually less methylated compared to repetitive regions, which were mostly removed during library construction in the *GeneThresher* program.

The oil palm genome sequence released by MPOB provides an accurate record of genes, predicted proteins, and other genomic elements and is a valuable reference for crop improvement. The sequence is expected to provide a framework for faster and more effective breeding methods, and this was very quickly realized as described below.

9.3.5 Realization of Marker-Assisted Selection to Improve Breeding Efficiency

One of the most important economic traits of the oil palm relates to how the thickness of the shell correlates to fruit size and oil yield. The genome sequence facilitated the identification of the SHELL gene, responsible for the three fruit forms: *dura* (thick-shelled), *pisifera* (shell-less), and *tenera* (thin-shelled). The inheritance of the shell gene was assessed in a population derived from the self-pollination of the *tenera* palm, T128, from MPOB's Nigerian germplasm collection. The presence and absence of shell is the most important monofactorial trait in oil palm, because of the increase in oil yield from palms carrying the heterozygous locus (*tenera*) of the gene, which is the preferred commercial planting material. One homozygote (*dura*) has on average 30% less oil yield than the *tenera* palms and the other homozygote (*pisifera*) is often female sterile, producing no oil yield. Using information derived from the genome build as well as homozygosity mapping of the *AVROS pisifera* lines over five generations, the shell gene responsible for the different fruit forms was identified (Singh et al. 2013b). Two independent mutations were found in the *Shell* gene, which codes for a homolog of the MADS box gene *SEEDSTICK*. The mutations observed correlate almost perfectly with the different fruit forms and allowed the development of a molecular diagnostic assay known as *SureSawit*Shell to differentiate the fruit forms early in the nursery before field planting. This enables significantly enhanced breeding operations. Currently, it can take four to six years to confirm whether an oil palm tree possesses the desired fruit form. The *Shell* marker is an important quality control tool in commercial seed production, as its application could avoid planting of the low-yielding *dura* in the commercial plantations.

More importantly, the diagnostic assay for the first time provides breeders with a tool to implement MAS in their breeding program, especially in the development of the male (*pisifera*) parental lines. Owing to female sterility of the *pisiferas*, the male parental populations are usually produced using a *tenera* × *tenera* or *tenera* × *pisifera* crossing scheme. All three fruit forms (*tenera*, *dura*, and *pisifera*) are offspring from the former, whereas only *tenera* and *pisifera* palms are offspring from the latter. Oil palm breeders are now, for the first time, equipped with a tool to selectively identify *pisifera* palms in the nursery and focus on planting only the *pisiferas* in the field. The unwanted *dura* palms (up to 25% in the *tenera* × *tenera* cross) can be removed thus, saving land and other resources in the progeny-testing exercise. This significantly improves breeding efficiency and can fast-track the seed production for commercial planting and the development of new and improved varieties.

The unraveling of the oil palm genome sequence also paved the way for the discovery of the gene influencing another important agronomic trait, the fruit color. The fruit of the oil palm can vary considerably based on external appearance, most notably the color of the fruit. The most common fruit is known as *nigrescens*, which is deep violet to black at the apex and yellow at the base when unripe, changing to dark red upon ripening. The other major fruit type is *virescens*, which is green before ripening and exhibits a striking color change to reddish-orange at maturity. Both fruit phenotypes are found in nature, although *virescens* are found at lower frequency, despite the trait being dominant to *nigrescens*. However, *virescens* palms may be more desirable to planters as the marked difference in color between ripe and unripe bunches makes it potentially easier to identify ripe bunches, particularly in tall palms, where bunches can be obscured by fronds. The inheritance of the exocarp color of oil palm fruits was initially assessed in the same population derived from the self-pollination of the *tenera* palm, T128, from MPOB's Nigerian germplasm collection. The data, as expected, suggested that one dominant Mendelian locus controlled the trait. Subsequently using the reference genome of *E. guineensis*, an R2R3 MYB transcription factor was found to be responsible for controlling the exocarp color of the fruit (Singh et al. 2014). Surprisingly, five independent mutant alleles were identified in over 400 palms from MPOB's *E. guineensis* germplasm collection from Africa. The identification of the gene has potentially important implications in improving harvesting standards and hence oil yield.

For the two traits described above, a single gene explains the differences in the traits observed. However, a number of the more desirable agronomic traits such as yield components, oil quality, and vegetative parameters, including height, are likely under polygenic control (Singh and Cheah 2005). Advances in sequencing technology will also assist in dissecting complex traits in oil palm. Using genotype-by-sequencing (GBS) implemented via NGS technology, namely, ion torrent, Pootakham et al. (2015) identified over 3000 informative SNP markers. Of these, 1086 were placed on a genetic map constructed from the selfing of a cloned *tenera* palm. Interestingly, the group identified several QTLs linked to height and of particular interest was a QTL on linkage group 14. The sequence of the SNP markers flanking the QTL region were mapped back to the publicly available genome sequence (Singh et al. 2013a), which allowed the identification of two candidate genes in the region, namely, the Della protein GAI1 and a gibberellin 2-oxidase, both of which have been implicated in regulating plant height in other crops (Sun and Gubler 2004). Admittedly, additional work has to be done before concluding that these two genes are actually involved in regulating plant height in oil palm. This is even more the case after another group using a different genetic background reported a distinct QTL region linked to oil palm stem height (Lee et al. 2015). The researchers also mapped the QTL region to the publicly available genome sequence and uncovered an interesting candidate gene, namely, an asparagine synthase-related protein that has been associated with the dwarf phenotype (Sivori and Alaniz 1973; Alonso et al. 2009). This points to the fact that unraveling polygenic-based traits is going to be a challenge with many candidate genes being uncovered for further analysis. Nevertheless, the work does indicate the possibilities offered by the advent of NGS technology and the availability of a genome sequence in deciphering the molecular basis of complex traits in oil palm. This was clearly demonstrated recently where the availability of the genome build allowed the epigenome of oil palm to be unraveled and with it the examination of DNA methylation alterations in clonal palms. As a result, the molecular mechanism responsible for clonal abnormality related to mantled fruits was uncovered (Ong-Abdullah et al. 2015), making large-scale tissue culture of oil palm feasible. The mantled abnormality was the subject of intense research and speculation over the last three decades and the advent of NGS technology and with it highly refined genome and epigenome builds was instrumental in solving the puzzle.

9.3.6 Moving to the Post-Genomics Era: Proteomics and Metabolomics Applications in Oil Palm Improvement

In parallel with the current progress in oil palm genomics, proteomics and metabolomics are recognized to be very important for post-genome studies. Recent developments in mass spectrometry have made it possible to acquire comprehensive analyses of proteins from the oil palm, and in particular, to elucidate the precise functions of individual proteins differentially expressed in cells. In addition to genomics, transcriptomics, and proteomics, metabolomics are also essential to obtain a further understanding of each physiological and biological function of proteins.

The rapid growth of the palm oil industry has brought about many new challenges to sustainability and competitiveness. These challenges encourage aggressive research efforts into proteomics and metabolomics toward value addition and improved productivity through the identification of novel compounds and biomarkers for specific traits, including oil palm disease. An improved understanding of oil palm fruit maturation would provide indications for future marker-assisted selection breeding programs aimed to improve oil palm fruit quality. Protein profiling was developed based on multiple platforms, including gel and liquid chromatography combined with tandem mass spectrometry, to profile the oil palm proteome from various tissues, including mesocarp during fruit ripening (Hasliza et al. 2014; Hasliza and Ramli 2015; Lau et al. 2015, 2016). The established procedures would provide a solid foundation for further functional studies, including fatty acid and lipid biosynthetic expression profiling and evaluation of regulatory function. For instance, the identification of differentially expressed proteins associated with physiological processes during oil palm fruit ripening provides information toward understanding the metabolic process and molecular mechanisms involved in lipid biosynthesis affecting yield and oil production. The proteomics studies also generate large volumes of raw experimental data and inferred biological results. These data will be deposited into centralized data repositories developed for oil palm proteome to facilitate the dissemination and access.

The palm oil industry generates several by-products. More than half of the dry weight of the waste is oil palm leaf, a tissue that is underutilized. Metabolomics promises a powerful tool in discovering potential bioactive compounds from oil palm fronds. Derivatives of flavones conjugated with hydroxymethylglutaric (HMG) acid were identified from the oil palm leaves using LC-MS/MS and this finding will further promote efficient utilization of this agricultural by-product (Tahir et al. 2012, 2013).

Metabolomics studies on fungal plant pathogens are only just beginning, to dissect many facets of the pathogen and disease (Tan et al. 2009). Similarly, in oil palm, these techniques have been applied toward understanding disease mechanisms and for biomarker discoveries for early detection of basal stem rot disease largely caused by the fungus *Ganoderma* f.sp. (Idris et al. 2004; Paterson et al. 2008; Cooper et al. 2011). Managing the basal stem rot disease in oil palm using chemical (Susanto et al. 2005) and biological control agents such as *Trichoderma harzianum* and endophytic bacteria have been reported to have some beneficial effects (Sundram et al. 2008; Sapak et al. 2008). However, oil palm disease management is complicated by the presence of multiple *Ganoderma* species, which exhibit different characteristics and aggressiveness, and the dominant species for BSR in oil palms can differ based on locality (Wong et al. 2012). Proteomics analyses detected plant defense and stress response-associated proteins such as β-1,3-glucanase, nucleoside diphosphate kinase, glutathione-*S*-transferase, early flowering protein 1, ferritin, and thioredoxin H2 in infected standard (Deli × AVROS) palm roots at 1-week post-infection

(Syahanim et al. 2013). β-1,3-Glucanase, a pathogenesis-related protein class 2, is widely known to be induced during fungal response in many plants such as wheat (Menu-Bouaouiche et al. 2003) and tomato (Benhamou et al. 1989). The activity of glucanases and chitinases was higher in the diseased tissues infected by *Ganoderma* sp. than in healthy tissues of oil palm (Siswanto and Darmono 1998). Preliminary primary metabolic profiling at early weeks of infection revealed accumulation of metabolites, including chelidonic acid, apigenindiglycoside and triglycoside, luteolintriglycoside, and chelidonic–vanilloyl conjugate. Longer infection showed higher abundance of chelidonic acid in standard (Deli × AVROS) palm seedlings treated with rubber wood blocks fully colonized by *Ganoderma boninense* (Shahirah Balqis et al. 2015).

The first demonstration on the potential use of metabolite profiling in oil palm to distinguish highly tolerant and susceptible parental palms was also recently reported (Nurazah et al. 2013). Figure 9.6 shows an LC-MS base peak chromatogram (BPC) of oil palm root extracts from the highly tolerant and susceptible parental palms. Nine distinctive peaks based on peak height were identified. Identification of the compounds using tandem mass spectrometry and fold increment of ion intensity in the tolerant and susceptible parental palms were also reported (Nurazah et al. 2013). Monitoring proteome and metabolome changes among standard, tolerant, and susceptible progenies will also reveal if progenies resistant to basal stem rot disease produce unique proteins and metabolites and how they overcome the infection. Results will provide information that can be employed to develop metabolo-proteomics techniques for early detection of the disease. The information can also be used by breeders in their effort at developing *Ganoderma*-resistant oil palm (Figure 9.6).

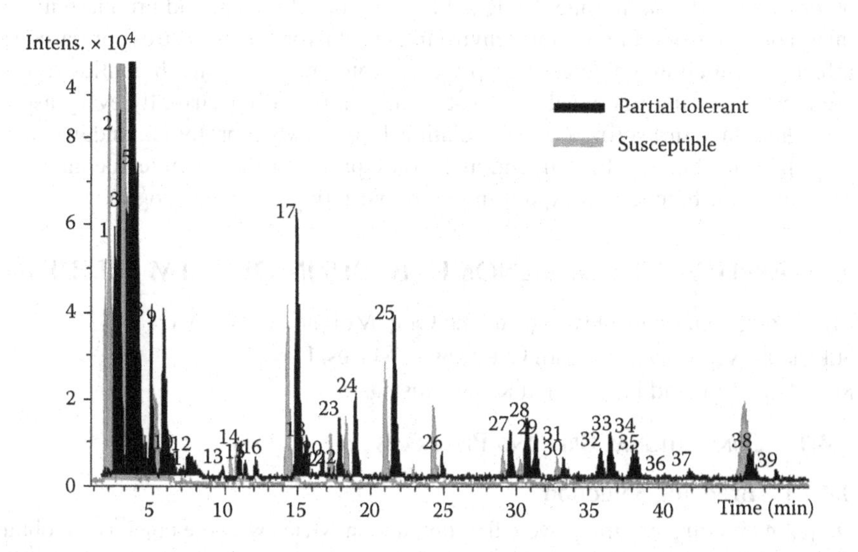

FIGURE 9.6 Liquid chromatography-mass spectrometry (LC-MS) base peak chromatogram (BPC) of partially tolerant and susceptible (shaded) parental palm root extracts. Identity of compounds that show qualitative difference based on peak height were identified by mass spectrometry.

9.3.7 Conclusions

Breeding progress in oil palm and the crop's perennial nature has provided a competitive advantage over the other oil crops. Nevertheless, to remain competitive, oil palm breeders will need to apply newer strategies incorporating modern diagnostics assays, in order to develop new and improved varieties at a much faster pace. The availability of the oil palm genome sequence has already paved the way for this effort. The identification of genes responsible for two important monogenic traits—shell and fruit color—was largely possible due to the availability of the high-quality genome build. This will have a profound effect on oil palm breeding. The development of the molecular diagnostic assay for *Shell* known as *SureSawit*Shell has for the first time made marker-assisted selection a reality for oil palm. The discovery of the fruit color gene also paves the way for the development of new and improved *virescens* planting materials. The genome sequence of oil palm will be a rich source of information for oil palm breeders and producers as well as geneticists and evolutionary biologists. It will also allow the deciphering of the molecular basis of more complex traits. A case in point is that it facilitated the mapping of epigenetic alterations that result in the fruit abnormalities, which currently restrict the use of clones in commercial plantings. The dense representation of sequenced scaffolds on the genetic map will facilitate the identification of genes responsible for other yield and quality traits such as fatty acid composition, plant stature, and disease resistance. The well-characterized oil palm germplasm could be the source for tagging of the alleles or genes linked to these traits. This will allow the incorporation of desirable genotypes into advanced breeding lines to enhance the genetic base and improve yield and other traits at the same time. Undeniably, plant development and productivity are inevitably controlled by various environmental factors such as stress or pathogen infection, which may decrease crop yield. Proteomics and metabolomics studies are now beginning to contribute substantially in revealing virtually every aspect of plants, thus unraveling possible relationships between protein abundance and/or metabolic changes. Protein and metabolite profiling will in time become powerful tools for breeders to exploit in the oil palm improvement program.

9.4 APPLICATION OF GENOMIC TOOLS IN OIL PALM BREEDING

Chee Keng Teh, Qi Bin Kwong, Ai Ling Ong, Mohaimi Mohamed, Sukganah Apparow, Fook Tim Chew, Sean Mayes, David Ross Appleton, and Harikrishna Kulaveerasingam

9.4.1 Conventional Breeding Programs

9.4.1.1 Breeding Selection

Oil palm breeding programs were first initiated in Africa with the objective to obtain high-yielding materials for commercial planting. In the early years, apart from the Department of Agriculture in Malaysia, almost all the breeding programs were carried out by private plantation companies. One of the major challenges for the breeding of perennial crops is to achieve sufficient selection intensity and reasonable breeding gains while maintaining genetic variability for selection gains in the development of future generations. Owing to the poor quality of the available *dura* origins in Africa, *tenera* thin-shelled fruiting palms were adopted, making way to a series of carefully conducted crossing experiments that led to the discovery of the single gene inheritance of the shell thickness trait in 1941, from various test crosses made between *dura*, *tenera*, and *pisifera* (Beirnaert and Vanderweyen 1941). The study also found that the cross between *dura* and *pisifera* produced 100% *tenera*, which was the best combination for high oil yield due to a higher bunch number and thicker mesocarp. To date, the same approach is still being practiced for selecting the elite *dura* and *pisifera* parents, based on the performance of their *dura* × *pisifera* (*tenera*) progeny.

9.4.1.2 Constraints

Two selection schemes, that is, reciprocal recurrent selection (RRS) and family-and-individual palm selection (FIPS) are generally deployed in oil palm breeding. Figure 9.7 illustrates current breeding practice in oil palm. FIPS predominantly relies on the selection of individual palms with the best performance, while RRS is based on the breeding values determined by progeny testing (Corley and Tinker 2003). Each scheme indeed has its *pros* and *cons* and many plantations in effect practice a combination of the pure approaches, partly due to the female sterility present in many *pisifera* origins. Unlike the African-based populations, the Deli palms derived from four *dura* palms planted at Bogor Botanical Garden in 1848 (Pamin 1998) exhibit better trait uniformity, thicker mesocarp, high bunch number, and high mesocarp oil yield. The oil palm breeding programs in Malaysia thus emphasized the genetic

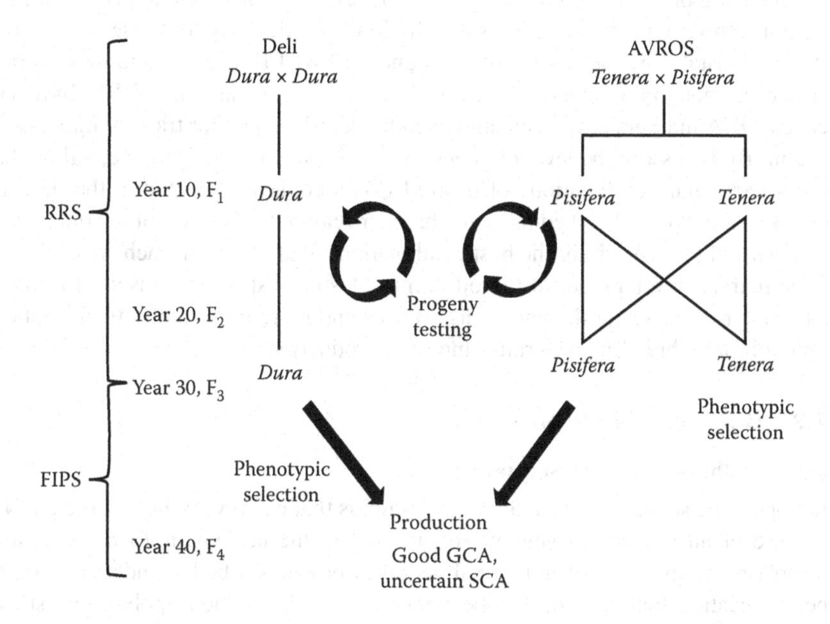

FIGURE 9.7 Conventional selection schemes in oil palm breeding. RRS—reciprocal recurrent selection; FIPS—family-and-individual palm selection; GCA—general combining ability; SCA—specific combining ability.

improvement of the Deli *dura* population through pedigree selection to produce high-yielding commercial *dura* materials until the 1960s. At that time, FIPS was the best method to improve yield and bunch component traits by selecting *dura* palms based on their phenotypic performance with no requirement in *dura* parents to evaluate their combining abilities. The *dura* × *pisifera* planting material was only accepted in Southeast Asia after its success in Africa. The commercial *dura* × *pisifera* breeding programs kept the Deli *dura* populations and AVROS *tenera*/*pisifera* populations separate. Owing to the direct evaluation of breeding values through progeny testing, RRS permits the exploration of specific combining ability (SCA) as well as general combining ability (GCA). However, the scheme requires a large progeny-testing population (Rosenquist 1990) derived from a small number of parents. Large-scale combinatory crossing for the evaluation of individual *dura* and *pisifera* combinations is not feasible due to resource limitation. This may then result in the best combination for GCA and SCA not being detected, which can hinder genetic improvement (Hardon 1970). RRS requires two generations of more than 20 years to complete a selection cycle. One generation is to propagate the Deli parents and the AVROS parents, respectively, followed by another generation of progeny testing between the parents. The breeders may lose some of the elite Deli and AVROS parents due to aging, disease infection, etc. after the completion of a selection cycle. This has become a more significant issue recently with the growing problem of *Ganoderma* infection and basal stem rot losses within palm stands. As a solution, selected palm sibs are often used to replace lost palms for the next cycle based on GCA, which unfortunately dilutes the power of RRS. Most breeding programs in Malaysia have shifted back to FIPS, focusing on GCA to reduce the dependency on progeny testing. However, the selection response for this method is usually lower particularly for those phenotypes with poor heritability, such as fresh fruit bunch (FFB). FIPS requires more selection cycles, even though the interval per cycle is shorter in comparison to RRS. In recent decades, DNA markers have been employed to identify QTLs for traits of interests in oil palm. By accessing the level of genetic polymorphism in the genome, palms that possess particular combinations of desired QTLs could be selected at the nursery stage. By using the trait-linked markers, breeders can reduce conventional phenotyping cycles and also enrich for the best combinations of alleles from each cross. So far, genetic marker development in the oil palm industry is still in its discovery phase. This chapter discusses both genetic discoveries and more importantly their applications to oil palm breeding programs and seed production.

9.4.2 Boom in DNA Sequencing

9.4.2.1 Whole-Genome Sequence

Whole-genome sequencing is a laboratory process that can reveal the complete DNA repertoire of an organism's genome, distributed in the nucleus, mitochondria, and chloroplasts. Sequencing of individual genomes provides a better understanding of genetic variation both within and between species. The genetic polymorphism in populations of an organism becomes extremely important when the inherited differences in genic or other portions of the genome are found to be directly responsible for phenotypic variants, influencing an individual's risk of disease and or affecting their fitness in the environment. Hence, whole-genome sequencing is the essential step to localize DNA variants that may contribute to both major and minor genetic effects of the targeted phenotypes, followed by a systematic and complete functional characterization of the genome. The three billion DNA base pairs (bp) of the human genome were sequenced and completed in 2003 under the Human Genome Project (HGP), an international public project led by the United States. With the published genomes, the HGP has facilitated the discovery of more than 1800 disease genes. The scientific communities are seeking to duplicate the same achievements in agricultural crops. As for commercial crop genomes, rice (*Oryza sativa* L.) was first sequenced and published by The International Rice Genome Sequencing Project (IRGSP) in 2000. A series of genome improvements were subsequently achieved, by correcting the reassembly, with a reduction in redundancy and an increase in genome completeness. The current release of the rice genome 7.0 was published in the Rice Annotation Project Database (http://rice.plantbiology.msu.edu/annotation_pseudo_current.shtml) in 2012. A total of 373.2 Mbp of non-overlapping rice genome sequence and 55,986 genes (loci) from the 12 rice chromosomes were reported. This represents only a slight increment in genome size and number of identified genes compared to release 4.0 (372.1 Mbp and 42,653 genes) (Ouyang et al. 2007), suggesting that saturating coverage of the genome has been achieved. The oil palm, as the most traded and dominant source of vegetable oil production in the world, was only sequenced and became publicly available in 2013. As for the 1.8-Gbp oil palm genome, 1.535 Gbp of assembled sequence and transcriptome data from 30 tissue types were used to predict at least 34,802 genes (Singh et al. 2013b). About 53% of the genomic scaffolds are represented in a physical map with 16 linkage groups, which are equivalent to the chromosome pairs (diploid number, 2n = 32) in oil palm. The genome appears to be of medium size among monocotyledons. It is larger than banana (0.60 Gbp) (Sagi et al. 2005) and soybean (1.15 Gbp) (Swaminathan et al. 2007), but smaller than barley (5.00 Gbp) (The International Barley Sequencing Consortium, 2012) and maize (2.30 Gbp) (Messing et al. 2004). The published genome sequence has become an important reference for genetic discovery programs that are critical to expedite breeding progress in oil palm.

9.4.2.2 Resequencing of Various Palm Populations

With the existence of the reference genome, whole-genome resequencing is the most comprehensive and cost-effective method to discover or capture polymorphism in the genome (Feuillet et al. 2011). The polymorphism can be categorized as either large or small variations caused by different mutation mechanisms. The popular SNPs and SSRs are widely utilized as molecular markers due to their high abundance, polymorphism, reproducibility, and codominance. Resequencing of pooled individuals provides a quick genome-wide polymorphism survey at very moderate cost and is also useful for the estimation of allelic frequencies, which is essential for population genetic studies. However, this approach does not permit LD analysis, compared to individual resequencing. At the Sime Darby Plantation R&D Centre, 132 individual oil palms (*E. guineensis* and *E. oleifera*) belonging to 59 diverse populations were pooled for resequencing, followed by further individual sequencing of 55 representative palms (one palm per population). The pooled sequencing data yielded more than seven million raw SNPs. About one million of the *E. oleifera*-specific variants were then separated out, based on the individual sequence data. After considering good sequencing depth from the pooled sequencing data, about three million high-quality SNPs were identified for further analysis and validation. The genome and SNP sequence data were then available to facilitate an understanding of the genetic basis of underlying QTLs of important agronomic traits in oil palm.

9.4.3 Development of DNA Markers

DNA markers, such as RFLP, AFLP, and RAPD, were initially developed to evaluate the genetic diversity of oil palm germplasm (Shah et al. 1994; Barcelos et al. 2002) and to construct linkage maps (Mayes et al. 1997). Attention then shifted rapidly to multiallelic SSR markers, which were derived from genomic libraries and EST databases (Billotte et al. 2001; Ting et al. 2010), but this method only provides medium mapping resolution and throughput. The publication of the oil palm genome, together with the reduced cost of high-throughput resequencing provided a better opportunity for high-density SNP and SSR detection in oil palm. In the last decade, high-density genome-wide SNP arrays have been developed for many crops and livestock species, including rice, corn, cattle, horse, and salmon. The first SNP genotyping array for oil palm was a 4.5K custom Illumina SNP array, which has been proven to be useful in the construction and comparison of linkage maps (Ting et al. 2014). Nevertheless, the density of genome coverage of the array for association analysis may be inadequate, especially in highly heterogeneous populations. Hence, a higher-density Illumina OP200K SNP array consisting of 200,000 SNPs was designed from the three million loci detected in the resequencing data and successfully validated in various populations using different genetic analyses (Kwong et al. 2016; Teh et al. 2016). According to the minimal LD decay of 19.5 Kbp observed among the oil palm populations examined, the OP200K SNP array with one SNP every 11 Kbp, on average provided excellent mapping resolution for linkage analysis and genome-wide association study (GWAS).

9.4.4 Genetic Dissection

9.4.4.1 Linkage Mapping

Methods of genetic mapping were developed to localize genes responsible for traits on the basis of a correlation between phenotype and DNA variation in an unbiased fashion. The first genetic map of fruit flies was successfully constructed through linkage analysis in 1913 (Sturtevant 1913). The method involves characterizing the meiotic recombination events in a family (mapping population) through analysis of the combination of markers (phenotypic or genetic) transmitted from the parents. DNA markers that show strong correlation to the trait variation (showing "linkage") are expected to reside close to the genes controlling variation in the trait. The QTL localization for heritable traits is still mainly dependent on family-based mapping methods. A linkage map with 576 SSR markers and an additional 102 diversity arrays technology (DArT) markers was constructed at Sime Darby for a commercial cross of Deli *dura* × AVROS *pisifera* (branded as Calix 600) using the regression model implemented in JoinMap4 (van Ooijen 2006) at a LOD = 5.0 threshold, using

Kosambi's distances. The map consisted of 23 linkage groups, spanning 1308 cM with a 2.1 cM average mapping interval (unpublished), which was one of the densest maps for oil palm at the time. However, the mapping resolution, particularly for complex traits (e.g., oil yield and disease tolerance), was still limited by insufficient recombination within these small populations (Billotte et al. 2010; Montoya et al. 2013; Jeennor and Volkaert 2014). More than 1000 seeds are usually obtained from a single cross of oil palm, but only 16–96 palms are planted and trait-recorded in field trials. The same problem with limited family sizes was also a limitation of linkage analysis in humans, especially for non-Mendelian diseases in the twentieth century (Altshuler et al. 2008).

9.4.4.2 GWAS and GS

The reference oil palm genomes (Singh et al. 2013b) and high-density OP200K genotyping array (Kwong et al. 2016) has enabled the first GWAS of mesocarp oil content (%) in oil palm (Teh et al. 2016). Unlike linkage analysis, GWAS provides access to the total meiotic recombination accumulated in a large population during evolution to increase the mapping resolution. The GWAS was carried out on 2045 *tenera* palms using the OP200K array. In order to maximize the mapping resolution, the discovery population was selected based on short-range LD distance, which is inherited with long breeding cycles and heterogeneous breeding populations. Figure 9.8 shows a genetic stratification and association study for mesocarp oil content. Deli × AVROS

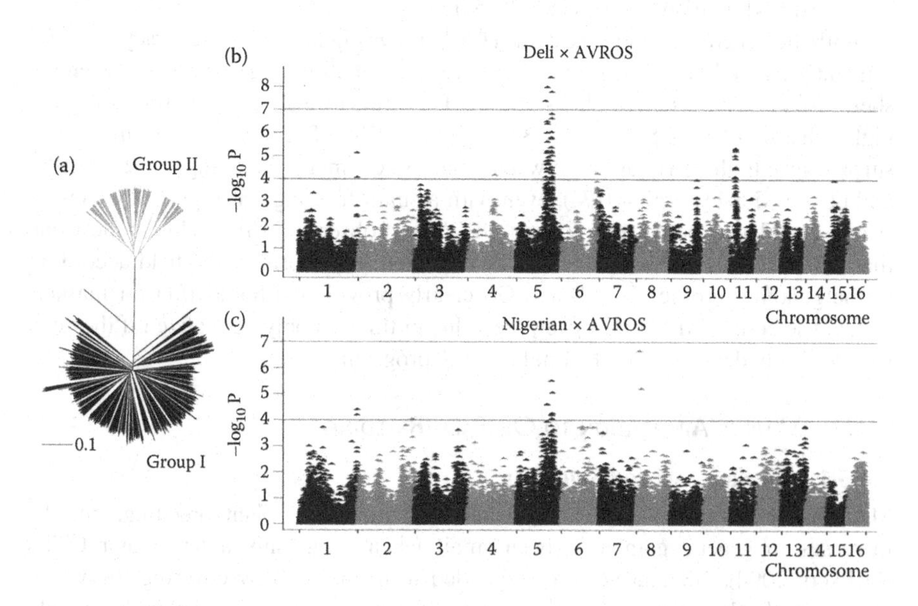

FIGURE 9.8 Genetic stratification and genome-wide association study for mesocarp oil content (%) of 2045 individuals representing Deli × AVROS and Nigerian × AVROS. (a) Neighbor-joining tree (NJ) constructed based on genetic distance. Group I and group II are Deli × AVROS and Nigerian × AVROS. (b) Manhattan plot of compressed MLM model for Deli × AVROS. (c) Manhattan plot of compressed MLM model for Nigerian × AVROS. (Based on Teh et al. 2016.)

and Nigerian × AVROS palms were grouped through genetic clustering analysis (Figure 9.8a) and subjected to phenotype–genotype association analysis independently using a simple linear model, but genomic inflation was still unsatisfactorily high. The false positives were then significantly reduced in Deli × AVROS (GIF = 1.1) and Nigerian × AVROS (GIF = 1.9) populations using a compressed mixed linear model with population parameters previously determined (P3D) (Zhang et al. 2010). A total of 62 and 18 significant association signals for mesocarp oil content (%) were detected in Deli × AVROS (Figure 9.8b) and Nigerian × AVROS (Figure 9.8c), respectively, according to whole-genome significance thresholds ($\log_{10}$ p-value ≥4.0; Bonferroni's $\log_{10}$ p-value ≥7.0) (Teh et al. 2016).

The major association peaks observed in Chromosome 5 had not been reported in any previous QTL linkage studies. The GWAS results also revealed potential genes responsible for variation in mesocarp oil content (%) of the oil palm populations. GWAS effectively identifies only the major QTLs that exceed the genome-wide significance threshold. The predictability of breeding value remains poor if these QTLs only contribute small effects to the complex traits. Hence, GS was then investigated to evaluate the total genetic effects of the entire genome relative to a trait, whereas all QTLs are expected to be in LD with at least one marker (Goddard and Hayes 2007). In order to achieve this, the GS accuracy in an oil palm population using a low density of SSR markers was first reported (Cros et al. 2015b) and the work was continued with reciprocal recurrent genomic selection (RRGS) based on simulated SNP markers (Cros et al. 2015a).

With the generated genotypic data, a further investigation of GS accuracy was subsequently carried out for the quantitative residual shell thickness variation (percent shell-to-fruit within *tenera* shell-types, S/F%) trait in 312 *tenera* palms using the high-density OP200K SNP array (Kwong et al. 2016). The prediction accuracy was surprisingly high (correlation between genomic estimated breeding value [GEBV] and observed S/F% trait = 0.74), even with a small training set. Figure 9.9 shows a strong correlation between predicted GEBV and observed S/F%. This is the value that tells a breeder how a palm will be expected to perform in the field according to the genomic profile. GWAS and GS clearly provide the foundation for marker-assisted selection (MAS); however, the value of this can only be realized if they can be effectively deployed in oil palm breeding programs.

9.4.5 Marker Application to Oil Palm Breeding

9.4.5.1 Marker-Assisted Selection

MAS has been extensively implemented in animal and plant breeding, but the increases in genetic gain have been small when using only a few major QTLs (Dekkers 2004). GS indeed is an expanded form of MAS by covering the whole genome so that breeding values of complicated traits can be predicted more accurately (Goddard and Hayes 2007). As a cost-optimized solution, both methods could be combined to optimize marker density for routine genotyping. The identified QTLs and GS model can break the cycle time and land use bottlenecks of RRS and FIPS by early selection at the nursery stage leading to a reduction in the number of progeny testing in the field. In Sime Darby Plantation R&D Centre, a large F_2 progeny test population

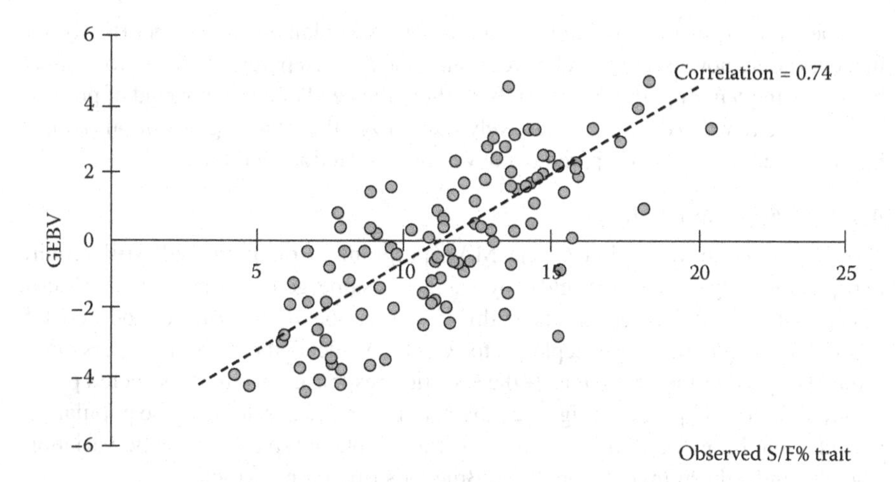

FIGURE 9.9 Representative regression plot of genomic estimated breeding value (GEBV) and observed shell thickness trait variation within *tenera* for the validation set. The correlation coefficient between GEBV and the observed trait was 0.74 with 55% of the variation explained for this representative case. (Based on Kwong et al. 2016.)

(~1000 palms/trial) derived from multiple Deli *dura* × AVROS *pisifera* crosses was used as a training set for GWAS and GS modeling of oil yield traits. Figure 9.10 shows the scheme for selection of the best palms from current commercial material. All germinated Deli *dura* and AVROS *pisifera* seedlings in the F_3 were screened and selected based on the genetic profiles of the training set; thus allele dropout should be minimal. Only those *dura* and *pisifera* seedlings that potentially contribute the best enrichment

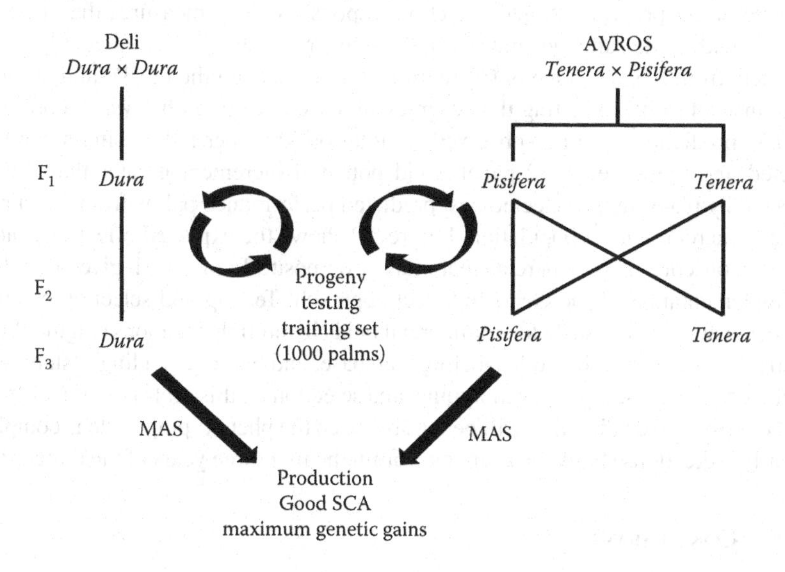

FIGURE 9.10 Marker-assisted selection (MAS) in oil palm selection schemes. SCA—specific combining ability.

of beneficial alleles to their future progeny were field planted. More importantly, the breeders could now assess SCA between *dura* and *pisifera* parents in the F_3 to produce high-yielding *tenera* planting materials, without the need of another round of progeny testing. The new MAS model effectively maximizes the genetic gains of oil palm in half the selection cycle compared to conventional selection methods.

9.4.5.2 "Genome Select"

The retrospective validation of the MAS program in oil palm will still require two phenotyping cycles. However, by using the same genetic findings, short-term selection response is measurable within a generation of *tenera*. A modified GS model for oil yield traits was deployed to "Genome Select" the highest yielders out of Sime Darby's commercial seeds. If the selection response is high, the selected palms should have a yield potential significantly better than the unselected base population, provided the base population possesses sufficient phenotypic and genetic variation. The idea indeed originated from "The Breeder's Equation," which is

$$R = h^2 S$$

S is known as selection differential, indicating the difference between population mean of all potential parents before selection and the population mean of the actual selected parents that will be used to generate the next generation. The h^2 is heritability for a trait and R is the selection response, which is the change in means between the population before selection and the population in the next generation. In this relationship, the h^2 of a trait is the link between the within-generation change S and the between-generation change R. However, the commercial *tenera* palms are basically the end products of breeding and hybridization, so h^2 may be less relevant to the Genome Select program. A new selection response R_a only measures the mean differences within the same generation. In this scenario, the main component by which R_a differs from S is the slope of GS regression line between the GEBV and observed traits, instead of h^2, reflecting the accuracy of selection using GEBVs. According to the 0.65 prediction accuracy observed for total oil yield per palm using a carefully selected and optimized SNP panel, yield potential increment greater than 16% is expected by planting the selected best predicted performance palms from the current best commercial *tenera* population. Figure 9.9 shows the expected effect of genome selection on current commercial materials. Composite Figure 9.11 gives details of the implementation of the Genome Select approach. Testing and selection has been conducted at sufficient scale for commercial fields in multiple locations (Figure 9.12a). Practical tools, such as barcode labeling, database, and barcode reading system, were developed to allow sampling, genotyping, and selection on this scale (Figure 9.12b and c). The prospective validation will be feasible once the phenotyping cycle is complete, but early indications should be available within the first three years of yield recording.

9.4.6 Conclusions

The developments in molecular technology during the past 10 years have enabled significant improvements in throughput combined with a reduction in costs for

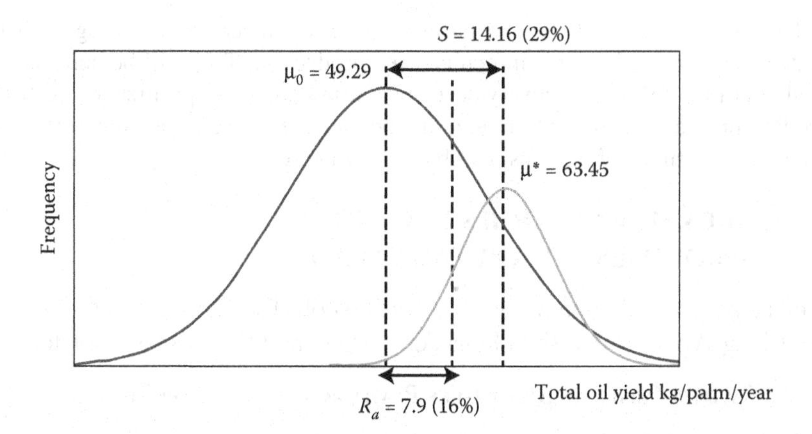

FIGURE 9.11 Example of "selection response" of the genome selected materials compared to the unselected base population. μ_0 is the yield mean of the unselected base population (dark curve). μ^* is the yield mean of the top-30% genome selected population (gray curve). S is selection differential. R_a is selection response within generation using GS.

genome sequencing as well as genetic marker assays. Concurrent developments in bioinformatics tools that allow detailed quantitative genetic analysis of large datasets means that the oil palm industry now has access to the tools to develop and deploy molecular markers for the selection of the best planting materials. The critical success factors for this technology to provide the much needed yield improvements

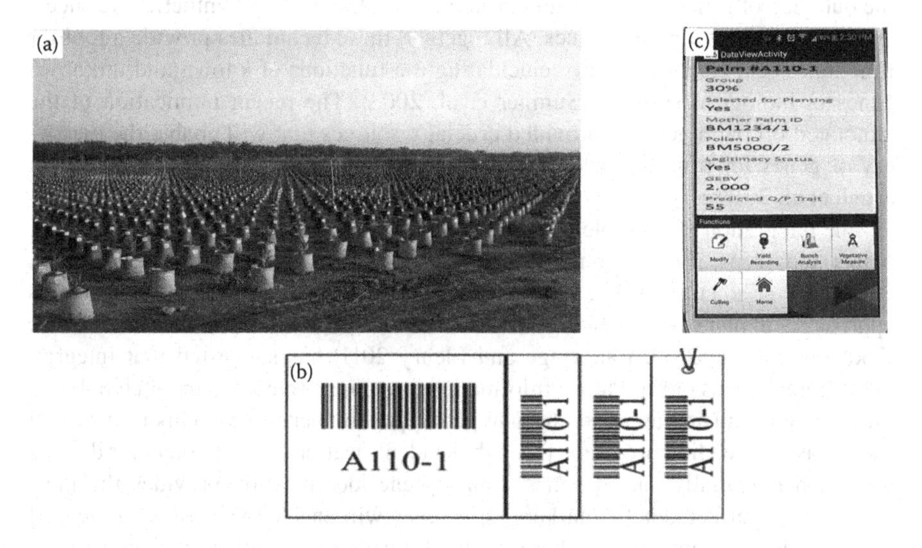

FIGURE 9.12 Deployment of Genome Select on the current best commercial *tenera* population. (a) Tested and barcoded seedlings were planted at the nursery; the genome selected individuals were transplanted to the commercial field. (b) Barcode labeling system was introduced. (c) Barcode reading system was developed as a smartphone app to retrieve palm biodata at nursery for selection purposes, including the predicted GS value of the individual palm.

in the coming years will be the harmonization of molecular breeding with the planted breeding trials and commercial seed production. This will be necessary to enable both the marker discovery/development process and optimization of future breeding programs in order to realize the potential increase in breeding efficiency and desirable trait combinations possible through MAS.

9.5 OMICS TECHNOLOGIES TOWARD IMPROVEMENT OF OIL PALM YIELD

Huey Fang Teh, Katharina Mebus, Bee Keat Neoh, Tony Eng Keong Ooi, Yick Ching Wong, Harikrishna Kulaveerasingam, and David Ross Appleton

9.5.1 Twenty-First Century Omics Platforms in Oil Crop Studies

As food demands increase and available farmland dwindles, it has become critical to not only increase edible oil production, but to do so more efficiently. Omics technologies, including whole-genome resequencing, transcriptomics, proteomics (Jorrín-Novo et al. 2015), metabolomics (Saito and Matsuda 2010), ionomics (Salt et al. 2008), and phenomics (Kumar et al. 2015) enable the dissection of underlying molecular mechanisms in a complex environment. Furthermore, integrating data from the genome, transcriptome, proteome, and metabolome into a single dataset can greatly facilitate the discovery of biomarkers for traits of interest. Transcriptomics studies the set of RNA transcripts produced by the genome at a specific point in time and tissue, while proteomics analyzes the translated proteins in a biological system. Metabolomics or the large-scale analysis of metabolites provides insight on the outcome of gene expression and protein translation into biosynthetic intermediates, end products, and hormones. All together, these techniques provide a tool for functional genomics to assist in elucidating the functions of known and unknown genes in biological systems (Sumner et al. 2003). The recent publication of the annotated oil palm genome provided crucial resources that will enable the discovery of genes for traits important to the oil palm industry (Abdullah et al. 2012; Singh et al. 2013b).

The integration of multiple omics techniques has been utilized in several higher plants and crops, such as tomato (Osorio et al., 2011), aspen tree (Bylesjö et al. 2008), rice (Cho et al. 2008), and maize (Amiour et al. 2012). Evaluation of omics approaches in plant system biology (Fiehn et al. 2001; Fridman and Pichersky 2005; Fukushima et al. 2009; Langridge and Fleury 2011) has suggested that integrating different levels of regulatory information (genome, proteome, and metabolome) might be a promising approach to map genotypes to phenotypes. This in turn will greatly assist breeding programs through the identification of biomarkers for desired traits and potentially causal genes. Causal gene identification provides the most accurate or "perfect genetic markers" since they will enable the most accurate trait selection across populations where polymorphisms exist. Integrative omics have proved useful in a number of studies targeting various traits. For example, multiple omics analyses have been used to study abiotic stress signaling, which has allowed for a more robust identification of molecular targets for future biotechnological applications in cereals (Cramer et al. 2011; Singh et al. 2015). Multiple omics have also been used for the improvement of seed-based nutrition (Galland et al. 2012). Bylesjö et al. (2008) applied a systems biology approach for integrated modeling of transcriptomics, proteomics, and metabolomics data in relation to genotype-specific perturbations affecting lignin biosynthesis and growth of hybrid aspen trees. A comprehensive study of rice seed aging using omics technologies has been carried out to assess the impact of aging on rice seed nutritional value (Galland et al. 2012). Multiple omics have also been employed to evaluate genetically modified crops for potential unintended effects (Ricroch et al. 2011; Decourcelle et al. 2015). A combination of targeted terpenoid analysis, and nontargeted metabolomic, proteomic, and transcriptomic profiling in maize transgenic lines has successfully revealed changes in carbohydrate metabolism to enhance carotenoid synthesis (Decourcelle et al. 2015).

Different omics approaches such as genomics (Cregan et al. 1999; Jarquín et al. 2014a,b), transcriptomics, metabolomics, proteomics (Hajduch et al. 2011; Natarajan et al. 2012; Xu et al. 2015), and phenomics and their integrated tools are being used for oil crops such as soybean, (Chaudhary et al. 2015), maize, rapeseed, castor, and jatropha for seed composition improvement and breeding programs. These omics technologies are revealing the information about dynamic changes taking place at these functional levels in oil crops. Using omics technologies, scientists aim to enhance seed oil content in several groups of oil crops through a number of different approaches, such as increasing the quantity of oil per seed (oil content), increasing the size of seeds, or increasing the number of seeds per plant (Syrenne et al. 2012). In soybeans, a combination of these approaches has led to successful discoveries, for instance, Kovinich et al. (2011) combined gene expression and metabolite data to elucidate the control of the R locus and allow identification of pigment biosynthesis genes. In a similar study, metabolic and transcriptional changes were assessed in developing soybean seeds to identify metabolic engineering targets. The study concluded that transcriptional activation and the involvement of signaling molecules were at much higher levels during seed maturation and dormancy (Collakova et al. 2013). The integration of expression QTL (eQTL) and phenotypic QTL (pQTL) has helped to identify 11 potential candidate genes related to isoflavone content in soybean seeds (Wang et al. 2014). Recently, ionomics and metabolomics were coupled for a comprehensive assessment of GM and non-GM soybean lines (Kusano et al. 2015) and Li et al. (2015) found the highest metabolic flux during early seed filling by integrating metabolomics and transcriptomics analysis. Furthermore, the metabolic flux was found to be consistent with regard to the transcript and metabolite level changes during the seed development stages. All of these studies clearly illustrated that an integrated "omics" approach needs to be applied for better understanding of seed composition traits in soybean.

In oil palm to date, transcriptomics, proteomics, and metabolomics have been used separately to study ganoderma disease (Lim et al. 2010; Alizadeh et al. 2011; Jeffery Daim et al. 2015), yield (Loei et al. 2013; Ooi et al. 2015; Guerin et al. 2016), fruit development (Bourgis et al. 2011; Tranbarger et al. 2011; Al-Shanfari et al. 2012; Neoh et al. 2013; Teh et al. 2014; Wong et al. 2014), and stress response (Ebrahimi et al. 2015; Azzeme et al. 2016). In the following section, we describe the various "omics"-based approaches, employed in recent years to understand oil biosynthesis metabolism. We restrict our discussion to oil crops and oil palm yield enhancement. The pitfalls of omics technologies and challenges in conducting omics research in a field crop are also being discussed here.

9.5.2 Omics Platforms in Oil Palm Yield Studies

Oil palm is the most productive oil-producing crop, with oil making up a remarkable 90% of its fruit mesocarp dry mass (Murphy 2014). For the past decades, oil palm breeders selected and crossed oil palms based on desired traits in order to improve yield (Corley and Tinker 2016). However, these conventional breeding methods are most often unable to determine specific gene–trait associations. New omics technologies may complement plant breeding and molecular breeding in oil palm plantations by providing the link between gene function and physiological traits (Mayes et al. 2008; Weselake et al. 2009; Appleton et al. 2014). Omics technologies, in a complementary approach, have been used to profile differences in transcripts, proteins, and metabolites that distinguish "high-yielding individual" palms (HY: 10–12 MT oil/ha) from palms with low-average yield (4–7 MT oil/ha), termed "low-yielding" (LY) within a group of siblings planted in the same location (Teh et al. 2013). Fruit bunches were harvested at different developmental stages preceding, during, and after the major oil biosynthesis period at 12, 14, 16, 18, 20, and 22 weeks after anthesis (WAA). All mesocarp samples were characterized using parallel profiling techniques for transcript (cDNA microarrays), protein/peptide (UPLC/MS), and metabolite (GC/MS, CE/MS, LC/MS) levels. A custom oil palm transcriptome microarray was developed using the Agilent platform (Wong et al. 2014) as a tool to study gene expression profiles of mesocarp tissue. Biochemical changes before, during, and after oil biosynthesis in the fruit mesocarp were compared in order to gain a deeper understanding of the control of oil biosynthesis in the oil palm (Neoh et al. 2013). The three datasets were then simplified by categorizing the genes, proteins, and metabolites into their cellular pathways and looking for concordance as a method of identifying the most important differences associated with oil palm yield. The data analysis was able to identify more than 2000 transcripts that were differentially expressed between the high- and low-yielding sample sets, and therefore may contribute to significant yield differences among progenies. Microarray analysis revealed a total of 1331 unique annotated gene candidates that were found to be differentially expressed between high- and low-yielding palms throughout mesocarp development (Wong, unpublished data). Among these candidates, many genes were found to be involved in the oil biosynthesis, triacylglycerol, glycolysis, and tricarboxylic acid (TCA) pathway, as well as N-assimilation, from which 127 were prioritized. Figure 9.13 shows the distribution of gene classes observed.

However, identification of the critical ("causal" vs. "effect," or "association") genes is not possible using transcriptomics analysis alone. A more complete picture can only be obtained by studying gene expression in combination with the analysis of the "outcome"—that is, proteins and metabolites. Protein levels along with metabolite changes were consistent in many cases between 12 and 22 WAA. The three datasets were simplified by categorizing the genes, proteins, and metabolites into their cellular pathways and analyzing areas of concordance along with known biochemical changes during mesocarp development and oil biosynthesis. In order

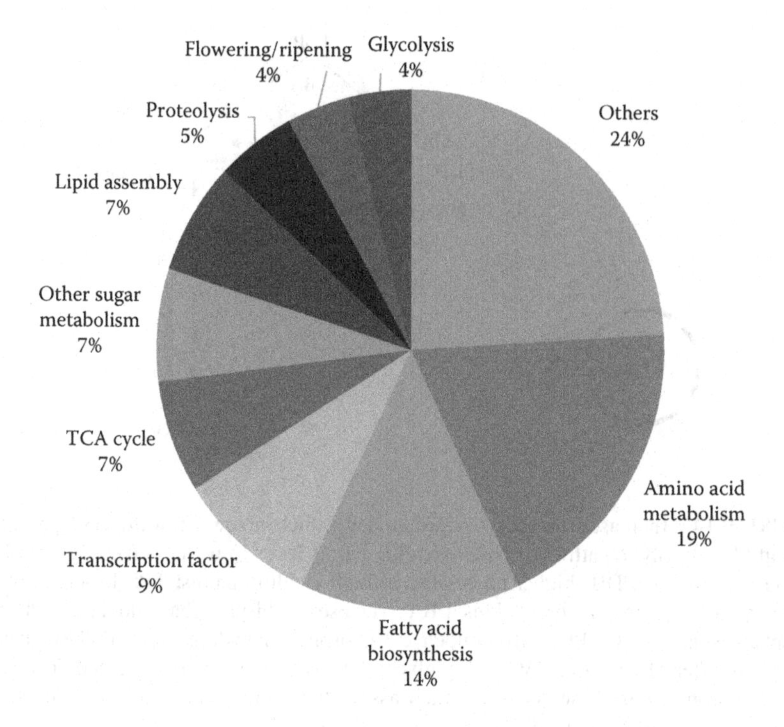

FIGURE 9.13 Biological classifications of 127 selected oil yield differentially expressed genes of interest identified through omics analysis.

to identify the best candidates for overall control, focus could be centered on three main biosynthetic differentials:

1. Rate-limiting steps where metabolite intermediates are depleted in subsequent biosynthetic steps
2. Branch points in pathways
3. Changes in global regulation of a biosynthetic process

Specific metabolite levels in the glycolysis pathway and TCA cycle were found to be high in the high-yielding palms. Glycolysis is the main pathway that converts sugar molecules transported from the leaves to the carbon building blocks of lipids in the mesocarp. Transcript profiling showed that four key enzymes centered around a branch point in the glycolytic metabolism were differential, namely, fructose-1,6-biphosphate aldolase, triosephosphate isomerase, glycerol-3-phosphate dehydrogenase, and glyceraldehyde-3-phosphate dehydrogenase. Isobaric tags for relative and absolute quantitation and multiple reaction monitoring data confirmed that in oil palm, several key glycolytic enzymes were differentially expressed in high-yielding palms (Loei et al. 2013; Ooi et al. 2015). In particular, higher expression of fructose-1,6-biphosphate aldolase and glyceraldehyde-3-phosphate dehydrogenase, combined with the reduced expression of triose phosphate isomerase may be indicative of important carbon flux balance changes in the glycolysis pathway in high-yielding palms.

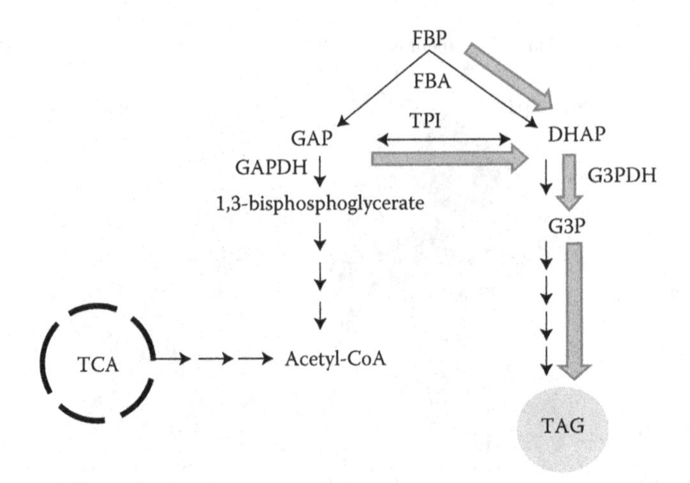

FIGURE 9.14 Increased balance of carbon flow (thick arrows) toward G3P production through the glycolytic pathway in high-yielding palms leads to increase triacylglycerol biosynthesis (FBA, G3PDH: higher expression in high-yielding palms; TPI: lower expression in high-yielding palms). Abbreviations: FBA, fructose-1,6-bisphosphate aldolase; TPI, triose phosphate isomerase; GAPDH, glyceraldehye-3-phosphate dehydrogenase; G3PDH, glycerol-3-phosphate dehydrogenase; TAG, triacylglycerol. Current work is now focused on studying the polymorphisms in these genes and their association to oil palm yields as well as functional characterization in model systems.

Figure 9.14 shows the observed flux changes in the glycolytic pathway. The absence of detectable dihydroxyacetone phosphate and glyceraldehyde-3-phosphate in metabolite analysis also indicates that fructose-1,6-biphosphate aldolase sits at a rate-limiting step in this pathway. In the glycolytic pathway of high-yielding palms, there seems to be greater carbon flux toward the production of glycerol-3-phosphate, the precursor for lipid biosynthesis. Glycerol-3-phosphate was observed to be higher in the HY group compared to the LY group throughout the last stages of fruit development. Glycerol-3-phosphate is the building block that is acylated with fatty acids to produce lipid/triacylglycerol molecules and has been linked to significantly higher yields in other oil crops (Vigeolas and Geigenberger 2004; Vigeolas et al. 2007; Weselake et al. 2009). Hence, an increment of glycerol-3-phosphate could be a significant driver of increased lipid production in oil palm mesocarp. Concordance between protein, transcriptome, and metabolite levels was clearly evident and provided confidence in the significance of this group of enzymes. Furthermore, overexpression of oil palm FBA and G3PDH genes in yeast has proven to increase oil accumulation (unpublished data).

Significant differentials were also observed in certain metabolites in the TCA cycle. Figure 9.15 shows the ratio of malate to citrate, which changed markedly with high-yielding palms exhibiting a much higher ratio from 12 to 18 WAA, a result similar to that found in a study investigating lipid biosynthesis in the mesocarp of olive (Donaire et al. 1975). The energy carrier molecule ATP had lower levels in high-yielding palms during oil biosynthesis (16–20 WAA). The lower concentrations are most likely a result of increased energy demand required to produce more lipids

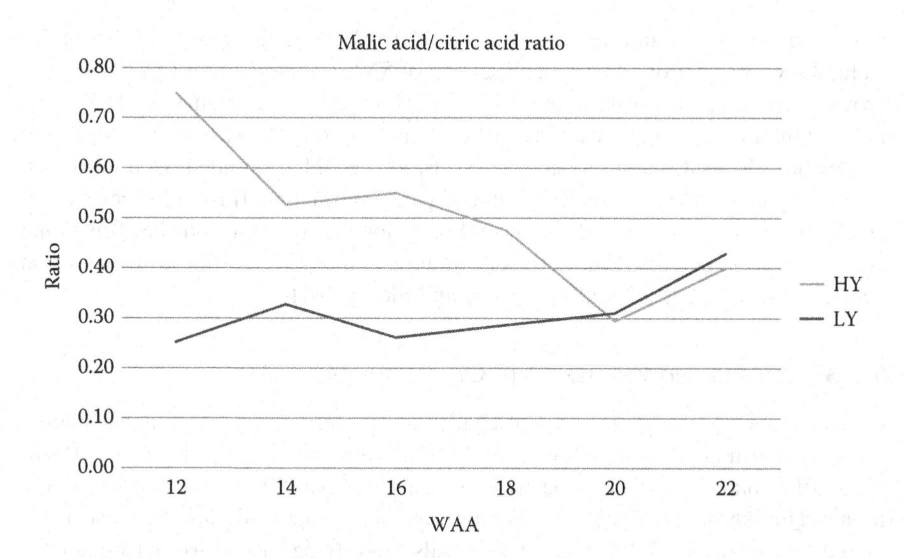

FIGURE 9.15 Differences in malic acid/citric acid ratios in HY and LY oil palm mesocarp during lipid biosynthesis (12–22 WAA, n = 8).

in the HY palms. Interestingly, isobaric tags for relative and absolute quantitation data showed that the abundance of the beta-subunit of the ATP synthase complex elevates during fruit maturation in high-yielding palms but is reduced in the low-yielding palms (Loei et al. 2013). This indicates that increased energy supply may be critically required for higher oil accumulation.

Differential metabolite concentrations were also observed in nitrogen assimilation for oil production. Concentration of amino acids is high at early stages in the mesocarp of high oil-yielding palms. Similarly, protein and gene expression analysis also showed several genes involved in the amino acid metabolism to be differentially expressed between high- and low-yielding palms. High-yielding palms were found to accumulate higher levels of these N-containing building blocks of proteins, which is possibly related to future cell expansion and mesocarp growth. Furthermore, nucleotides, which are another major class of N-containing metabolites found in the mesocarp, were found to be increased along with lipid accumulation in high-yielding palms. Levels of nucleotides such as adenine, cytidine, guanosine, uracil, and uridine are clearly higher in HY during the later stages of fruit development (18–22 WAA) compared to LY fruits. These data appear to indicate an important role for N assimilation in the development of fruit tissue for oil yield production (Teh et al. 2013).

Taken together, omics analysis of oil palm mesocarp enabled the identification of genes that play important roles leading to increased oil production. The patterns observed were in agreement with phenotypic data obtained through years of yield recording prior to the biochemical analysis. Data for yield revealed simultaneous increases in the number of bunches produced per year, the average mesocarp mass per bunch, and the oil content of the mesocarp, while omics analysis indicated changes in glycolytic flux, nitrogen assimilation, and energy management. Many key traits of plants such as yield and disease resistance are often controlled by a large number of genomic regions. For future work, combining GWAS and omics technology information allows identification of SNPs related to molecular traits and provide strong candidate loci for yield traits (Langridge and Fleury 2011). GWAS in combination with gene expression QTL analysis (eQTL) (Liu et al. 2011) and metabolic QTL (mQTL) (Scossa et al. 2016) also enabled the identification of key controlling genes of particular biosynthetic pathways in a multitude of species of oil crops (Holloway et al. 2011; Li et al. 2013a,b; Song et al. 2014). Further functional characterization of such identified loci can unravel the effects of these variations on gene function and expression (Langridge and Fleury 2011).

9.5.3 Pitfalls and Challenges in Omics Studies

Omics studies typically face several challenges and criticisms, those being mainly issues concerning sampling (Lay et al. 2006), reproducibility of data (Jorrín-Novo et al. 2015), bias, statistical challenges (Choi and Pavelka 2011), as well as the search for novel markers in complex samples using discovery-mode omics in which one hopes to find something interesting and yet often falls short (Ning and Lo 2010). Unlike sampling for model plants such as *Arabidopsis* or those that can be cultivated in controlled environments, sampling of field crops such as oil palm can be quite challenging for a number of reasons: First, oil palm is grown exclusively in the field. This provides an inherent lack of environmental control in any omics study, including sunlight changes during the day of sampling, significant soil differences across the sampling location, different stresses, and the pest and disease states of sampled palms (Appleton et al. 2014). Field sampling also provides challenges in sample preservation and transport to the laboratory. Second, the size of mature oil palms and geographic spread also results in extremely labor-intensive sample collection that must be completed in the shortest time possible due to the points noted above. The molecular responses of plants to their environment are very complex and the field environment cannot be controlled easily (Cramer et al. 2011). Careful selection is critical to avoid sampling bias (Alexandersson et al. 2014). The logistics of sampling in several hectares of oil palm estate and the preservation of the biochemical status in the sample at the point of sampling are key challenges, as harvesting might activate enzymatic reactions that change the composition in the sample. One way to overcome these problems is to immediately freeze the samples in the field using liquid nitrogen to quench metabolism of the plant tissues and enzymatic activity. Weckwerth et al. (2004) proposed an integrated extraction for omics techniques. He described a method by which RNA, proteins, and metabolites are sequentially extracted from the same sample to provide convenient and unbiased analysis. However, this integrated extraction approach might pose difficulties in that each of the solvents might influence the quality of extraction. For example, according to Fiehn et al. (2008), extraction parameters have a direct impact on metabolic data as the metabolite profiles will differ according to the extraction protocol used. Although rapid progress in the use of omics tools have been demonstrated, data mining and analyses are still challenging tasks (Zu Castell and Ernst 2012). Another demanding task is to extract biological meaning from this enormous amount of data. Deriving knowledge from tables, revealing patterns among biological samples, and identifying discriminatory variables are challenging (Boccard et al. 2010). These omics technologies are at different stages of development and present unique advantages as well as limitations (Joyce and Palsson 2006; Gomez-Cabrero et al. 2014). It is therefore crucial to carefully select the suitable profiling technology and analytical strategy with a proper experiment design to eliminate a potentially high number of analytical errors.

9.5.4 Conclusions and Future Perspectives

Yield enhancement through identification of genetic markers has been the focus of recent research efforts. However, with a complex trait, such as yield, a number of factors, including environmental effects and epistasis, need to be taken into account and can complicate the association between yield traits and genetic markers. Here, we have reviewed how the use of omics technologies has helped in a complementary approach to enhance seed oil content and improve oil yield in oil crops, including oil palm. The deployment of omic technologies is providing valuable insight into the gene regulatory networks of investigated oil crops. The knowledge of oil biosynthesis pathways and lipids accumulation has grown in leaps and bounds since the emergence and subsequent developments of omics technologies. This allows further manipulation of oil crop composition with the understanding of posttranscriptional and metabolic regulation via omic technologies. Furthermore, omics have the potential to provide additional measurable molecular phenotypes. These can serve as indicators and help us to understand complex traits through closing the gap between observed traits and the understanding of their molecular and genetic basis. This can be applied to many other traits besides yield as well. Therefore, integrating omics techniques into the oil palm breeding program promises to allow for more accurate selection of palms, which could improve breeding efficiency. Using these data to complement molecular marker trait association studies will further enable identification of causal gene variants for specific desirable traits. To this end, omics technologies can help plant biologists to select the important genes for a trait of interest.

9.6 GENOMIC SELECTION FOR OIL PALM

David Cros

In oil palm, conventional breeding usually relies on RRS between two parental groups A (Deli and Angola populations) and B (other African populations). The genetic trials give reliable estimates of parental GCAs with accuracy around 0.9 for yield components (Cros et al. 2015b) when analyzed with traditional pedigree-based mixed models. As with other perennials, the major drawbacks of oil palm breeding are the long generation interval, currently around 20 years, and the low selection intensity, with usually less than 200 individuals progeny tested per generation and parental group. Two different approaches of GS have been studied so far to increase oil palm yield: a single-crosses approach (Wong and Bernardo 2008) and a population-level approach (Cros et al. 2015a,b). These studies are summarized and discussed here.

9.6.1 GS Studies Based on Simulations

Wong and Bernardo (2008) compared conventional phenotypic selection, MARS (marker-assisted recurrent selection, based on QTL detection) and a single-crosses GS approach. Their simulation started with an initial population originated from the selfing of a cross between two inbred palms. In conventional selection, this population was crossed to an unrelated tester and the selected individuals were recombined to produce progenies used in a similar cycle. In MARS and GS, the initial population was also genotyped and used to estimate marker effects. For three consecutive cycles, the selection was made on markers only. The conventional and marker-assisted breeding strategies thus extended over a different number of generations but a comparable number of years. In MARS, the genotyping was made with 100 SNPs and only the SNPs with a significant effect were used for selection. In GS, 140 SNPs were used. The population sizes were defined as the number of test-crossed full-sibs in the initial population and of individuals in the subsequent generations. Their simulation included test-cross trials with two replicates instead of the usual number of four and populations twice as large as the populations resulting from the usual experimental designs, in order to achieve a wide range of population sizes (i.e., from 15 to 70) while maintaining the same phenotyping effort. They found that GS and conventional selection consistently outperformed MARS in terms of selection response. With population sizes from 50 to 70, GS always outperformed conventional selection with a response from 4% to 25% higher, depending on the population size, heritability, and number of QTLs. The authors also considered the relative costs of the different breeding strategies. To our knowledge, this is the only study giving estimates of the costs of conventional breeding in oil palm. From the estimates obtained from a private oil palm research company, they concluded that with genotyping at US $1.50 per data point, the cost per unit gain was 26%–57% lower with GS than with conventional selection; while with genotyping at US $0.15 per data point, the cost per unit gain was 35%–65% lower.

To extend these initial results, Cros et al. (2015a) simulated two breeding populations of 120 individuals each, matching with the actual current Deli and La Mé populations, and compared over four generations conventional RRS with a population-level GS approach termed RRGS. They focused on bunch weight, bunch number, and the resulting bunch production. For RRGS, the progeny tests of the initial cycle of conventional RRS were used to train the GS model and the following cycles were used either to implement selection on markers alone or to retrain the GS with new progeny tests. The parental GEBVs were obtained using GBLUP with hybrid phenotypes as data records and 2500 SNPs. The genotypes were those from the progeny-tested individuals (RRGS_PAR strategy), or also from hybrid individuals (RRGS_HYB). For RRGS, they also studied the effect of the frequency of progeny tests used for model calibration (in every generation, in generations 1 and 3, and in generation 1 only), the number of candidates (120 or 300) and in RRGS_HYB the number of genotyped hybrids (up to 1700). They concluded that RRGS could increase the annual selection response compared to RRS by decreasing the generation interval and by increasing the selection intensity. The best breeding strategy was RRGS_HYB with 1700 genotyped hybrids, calibration only in generation 1 and 300 candidates per generation and population, which led to an annual selection response 72% higher than RRS. RRGS_PAR with calibration of the GS model in generations 1 and 3 and 300 candidates was a relevant alternative, with a high increase in the annual response compared to RRS (+46%) and lower risk around the expected response, lower increased inbreeding and lower cost than RRGS_HYB.

9.6.2 GS Studies Based on Empirical Data

Cros et al. (2015b) investigated a population-level GS strategy for Deli and Group B populations (mostly La Mé and Yangambi here). The two populations comprised 131 progeny-tested individuals each, genotyped with 265 SSR. Using a cross-validation approach, they estimated within population GS accuracies when predicting breeding values of non-progeny-tested individuals. They considered eight yield traits, with three methods to sample training sets and five statistical methods to estimate GEBV. In Group B, they found that the GEBV of the candidate individuals could take into account both the mean value of their family and their individual value (Mendelian sampling term), as indicated by the significant accuracies that could be observed in full-sib families. In Deli, only family effects could be estimated. The authors hypothesized that this difference between populations originated from their contrasting breeding histories. The Deli population had a narrower genetic base and a longer history of artificial selection, inbreeding and drift than Group B. As a result, the Mendelian sampling terms of individuals not progeny tested had a higher magnitude in Group B than in Deli (i.e., Group B had a higher within family additive variance), and were therefore easier to estimate with GS. For the same reason, the markers were more informative in Group B. The GS accuracy ranged from –0.41 to 0.94 and was positively correlated with the relationship between training and test sets. Good accuracies (>0.60) could be obtained for selection candidates highly related to the training set (in particular full-sibs and progeny). Training sets optimized with the CD mean criterion (Rincent et al. 2012) gave the highest accuracies, ranging from 0.49 (pulp to fruit ratio in Group B) to 0.94 (fruit weight in Group B). The statistical methods did not affect the accuracy. Finally, Group B could be preselected for progeny tests by applying GS to key yield traits, therefore increasing the selection intensity.

9.6.3 Conclusion and Perspectives

The results obtained so far point out that GS has the potential to make oil palm breeding more efficient, although other studies are necessary to optimize its implementation and obtain a broader view of its possibilities, depending on the population, trait, etc. An efficient implementation of GS in practice will require an even more integrated approach to breeding. The definition of relevant training and candidate sets, the high-throughput genotyping on a large-scale basis (with fast-evolving technologies), the management of the large datasets of molecular data generated, and the joint analysis of phenotypic, molecular, and possibly other types of information in complex mathematical models will require a strengthened collaboration between breeders, molecular geneticists, bioinformaticians, and biostatisticians, and as usual with the specialists of the traits of interests (phytopathologists, agronomists, etc.).

9.6.3.1 Using GS to Increase the Rate of Genetic Gain in Oil Palm Yield

The single-crosses GS approach of Wong and Bernardo (2008) can be implemented within full-sib families of the parental populations of commercial hybrids, using an individual of another heterotic population as tester. The test-crossed individuals of a full-sib family would make a training set efficient to implement GS among their non-test-crossed full-sibs and among their progenies, to select for individuals that would produce elite hybrids with the tester (or with the population of the tester). However, this appealing approach has some drawbacks. In the simulation, inbred palms were used to generate the training set. However, the practical application of the single-crosses GS approach would use individuals heterozygous to some extent, as inbred palms are not available. This would likely lead to a lower LD compared to the simulation, and therefore GS would require higher marker density and larger training size to achieve the simulated results. Also, as the training set is made of a single cross, this will likely be inefficient to predict the genetic value of individuals of other crosses, even from the same population. This approach must therefore be implemented for each cross. In addition, in the single-crosses approach as described here, it would probably not be possible to compare the GEBV of individuals of different crosses. An empirical study appears necessary.

The simulation by Cros et al. (2015a) confirmed the results of Wong and Bernardo (2008) regarding the usefulness of GS in oil palm, and indicated that a population-level GS approach (RRGS) was an interesting alternative to the single-crosses GS approach, as it could give an even higher annual selection response. RRGS also appeared valuable from empirical results (Cros et al. 2015b), which showed that it could take into account both the average family value and the individual value of selection candidates, thus allowing selecting within full-sib families of parental groups. However, this was only the case for some traits in Group B. In such conditions, breeders can make a preselection before progeny tests, thus increasing selection intensity compared to conventional selection, but progeny tests would still be required. This is contradictory to the simulation results of the same authors, but this may have been the consequence of a too small training set in the empirical study. Indeed, this only used 80% of the progeny-tested individuals, in order to keep a validation set. In real conditions, all the progeny-tested individuals would be used to calibrate the GS model, as it was the case in the simulations. We must also stress that the simulations studies rely on simplifying assumptions, for example, in the two studies presented here: biallelic QTLs, additive genetic determinism, no genotyping errors, and no missing molecular data. This may bias the results compared to real situations. In order to clarify the discrepancy between empirical and simulation results regarding the population-level GS approach, it is necessary to obtain empirical estimates of accuracy along generations after model calibration, for the two parental groups and all yield components, when calibrating the model with all the individuals that were progeny tested in a single breeding cycle.

Compared to the single-crosses GS approach, RRGS will likely be more difficult to implement. In particular, working with a whole breeding population implies a larger training set to achieve good accuracy. The results obtained so far suggest that it should be made of roughly >120 individuals progeny tested per parental population, evaluated according to mating and experimental designs well designed enough to insure in particular homogenous phenotypic data and connections between trials and progeny-tested individuals. This might not be currently available to all breeding companies, and if not, this will take some time to achieve. The population-level GS approach also requires higher marker density, but with the small N_e of oil palm breeding populations (Cros et al. 2014) and the currently available high-throughput genotyping strategies, marker density should not be a limiting factor for GS in oil palm. Although further studies are needed, it seems that with a good relationship between the training set and the selection candidates (full-sibs, progenies), and a training set large enough (probably at least 50 individuals for single crosses and 120 per parental group for RRGS), GS should allow the selection, with markers, of individuals with the best GCA to produce elite hybrid crosses. When compared to conventional breeding, the decrease in generation interval and increase in selection intensity allowed by GS would more than balance out the likely associated decrease in selection accuracy. Figure 9.16 illustrates a possible RRGS breeding scheme, where one breeding cycle with progeny tests alternates with one cycle with selection on markers alone. First, considering that two cycles with progeny tests require 38–40 years against 24 when selecting only on markers (Wong and Bernardo 2008; Cros et al. 2015b), a reduction in the mean generation interval of around 40% is obtained. Second, the selection intensity is increased. Indeed, in conventional breeding, the selection can only be made among the progeny-tested individuals, while with GS it will be made among all the genotyped individuals, progeny tested or not. Third, as a result of the ability to select on markers alone, GS will increase the breeding value of the seed garden. In conventional breeding, each new generation is made of

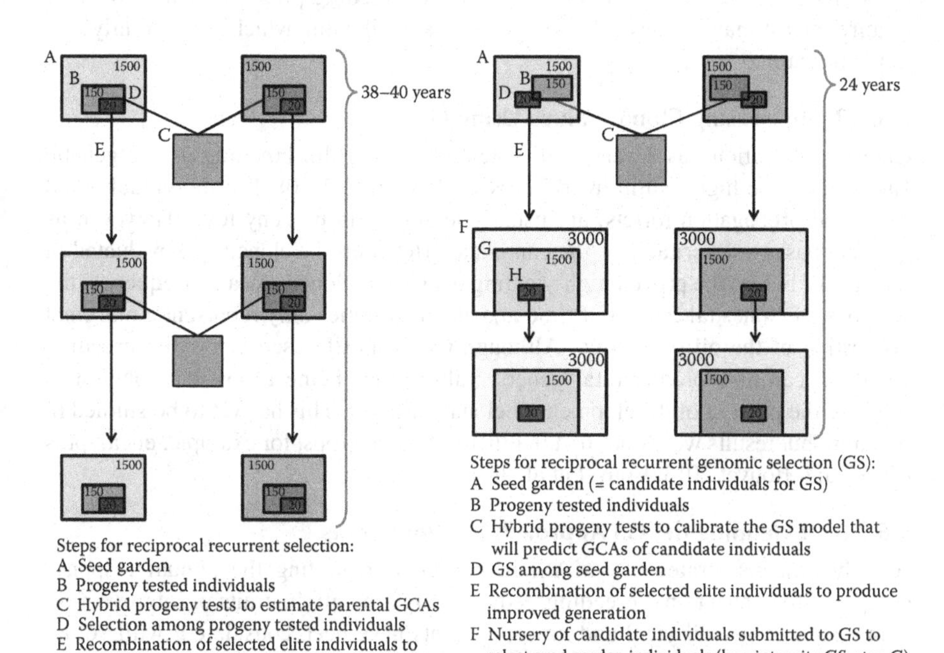

FIGURE 9.16 **(See color insert.)** Integrating genomic information into oil palm breeding to increase annual response: example breeding scheme of the population-level GS approach (reciprocal recurrent genomic selection, RRGS; right) versus conventional phenotypic breeding (left). GCAs—general combining abilities.

a random sample of individuals obtained by recombining the elite individuals of the previous cycle, which constitute the improved seed garden. On average, assuming random mating of the elite individuals and a large enough number of progeny individuals, the average breeding value of this new seed garden is equal to the average breeding value of the selected parents. However, not all the seedlings produced when recombining the elite individuals are planted in the field. With GS, all the seedlings can be genotyped and their GCAs estimated at the nursery stage, allowing planting only the best ones. In this case, the average breeding value of the new seed garden will be higher than that of the selected parents, leading to further increase in the performance of the commercial hybrids.

GS generates new costs due to the genotyping of the training and candidate individuals. However, it is less expensive to genotype than to progeny test an individual, and this should amplify with declining whole-genome genotyping cost and increasing phenotyping cost (in particular due to labor costs). Each breeding company must carry out a cost–benefit study, taking into account the cost of progeny test per individual, the cost of genotyping per individual, the current number of progeny-tested individuals per cycle, the proportion of cycles with progeny tests versus cycles with selection on makers alone and the number of genotyped individuals, in order to estimate the cost of a genomic version of its breeding program. This obviously requires an estimate of the cost of progeny tests in oil palm, which are currently rare in the literature.

9.6.3.2 Increasing Clonal Values Using GS

Clonal propagation has a very high potential in oil palm breeding as the genetic variance can be high within hybrid crosses (Soh et al. 2010). The individuals used for clonal propagation (ortets) are usually identified in progeny tests. However, as oil yield has a low heritability, the candidate ortets must be cloned and evaluated in field trials, before the proven high-yielding ones are recloned. As a consequence, the selection of clones takes a lot of time and clonal varieties only represent a marginal proportion of the oil palm area. Although GS is mostly used to predict breeding values, it can also predict total genetic values; thus being a possible solution to improve the process of development of clonal varieties. This has yet to be studied in oil palm, but results were obtained in other perennial crops, for example, eucalyptus (Denis and Bouvet 2013; Bouvet et al. 2016).

9.6.3.3 Extending the GS Approach to Other Traits

A fully efficient strategy of MAS, in particular regarding the annual response compared to conventional breeding, requires targeting all the traits of interest of oil palm. So far, only oil yield and its components have been considered, while other traits are also of major importance, in particular resistance to diseases. Good results have already been obtained for resistance to diseases using GS in plant species, for example, with fusarium head blight in barley (Lorenz et al. 2012), stem rust in wheat (Rutkoski et al. 2014), lethal necrosis and northern corn leaf blight in maize (Technow et al. 2013; Gowda et al. 2015), and fusiform rust in loblolly pine (Resende et al. 2012b). Studies regarding the performance of GS for disease resistance in oil palm are likely to be implemented in the near future. Ongoing molecular and genetic studies regarding the resistance to the major diseases (Tee et al. 2013; Hama-Ali et al. 2015; Yusof et al. 2015) should help in developing efficient MAS strategies. Before this is achieved, the practical implementation of GS for oil yield can still start in candidate populations made of elite families in terms of resistance to diseases. Similarly, the potential of GS for oil quality, vegetative growth, etc. requires to be investigated.

9.6.3.4 Integrating Multiple Omics and Genetic Data in GS

GS uses neutral markers, but *a priori* information about the effects of some markers can be available, for example, from GWAS or functional genomic studies. Several approaches that combine the GS methodology and *a priori* information about markers are already available. For instance, Zhang et al. (2014) developed the BLUP conditional on the genetic architecture. This is an extension of GBLUP where markers are weighted when used to compute the **G** matrix of realized relationships. The weights can be the effects or variances estimated at each marker with a GS model, or at some significant markers with a GWAS approach, etc. Bernardo (2014) and Rutkoski et al. (2014) showed that including markers strongly associated with the trait ($r^2 > 10\%$) as fixed effects in the GBLUP increased the accuracy. Speed and Balding (2014) extended the RR-BLUP with MultiBLUP, taking into account marker classes that differ in terms of expected magnitude of effects. The classes can be defined in any way: location of markers (exons, introns, QTLs, etc.), functional or neutrals markers, etc. These sophisticated GS models should be investigated in oil palm, where *a priori* information about markers are becoming available at a rapid pace, from transcriptomic studies (Dussert et al. 2013; Azni et al. 2014; Xiao et al. 2014), QTL detections, and the genome sequence (Singh et al. 2013b). Data acquired using other omic approaches (transcriptomics, proteomics, methylomics, metabolomics) can be included in the GS model in addition to genomic data, in order to increase accuracy. Promising results in this recent field of GS were obtained in maize (Riedelsheimer et al. 2012) and loblolly pine (Whetten et al. 2015). The latter study showed that adding the levels of 199 transcripts and 382 metabolites to the genomic data of 3562 SNPs increased the GS accuracy.

9.6.3.5 Interactions with the Environment

Oil palm yield is affected by genotype × abiotic environment interactions, in particular under the effect of variations in water balance. Taking these interactions into account could allow the prediction of combining ability of selection candidates in environments where they were not progeny tested. This requires models combining phenotypic data from various environments, genomic data, and environmental covariables. Some promising results were obtained by the few authors that investigated this aspect of GS (Heslot et al. 2014; Jarquín et al. 2014a,b; Technow et al. 2015).

REFERENCES

Abdullah, M.O., Madon, M., Low, E.T.L., Ithnin, M., and Singh, R. 2012. Oil palm genomics. In: Oi-Ming Lai, Ching-Ping Tan, and Casinir C. Atoh (eds) *Palm Oil: Production, Processing, Characterization, and Uses*. AOCS Press, Urbana, Illinois, p 59.

Alexandersson, E., Jacobson, D., Vlivier, M.A., Weckwerth, W., and Andreaszon, E. 2014. Field-omics—Understanding large-scale molecular data from field crops. *Frontiers in Plant Science* 5: 286.

Alizadeh, F., Abdullah, S.N.A., Khodavandi, A., Abdullah, F., Yusuf, U.K., and Chong, P.P. 2011. Differential expression of oil palm pathology genes during interactions with *Ganoderma boninense* and *Trichoderma harzianum. Journal of Plant Physiology* 168: 1106–1113.

Alonso, R., Onate-Sanchez, L., Weltmeir, F., Ehlert, A., Diaz, I., Deitrich, K., Vicente-Carbojosa, J., and Droge-Laser, W. 2009. A pivotal role of the basic leucine zipper transcription FactorbZIP53 in the regulation of arabidopsis seed maturation gene expression based on heterodimerization and protein complex formation. *Plant Cell* 21: 1747–1761.

Al-Shanfari, A.B., Abdullah, S.N.A., Saud, H.M., Omidvar, V., and Napis, S. 2012. Differential gene expression identified by suppression subtractive hybridization during late ripening of fruit in oil palm (*Elaeis guineensis* Jacq.). *Plant Molecular Biology Reporter* 30: 768–779.

Altshuler, D., Daly, M.J., and Lander, E.S. 2008. Genetic mapping in human disease. *Science* 322: 881–888.

Amiour, N., Imbaud, S., Clement, G., Agier, N., Zivy, M., Valot, B., Baliau, T., Armengaud, P., Quillere, I., and Canas, R. 2012. The use of metabolomics integrated with transcriptomic and proteomic studies for identifying key steps involved in the control of nitrogen metabolism in crops such as maize. *Journal of Experimental Botany* 63: 5017–5033.

Appleton, D., Teh, H.F., Neoh, B.K., Ooi, E.K., Wong, Y.C., Kwong, Q.B., Yusof, H.M., Chew, F.T., and Harikrishna, K. 2014. Omics: Mesocarp biochemistry provides insight into increased oil palm yield. *Planter* 90: 241–254.

Azni, I., Namasivayam, P., Ling, H.C. et al. 2014. Differentially expressed trancripts related to height in oil palm. *Journal of Oil Palm Research* 26: 308–316.

Azzeme, A.M., Abdullah, S.N.A., Aziz, M.A., and Wahab, P.E.M. 2016. Oil palm leaves and roots differ in physiological response, antioxidant enzyme activities and expression of stress-responsive genes upon exposure to drought stress. *Acta Physiologiae Plantarum* 38: 1–12.

Barcelos, E., Amblard, P., Berthaud, J., and Seguin, M. 2002. Genetic diversity and relationship in American and African oil palm as revealed by RFLP and AFLP molecular markers. *Pesquisa Agropecuária Brasileira* 37: 1105–1114.

Basiron, Y. 2000. Techno-economic aspects of research and development in the Malaysian oil palm industry. In: Basiron, Y., Jalani, B.S. and Chan, K.W. (eds) *Advances in Oil Palm Research.* Malaysian Palm Oil Board (MPOB), Bangi, Selangor, Volume 1, pp 1–18.

Beirnaert, A. and Vanderweyen, R. 1941. Contribution à l'Etude Génétique et Biométrique des Variétiés d'Elaeis guineensis Jacq. *Institut National pour l' Etude Agronomique du Congo Serie Scientific.*

Benhamou, N., Grenier, J., Asselin, A., and Legrand, M. 1989. Lmmunogold localization of β-I,3-glucanases in two plants infected by vascular wilt fungi. *Plant Cell* 1: 1209–1221.

Bernardo, R. 2014. Genomewide selection when major genes are known. *Crop Science* 54: 68–75.

Billotte, N., Jourjon, M.F., Marseillac et al. 2010. QTL detection by multi-parent linkage mapping in oil palm (*Elaeis guineensis* Jacq.). *Theoretical and Applied Genetics* 120: 1673–1687.

Billotte, N., Marseillac, N., Risterucci, A.M. et al. 2005. Microsatellite-based high density linkage map in oil palm (*Elaeis guineensis* Jacq.). *Theoretical and Applied Genetics* 110: 754–765.

Billotte, N., Risterucci, A.M., Barcelos, E., Noyer, J.L., Amblard, P., and Baurens, F.C. 2001. Development, characterisation and across-taxa utility of oil palm (*Elaeis guineensis* Jacq.) microsatellite markers. *Genome* 44: 413–425.

Boccard, J., Veuthey, J.L., and Rudaz, S. 2010. Knowledge discovery in metabolomics: An overview of MS data handling. *Journal of Separation Science* 33: 290–304.

Bourgis, F., Kilaru, A., Cao, X., Ngando-Ebongue, G.-F., Drira, N., Ohlrogge, J.B., and Arondel, V. 2011. Comparative transcriptome and metabolite analysis of oil palm and date palm mesocarp that differ dramatically in carbon partitioning. In: *Proceedings of the National Academy of Sciences* 108: 12527–12532.

Bouvet, J.-M., Makouanzi, G., Cros, D., and Vigneron, P. 2016. Modeling additive and non-additive effects in a hybrid population using genome-wide genotyping: Prediction accuracy implications. *Heredity* 116: 146–157.

Budiman, M.A., Rajinder, S., Low, E.-T.L., Nunberg, A., Citek, R., Rohlfing, T., Bedell, J.A., Lakey, N.D., Martienssen, R.A., and Cheah, S.C. 2005. Sequencing of the oil palm genespace. In: *MPOB International Palm Oil Conference*, PIPOC 2005, September 25–29, 2005, Malaysian Palm Oil Board, Kuala Lumpur Conference Centre, pp 628–639.

Bylesjö, M., Nilsson, R., Srivasta, V., Gronlund, A., Johansson, A.I., Jansson, S., Karlsson, J., Moritz, T., Wingsle, G., and Trygg, J. 2008. Integrated analysis of transcript protein and metabolite data to sudy lignin biosynthesis in hybrid aspen. *Journal of Proteome Research* 8: 199–210.

Castilho, A., Vershinin, A.V., and Heslop-Harrison J.S. 2000. Repetitive DNA and the chromosomes in the genome of oil palm (*Elaeis guineensis*). *Annals of Botany* 856: 837–844.

Chan, P.L., Sun, M.L., Low, E.-T.L., Shariff, E.M., Ooi, L.C.-L., Cheah, S.-C., and Singh, R. 2010. Normalized embryoid cDNA library of oil palm (*Elaeis guineensis*). *Electronic Journal of Biotechnology* 13: 1–17.

Chaudhary, J., Patil, G.B., Sonah, H., Deshmukh, R.K., Vuong, T.D., Valliyodan, B., and Nguyen, H.T. 2015. Expanding omics resources for improvement of soybean seed composition traits. *Frontiers in Plant Science* 6: 1021.

Cheah, S.C., Madon, M., and Singh, R. 2000. Oil palm genomics. In: Yusof, B, Jalani, B.S and Chan, K.W. (eds.). *Advances in Oil Palm Research Volume 1.* Malaysian Palm Oil Board, Kuala Lumpur, Malaysia.

Cho, K., Shibato, J., Agrawal, G.K., Jung, Y.-H., Kubo, A., Jwa, N.-S., Tamogami, S., Satoh, K., Kikuchi, S., and HigashiI, T. 2008. Integrated transcriptomics, proteomics, and metabolomics analyses to survey ozone responses in the leaves of rice seedlings. *Journal of Proteome Research* 7, 2980–2998.

Choi, H. and Pavelka, N. 2011. When one and one gives more than two: Challenges and opportunities of integrative omics. *Frontiers in Genetics* 2: 105.

Collakova, E., Aghamirzaie, D., Fang, Y., Klumas, C., Tabataba, F., Kakumanu, A., Myers, E., Heath, L.S., and Grene, R. 2013. Metabolic and transcriptional reprogramming in developing soybean (*Glycine max*) embryos. *Metabolites* 3: 347–372.

Cooper, R.M., Flood, J., and Rees, R. 2011. *G. boninense* in oil palm plantations: Current thinking on epidemiology, resistance and pathology. *Planter* 87: 515–526.

Corley, R.H.V. and Lee, C.H. 1992. The physiological basis for genetic improvement of oil palm in Malaysia. *Euphytica* 60: 179–184.

Corley, R.H.V. and Tinker, P.B. 2003. Selection and breeding. *The Oil Palm* 3rd edn. Blackwell Publishing, Oxford, pp. 133–199.

Corley, R.H.V. and Tinker, P.B. 2016. Selection and breeding. *The Oil Palm*, Fifth edn. John Wiley & Sons, Ltd, UK.

Cramer, G.R., Urano, K., Delrot, S., Pezzotti, M., and Shinozaki, K. 2011. Effects of abiotic stress on plants: A systems biology perspective. *BMC Plant Biology* 11: 163.

Cregan, P., Jarvik, T., Bush, A., Shoemaker, R., Lark, K., Kahler, A., Kaya, N., Vantoai, T., Lohnes, D., and Chung, J. 1999. An integrated genetic linkage map of the soybean genome. *Crop Science* 39: 1464–1490.

Cros, D., Sánchez, L., Cochard, B. et al. 2014. Estimation of genealogical coancestry in plant species using a pedigree reconstruction algorithm and application to an oil palm breeding population. *Theoretical and Applied Genetics* 127: 981–994.

Cros, D., Denis, M., Bouvet, J.-M., and Sanchez, L. 2015a. Long-term genomic selection for heterosis without dominance in multiplicative traits: Case study of bunch production in oil palm. *BMC Genomics* 16: 651.

Cros, D., Denis M., Sánchez, L. et al. 2015b. Genomic selection prediction accuracy in a perennial crop: Case study of oil palm (*Elaeis guineensis* Jacq.). *Theoretical and Applied Genetics* 128: 397–410.

Daim, L.D.J., Ooi, T.E.K., Ithnin, N., Yusof, H.M., Harikrishna, K., Majid, N.A., and Karsani, S.A. 2015. Comparative proteomic analysis of oil palm leaves infected with *Ganoderma boninense* revealed changes in proteins involved in photosynthesis; carbohydrate metabolism; and immunity and defence. *Electrophoresis* 36: 1699–1711.

Decourcelle, E.M., Perez-Fons, L., Bulande, S. et al. 2015. Combined transcript, proteome and metabolite analysis of transgenic maize seeds engineered for enhanced carotenoid synthesis reveals pleotropic effects in core metabolism. *Journal of Experimental Botany* 66: 3141–3150.

Dekkers, J.C.M. 2004. Commercial application of marker- and gene-assisted selection in livestock: Strategies and lessons. *Journal of Animal Science* 82: E-Suppl:E313–E328.

Denis, M. and Bouvet, J-M. 2013. Efficiency of genomic selection with models including dominance effect in the context of eucalyptus breeding. *Tree Genetics and Genomes* 9: 37–51.

Donaire, J.P., Sanchez, A.J., Lopez-Gorge, J., and Recalde, L. 1975. Metabolic changes in fruit and leaf during ripening in the olive. *Phytochemistry* 14: 1167–1169.

Dunwell, J.M., Wilkinson, M.J., Nelson, S. et al. 2010. Production of haploids and doubled haploids in oil palm. *BMC Plant Biology* 10: 218.

Dussert, S., Guerin, C., Andersson, M. et al. 2013. Comparative transcriptome analysis of three oil palm fruit and seed tissues that differ in oil content and fatty acid composition. *Plant Physiology* 162: 1337–1358.

Ebrahimi, M., Abdullah, S., Aziz, M., and Namasivayam, P. 2015. A novel CBF that regulates abiotic stress response and the ripening process in oil palm (*Elaeis guineensis*) fruits. *Tree Genetics and Genomes* 11: 1–16.

Feuillet, C., Leach, J.E., Rogers, J., Schnable, P.S., and Eversole, K. 2011. Crop genome sequencing: Lessons and rationales. *Trends in Plant Science* 16: 77–88.

Fiehn, O., Kloska, S., and Altmann, T. 2001. Integrated studies on plant biology using multiparallel techniques. *Current Opinion in Biotechnology* 12: 82–86.

Fiehn, O., Wohlgemuth, G., Scholz, M., Kind, T., Lee, D.Y., Lu, Y., Moon, S., and Nikolau, B. 2008. Quality control for plant metabolomics: Reporting MSI-compliant studies. *The Plant Journal* 53: 691–704.

Figueroa, D.M. and Bass, H.W. 2012. Development of pachytene FISH maps for six maize chromosomes and their integration with other maize maps for insights into genome structure variation. *Chromosome Research* 20: 363–380.

Fonseca, A., Ferreira, J., Barros dos Santos, T. et al. 2010. Cytogenetic map of common bean (*Phaseolus vulgaris* L.). *Chromosome Research* 18: 487–502.

Fridman, E. and Pichersky, I.E. 2005. Metabolomics, genomics, proteomics, and the identification of enzymes and their substrates and products. *Current Opinion in Plant Biology* 8: 242–248.

Fukushima, A., Kusano, M., Redestig, H., Arita, M., and Saito, K. 2009. Integrated omics approaches in plant systems biology. *Current Opinion in Chemical Biology* 13: 532–538.

Galland, M., Lounifi, I., Cueff, G., Baldy, A., Morin, H., Job, D., and Rajjou, L. 2012. A role for "omics" technologies in exploration of the seed nutritional quality. In: Agrawal, G.K. and Rakwal, R. (eds) *Seed Development: OMICS Technologies toward Improvement of Seed Quality and Crop Yield*. Springer, Dordrecht, Netherlands.

Ghesquiere, M. 1984. Enzyme polymorphism in oil palm I—Genetic control of nine enzyme systems. *Oleagineux* 39: 561–574.

Goddard, M.E. and Hayes, B.J. 2007. Genomic selection. *Journal of Animal Breeding and Genetics* 124: 323–330.

Gomez-Cabrero, D., Abugessaisa, I., Maier, D., Teshendofff, A., Merkenschlager, M., Gisel, A., Ballestar, E., Bongcam-Rudloff, E., Conesa, A., and Tegner, J. 2014. Data integration in the era of omics: Current and future challenges. *BMC Systems Biology* 8: 1–10.

Gowda, M., Das, B., Makumbi, D. et al. 2015. Genome-wide association and genomic prediction of resistance to maize lethal necrosis disease in tropical maize germplasm. *Theoretical and Applied Genetics* 128: 1957–1968.

Guerin, C., Joët, T., Serret, J. et al. 2016. Gene coexpression network analysis of oil biosynthesis in an interspecific backcross of oil palm. *The Plant Journal* 87(5): 423–441. doi: 10.1111/tpj.13208.

Hajduch, M., Matusova, R., Houston, N.L., and Thelen, J.J. 2011. Comparative proteomics of seed maturation in oilseeds reveals differences in intermediary metabolism. *Proteomics* 11: 1619–1629.

Hama-Ali, E., Panandam, J., Tan, S. et al. 2015. Association between basal stem rot disease and simple sequence repeat markers in oil palm, *Elaeis guineensis* Jacq. *Euphytica* 202: 199–206.

Hardon, J.J. 1970. Inbreeding in populations of the oil palm (*Elaeis guineensis* Jacq.) and its effects on selection. *Oleagineux* 25: 449–456.

Hasliza, H., Lau, B.Y.C., and Ramli, U.S. 2014. Extraction method for analysis of oil palm leaf and root proteins by two-dimensional gel electrophoresis. *Journal of Oil Palm Research* 26: 54–61.

Hasliza, H. and Ramli, U.S. 2015. Profiling of oil palm mesocarp proteome using gel-LC-MS/MS analysis. In: *Proceedings of PIPOC 2015*, October 6–8, 2015, Kuala Lumpur Convention Centre, Kuala Lumpur, pp 398–403.

Heslop-Harrison, J.S. 1991. The molecular cytogenetics of plants. *Journal of Cell Science* 100: 5–21.

Heslop-Harrison, J.S. and Schwarzacher, T. 2011. Organisation of the plant genome in chromosomes. In: *Plant Journal* 66: 18–33.

Heslop-Harrison, J.S. and Schwarzacher, T. 2013. Nucleosomes and centromeric DNA packaging. *Proceedings of the National Academy of Sciences USA* 150: 19974–19975.

Heslot, N., Akdemir, D., Sorrells, M., and Jannink, J.-L. 2014. Integrating environmental covariates and crop modeling into the genomic selection framework to predict genotype by environment interactions. *Theoretical and Applied Genetics* 127: 463–480.

Ho, C.L., Kwan, Y.Y., Choi, M.C. et al. 2007. Analysis and functional annotation of expressed sequence tags (ESTs) from multiple tissues of oil palm (*Elaeis guineensis* Jacq.). *BMC Genomics* 8: 381–393.

Holloway, B., Luck, S., Beatty, M., Rafalski, J.-A., and Li, B. 2011. Genome-wide expression quantitative trait loci (eQTL) analysis in maize. *BMC Genomics* 12: 1–14.

Idris, A.S., Kushairi, A., Ismail, S., and Ariffin, D. 2004. Selection of partial resistance in oil palm progenies to *Ganoderma* basal stem rot. *Journal of Oil Palm Research* 16: 12–18.

Jack, P.L., Dimitrijevic, T.A., and Mayes, S. 1995. Assessment of nuclear, mitochondrial and chloroplast RFLP markers in oil palm (*Elaeis guineensis* Jacq.). *Theoretical and Applied Genetics* 90: 643–649.

Jaligot, E., Adler, S., Debladis É., Beulé, T., Richaud, F., Ilbert, P., Finnegan, E.J., and Rival, A. 2011. Epigenetic imbalance and the floral developmental abnormality of the in vitro-regenerated oil palm *Elaeis guineensis*. *Annals of Botany* 108: 1453–1462.

Jarquín, D., Crossa, J., Lacaze, X. et al. 2014a. A reaction norm model for genomic selection using high-dimensional genomic and environmental data. *Theoretical and Applied Genetics* 127: 595–607.

Jarquín, D., Kocak, K., Posadas, L., Hyma, K., Jedlicka, J., Graef, G., and Lorenz, A. 2014b. Genotyping by sequencing for genomic prediction in a soybean breeding population. *BMC Genomics* 15: 1–10.

Jeennor, S. and Volkaert, H. 2014. Mapping of quantitative trait loci (QTLs) for oil yield using SSRs and gene-based markers in African oil palm (*Elaeis guineensis* Jacq.). *Tree Genetics and Genomes* 10: 1–14.

Jeffery Daim, L.D., Ooi, T.E.K., Ithnin, N., Mohd Yusof, H., Kulaveerasingam, H., Abdul Majid, N., and Karsani, S.A. 2015. Comparative proteomic analysis of oil palm leaves infected with *Ganoderma boninense* revealed changes in proteins involved in photosynthesis, carbohydrate metabolism, and immunity and defense. *Electrophoresis* 36: 1699–1710.

Jorrín-Novo, J.V., Pascual, J., Sánchez-Lucas, R., Romero-Rodríguez, M.C., Rodríguez-Ortega, M.J., Lenz, C., and Valledor, L. 2015. Fourteen years of plant proteomics reflected in proteomics: Moving from model species and 2DE-based approaches to orphan species and gel-free platforms. *Proteomics* 15: 1089–1112.

Jouannic, S., Argout, X., Lechauve, F., Fizames, C., Borgel, A., Morcillo, F., Aberlenc-Bertossi, F., Duval, Y., and Tregear, J. 2005. Analysis of expressed sequence tags from oil palm (*Elaeis guineensis*). *FEBS Letters* 579: 2709–2714.

Joyce, A.R. and Palsson, B.Ø. 2006. The model organism as a system: Integrating "omics" data sets. *Nature Reviews Molecular Cell Biology* 7: 198–210.

Kao, F., Cheng, Y.Y., Chow, T.Y. Chen, H.H., Liu, S.M., Cheng, C.H., and Chung, M.C. 2006. An integrated map of *Oryza sativa* L. chromosome 5. *Theoretical and Applied Genetics* 112: 891–902.

Kim, J.S., Islam-Faridi, M.N., Klein, P.E. et al. 2005. Comprehensive molecular cytogenetic analysis of sorghum genome architecture: Distribution of euchromatin, heterochromatin, genes and recombination in comparison to rice. *Genetics* 1714: 1963–1976.

Kovinich, N., Saleem, A., Arnason, J.T., and Miki, B. 2011. Combined analysis of transcriptome and metabolite data reveals extensive differences between black and brown nearly-isogenic soybean (*Glycine max*) seed coats enabling the identification of pigment isogenes. *BMC Genomics* 12: 381–399.

Kubis, S.E., Castilho, A.M., Vershinin, A.V., and Heslop-Harrison, J.S. 2003. Retroelements, transposons and methylation status in the genome of oil palm (*Elaeis guineensis*) and the relationship to somaclonal variation. *Plant Molecular Biology* 52: 69–79.

Kumar, J., Pratap, A., and Kumar, S. 2015. Plant phenomics: An overview. In: Kumar, J., Pratap, A. and Kumar, S. (eds) *Phenomics in Crop Plants: Trends, Options and Limitations.* Springer India, New Delhi.

Kusano, M., Baxter, I., Fukushima, A., Okazaki, Y., Nakabayashi, R., Bouvrette, D.J., Achard, F., Jakubowski, A.R., and Ballam, J.M. 2015. Assesing metabolomics and chemical diversity of a soybean lineage representing 35 years of breeding. *Metabolomics* 11: 261–270.

Kwong, Q.B., Teh, C.K., Ong, A.L. et al. 2016. Development and validation of a high density SNP genotyping array for African oil palm. *Molecular Plant.* doi: 10.1016/j.molp.2016.04.010.

Langer, S.M., Longin, F.H., and Wurschum, T. 2014. Flowering time control in European winter wheat. *Frontiers in Plant Science* 5: 537.

Langridge, P. and Fleury, D. 2011. Making the most of "omics" for crop breeding. *Trends in Biotechnology* 29: 33–40.

Lau, B.Y.C., Clerens, S., Morton, J.D., Dyer, J.M., Deb-Choudhury, S., and Ramli, U.S. 2015. Method developments to extract proteins from oil palm chromoplast for proteomic analysis. *Springer Plus* 4: 791.

Lau, B.Y.C., Clerens, S., Morton, J.D., Dyer, J.M., Deb-Choudhury, S., and Ramli, U.S. 2016. Application of mass spectrometry approach to detect the presence of fatty acid biosynthetic phosphopeptides. *Protein Journal* 35: 163–170.

Lay, J.R., Liyanage, R., Borgmann, S., and Wilkins, C.L. 2006. Problems with the "omics." *TrAC Trends in Analytical Chemistry* 25: 1046–1056.

Lee, M., Xia, J.H., Zou, Z. et al. 2015. A consensus linkage map of oil palm and a major QTL for stem height. *Scientific Reports* 5: 8232. doi: 10.1038/srep08232.

Li, G., Yu, M., Fang, T., Cao, S., Carver, B.F., and Yan, L. 2013a. Vernalization requirement duration in winter wheat is controlled by TaVRN-A1 at the protein level. *Plant Journal* 76: 742–753.

Li, H., Peng, Z., Yang, X., Wang, W., Fu, J., Wang, J., Han, Y., Chai, Y., Guo, T., and Yang, N. 2013b. Genome-wide association study dissects the genetic architecture of oil biosynthesis in maize kernels. *Nature Genetics* 45: 43–50.

Li, L., Hur, M., Lee, J.-Y., Zhou, W., Song, Z., Ransom, N., Demirkale, C.Y., Nettleton, D., Westgate, M., and Arendsee, Z. 2015. A systems biology approach toward understanding seed composition in soybean. *BMC Genomics* 16: S9.

Liu, P., Wang, C.M., Li, L., Sun, F., and Yue, G.H. 2011. Mapping QTLs for oil traits and EQTLs for oleosin genes in jatropha. *BMC Plant Biology* 11: 1–9.

Lim, K.-A., Shamsuddin, Z., and Ho, C.-L. 2010. Transcriptomic changes in the root of oil palm (*Elaeis guineensis* Jacq.) upon inoculation with *Bacillus sphaericus* UPMB10. *Tree Genetics and Genomes* 6: 793–800.

Loei, H., Lim, J., Tan, M., Lim, T.K., Lin, Q.S., Chew, F.T., Kulaveerasingam, H., and Chung, M.C. 2013. Proteomic analysis of the oil palm fruit mesocarp reveals elevated oxidative phosphorylation activity is critical for increased storage oil production. *Journal of Proteome Research* 12: 5096–5109.

Lorenz, A.J., Chao, S., Asoro, F.G. et al. 2011. Genomic selection in plant breeding: Knowledge and prospects. *Advances in Agronomy* 110: 77–123.

Low, E.T., Rosli, R., Jayanthi, N., Mohd-Amin, A.H., Azizi, N., Chan, K.L., Maqbool, N.J., Maclean, P., Brauning, R., McCulloch, A., Moraga, R., Ong-Abdullah, M., and Singh, R. 2014. Analyses of hypomethylated oil palm gene space. *PLoS One* 9: 86728.

Madon, M., Heslop-Harrison, J.S., Schwarzacher, T., and Hashim, A.T. 2012. Analysis of oil palm calli and regenerants using flow and image cytometry and 18S-25S ribosomal DNA fluorescence *in situ* hybridisation (FISH). *Journal of Oil Palm Research* 24: 1318–1329.

Madon, M., Clyde, M.M., and Cheah, S.C. 1995. Cytological analysis of *Elaeis guineenis* (*tenera*) chromosomes. *Elaeis* 7: 122–134.

Madon, M., Heslop-Harrison, J.S., Schwarzacher, T., Mohd Rafdi, M.H., and Clyde, M.M. 2005. Cytological analysis of oil palm pollen mother cells (PMCs). *Journal of Oil Palm Research* 17: 176–180.

Mayes, S., Jack, P.L., Corley, R.H.V., and Marshall, D.F. 1997. Construction of a RFLP genetic linkage map for oil palm (*Elaeis guineensis* Jacq.). *Genome* 40: 116–122.

Mayes, S., Hafeez, F., Price, Z., MacDonald, D., Billotte, N., and Roberts, J. 2008. Molecular research in oil palm, the key oil crop for the future. In: Moore, P.H. and Ming, R. (eds) *Genomics of Tropical Crop Plants.* Springer, New York.

Menu-Bouaouiche, L., Vriet, C., Peumans, W.J., Barre, A., Van Damme E.J.M., and Rouge, P. 2003. A molecular basis for the endo-β-1,3-glucanase activity of the thaumatin-like proteins from edible fruits. *Biochimie* 85: 123–131.

Messing, J., Bharti, A.K., Karlowski, W.M., Gunlach, H., Kim, H.R., Yu, Y., Wei, F., Fuks, G., Soderlunds, C.A., Mayer, K.F.X., and Wing, R.A. 2004. Sequence composition and genome organization of maize. In: *Proceedings of the National Academy of Sciences of the United States of America* 101: 14349–14354.

Montoya, C., Lopes, R., Flori, A. et al. 2013. Quantitative trait loci (QTLs) analysis of palm oil fatty acid composition in an interspecific pseudo-backcross from *Elaeis oleifera* (H.B.K.) cortés and oil palm (*Elaeis guineensis* Jacq.). *Tree Genetics and Genomes* 9:1207–1225.

Murphy, D.J. 2014. The future of oil palm as a major global crop: Opportunities and challenges. *Journal of Oil Palm Research* 26: 1–24.

Natarajan, S.S., Xu, C., Garrett, W.M., Lakshman, D., and Bae, H. 2012. Assessment of the natural variation of low abundant metabolic proteins in soybean seeds using proteomics. *Journal of Plant Biochemistry and Biotechnology* 21: 30–37.

Neoh, B.K., Teh, H.F., Ng, T.L.M., Tiong, S.H., Thang, Y.M., Ersad, M.A., Mohamed, M., Chew, F.T., Kulaveerasingam, H., and Appleton, D.R. 2013. Profiling of metabolites in oil palm mesocarp at different stages of oil biosynthesis. *Journal of Agricultural and Food Chemistry* 61, 1920–1927.

Ning, M. and Lo, E. 2010. Opportunities and challenges in omics. *Translational Stroke Research* 1, 233–237.

Nurazah, Z., Idris, A.S., Kushairi, A.D., and Ramli, U.S. 2013. Metabolite profiling of oil palm towards understanding basal stem rot (BSR) disease. *Journal of Oil Palm Research* 25: 58–71.

Ong-Abdullah, M., Ordway, J.M., Jiang, N. et al. 2015. Loss of karma transposon methylation underlies the mantled somaclonal variant of oil palm. *Nature* 525: 533–537.

Ooi, T.E.K., Yeap, W.C., Jeffery Daim, L.D., Ng, B.Z., Lee, F.C., Othmann, A.M., Appleton, D.R., Chew, F.T., and Kulaveerasingam, H. 2015. Differential expression analysis of mesocarp protein from high- and low-yielding oil palms associates non-oil biosynthetic enzymes to lipid biosynthesis. *Proteome Science* 13: 28–43.

Osorio, S., Alba, R., Damasceno, C.M., Lopez-Casado, G., Lohse, M., Zanor, M.I., Tohge, T., Usadel, B., Rose, J.K., and Fei, Z. 2011. Systems biology of tomato fruit development: Combined transcript, protein, and metabolite analysis of tomato transcription factor (nor, rin) and ethylene receptor (Nr) mutants reveals novel regulatory interactions. *Plant Physiology* 157: 405–425.

Ouyang, S., Zhu, W., Hamilton, J. et al. 2007. The TIGR rice genome annotation resource: Improvements and new features. *Nucleic Acids Research* 35: D883–D887.

Paesold, S., Borchardt, D., Schmidt, T., and Dechyeva, D. 2012. A sugar beet (*Beta vulgaris* L.) reference FISH karyotype for chromosome and chromosome-arm identification, integration of genetic linkage groups and analysis of major repeat family distribution. *Plant Journal* 72: 600–611.

Pamin, K.A. 1998. A hundred and fifty years of oil palm in Indonesia: From the Bogor Botanical Garden to the industry. In: *International Oil Palm Conference.* "Commodity of the Past, Today and the Future", September 23–25, 1998, Sheraton Nusa Indah Hotel Bali, Indonesia, pp 3–23.

Paterson, R.R.M., Sariah, M., Abidin, M.A.Z., and Lima, N. 2008. Prospects for inhibition of lignin degrading enzymes to control *Ganoderma* white rot of oil palm. *Current Enzyme Inhibition* 4: 172–179.

Pedrosa-Harand, A., Kami, J., Gepts, P., Gefffroy, V., and Schweizer, D. 2009. Cytogenetic mapping of common bean chromosome reveals a less compartmentalized small-genome plant species. *Chromosome Research* 17: 405–417.

Pootakham, W., Jomchai, N., Ruang-Areerate, P., Shearman, J.R., Sonthirod, C., Sangsrakru, D., Tragoonrung, S., and Tangphatsornruang, S. 2015. Genome-wide SNP discovery and identification of QTL associated with agronomic traits in oil palm using genotyping-by-sequencing (GBS). *Genomics* 105: 288–295.

Rabinowicz, P.D., Schutz, K., Dedhia, N., Yordan, C., Parnell, L.D., Stein, L., Mccombie, W.R., and Martienssen, R.A. 1999. Differential methylation of genes and retrotransposons facilitates shotgun sequencing of the maize genome. *Nature Genetics* 23: 305–308.

Rajanaidu, N., Kushairi, A., Rafii, M., Mohd Din, A., Maizura, I., and Jalani, B.S. 2000. Oil palm breeding and genetic resources. In: Basiron, Y., Jalani, B.S. and Chan, K.W. (eds) *Advances in Oil Palm Research*. Malaysian Palm Oil Board (MPOB), Bangi, Selangor, Volume 1, pp 171–237.

Rance, K., Mayes, S., Price, Z., Jack, P.L., and Corley, R.H.V. 2001. Quantitative trait loci for yield components in oil palm (*E. guineensis* Jacq.). *Theoretical and Applied Genetics* 103: 1302–1310.

Resende, M.D.V., Resende M.F.R., Sansaloni, C.P. et al. 2012a. Genomic selection for growth and wood quality in eucalyptus: Capturing the missing heritability and accelerating breeding for complex traits in forest trees. *New Phytologist* 194: 116–128.

Resende, M.F.R., Muñoz, P., Resende, M.D.V. et al. 2012b. Accuracy of genomic selection methods in a standard data set of loblolly pine (*Pinus taeda* L.). *Genetics* 190: 1503–1510.

Riedelsheimer, C., Czedik-Eysenberg, A., Grieder, C. et al. 2012. Genomic and metabolic prediction of complex heterotic traits in hybrid maize. *Nature Genetics* 44: 217–220.

Riju, A., Chandrasekhar, A., and Arunachalam, V. 2007. Mining for single nucleotide polymorphism and insertions/deletions in expressed sequence tag libraries of oil palm. *Bioinformation* 2: 128–131.

Rival, A., Beule, T., Phippe, B., and Noirot, M. 1997. Comparative flow cytometric estimation of nuclear DNA content in oil palm (*Elaeis guineensis* Jacq.) tissue cultures and seed-derived plants. *Plant Cell Reports* 16: 884–887.

Ricroch, A.E., Berge, J.B., and Kuntz, M. 2011. Evaluation of genetically engineered crops using transcriptomic, proteomic, and metabolomic profiling techniques. *Plant Physiology* 155: 1752–1761.

Rincent, R., Laloë, D., Nicolas, S. et al. 2012. Maximizing the reliability of genomic selection by optimizing the calibration set of reference individuals: Comparison of methods in two diverse groups of maize inbreds (*Zea mays* L.). *Genetics* 192(2): 715–728. doi: 10.1534/genetics.112.141473.

Rosenquist, E.A. 1990. An overview of breeding technology and selection in *Elaeis guineensis*. In: *International Oil Palm Development Conference—Agriculture*. Palm Oil Research Institute Malaysia, Kuala Lumpur.

Rutkoski, J.E., Poland, J.A., Singh, R.P. et al. 2014. Genomic selection for quantitative adult plant stem rust resistance in wheat. *The Plant Genome* 7(3): 1–10. doi: 10.3835/plantgenome2014.02.0006.

Sagi, L., Remy, S., Coemans, B., Thiry, E., Santos, E., Matsumara, H., Terushi, R., and Swennen, R. 2005. Functional analysis of the banana genome by gene tagging and SAGE. In: *Plant and Animal Genome XIII Conference*. San Diego.

Saito, K. and Matsuda, F. 2010. Metabolomics for functional genomics, systems biology, and biotechnology. *Annual Review of Plant Biology* 61: 463–489.

Salt, D.E., Baxter, I., and Lahner, B. 2008. Ionomics and the study of the plant ionome. *Annual Review of Plant Biology* 59: 709–733.

Sapak, Z., Meon, S., and Ahmad, Z.A.M. 2008. Effect of endophytic bacteria on growth and suppression of *Ganoderma* infection in oil palm. *International Journal of Agricultural Biology* 10: 127–132.

Sato, D. 1949. Karyotype alteration and phylogeny, VI karyotype analysis in palmae. *Cytologia* 12:174–186.

Schwarzacher, T. and Heslop-Harrison, J.S. 2000. *Practical In Situ Hybridization*. Bios, Oxford. 203+xii pp.

Schwarzacher, T. 2016. Preparation and fluorescent analysis of plant metaphase chromosomes. In: Caillaud, M.C. (ed) *Plant Cell Division: Methods and Protocols, Methods in Molecular Biology 1370*. Humana Press, Springer, New York, Chapter 7.

Scossa, F., Brotman, Y., De Abreu E Lima, F., Willmitzer, L., Nikoloski, Z., Tohge, T., and Fernie, A.R. 2016. Genomics-based strategies for the use of natural variation in the improvement of crop metabolism. *Plant Science* 242: 47–64.

Shah, F.H. and Rashid, O. 1996. Nucleotide sequence of a complementary DNA clone encoding stearoyl-acyl-carrier protein desaturase from *Elaeis guineensis* var *tenera* (PGR96-110). *Plant Physiology* 112: 1399.

Shah, F.H., Rashid, O., Simons, A.J., and Dunsdon, A. 1994. The utility of RAPD markers for the determination of genetic variation in oil palm (*Elaeis guineensis*). *Theoretical and Applied Genetics* 89: 713–718.

Shahirah Balqis, D., Abrizah, O., Ramli, U.S., Tahir, N.I., Syahanim, S., Nurazah, Z., Mohamad Arif, A.M., Idris, A.S., Mohd Din, A., and Sambanthamurthi, R. 2015. Identification of chelidonic acid in oil palm spear leaf artificially infected with *Ganoderma boninense* using liquid chromatography mass spectrometry. In: *Proceedings of PIPOC 2015*, October 6–8, 2015, Kuala Lumpur Convention Centre, Kuala Lumpur, pp 386–391.

Shariff, E.M.D., Low, E.-T.L., Alias, H., Suan-Choo, C., and Singh, R. 2008. Identification of genes expressed in the embryoid tissue of oil palm (*Elaeis guineensis* Jacq.) tissue culture via expressed sequence tag analysis. *Journal of Oil Palm Research: Special Issue on Malaysia-MIT (Biotechnology Partnership Programme: Volume 1—Oil Palm Tissue Culture* 51–63.

Sharma, A.K. and Sarkar, S.K. 1956. Cytology of different species of palms and its bearings on the solutions of the problems of phylogeny and speciation. *Genetica* 28: 361–488.

Shary, S. and Guha-Mukherjee, S. 2004. Isolation and expression studies of differentiation-specific genes in tobacco dihaploids using PCR-based subtractive hybridization method. *Plant Science* 166: 317–322.

Singh, B., Bohra, A., Mishra, S., Joshi, R., and Pandey, S. 2015. Embracing new-generation "omics" tools to improve drought tolerance in cereal and food-legume crops. *Biologia Plantarum* 59: 413–428.

Singh, R., Cheah, S.C., and Rahman, R.A. 1998. Generation of molecular markers in oil palm (*Elaeis guineensis*) Using AFLP analysis. *FOCUS* 20: 26–27.

Singh, R. and Cheah, S.C. 2005. Potential application of marker-assisted selection in oil palm. *MPOB Bulletin* 51: 1–9.

Singh, R., Tan, S.G., Panandam, J., Rahimah, A.R., and Suan-Choo, C. 2008a. Identification of cDNA-RFLP markers and their use for molecular mapping in oil palm (*Elaeis guineensis*). *Asia Pacific Journal of Molecular Biology and Biotechnology* 16: 53–63.

Singh, R., Zaki, N.M., Ting, N.C., Rosli, R., Tan, S.-G., Low, E.-T.L., Ithnin, M., and Suan-Choo, C. 2008b. Exploiting an oil palm EST database for the development of gene-derived SSR markers and their exploitation for assessment of genetic diversity. *Biologia* 63: 227–235.

Singh, R., Tan, S.G., Panandam, J., Rahimah, A.R., Ooi, L.C.-L., Low, E.-T.L., Sharma, M., Jansen, J., and Suan-Choo, C. 2009. Mapping quantitative trait loci (QTLs) for fatty acid composition in an interspecific cross of oil palm. *BMC Plant Biology* 9: 114.

Singh, R., Madon, M., Low, E.-T.L., Ooi, L.C.-L., Chan, P.L., Rosli, R., Ting, N.C., and Ithnin, M. 2011. Oil palm genomics: A foundation for improved agricultural productivity. In: Mohd. Basri B. Wahid, Yuen May Choo (Datuk.), Kook Weng Chan (eds) *Further Advances in Oil Palm Research (2000–2010)*. Malaysian Palm Oil Board, Kuala Lumpur, Malaysia, pp 202–251.

Singh, R., Low, E.-T.L., Ooi, L.C.-L. et al. 2013a. The oil palm SHELL gene controls oil yield and encodes a homologue of SEEDSTICK. *Nature* 500: 340–344.

Singh, R., Ong-Abdullah, M., Low, E.-T.L. et al. 2013b. Oil palm genome sequences reveals divergence of interfertile species in old and new worlds. *Nature* 500: 335–339.

Singh, R., Low, E.-T.L., Ooi, L.C.-L. et al. 2014. The oil palm VIRESCENS gene controls fruit colour and encodes a R2R3-MYB. *Nature Communications* 5: Article No. 4106. doi: 10.1038/ncomms5106.

Siswanto, S.D. and Darmono, T.W. 1998. Chitinase and β-1,3-glucanase activities against *Ganoderma* sp. in oil palm. In: Tahardi, J.S., Darmono, T.W., Siswanto, S.D., Nataatmadja, R. (eds) *Proceedings of the BTIG Workshop on Oil Palm Improvement through Biotechnology*. Bogor, Indonesia, pp 104–114.

Sivori, E.M. and Alaniz, J.R. 1973. Relationships between the dwarfing of *Tropaeolummajus* and asparagine synthesis. *Plant Cell Physiology* 14: 653–659.

Soh, A., Wong, C., Ho, Y., and Choong, C. 2010. Oil palm. In: Vollmann, J., Rajcan, I. (eds) *Oil Crops*. Springer, New York, pp 333–367.

Song, H., Yin, Z., Chao, M., Ning, L., Zhang, D., and Yu, D. 2014. Functional properties and expression quantitative trait loci for phosphate transporter GmPT1 in soybean. *Plant, Cell and Environment* 37: 462–472.

Speed, D. and Balding, D.J. 2014. MULTIBLUP: Improved SNP-based prediction for complex traits. *Genome Research* 24(9): 1550–1557. doi: 10.1101/gr.169375.113.

Sturtevant, A.H. 1913. The linear arrangement of six sex-linked factors in drosophila, as shown by their mode of association. *Journal of Experimental Zoology* 14: 43–59.

Sumner, L.W., Mendes, P., and Dixon, R.A. 2003. Plant metabolomics: Large-scale phytochemistry in the functional genomics era. *Phytochemistry* 62, 817–836.

Sun, J., Zhang, Z., Zong, X., Huang, S., Li, S., and Han, Y. 2013. A high-resolution cucumber cytogenetic map integrated with the genome assembly. *BMC Genomics* 14: 461.

Sun, T.P. and Gubler, F. 2004. Molecular mechanism of gibberellin signaling in plants. *Annual Review of Plant Biology* 55: 197–223.

Sundram, S., Abdullah, F., Ahmad, Z.A.M., and Yusuf, U.K. 2008. Efficacy of single and mixed treatments of *Trichoderma harzianum* as biocontrol agents of *Ganoderma* basal stem rot in oil palm. *Journal of Oil Palm Research* 20: 470–483.

Susanto, A., Sudharto, P.S., and Purba, R.Y. 2005. Enhancing biological control of basal stem rot disease (*Ganoderma boninense*) in oil palm plantations. *Mycopathologia* 159: 153–157.

Swaminathan, K., Varala, K., and Hudson, M. 2007. Global repeat discovery and estimation of genomic copy number in a large, complex genome using a high-throughput 454 sequence survey. *BMC Genomics* 8: 132.

Syahanim, S., Abrizah, O., MohamadArif, A.M., Idris, A.S., and Mohd Din, A. 2013. Identification of differentially expressed proteins in oil palm seedlings artificially infected with *Ganoderma*: A proteomics approach. *Journal of Oil Palm Research* 25: 298–304.

Syrenne, R.D., Shi, W., Stewart, C.N., and Yuan, J.S. 2012. Omics platforms: Importance of twenty-first century genome-enabled technologies in seed developmental research for improved seed quality and crop yield. In: Agrawal, K.G. and Rakwal, R. (eds) *Seed Development: OMICS Technologies toward Improvement of Seed Quality and Crop Yield: OMICS in Seed Biology*. Springer Netherlands, Dordrecht.

Szinay, D., Chang, S.-B., Khrustaleva, L. et al. 2008. High-resolution chromosome mapping of BACs using multi-colour FISH and pooled-BAC FISH as a backbone for sequencing tomato chromosome 6. *Plant Journal* 56: 627–637.

Tahir, N.I., Shaari, K., Abas, F., Ahmad Parveez, G.K., Ahmad Tarmizi, H., and Ramli, U.S. 2013. Identification of oil palm (*Elaeis guineensis*) spear leaf metabolites using mass spectrometry and neutral loss analysis. *Journal of Oil Palm Research* 25: 72–83.

Tahir, N.I., Shaari, K., Abas, F., Ahmad Parveez, G.K., Ishak, Z., and Ramli, U.S. 2012. Characterization of apigenin and luteolin derivatives from oil palm (*Elaeis guineensis* Jacq.) leaf using LC-ESI-MS/MS. *Journal of Agricultural and Food Chemistry* 60: 11201–11210.

Tan, K.C., Ipcho, S.V., Trengove, R.D., Oliver, R.P., and Solomon, P.S. 2009. Assessing the impact of transcriptomics, proteomics and metabolomics on fungal phytopathology. *Molecular Plant Pathology* 10: 703–715.

Tang, X., Szinay, D., Lang, C. et al. 2008. Cross-species BAC-FISH painting of the tomato and potato chromosome 6 reveals undescribed chromosomal rearrangements. *Genetics* 180: 1319–1328.

Technow, F., Bürger, A., Melchinger, A.E. 2013. Genomic prediction of northern corn leaf blight resistance in maize with combined or separated training sets for heterotic groups. *G3: Genes|Genomes|Genetics* 3: 197–203.

Technow, F., Messina, C.D., Totir, L.R., and Cooper, M. 2015. Integrating crop growth models with whole genome prediction through approximate Bayesian computation. *PLoS One* 10: e0130855.

Tee, S.-S., Tan, Y.-C., Abdullah, F. et al. 2013. Transcriptome of oil palm (*Elaeis guineensis* Jacq.) roots treated with *Ganoderma boninense. Tree Genetics and Genomes* 9: 377–386.

Teh, C.K., Ong, A.L., Kwong, Q.B., Apparow, S., Chew, F.T., Mayes, S., Mohamed, M., Appleton, D., and Kulaveerasingam, H. 2016. Genome-wide association study identifies three key loci for high mesocarp oil content in perennial crop oil palm. *Scientific Reports* 6: 19075. doi: 10.1038/srep19075.

Teh, H.F., Neoh, B.K., Hong, M.P.L., Low, J.Y.S., Ng, T.L.M., Ithnin, N., Thang, Y.M., Mohamed, M., Chew, F.T., and Yusof, H.M. 2013. Differential metabolite profiles during fruit development in high-yielding oil palm mesocarp. *PloS One* 8: e61344.

Teh, H.F., Neoh, B.K., Wong, Y.C., Kwong, Q.B., Ooi, T.E.K., Ng, T.L.M., Tiong, S.H., Low, J.Y.S., Danial, A.D., and Ersad, M.A. 2014. Hormones, polyamines, and cell wall metabolism during oil palm fruit mesocarp development and ripening. *Journal of Agricultural and Food Chemistry* 62: 8143–8152.

The International Barley Sequencing Consortium. 2012. A Physical, Genetic and Functional Assembly of the Barley Genome. *Nature* 491: 711–716.

Ting, N.C., Jansen, J., Mayes, S. et al. 2014. High density SNP and SSR-based genetic maps of two independent oil palm hybrids. *BMC Genomics* 15: 309. doi: 10.1186/1471-2164-15-309.

Ting, N.C., Jansen, J., Nagappan, J., Ishak, Z., Chin, C.W., Tan, S.G., Cheah, S.C., and Singh, R. 2013. Identification of QTLs associated with callogenesis and embryogenesis in oil palm using genetic linkage maps improved with SSR markers. *PloS One* 8: e53076.

Ting, N.C., Noothariza, M.Z., Rozana, R., Low, E.T., Maizura, I., Cheah, S.C., Tan, S.C., and Singh, R. 2010. SSR mining in oil palm EST database: Application in oil palm germplasm diversity studies. *Journal of Genetics* 89: 135–145.

Tisné, S., Denis, M., Cros, D. et al. 2015. Mixed model approach for IBD-based QTL mapping in a complex oil palm pedigree. *BMC Genomics* 16: 798–810.

Tranbarger, T.J., Dussert, S., Joet, T., Argout, X., Summo, M., Champion, A., Cros, D., Omore, A., Nouy, B., and Morcillo, F. 2011. Regulatory mechanisms underlying oil palm fruit mesocarp maturation, ripening, and functional specialization in lipid and carotenoid metabolism. *Plant Physiology* 156: 564–584.

Van Ooijen, J.W. 2006. *JoinMap 4, Software for the Calculation of Genetic Linkage Maps in Experimental Populations*. Kyazma B.V., Wageningen, Netherlands.

Vigeolas, H. and Geigenberger, P. 2004. Increased levels of glycerol-3-phosphate lead to a stimulation of flux into triacylglycerol synthesis after supplying glycerol to developing seeds of *Brassica napus* L. *Planta* 219: 827–835.

Vigeolas, H., Waldeck, P., Zank, T., and Geigenberger, P. 2007. Increasing seed oil content in oil-seed rape (*Brassica napus* L.) by over-expression of a yeast glycerol-3-phosphate dehydrogenase under the control of a seed-specific promoter. *Plant Biotechnology Journal* 5: 431–441.

Wai, C.M., Moore, P.H., Paull, R.E., Ming, R., and Yu, Q. 2012. An integrated cytogenetic and physical map reveals unevenly distributed recombination spots along the papaya sex chromosomes. *Chromosome Research* 20: 753–767.

Wallace, J.G. 2014. Association mapping across numerous traits reveals patterns of functional variation in maize. *PLoS Genetics* 10: e1004845.

Walling, J.G., Shoemaker, R., Young, N., Mudge, J., and Jackson, S. 2006. Chromosome-level homeology in paleopolyploid soybean (*Glycine max*) revealed through integration of genetic and chromosome maps. *Genetics* 172: 1893–1900.

Wang, C.J.R., Harper, L. and Cande, W.Z. 2006. High-resolution single copy gene fluorescence in situ hybridization and its use in the construction of a cytogenetic map of maize chromosome 9. *Plant Cell* 18: 529–544.

Wang, Y., Han, Y., Teng, W., Zhao, X., Li, Y., Wu, L., Li, D., and Li, W. 2014. Expression quantitative trait loci infer the regulation of isoflavone accumulation in soybean (*Glycine max* L. Merr.) seed. *BMC Genomics* 15: 680.

Weckwerth, W., Wenzel, K., and Fiehn, O. 2004. Process for the integrated extraction, identification and quantification of metabolites, proteins and RNA to reveal their co-regulation in biochemical networks. *Proteomics* 4: 78–83.

Weselake, R.J., Taylor, D.C., Rahman, A.M.H., Shah, S., Laroche, A., Mcvetty, P.B.E., and Harwood, J.L. 2009. Increasing the flow of carbon into seed oil. *Biotechnology Advances* 27: 866–878.

Whetten, R., Lu, M., Krutovsky, K.V. et al. 2015. Predictive accuracy of models based on multiple genomic attributes in *Pinus taeda* L. In: *Plant and Animal Genome XXIII Conference*. Plant and Animal Genome, San Diego, CA, USA.

Wong, C.K. and Bernardo, R. 2008. Genomewide selection in oil palm: Increasing selection gain per unit time and cost with small populations. *Theoretical and Applied Genetics* 116: 815–824.

Wong, C.L., Bong, J.F.C., and Idris, A.S. 2012. *Ganoderma* species associated with basal stem rot disease of oil palm. *American Journal of Applied Science* 9: 879–885.

Wong, Y.C., Kwong, Q.B., Lee, H.L., Ong, C.K., Mayes, S., Chew, F.T., Appleton, D.R., and Kulaveerasingam, H. 2014. Expression comparison of oil biosynthesis genes in oil palm mesocarp tissue using custom array. *Microarrays* 3: 263–281.

Xiao, Y., Zhou, L., Xia, W. et al. 2014. Exploiting transcriptome data for the development and characterization of gene-based SSR markers related to cold tolerance in oil palm (*Elaeis guineensis*). *BMC Plant Biology* 14: 1–13.

Xu, X.P., Liu, H., Tian, L., Dong, X.B., Shen, S.H., and Qu, L.Q. 2015. Integrated and comparative proteomics of high-oil and high-protein soybean seeds. *Food Chemistry* 172: 105–116.

Yusof, Z.N.B., Borhan, F.P., Mohamad, F.A., Rusli, M.H. 2015. The effect of ganoderma boninense infection on the expressions of thiamine (vitamin b1) biosynthesis genes in oil palm. *Journal of Oil Palm Research* 27: 12–18.

Zhang, Z., Ersoz, E., Lai, C.Q. et al. 2010. Mixed linear model approach adapted for genome-wide association studies. *Nature Genetics* 42: 355–360.

Zhang, Z., Ober, U., Erbe, M. et al. 2014. Improving the accuracy of whole genome prediction for complex traits using the results of genome wide association studies. *PLoS One* 9: e93017.

Zu Castell, W. and Ernst, D. 2012. Experimental 'omics' data in tree research: Facing complexity. *Trees* 26: 1723–1735.

10 *Elaeis oleifera* × *Elaeis guineensis* Interspecific Hybrid Improvement

Aik Chin Soh, Sean Mayes, Jeremy Roberts, Edson Barcelos, Philippe Amblard, Amancio Alvarado, Jeremy Henry Alvarado, Ricardo Escobar, Kandha Sritharan, Mohan Subramaniam, and Xaviar Arulandoo

CONTENTS

10.1 Introductory Overview (*Aik Chin Soh, Sean Mayes, and Jeremy Roberts*) 283
10.2 Utilization of *E. oleifera* Genetic Resources in Breeding Programs in CIRAD (*Edson Barcelos and Philippe Amblard*) 284
10.2.1 Background 284
10.2.2 F_1 Hybrids 284
10.2.3 *E. oleifera* Introgression into *E. guineensis* 284
10.3 ASD's O × G Interspecific Hybrid Breeding and Compact Clone Program (*Amancio Alvarado, Jeremy Henry, and Ricardo Escobar*) 285
10.3.1 Background 285
10.3.2 Genetic Diversity 286
10.3.3 Backcrossing Program 286
10.3.4 Tissue Culture and Clonal Performance 288
10.3.5 Use of *E. oleifera* to Breed for OG Varieties 288
10.4 Interspecific Hybrid Improvement at United Plantations Berhad (*Kandha Sritharan, Mohan Subramaniam, and Xaviar Arulandoo*) 289
10.4.1 Background 289
10.4.2 *Elaeis oleifera* 290
10.4.3 *E. oleifera* × *E. guineensis* Hybrids 291
10.5 Conclusions (*Aik Chin Soh, Sean Mayes, and Jeremy Roberts*) 294
References 295

10.1 INTRODUCTORY OVERVIEW

Aik Chin Soh, Sean Mayes, and Jeremy Roberts

In Chapter 3, some aspects of the utilization of *E. oleifera (EO)* in intra- and inter-hybrids with *E. guineensis* (*EG*) in breeding programs were introduced. This chapter examines in more detail three focused *EO* and *EO* × *EG* or *OG* breeding and cloning programs in Cirad, ASD Costa Rica, and United Plantations, Malaysia, and reflects on the breeding progress made in this direction and the future prospects of such work.

10.2 UTILIZATION OF *E. OLEIFERA* GENETIC RESOURCES IN BREEDING PROGRAMS IN CIRAD

Edson Barcelos and Philippe Amblard

10.2.1 Background

A strategy for the breeding of *EO* was defined earlier by Meunier (1975). It comprised

- Testing of combining ability between *EO* origins, to determine which populations combined well together
- Testing of individual combining ability to identify the best parents within *EG* and within *EO* populations

Subsequent results coming from the assessment of F_1 hybrids between *EO* from Latin America and *EG* showed that the Costa Rica Pacific Centre *EO* population and La Mé *EG* population combined best for bunch production (Le Guen et al. 1991). Since then, new combinations between Brazilian *EO* and La Mé *EG* have been found to give even better bunch yields. However, as the interspecific F_1 oil yield is substantially below that of the *EG*, it was decided to implement a backcross (BC) program (Le Guen et al. 1991) based on the initial success with hybrids.

10.2.2 F_1 Hybrids

Interspecific hybrids are the only option in areas where the Bud Rot complex exists. The main commercial hybrids are produced using Brazilian populations (Coari, Manicoré Manaus from the Amazonian Basin). More recently, the promising Taisha origin has been used. Although bunch yields were excellent, oil to mesocarp (O/M) was low with oil extraction rate (OER) below 18% for the best crosses. In order to increase the interspecific F_1 oil yield, it was decided to complete the original strategy by improving *EO* itself first. The most interesting *EO* populations were then test-crossed with *EG* to identify the best parents to produce hybrids and to be used in the BC program.

10.2.3 *E. oleifera* Introgression into *E. guineensis*

The BC program at Cirad was based on the results of F_1 *OG* hybrid tests. The best *OG* hybrid palms were introgressed into the best *EG* parents. The selection criteria were bunch yield, IV, height, and vegetative growth (Le Guen et al. 1991). In parallel with the BC program, tests were carried out to evaluate resistance to diseases. In Latin America, BC programs have been mostly focused on the improvement of resistance to Bud Rot. In Africa, programs were also focused on Fusarium Wilt and the leaf miner *Coelaenomenodera minuta* pest damage.

The first BCs were based on Brazilian *EO* origins; Brazil × La Mé and Brazil × Yangambi *OG* hybrids. A second series of BCs was developed with Central American *EO* (Le Guen et al. 1991). Currently, the BC program is mainly based on the Coari (Brazil Amazon) origin. An analysis for quantitative trait loci (QTL) was initiated in the BCs for the Bud Rot resistance trait and also to study the genetic determination of fatty-acid composition (FAC). QTL and intragenic single-nucleotide polymorphism (SNPs) were identified (Montoya et al. 2013).

10.3 ASD'S O × G INTERSPECIFIC HYBRID BREEDING AND COMPACT CLONE PROGRAM

Amancio Alvarado, Jeremy Henry, and Ricardo Escobar

10.3.1 Background

For 50 years, the ASD Costa Rica oil palm breeding program has used the genetic diversity of the two species, *EG* and *EO* to obtain commercial varieties. Emphasis has been placed on the development of varieties with reduced vegetative growth through BC programs, which have helped stabilize compact populations with short trunks, short leaves, and high oil production. The possibility of reproducing elite individual palms through *in vitro* culture from the best families has enabled the option of producing new commercial compact clonal varieties for high density planting with expected oil yields (OY) of 10–12 t/ha/yr. They also provide a longer economic life span to the plantations as compared to normal oil palms which have to be replanted at 20–30 years when they have grown too tall for economic harvesting.

The development of the compact varieties has been the outcome of an extensive cycle of backcross (BC) breeding, starting with a wild *OG* compact palm with palms from a recurrent *EG* gene pool which has high production potential, with backcrossing over four cycles, (Table 10.1).

TABLE 10.1
Chronology of the ASD Backcrossing Program to Develop the "Compact" Populations, with the Theoretical Proportion of *E. oleifera* Genes at Each Cycle of the BC

Year	Population	*E. oleifera* Genes (%)
1966	Hybrid O × G (wild, open pollination)	50
1970	First backcross (BC_1): OCP	25
1978	Second backcross (BC_2)	12.5
1985	Sub population: BC_2F_1	12.5
1994	Sub population: BC_2F_2	12.5
2008	Sub population: BC_2F_3	12.5
1986	Third backcross (BC_3)	6.25
1995	Sub population: BC_3F_1	6.25
2008	Sub population: BC_3F_2	6.25
1995	Fourth backcross (BC_4)	3.125
2008	Sub population: BC_4F_1	3.125

Specific combinations of *dura* (D, with thick shells) or *pisifera* (P, without shell) palms from "compact" populations with *EG* palms, resulted in commercial varieties referred to as "high density" types. These can be planted at densities of 160–180 palms per hectare (p/ha), depending on soil and growing conditions, making it possible to achieve average oil yields of 9 or more t/ha/yr.

The following summarizes the main aspects of the development of varieties and clones for planting at high density by the ASD Costa Rica breeding program. It gives the origins of the program, a summary of the genetic diversity of *EO* as well as *EG* and the development of the BC programs, together with their results in obtaining compact varieties and clones. Finally, we highlight the use of *EO* for the development of an *OG* hybrid with particular characteristics with the use of P*s* from the "compact" population, which translates into moderate leaf length and a low trunk height growth.

10.3.2 Genetic Diversity

Genetic diversity has been the main strength of ASD Costa Rica's oil palm breeding program. The basis of the program is related to an extensive germplasm collection of *EO* acquired in the 1960s and 1970s. Collections were made in diverse regions of Honduras, Nicaragua, Costa Rica, Panama, Colombia, Suriname, and Brazil (Escobar 1981; Sterling et al. 1999). About 40 localities were prospected with more than 350 accessions collected, some of which were evaluated in *OG* trials planted in 1978. Most of the *EO* collections were planted in the southern zone of Costa Rica. During the 1970s, some accessions were exchanged for advanced generation *EG* material from experimental stations in Africa and Asia. The *EG* genetic materials obtained were Deli D, sourced from the Chemara, Harrison & Crossfield, Banting, SOCFIN, and MARDI (now MPOB) (Malaysia) stations and from Dami (Papua, New Guinea). Also introduced were AVROS P palms from Harrisons & Crossfield (Malaysia), Ekona from Unilever (Cameroon), Ghana and Nigeria from the Kade Station (Ghana) and the NIFOR station (Nigeria), and La Mé and Yangambi from IRHO (Ivory Coast). These exchange programs were strengthened with the subsequent introduction of seeds from wild palms of the Bamenda highlands (Cameroon) and Tanzania, and from several regions in Sierra Leone, Uganda, Zambia, and Malawi (Alvarado et al. 2009).

The development of *EG* commercial varieties was initiated in the 1970s using Deli D and AVROS P. At the same time, several *EG* sources were used in *OG* trials, as well as in a BC program based on a wild *OG* hybrid, the seeds of which were collected in the Central Pacific region of Costa Rica during the initial prospection period for *EO*.

10.3.3 Backcrossing Program

The discovery of an outstanding *OG* hybrid in 1966 with shorter leaves compared with standard *OG* palms and the identification of a descendant BC_1 palm with short stems and leaves, named "original compact palm" (OCP), led to the start of the BC program at ASD. Given that the OCP had poor bunch quality, there was a need to introgress compact genes into other *EG* backgrounds (Sterling et al. 1987). Two different breeding strategies were adopted: successive BCs to *EG* to improve oil yield; and development of a recombinant population in each BC cycle to stabilize growth traits (Table 10.1).

In each BC cycle, specific crosses were made to concentrate and stabilize the yield and growth traits in advanced generations. The BC_2F_3 and BC_3F_2 populations, segregating Ds and Ts, were planted in 2008. The weaknesses of the first population (BC_2F_3), were its lower M/F and O/B which were overcome in the BC_3F_2 population. However, the BC_2F_3 D population was superior in fresh fruit bunch (FFB) yield, and both BC_2F_3 Ds and Ts, had shorter leaf length (LL) when compared with the subsequent BC_3F_2 generation. Based on the performance of the two maternal populations, it was thought that the development of the new varieties for high planting density planting should concentrate on the descendants of the BC_2 (Table 10.2).

TABLE 10.2
FFB and Oil Yield and Growth Characteristics of *dura* and *tenera* "Compact" Populations Compared to Deli *dura* and One *E. guineensis* D × P Variety, Planted in 2008

Origin	FFB	TH	LL	M/F	O/M	O/B
Compact BC_2F_3, D × D						
Average (10 families, 912 palms)	120.0	104	361	60.4	41.7	17.2
Standard deviation	29.8	21	44	4.5	6.2	3.2
Compact BC_3F_2, D × D						
Average (25 families, 1987 palms)	103.1	127	433	62.9	49.8	21.1
Standard deviation	30.6	23	37	4.8	5.9	4.0
Deli dura						
Average (79 families, 6207 palms)	106.7	143	535	67.4	44.0	20.5
Standard deviation	35.2	33	67	4.1	5.3	3.6
Compact BC_2F_3, T × T						
Average (8 families, 482 palms)	75.4	100	386	81.6	46.5	23.8
Standard deviation	37.8	28	53	4.7	6.0	4.9
Compact BC_3F_2, T × T						
Average (11 families, 532 palms)	93.3	129	449	87.8	49.3	25.3
Standard deviation	40.7	25	46	4.6	5.1	5.7
Deli Nigeria, D × P						
Average (3 families, 160 palms)	203.1	158	557	86.0	48.9	26.6
Standard deviation	40.2	20	39	3.7	4.7	4.2

Note: *BC_2F_3*, third filial generation from second backcross; *BC_3F_2*, second filial generation from third backcross; *FFB*, fresh fruit bunches, kg/palm/yr (4 years of evaluation); *TH*, trunk height measured at the base of the petiole of leaf 6, 48 months old, cm; *LL*, leaf length, 48 months old, cm; *M/F*, mesocarp in the fruit, %; *O/M*, oil in the mesocarp, %; *O/B*, oil in the bunch, %; bunch analysis data for *dura (D)* palms in D × D families and for *tenera (T)* palms in T × T families.

10.3.4 Tissue Culture and Clonal Performance

Since the 1980s, ASD has developed a reliable and unique protocol for cloning compact palms by using explants from inflorescence tissue, with minimal somaclonal abnormalities (below 1% mantled fruit palms experienced in the field). The low rate of abnormalities could be attributed to ASD's unique cloning protocol based on the use of immature inflorescence (Alvarado et al. 2006; Guzmán 1999).

With this ASD is able to fix and propagate compact high yielding palms segregated from the BC populations as commercial varieties. The ASD Tissue Culture program has been using this method to clone compact ortets (clonal parent) from the BC_3 population. In a semi-commercial 35 ha plot in Costa Rica planted in 2004 on highly fertile alluvial soils at 190 p/ha, two clones achieved average OY of 8–10 t/ha/yr from the fourth to the 10th year of harvesting. The *EG* standard check variety planted at 143 palms/ha yielded 6.5–8.0 t/ha/yr during the same period. In addition, in year 8, the compact clones showed 28 cm annual height increment and 544 cm of LL as compared to 54 and 710 cm, respectively, for the *EG* check. The next selection cycle of "high density" ortets are expected to produce clones with a good combination of total oil yield and short leaf traits when planted at densities of 160–180 p/ha (Table 10.3).

10.3.5 Use of *E. oleifera* to Breed for OG Varieties

ASD initiated a parallel program for the development and evaluation of *OG* hybrids based on its *EO* collection. This began in 1978 with the establishment of a progeny trial that evaluated 236 *OG* combinations: 105 *EO* palms from 13 regions of C. America and Colombia. *EG* palms of six origins (AVROS, Ekona, Ulu Remis, Yangambi, WAIFOR, and a Deli/Yangambi composite) were used as pollen sources. In subsequent stages in the 1990s, the diversity of sources of the two species was expanded, including native palms from Brazil and parents from the high density

TABLE 10.3
Fresh Fruit Bunch and Oil Yield and Leaf Length for a Sample of New Compact Ortets for Tissue Culture, Planted in 2008

Origin	N	FFB (kg/p/yr)	LL (cm)	O/B (%)
"Compact" BC_2F_3 × *E. guineensis*	10	191	544	27.6
"Compact" BC_3F_2	5	211	556	31.1
Deli × "Compact" BC_3F_2	6	212	544	30.4
Average	21	205	548	29.7
Tornado (current commercial clone)		174	475	31.6
E. guineensis D × P check variety		201	680	26.0

Note: *n*, number of ortets; *FFB*, fresh fruit bunch yield (average of 4 years of evaluation); *LL*, leaf length, measured at year 8; *O/B*, oil to bunch.

population. The outcome of these trials stimulated the development of a compact growth interspecific hybrid program.

The ASD Amazon compact hybrid was derived from *EO* mother palms of Manaus origin (Brazil, F_1 population), characterized by a high O/M (30% vs. 10%–20% in other *EO* populations) crossed with P*s* from the BC_2F_2 compact population. The shorter LL transmitted by the paternal origin, in comparison with other hybrids, has allowed the commercial planting of this hybrid at densities of 128–143 p/ha.

This *OG* cross has an estimated 56.2% of *EO* genes: 50% from the mother palm and 6.2% from the compact pollen-source palm, giving it particular attributes such as high tolerance to spear rots, significant opening of the spathes, attraction of insect pollinators, partial self-compatibility of the pollen, and the ability to achieve high rates of fruit set, even under natural conditions without assisted pollination in locations with a high concentration of *EG* pollen from neighboring plots. In commercial plantations in Colombia (Tumaco), the Amazon hybrid has achieved an average 36 t/ha FFB yield (50 t/ha for the best plots) at year 6. Several families from a 10 ha experimental plot in Costa Rica achieved 20–25 t/ha FFB yield at year 3, with average values for O/B of 23.6%.

10.4 INTERSPECIFIC HYBRID IMPROVEMENT AT UNITED PLANTATIONS BERHAD

Kandha Sritharan, Mohan Subramaniam, and Xaviar Arulandoo

10.4.1 Background

United Plantations Berhad (UPB) has been actively involved in plant breeding activities since the early 1950s. Since its inception, the Research Department of United Plantations had earnestly participated in germplasm exchange programs coordinated by the Department of Agriculture, Malaysia, such as the Combined Breeding Scheme (CBS) (Martin 2003). Improving the profitability of oil palm does not have to be confined to increasing yields. The value of palm oil can be amplified with improved quality. With this in mind, breeders ventured into the possibility of interspecific hybrids between *EG* and *EO*, the latter originating from Latin America is a species known for its high oil quality but is not commercially planted due to its low OY. Introgression of this species into *EG* was envisaged to produce planting material with both high yields and premium quality oil (Rajanaidu et al. 1986). *E. oleifera* germplasm obtained in the first material transfer programs (Gen-1) was evaluated for yield traits, bunch components, and oil quality mainly for FAC determined by its iodine value (IV, percent oil unsaturation). Outstanding individuals within superior families were selected and crossed with Yangambi and La Mé-based *EG* lines. *OG* crosses obtained were evaluated and then backcrossed to Yangambi and La Mé lines to create a BC_1 (UPB Annual Report 2000).

E. oleifera and interspecific *OG* hybrid breeding programs in UPB began in 1959 with the first interspecific cross created between a Deli *dura* and the KLM (Kuala Lumpur Melanococca) *oleifera* believed to be of Brazilian origin. Subsequently in the early 1970s pollen of *EO* (Brazilian) and seeds of F1 *OG* hybrid crosses from IRHO (now Cirad) were obtained. The poor performance of these materials especially in

TABLE 10.4
***E. oleifera* Germplasm in UPB**

Place of Collection	Country
San Alberto	Colombia
Monteria	Colombia
Chimichagua	Colombia
Rio Magdalena	Colombia
Caucasia	Colombia
Las Brisas	Colombia
Cereta	Colombia
Acandi	Colombia
La Dorado	Colombia
Valencia	Colombia
Quepos	Costa Rica
Dominical	Costa Rica
Limon	Costa Rica
Venecia	Costa Rica
Nicoya	Costa Rica
Puntarenas	Costa Rica
Sona Santiago	Costa Rica
Coto	Costa Rica
San Carlos	Nicaragua
Guabito	Panama
Colon	Panama
Chepo	Panama
David	Panama
Chitre	Panama
Guabala Sona	Panama
Unknown	Suriname
Unknown	Peru
Taisha	Ecuador
TeFe	Brazil

terms of O/B led to prospection of *EO* germplasm in Latin America either by UPB initiatives or through MPOB's collaborative efforts (Sharma and Tan 1990). Over the past 56 years, the researchers have been successful in collecting a wide array of *EO* germplasm (Table 10.4). From the pure collections, *OG* hybrids, and BC_1, BC_2, and more recently BC_3 have been created to enable selection for oil quality, lower height increment, and increased disease resistance.

10.4.2 *Elaeis oleifera*

FFB yields of pure *EO* germplasm over 6 years averaged 21.1 t/ha/yr⁻ (Table 10.5). Progeny 44-171, a Brazilian accession, recorded the highest FFB yields with a 6-year mean of 26.8 t/ha yr. Bunch analysis was carried out by separating analysis

TABLE 10.5
Yield and Oil Yield Performance (6-year mean) of Pure *EO* Germplasm (Generation 3)

	Yield (6-year mean)				OB	OER		
Progeny	FFB (kg/p/yr)	BNO (no/p/yr)	ABW (kg)	FFB (t/ha/yr)	(NF + PF) (%)	(NF + PF) (%)	Estimated Oil Yield (t/ha/yr)	IV
44-171	133.4	13.2	10.1	26.8	11.3	9.7	2.6	83.1
44-168	109.0	14.6	7.4	21.9	8.8	7.6	1.7	86.3
44-173	104.4	14.1	7.6	21.0	8.1	6.9	1.5	82.4
44-175	119.8	14.5	8.3	24.1	7.2	6.2	1.5	86.5
44-182	111.1	15.4	7.4	22.3	8.1	6.9	1.5	85.7
44-167	119.5	16.4	7.5	24.0	6.8	5.8	1.4	88.1
44-164	85.3	12.6	6.8	17.1	8.6	7.4	1.3	84.1
44-165	81.9	14.1	6.0	16.5	9.6	8.2	1.3	84.1
44-180	90.4	12.1	7.4	18.2	8.6	7.4	1.3	82.7
44-185	126.6	16.3	7.7	25.4	6.1	5.2	1.3	88.2
44-169	111.3	16.9	6.9	22.4	6.3	5.4	1.2	89.8
44-172	115.2	16.0	7.3	23.1	5.9	5.0	1.2	88.4
44-176	122.4	15.2	8.2	24.6	5.6	4.7	1.2	90.1
44-179	127.5	16.2	8.0	25.6	5.6	4.8	1.2	86.7
44-178	98.8	15.4	6.5	19.9	6.3	5.4	1.1	85.3
44-183	96.6	14.3	6.7	19.4	6.8	5.8	1.1	85.1
44-170	88.4	15.1	6.2	17.8	6.5	5.6	1.0	88.2
44-181	107.1	15.6	7.0	21.5	5.7	4.9	1.0	86.6
44-187	96.5	14.1	7.2	19.4	5.8	5.0	1.0	88.3
44-188	72.5	11.4	6.0	14.6	8.1	6.9	1.0	89.0
44-174	111.5	18.4	6.5	22.4	4.5	3.8	0.9	88.6
44-163	83.9	14.3	5.7	16.9	5.4	4.6	0.8	87.5
Mean	105.1	14.8	7.2	21.1	7.1	6.1	1.3	86.6
CV (%)	15.9	11.0	13.4	15.8	23.1	23.4	28.9	2.6

Note: *FFB*, fresh fruit bunch; *BNo*, bunch number; *ABW*, average bunch weight; *OB*, oil to bunch; *OER*, oil extraction rate (0.855 × OB); *IV*, iodine value; *NF*, normal fruit; *PF*, parthenocarpic fruit.

on fertile normal fruits and parthenocarpic fruits and combining results to produce total O/B%. Oil to bunch for *EO* progenies ranged from 4.5% in Progeny 44-174 to 11.3% in Progeny 44-171. OY per hectare was projected using FFB yields and OER (estimated mill OER = 0.855 × OB%). Mean OY averaged over 6 years was 1.3 t/ha/yr for all progenies. IVs averaged 86.6 for all progenies. Progeny 44-171 had the highest OY and was selected for creation of F_1 *OG* hybrids through crossings with Yangambi and La Mé *EG* parents.

10.4.3 *E. oleifera* × *E. guineensis* Hybrids

A total of five *OG* progenies were produced from a single *EO* accession. The average FFB production over eight years was 32.7 t/ha/yr (Table 10.6). This was

TABLE 10.6
Yield and Oil Yield Performance (8-year mean) of *E. oleifera* × *E. guineensis* Hybrid (Generation 2)

	Yield				OB	OER	Est. Oil	
Progeny	FFB (kg/p/yr)	BNO (no/p/yr)	ABW (kg)	FFB (t/ha/yr)	(NF + PF) (%)	(NF + PF) (%)	Yield (t/ha/yr)	IV
OG4	288.0	21.8	13.6	42.6	15.6	13.3	5.7	75.6
OG3	284.0	21.1	13.9	42.0	12.3	10.5	4.4	76.0
OG5	248.0	19.5	13.1	36.7	12.2	10.4	3.8	76.1
OG1	179.0	19.4	9.8	26.5	8.6	7.3	1.9	71.1
OG2	106.0	15.8	7.2	15.7	7.4	6.3	1.0	69.0
Mean	221.0	19.5	11.5	32.7	11.2	9.6	3.4	73.6
CV%	4.7	5.0	8.0	4.7	11.8	11.8	11.8	2.0

Note: *FFB*, fresh fruit bunch; *BNo*, bunch number; *ABW*, average bunch weight; *OB*, oil to bunch; *OER*, oil extraction rate (0.855 × OB); *IV*, iodine value; *NF*, normal fruit; *PF*, parthenocarpic fruit.

comparable to some of the high yielding pure *EG* lines. However, the *OG* crosses lost most of the potential for production through OY with a low O/B averaging at 11.2% due to poor bunch development (low fruit to bunch or F/B ratio). Oil yields across five progenies averaged at 3.4 t/ha/yr with progeny OG4 being the highest yielder at 5.7 t/ha/yr. IVs dropped to an expected range of 69.0–76.1 due to the dilution of unsaturated oils from *EG* genetic contribution. Selected palms from these crosses were backcrossed (BC_1) to Yangambi and La Mé which were then evaluated.

Average FFB production dropped to 23.6 t/ha/yr in the BC_1 crosses compared to 32.7 t/ha/yr in the *OG* F_1 crosses (Table 10.7). This might have been caused by severe segregation of traits seen among palms within individual crosses. Oil to bunch saw an increase from 11.2% in the F_1 to 20.7% in the BC_1s. Marked improvements were seen in fruit set and bunch development. Oil yields averaged lower than expected at 4.2 t/ha/yr due to lower FFB. IV was further diluted and reduced down to an average of 68.2. As much as this is an improvement from the standard IV of about 55 for *EG*, there is still a need for improvement in oil production.

Owing to unconvincing results from the BC_1 crosses, an initiative was taken to exploit the yield potential of individual outstanding palms among the F_1 *OG* population. Selections were made based on FFB and OB but not IV. Palms were assessed, cloned through tissue culture, field planted, and the clones assessed for a 6-year period (Table 10.8). Average FFB yields across progenies were 31.6 t/ha/yr with OB averaging at 24.1%. The highest yielder in terms of OY was OG8 with an average of 8.1 t/ha/yr. As the palms were selected for yield traits, IV remained at a moderate level with an average of 64.3.

TABLE 10.7
Yield and Oil Yield Performance (8-year mean) of Backcross 1 (Generation 1)

	Yield				OB	OER	Est. Oil	
Progeny	FFB (kg/p/yr)	B No (no/p/yr)	ABW (kg)	FFB (t/ha/yr)	(NF + PF) (%)	(NF + PF) (%)	Yield (t/ha/yr)	IV
BC3	182.0	19.9	9.1	26.9	20.9	17.8	4.8	64.4
BC6	166.0	18.0	9.2	24.6	22.7	19.4	4.8	65.8
BC2	142.0	16.3	8.7	21.0	23.9	20.4	4.3	72.9
BC7	176.0	19.8	8.9	26.0	18.9	16.1	4.2	67.7
BC5	160.0	18.8	8.5	23.7	20.6	17.6	4.2	70.1
BC4	144.0	17.5	8.2	21.3	20.2	17.3	3.7	66.8
BC1	146.0	16.8	8.7	21.6	18.1	15.5	3.4	69.4
Mean	159.4	18.2	8.8	23.6	20.7	17.7	4.2	68.2
CV%	10.1	7.8	4.0	10.1	9.7	9.7	12.6	4.2

Note: *FFB*, fresh fruit bunch; *BNo*, bunch number; *ABW*, average bunch weight; *OB*, oil to bunch; *OER*, oil extraction rate (0.855 × OB); *IV*, iodine value; *NF*, normal fruit; *PF*, parthenocarpic fruit.

TABLE 10.8
Yield and Oil Yield Performance (6-year mean) of Clonal *E. oleifera* × *E. guineensis* Hybrids (Generation 1)

	Yield (6-year mean)				OB	OER	Est. Oil	
Progeny	FFB (kg/p/yr)	BNO (no/p/yr)	ABW (kg)	FFB (t/ha/yr)	(NF + PF) (%)	(NF + PF) (%)	Yield (t/ha/yr)	IV
cOG8	216.1	25.4	8.5	32.0	29.4	25.2	8.1	64.1
cOG2	210.5	25.6	8.2	31.2	27.9	23.9	7.4	64.0
cOG1	217.3	25.0	8.7	32.2	25.8	22.1	7.1	67.7
cOG7	213.8	27.8	7.7	31.6	25.4	21.8	6.9	63.5
cOG10	236.5	25.9	9.1	35.0	22.6	19.4	6.8	64.2
cOG6	206.9	26.9	7.7	30.6	25.1	21.4	6.6	63.6
cOG9	214.8	27.8	7.8	31.8	24.0	20.5	6.5	63.0
cOG12	202.1	27.6	7.3	29.9	24.3	20.8	6.2	62.9
cOG3	199.8	27.6	7.2	29.6	22.9	19.6	5.8	65.7
cOG5	213.0	27.3	7.8	31.5	21.4	18.3	5.8	64.8
cOG4	208.8	24.2	8.6	30.9	21.4	18.3	5.7	64.4
cOG11	226.5	25.0	9.1	33.5	19.4	16.6	5.6	64.0
Mean	213.8	26.3	8.1	31.6	24.1	20.6	6.5	64.3
CV%	4.7	5.0	8.0	4.7	11.8	11.8	11.8	2.0

Note: *FFB*, fresh fruit bunch; *BNo*, bunch number; *ABW*, average bunch weight; *OB*, oil to bunch; OER, oil extraction rate (0.855 × OB); *IV*, iodine value; *NF*, normal fruit; *PF*, parthenocarpic fruit.

10.5 CONCLUSIONS

Aik Chin Soh, Sean Mayes, and Jeremy Roberts

It is undeniable that the American oil palm *EO* is an invaluable source of desirable traits unavailable or available at poorer trait levels in *EO*: high oil unsaturation, slow height growth, non-shedding and low lipase fruits, resistance to diseases, and adaptable to variable growing conditions. This was the rationale behind the rush in the 1970s and 1980s to collect pure or illegitimate *EO* and *OG* materials through private collection, for example, Martineux's collection, or exchange/collaboration with plantation companies in Latin America, for example, ASD, INDUPALMA, CENIPALMA by oil palm breeders from Malaysia, Indonesia, and Cirad, which still continues today. The bulk of initial F_1 hybrids especially with Colombian *EO* were disappointing with vigorous vegetative growth, short thick peduncles, persistent spathes, poor yields due to poor fruit set, low M/F and O/M but with higher IVs. Those with Brazilian and C. American O's gave better oil yields due to better fertility and oil content but lower IVs. Nevertheless, resulting from extensive F_1 *OG* hybrid testing, commercial varieties of *OG* hybrid seeds (e.g., INDUPALMA) essentially derived from Brazilian/Amazonian *EO* (Taisha, Coari, Manicore), and clones (UPB, mainly in-house) have become available. ASD has also "marketed" their BC_2 compact clones. Most of the early clones purchased and tested by clients did not appear promising except for a few, for example, the *Tornado* variety. Nevertheless ASD is confident of producing more promising clones as indicated here.

Commercial *OG* varieties whether as seeds or clones have yet to surpass good commercial *EG* D × P hybrids, although close in terms of oil yield. Assisted pollination is still needed in most cases. This could be due to inherent infertility, inappropriate pollinating weevils, or scent attractant. The *OG* cloning approach, while attractive for early commercial exploitation, may be even more limited than *Tenera* cloning. Interspecific hybridization in crops is known to result in genomic instability expressed as "genomic shock" affecting DNA methylation and gene expression (Mclintock 1984; Marfil et al. 2006; Anisova et al. 2009). The tissue culture cloning process can also induce such epigenetic changes. As such clonal propagation of *OG* hybrids may enhance or perpetuate such variant palm phenotypes although the phenomenon appears to occur at random. Although some groups have highlighted some promising normal clones of *OG* F_1 and BC hybrids there were also deviant clones. It would be interesting to see if the promising clones remain stable with recloning. As with clonal D × P production, clonal O × G production of promising hybrids is also a possible interim option. The BC breeding or introgression approach would be more promising and sustainable as the recurrent crossing and selection will help stabilize the genotype. A number of breeding groups, for example, AAR, FELDA (Chin et al. 2003; Mohd Din et al. 2003) besides the groups featured in this chapter have adopted this approach and are already in the BC_3 generation. Meunier, a pioneer in *OG* breeding, predicted earlier that future oil palm cultivars would have some *EO* genes in their genetic make-up and it is likely that this will be the case.

This, however, does not preclude the eventual development of pure *EO* cultivars possessing premium quality oils and agronomic traits with acceptable high yields.

After all the current commercial African oil palm came from similar modest genetic backgrounds not too long ago!

REFERENCES

Alvarado, A. Escobar, R., and Peralta, F. 2009. Avances en el Mejoramiento Genético de la Palma de Aceite en Centro América. In: *XVI Conferencia Internacional sobre Palma de Aceite*. Cartagena de Indias, Colombia, 22–25 de septiembre, 27 p.

Alvarado, A., Guzmán, N., Escobar, R., Peralta, F., and Chinchilla, C. 2006. Cloning program for oil palm compact varieties: Realities and commercial potential. In: *XV Conferencia Internacional sobre Palma de Aceite*, FEDEPALMA. Cartagena, Colombia, Septiembre 2006.

Anisova, I.N., Tumanova, L.G., Gavrilova, V.A., Dyageleva, L.I., Pasha, V.A., Mitin, V.A., and Timofeyova, G.I. 2009. Genomic instability in sunflower interspecific hybrids. *Russian Journal of Genetics* 45: 934.

Chin, C.W., Suhaumi, S., Mohd Nasarudin, M., and Ng, W.J. 2003. Selection of elite ortets from interspecific hybrids and backcrosses. In: *Proceedings of the Agricultural Conference on Palm Oil: The Power-house for the Global Oils and Fats Economy.* Malaysian Palm Oil Board, Kuala Lumpur, pp. 36–50.

Escobar, R. 1981. Preliminary results of the collection and evaluation of the American oil palm (*Elaeis oleifera*) in Costa Rica. In: *Proceedings of the International Conference on Oil Palm in Agriculture in the Eighties*. The Incorporated Society of Planters, Kuala Lumpur, 17–20 June, 1981, pp. 79–97.

Guzmán, N. 1999. Present status of clonal propagation of oil palm *Elaeis guineensis* Jacq. in Costa Rica by culture of immature inflorescences. In: *Seminar Worldwide Performance of D × P Oil Palm Planting Materials, Clones and Interspecific Hybrids*. Barranquilla, Colombia, 1995, N. Rajainadu, B.S. Jalani, Eds. *Proceedings*. PORIM. pp. 144–150.

Le Guen, V., Amblard, P., Omoré, A., Koutou, A., and Meunier, J. 1991. Le programme hybride interspecifique *Elaeis oleifera* × *Elaeis guineensis* de l'IRHO. *Oléagineux* 46: 479–487.

Marfil, C.F., Masuelli, R.W., Davison, J., and Comai, L. 2006. Genomic instability in *Solanum tuberosum* × *Solanum kurtzianum* interspecific hybrids. *Genome* 49: 104–113.

Martin, S.M. 2003. *The UP Saga*. Copenhagen, Denmark: NIAS Press.

Mclintock, B. 1984. The significance of responses of the genome to challenge. *Science* 226: 792–801.

Meunier, J. 1975. Le Palmier à Huile Américain *Elaeis melanococca. Oléagineux* 30: 51–61.

Mohd Din, A., Rajanidu, N., Kushari, A., Mohd Isa, Z.A., and Noh, A. 2003. Comparison of bunch performance of an interspecific hybrid *Elaeis oleifera* × *E. guineensis* (O × T) to its reciprocal (T × O) on peat soil. In: *Proceedings of the Agriculture Conference on Palm Oil: The Power-House for the Global Oils and Fats Economy.* Malaysian Palm Oil Board, Kuala Lumpur, pp. 24–28.

Montoya, C., Lopes, R., Flori, A., Cros, D., Cuellar, T., Summo, M., Espeout, S. et al. 2013. Quantitative trait loci (QTLs) analysis of oil palm fatty acid composition in an interspecific pseudo-backcross from *Elaeis oleifera* (HBK) Cortés and oil palm (*Elaeis guineensis* Jacq.). *Tree Genetics and Genomes* 9: 1207–1225.

Rajanaidu, N., Tan, Y.P., Ong, E.C., and Lee, C.H. 1986. The performance of inter-origin D × P planting material. In: *Workshop Proceedings of the Palm Oil Research Institute of Malaysia*. Institut Penyelidikan Minyak Kelapa Sawit Malaysia, Bangi.

Sharma, M. and Tan, Y.P. 1990. Performance of the *Elaeis oleifera* × *Elaeis guineensis* (O × G) hybrids and their backcrosses. In: B.S. Jalani et al. In: *Proceedings of 1989 International Palm Oil Development Conference Agriculture*, Palm Oil research Institute of Malaysia, Kuala Lumpur. pp. 40–43.

Sterling, F., Richardson, D.L., Alvarado, A., Montoya, C., and Chaves, C. 1999. Performance of O × G (*E. oleifera* Central American and Colombian biotype × *E. guineensis*) interspecific hybrids. In: N. Rajanaidu and B.S. Jalani, Eds., In: *Proceedings of the Seminar on Worldwide Performance of D × P Oil Palm Planting Materials, Cones and Interspecific Hybrids.* Palm Oil Research Institute of Malaysia, Kuala Lumpur. pp. 114–127.

Sterling, F., Richardson, D.L., and Chavez, C. 1987. Some phenotypic characteristics of the descendants of QB049, an exceptional hybrid of oil palm. In: A. Halim Hassan, P.S. Chew, B.J. Wood and E. Pushparajah, Eds., In: *Proceedings of the Oil Palm/Palm Oil Conference on Progress and Prospects.* PORIM, Kuala Lumpur. pp. 135–146.

11 Commercial Planting Material Production

Sean Mayes, Jeremy Roberts, Choo Kien Wong, Chin Nee Choo, Wei Chee Wong, Cheng Chua Tan, Abdul Razak Purba, and Aik Chin Soh

CONTENTS

11.1 Introductory Overview (*Aik Chin Soh, Sean Mayes, and Jeremy Roberts*) 298
11.2 Oil Palm Seed Companies, Seed and Ramet Market (*Abdul Razak Purba and Aik Chin Soh*) 298
11.3 Commercial Seed Production (*Choo Kien Wong and Aik Chin Soh*) 301
11.3.1 Commercial Seed: Mixed D × P, Near-Single Cross D × P, Semi/Biclonal D × P 301
11.3.2 Establishment of Seed and Pollen Gardens 301
11.3.3 Seed Production Processes 301
11.3.3.1 Controlled Pollination 301
11.3.3.2 Seed Processing and Germination 303
11.3.3.3 Innovations and Improvements 304
11.4 Commercial Clonal Plant Production (*Chin Nee Choo, Wei Chee Wong, Choo Kien Wong, Cheng Chua Tan, and Aik Chin Soh*) 306
11.4.1 Background 306
11.4.2 Principles and Procedures Applied in Ortet Selection 307
11.4.3 Commercial Oil Palm Tissue Culture Process 307
11.4.4 Cloning and Recloning Efficiencies 308
11.4.5 Acclimatization in the Ramet Nursery 310
11.4.6 Commercial Clonal Propagation Issues 310
11.4.6.1 Fruit Mantling 310
11.4.6.2 Cloning Inefficiency 311
11.4.7 Field Planting Issues 312
11.4.8 Clonal Field Planting Results 312
11.5 Seed Certification and Plant Variety Protection (*Choo Kien Wong, Wei Chee Wong, Abdul Razak Purba, and Aik Chin Soh*) 312
11.5.1 Seed Certification versus Plant Variety Protection 312
11.5.2 Application of DNA Fingerprinting in Genotype Identification and PVP 313
11.6 Conclusions (*Aik Chin Soh, Sean Mayes, and Jeremy Roberts*) 316
Appendix 11A.1 Clonal Field Trial Results 317
Appendix 11B.1 Commercial Clonal Field FFB Yield Results 318

Appendix 11C.1 Ramet Crop Mill OER or Oil Extraction Rate Results............. 318
Appendix 11D.1 Seed Certification... 319
Appendix 11E.1 Plant Variety Protection.. 322
References.. 324

11.1 INTRODUCTORY OVERVIEW

Aik Chin Soh, Sean Mayes, and Jeremy Roberts

Private seed companies (although some are government linked, e.g., FELDA, Sime Darby, IOPRI) are largely responsible for the development and expansion of the commercial oil palm seed market since the early days of the development of the crop. Almost all of these seed companies are plantation based. This arose first because plantations felt the need to be assured of the timely supply of top-quality seeds in their plantation development and expansion efforts. Second, the hybrid *Dura* × *Pisifera* (D × P) nature of the seeds ensured the propriety of production and continued commercial sale of the hybrid seed. Clients cannot save seed derived from their current planting for their next cycle of planting without a loss in productivity due to the segregation of the low-oil-yielding *Duras* (D), sterile *Pisiferas* (P), and reduced hybrid vigor of the *Teneras* (T). Third, although the original intention of setting up the oil palm breeding and seed production program is to ensure good-quality seeds for its own plantations, external sales of the seeds have been shown to be a lucrative business with profits two to three times that of production costs (even including research costs) that could be channeled to further plant breeding research or business profit.

Many seed companies have now progressed from producing heterogeneous mixed hybrids to more homogeneous near-true F_1 hybrids, clonal (semi-, bi-) hybrid seed and clones. Oil palm planting material production has become a lucrative and sophisticated business.

11.2 OIL PALM SEED COMPANIES, SEED AND RAMET MARKET

Abdul Razak Purba and Aik Chin Soh

The world oil palm annual seed market is about 400–500 million seeds worth US$300 million at around 70 US cents per seed. The main supply and demand has been in Indonesia and Malaysia. In the last couple of years, demand has dropped due to the moratorium on new plantings in Indonesia affecting both Indonesian and foreign (e.g., Malaysia, Singapore, China) investors who have been focusing their expansion efforts there. However, there is increasing demand for oil palm seeds from West Africa and South America where the Asian investors are making forays for new plantation development. The breakdown of the oil palm seed suppliers, many of which are plantation-based companies, and their estimated capacities for seed production are given in Table 11.1.

The largest seed demand is from Indonesia followed by Malaysia. The high crude palm oil or CPO prices boosted by its demand as a source of biofuel has prompted the opening of more land in Indonesia and other countries for oil palm cultivation, resulting in increased demand for oil palm seeds and consequently many seed companies have stepped up their production capacities.

TABLE 11.1
Oil Palm Seed and Clonal Plantlet (Ramet) Production Companies and Their Estimated Supply Capacities (per Annum)

Country	Company	Estimated Seed Production or Capacity (million)	Estimated Ramet Production or Capacity[a] (million)	Genetic Origin
Indonesia (2016)	Indonesian Oil Palm Research Institute (IOPRI)	50	<100,000 (semicommercial)	Deli, SP540/ AVROS, Yangambi, La Mé
	PT. Socfin Indonesia	50	NA (in-house use)	Deli, La Mé, Yangambi
	SUMBIO (PT. London Sumatera/IndoAgri)	20	NA (in-house use)	Deli, AVROS, Binga, Ekona
	PT. Bina Sawit Makmur (Sampoerna Agro)	20	NA (in-house use)	Deli, Ekona, Ghana, Calabar
	PT. Tunggal Yunus Estate (Asian Agri)	25	NA (in-house use)	Deli, Ekona, Ghana, Calabar
	PT. Dami Mas Sejahtera (Sinar Mas)	25	NA (in-house use)	Deli, AVROS
	PT. Tania Selatan (Wilmar)	10	NA (in-house use)	Deli, Ekona, Ghana, Calabar
	PT. Bakti Tani Nusantara	12		Deli, La Mé, AVROS
	Sarana Inti Pratama (IndoAgri)	6.5		Deli, Ekona, Ghana, Calabar
	Bakrie-ASD (Bakrie)	12		Deli, Ekona, Ghana, Calabar
	Sasaran Ehsan Mekarsari	6.5		Deli, La Mé
	Dura Inti Lestari (Darmex Agro)	6.5		Deli SP540/ AVROS, Yangambi, La Mé
	Gunung Sejahtera Ibu Pertiwi (Astra)	2.5	NA (in-house use)	Deli, SP540/ AVROS, Yangambi, La Mé
	PTPN IV	3.0		Deli, SP540, Yangambi, La Mé
Subtotal 1		**249**	NA	
Malaysia (2010)	Felda Agricultural Services S.B.	35.4	1.0	Deli, Yangambi, Nigerian (MPOB)
	Guthrie Research Chemara S.B. (currently Sime Darby Plantations)	20.3	0.75 (includes Guthrie, Golden Hope, and Ebor)	Deli, Yangambi, AVROS

(*Continued*)

TABLE 11.1 (*Continued*)
Oil Palm Seed and Clonal Plantlet (Ramet) Production Companies and Their Estimated Supply Capacities (per Annum)

Country	Company	Estimated Seed Production or Capacity (million)	Estimated Ramet Production or Capacity[a] (million)	Genetic Origin
	Golden Hope Research (currently Sime Darby Plantations)	6.8		Deli, AVROS
	HRU Sdn Bhd (currently Sime Darby Plantations)	5.0		Deli, Dumpy. AVROS
	Ebor Research, (currently Sime Darby Plantations)	3.0		Deli, AVROS
	United Plantations Bhd.	18.6	0.06	Deli, Yangambi
	Applied Agricultural Resources S.B.	9.1	0.6	Deli, Dumpy. Yangambi.AVROS
	IOI	6.3	1.0	Deli, AVROS
	Kulim	1.6	0.02	Deli, AVROS
	Sawit Kinabalu	9.5	0.2	Deli, AVROS
	IJM	3.2		Deli, AVROS
	Others	7.5		Deli, AVROS, Nigerian (MPOB), Socfin
	Agrocom		0.12	
	TSH		0.8	
	SEU		0.01	
Subtotal 2		**126.3**	4.56	
Costa Rica	ASD	20	NA	Deli, Ekona, Calabar, La Mé, AVROS, Yanganbi
Papua New Guinea	Dami	20	NA	Deli, AVROS
Africa	Ivory Coast, Zaire, Nigeria, Ghana, Benin	17	NA	Deli, AVROS, La Mé, NIFOR (e.g., Calabar, Aba)
Thailand	Univanvic	5	NA	Deli, AVROS
South America	Colombia, Brazil, Ecuador	14	NA	Deli, AVROS, La Mé, Yangambi
Subtotal 3		**76**		
Grand total		451.3	4.56[a]	

Note: Compiled from Soh et al. (2007), Kushairi et al. (2011), Purba et al. (2016, pers. comm.).
[a] Many produced for in-house use or research and development are unreported.

The supply of oil palm clonal planting material from about 20 tissue culture laboratories is still very small, constituting less than 1% (ca. 5 million) of planting material and may rise to 10 million in 5 years' time. The high price and limited supply can be attributed to the still inefficient commercial tissue culture process.

11.3 COMMERCIAL SEED PRODUCTION

Choo Kien Wong and Aik Chin Soh

11.3.1 Commercial Seed: Mixed D × P, Near-Single Cross D × P, Semi/Biclonal D × P

Defined by the status of parental palms used, commercial oil palm seed for field planting, in general, could be divided into three categories: mixed D × P, near-single cross D × P, and semi/biclonal D × P. The type of D × P produced depends primarily on the history of the breeding program. Oil palm breeding programs that adopted the Modified Recurrent Selection (MRS), or more familiarly referred to as the Family and Individual Palm Selection (FIPS), will commonly produce mixed D × P, owing to the selection technique that exploits General Combining Ability (GCA). The production of mixed D × P seeds usually involved a pool of D mothers from out-crossed or inbred populations, which meet a particular standard for the D performance per se crossed with well D × P progeny-tested Ps defined by GCA. While for the production of near-single cross D × P seeds using the Modified Reciprocal Recurrent Selection (MRRS) approach, the seed production parents used are selfed Ds and selfed Ps, derived from the respective D and T involved in the progeny test. This reproduction approach captures both the GCA of the original D and T parents and the Specific Combining Ability (SCA) of their cross (Jacquemard et al. 1981). With the successful development of Oil Palm Tissue Culture (OPTC) clonal propagation, the reproduction of the best D × P hybrids using the clones of the parents for the commercial production for biclonal (both clonal parents) or semi/monoclonal (one clonal parent) hybrid seeds has been made possible, with a consequent improvement in genetic uniformity and yield potential accordingly.

11.3.2 Establishment of Seed and Pollen Gardens

Dura × *dura* and T × T/P progenies planted in trials or progeny blocks are used to generate the D and P parents, respectively, for commercial D × P seed production. Mass selection is used in the choice of D mother palms (usually complying with minimum selection criteria or standards set up by the national seed certification body or using self-imposed standards), while the choice of the P parent is based on its D × P progeny test. Clonal D and P seed and pollen gardens derived from progeny-tested D and P parents can also be set up.

11.3.3 Seed Production Processes

11.3.3.1 Controlled Pollination

All oil palm seed producers generally adopt similar seed production procedures following early work due to the crop's physiology and morphology, and the need to produce an intercross of D × P for the sake of heterosis and to produce the tenera fruit form in the progeny. In the field, the heart of seed production centers on controlled pollination, which is critically detailed in several publications (Donough et al. 1993; Chin 1999; Rao and Kushairi 1999; Periasamy et al. 2002), with most efforts aimed to exclude foreign or illegitimate pollen brought in particularly by the ubiquitous pollinating weevil *Elaeidobius kamerunicus* (introduced to South East Asia in the 1980s), which is strongly attracted to the scent of the receptive oil palm female inflorescence. Mechanistically, controlled pollination involves isolating a prereceptive D female inflorescence using a fiber bag that is densely woven with pore size smaller than a single grain of pollen. Prior to bagging, the inflorescence is sanitized against foreign pollen with a formalin spray. An insecticide-infiltrated cotton wad collar is also inserted at the securely tied end of the bag to prevent the entry of weevil and other insects. Additional precautions are taken to avoid rodent damage, for example, placing rat baits and an additional wire net. When the isolated inflorescence becomes receptive—no earlier than seven days after inflorescence isolation as a precaution, so that all foreign pollen resting on the inflorescence would have become inviable—the preharvested and processed *P* pollen mixed with talcum powder (usually 1:10 or 1:20 ratio) is then blown onto the inflorescences through a punctured hole made in the plastic window of the still tightly intact bag. After this, the hole is sealed with cellophane tape. The blowing technique is important to obtain thorough coverage of the entire inflorescence while the talcum powder is to ensure even dispersal of the pollen onto the inflorescence. The bag is removed about a month after pollination in order to avoid illegitimate pollination from asynchronous flowers in the same bunch, when the pollinated bunch is sufficiently hardened and its normal growth unhindered. When the bunch matures, it is harvested for further processing to obtain the seeds. Some seed producers harvest the bunch at 150 days (although there is some variability in fruit ripeness, but with mature seed embryos) for ease of management and accountability, for example, loss of loose ripe fruits or potential contamination by loose fruit from other bunches.

The isolation procedure for collecting pollen from male inflorescences is similar to that for the female inflorescence. After the male inflorescence is isolated, the inflorescence is harvested when 70% of the florets have reached the anthesis stage with the bag still tightly secured. No insects should be present within the bag, otherwise the inflorescence is discarded. The freshly harvested inflorescence is usually moist and pollen is not easily dislodged from its anthers. Drying of the inflorescence, either by cool dry air (air-conditioning) or by heat (a seed heat-treatment chamber), is practiced prior to dislodging pollen within the bag. The partially dried pollen is shaken out, sieved, and further dried in filter paper envelopes in a 38–39°C oven or a desiccator until 6% moisture content is achieved. A viability test is done on the pollen by examining under a microscope pollen tube growth of a pin drop of pollen on a sucrose-borate solution after a few hours. Pollen with less than 60% viability is usually discarded. The acceptable pollen is then stored in specimen tubes in a freezer until used in controlled pollination. Such stored pollen can retain its viability from about six months to a year. Freeze-drying cum vacuum sealing is recommended for longer-term storage. The whole process is carried out under isolated conditions to avoid or minimize contamination by foreign pollen. During controlled pollination, the tube of the freezer-stored pollen is allowed to thaw in the refrigerator, the pollen is tested for acceptable viability (above 60%), before mixing with talcum and dispensing.

11.3.3.2 Seed Processing and Germination

In nature, oil palm seeds from rotted ripe fruit that have fallen onto the ground near the palm base or have been dispersed by animals will germinate when they have acquired sufficient heat and moisture to do so. Such germination is generally poor, sporadic, and unpredictable and indicates the undomesticated nature of oil palm in this regard. In the early days, this was emulated by planting the seeds in exposed or shaded sand beds with frequent watering. Such methods are unacceptable for a commercial process. With the advent of commercial seed production and the demand for legitimate D × P hybrid seeds, an efficient and controlled seed production process has been developed and improved (Rees 1962; Mok 1982; Hartley 1988; Periasamy et al. 2002). The process is outlined below:

After a controlled pollination bunch is harvested, the objective of the following process is to obtain oil palm seeds from the bunch. Previously, this involved chopping up the bunch to obtain the spikelets, followed by their retting (in moist gunny sacks) to loosen and separate out the fruits. The fruits were subsequently depericarped or depulped to obtain the seeds. This tedious process has mostly been simplified with customized machines whereby the entire bunch is treated with ethylene to promote fruit abscission and then placed within a spinning drum of a machine to dislodge the loose fruits. The mesocarp of the fruit is then removed using a modified upright ASD (ASD de Costa Rica S.A)-type depericarper (Escobar 1980) instead of the earlier messy horizontal wire-meshed rotating drum type. The depericarper functions by creating friction between the mesocarp and the corrugations within the spinning drum of the depericarper. Water is fed into the drum simultaneously during the spinning to facilitate the separation between the seeds and the mashed mesocarp. Generally, relatively clean seeds are obtained. Adhering fibers on some nuts are removed manually if necessary. The seeds are further washed with a detergent to make their surface oil-free; seeds at this stage are termed as fresh seeds. For some seed producers, in order to avoid imitation, fresh seeds are authenticated by marking either with indelible ink or laser engraving as practiced in applied agricultural resources (AAR).

Fresh seeds are usually stored at 22°C after adjusting the seeds' moisture to 19%–20%, although in Cirad's process, seeds are stored at 16%. The targeted moisture content of seeds is obtained either by further drying or soaking the seeds accordingly to the desired weight, which is estimated from the predetermined moisture status of seed samples taken from the bulk (whole bag). Precautions are taken to minimize moisture loss by packing the seeds either in suitable plastic bags or plastic containers, leaving sufficient air within the packing for respiration. Seed storage is not essential for germination although it has been observed that seeds stored for a month or so appear to give better early germination or otherwise a slightly longer period (60 days) of heat treatment is needed. Seeds from older mother palms also require a shorter period of heat treatment (40 days) Seeds stored for more than a year tend to give poorer germination. Nevertheless, for commercial reasons, it is preferable to store sufficient seeds to cater for demand especially during peak periods. Customers are often advised to place their seed order 10 weeks or more in advance.

As oil palm seeds are considered to be recalcitrant, dormancy breaking by heating the fresh seeds at 38–40°C for 40–60 days is required prior to soaking the seeds to raise the seeds' moisture content to 22% for germination in a cool ambient temperature room. The seeds are inspected at weekly interval for sprouts, which are then sorted to allow for further growth, the remainder set aside for subsequent germination. When the plumules and radicles have fully emerged, seeds with vigorous and balanced plumule:radicle growth are selected, the remainder regarded as abnormal are discarded. Four to six periods of sorting lasting four to six weeks are usually made, with the best-germination and best-quality seeds obtained from the first two to three sortings. Germination rates exceeding 90% with recovery of saleable seeds exceeding 75% from fresh seeds are achievable with good process control (Periasamy et al. 2002). The germinated seeds are then packed for delivery, taking precautions to minimize damage of the relatively tender plumule and radicle, for example, using bubble bags in packing and styrofoam beads/cork crumps/rubber crumps as padding in the packing box. Good sanitation is encouraged at all levels of seed production to minimize seed pest and disease damage. Fungicide treatment is commonly done prior to seed storage, seed setting for germination, and on the sprouts prior selection and delivery. Further prophylactic fungicide spraying is done to meet plant phytosanitary requirements for seed export.

Preheated seeds or seeds that have just undergone heat treatment are desired by clients in remote destinations who are prepared to germinate the seeds themselves. This is to avoid risk of damage to the tender germinating seedlings during transit. Stored preheated seeds are also used by some seed producers to respond to urgent seed demand. Usually, one to two weeks of additional heat treatment is applied to preheated seeds to restimulate the germination process. However, the germination and quality of seedlings usually deteriorate drastically within three months of storage with preheated seeds.

The process described above, called the "dry heat method," is commonly practiced. In the less frequently used "wet heat method," the fresh seed moisture content is raised to 21%–22%, and the seeds (without surface moisture) are then sealed into polythene bags and placed in a 39°C heat chamber for 80 days. After this, the seeds are placed in a cool chamber for germination. Germination will begin after a few days and is normally complete within 15–20 days (Tailliez 1970).

11.3.3.3 Innovations and Improvements

As commercial oil seed production is a lucrative business, reductions in production cost and improved efficiency and product (seed) quality are of prime concern. Commercial oil palm seed production was a tedious and inefficient process requiring much labor, space, and material (plastic bags, nets) and generating much waste (oily sludge, plastic, nuts) until the recent introduction of the ASD vertical drum depericarper and AAR's trolley and plastic tray-based system to replace the messy and labor-intensive traditional plastic bag system in the seed germination room. Nevertheless, there are still areas for improvement and innovation.

11.3.3.3.1 Pollen Supply

Sufficient supply of pollen from proven P parents has always been an issue despite routine stress treatments such as root pruning and trenching, severe leaf pruning, and induced drought, which have been shown to induce male inflorescence production. Cloning the P is one approach but with the attendant risk of killing it if using leaf explants; using inflorescence and root explants instead of young leaf explants is safer but less efficient. More efficient pollen usage, for example, 1:30 pollen to talc ratio or higher using a motorized puffer has been attempted. Proper storage and maintenance of viability of pollen is also important. The use of pollen of poor viability results in poor bunch and fruit formation and consequent high nursery culling rates, giving a poor image of the cultivar (Noiret and Ahizi 1970).

11.3.3.3.2 Controlled Pollination

After female inflorescence isolation, skilled workers are needed to monitor closely (sometimes daily) when the inflorescence becomes receptive. A temperature sensor attached to the isolated inflorescence that could "report back" the rise in temperature due to active respiration by the receptive flowers is being developed. Another possible reporting tool is an "artificial nose" that could detect the aniseed-like scent emitted by the inflorescence at the time of anthesis.

11.3.3.3.3 Illegitimate Seeds

Illegitimate seed contamination is a problem for seed producers. This is manifested by the occurrence of D (sometimes also P) palms in commercial T fields and was a serious issue in the 1980s. This was the negative side effect of the introduction of the pollinating weevil *Elaeidobius kamerunicus* from West Africa. The weevil is such a ubiquitous and powerful pollinator that it can seek out the slightest gap in the covered inflorescence. A weevil can carry more than 500 illegitimate pollen grains contaminating almost the whole bunch. Contaminated fields with up to 70% D palms were found resulting in high yield loss. Replanting was not an option then as it was expensive and the replants could also be contaminated. With more stringent bag isolation and quality control measures as described earlier, since the late 1990s, D contamination in field palms derived from legitimate seed producers is normally minimal but can occur due to unscrupulous suppliers especially during periods of high seed demand. *Dura* contamination in the field is detected using the laborious cut-fruit test of fruiting palms three years after field planting. The development of a molecular marker for shell, for example, Sure Shell Kit™, can be used as a quality control tool for detecting D contamination in seeds and seedlings of suppliers. DNA fingerprinting can also differentiate the seeds from different suppliers, both legitimate and illegitimate (Wong et al. 2015).

11.3.3.3.4 Seed Processing

Integrating fruit detachment, depericarping, fiber removal and seed cleaning, and seed sizing into one machine or a continuous process will greatly reduce manual handling.

11.3.3.3.5 Seed Germination

There have been various research attempts to circumvent the tedious heat treatment process with the application of plant growth regulators (PGR), for example, gibberellic acid or GA3 (Hussey 1958; Z.C. Alang, pers. comm.) and a combination of chemicals (sulfuric acid, hydrogen cyanamide) and PGRs (ethylene) (Herrera et al. 1998). None of these replicate the high germination rates achieved by the traditional method but nevertheless the latter's tedious process remains a research challenge.

11.3.3.3.6 Seed Inventory

Instead of manual count, fresh seeds could pass along a running belt, with continuous image capturing by camera and further computer analyses, to provide fresh seed number. Seed authentication by ink printing or engraving could be integrated into the process.

11.3.3.3.7 Seed Moisture Adjustment

In moisture adjustment, the seeds are frequently reweighed to obtain the required weight for the desired moisture content. A tool or gadget that could minimize transcription errors, for example, bar coding or RFID integrated with direct weight data capture would be useful to reduce this.

11.3.3.3.8 Seed Sorting and Selection

Riding on the same principle and development in detailed phenotyping in plant breeding, integrating imaging, robotics, and engineering, the sorting and selecting process could come under one process, whereby the germinating seeds are repeatedly observed until the defined phenotype is achieved and picked to be sold.

Most of the above ideas are still exploratory but their active pursuit is not only for the sake of improved efficiency but also for consumer demand of quality products based on both technical and aesthetic appeal.

11.4 COMMERCIAL CLONAL PLANT PRODUCTION

Chin Nee Choo, Wei Chee Wong, Choo Kien Wong,
Cheng Chua Tan, and Aik Chin Soh

11.4.1 Background

The advantage of clonal oil palms, the historical development of the OPTC plantlet regeneration process and clonal propagation technology, the attendant R&D efforts and progress made, and issues remaining to be resolved have been described in Chapter 8. As discussed previously, early commercial clonal propagation was halted due to the fruit mantling issue (Corley et al. 1986). Subsequently, with resurgence in confidence from the successful larger-scale clonal plantings with negligible or minimal mantling incidence, further interest has been led by AAR, Agrocom, and Felda in the late 1990s/early 2000s (Wong et al. 1997, 1999a). There are currently about 20 commercial OPTC labs or OPTCLs worldwide producing about 5 million ramets as compared to the oil palm seed market of 450 million. This underscores the potential size of the ramet market.

This section discusses the commercial applications of this new technology, the progress made to date, and the R&D issues still to be resolved at the lab and the field levels to make it a profitable and sustainable business. The results and experiences discussed are largely drawn from AAR who pioneered the large-scale commercial production of clonal palms or ramets in the largest OPTC lab in 2005 with an annual production capacity of 1.5 million ramets and had, and probably still has, the largest areas planted with clones in trials and in commercial blocks (Soh et al. 2011). AAR's experiences are reflected by some of the other groups' and comparisons have proven valuable to the Industry.

11.4.2 Principles and Procedures Applied in Ortet Selection

The principles and strategies for ortet selection have been discussed in detail in Chapter 8; this section illustrates how this is applied in practice.

The success of oil palm cloning begins with the selection of the "best" ortet material (palm from which tissues are taken for tissue culture). From the previous discussions, the best approach would be to select the best palms from the best families in progeny-test trials. Sometimes, palms from not the best families have also been selected provided they meet the selection criteria, especially when there are no statistical differences among the families. In ortet selection from field-tested clones for recloning, emphasis is placed on the mean clone performance in the trial. In Malaysia, SIRIM has drawn up the selection criteria and standards for ortet selection from progenies and clones (Appendix 11D1.1.3 and Appendix 11D1.1.4). The selection of ortets initially based on data is further confirmed by field observations by the breeder, sometimes together with the agronomist and the tissue culturist. Its pedigree authenticity (confirming that it is not illegitimate or a replacement palm) can be confirmed through a DNA sample for genetic fingerprinting, and the observed traits should reflect previous data, with the palm not having an unfair advantage over its neighbors (more vigorous or aggressive or next to empty space). The palm must also be free of any abnormalities and any pest or disease damage.

11.4.3 Commercial Oil Palm Tissue Culture Process

The OPTC process begins with the selection of suitable explants, such as immature leaves, young inflorescences, seed embryos, or tips of tertiary roots. The adoption of immature leaves is preferable because it does not require severe surface sterilization, being completely enclosed by the bases of older leaves and large amounts of explants are available. However, the excision of the immature leaves if not done properly to avoid injuring the apical meristem may jeopardize the survival of the ortet palm, which would otherwise recover within two to three years after the spear excision process. Adopting the somatic embryogenesis pathway, the OPTC process comprises five stages (see Figure 8.1, Chapter 8). The entire process may take 12 months to over two years to generate the first batch of conditioned ramets (Wong et al. 1997, 1999b).

Following the gel pathway, explants are first placed onto a modified Murashige and Skoog (1962) culture medium supplemented with sugar, vitamins, and growth regulators, usually 2,4-dichlorophenoxyacetic acid (2,4-D) or α-naphthalene acetic acid (NAA), to stimulate callus initiation. The explants are incubated in the dark to minimize the production of phenolic compounds, which inhibit the callus initiation process. Callus can be observed as early as one and a half months after inoculation. Most calluses grow slowly, forming a compact, nodular mass, but fast-growing calluses may develop spontaneously (Smith and Thomas 1973). For the induction of embryoids, the callus is transferred onto fresh media with reduced auxin concentration (Paranjothy and Rohani 1982). The first batch of embryoids can be obtained about five months after inoculation and the callus differentiation process will be extended over 18 months. Embryoid multiplication is carried out by subculturing onto fresh media every 8–12 weeks. At each subculture, embryoids are separated from the shoot clusters; the embryoids are subcultured onto embryoid proliferation media while plantlets are transferred to shoot development media. Shoots that had reached a height of 7 cm and above are excised for rooting on a root induction medium containing an auxin such as NAA or indole-3-butyric acid (IBA). According to Wooi (1990, 1995), root development can be stimulated by transfer to a medium with very low NAA concentration or by a short exposure to high NAA, followed by hormone-free basal medium. The advantages of using the double-layer technique have also been reported by Zamzuri (1999). Initial roots normally develop after two to three weeks in the root induction medium.

For the liquid suspension pathway, explant inoculation and callogenesis stages are the same as in the gel pathway (see Figure 8.1, also Wong et al. 1999a; Soh et al. 2011). However, the desirable embryogenic callus is picked out at the callus differentiation stage and put into the liquid medium in the shake-flask system for proliferation. The proliferating embryogenic suspension cultures (EC) are subcultured at monthly intervals using a fine sieve to separate the culture aggregates for further subcultures. When the EC are mature, they are plated back onto gel medium for embryoid formation and subsequent plantlet regeneration as in the gel pathway. The advantages of the liquid suspension pathway are the shorter gestation period to obtain the first plantlet (18 months compared to 29 months for the gel pathway, Figure 8.1); the very high levels of feasibility of uniform ramet production; the amenability of the technique to batch culture and automation; and the possibility of a great reduction in skilled worker requirement for culture selection and transfer. The disadvantages are the very low frequency of embryogenic cultures amenable to proliferation in liquid media, which appears to be genotype and worker skill dependent, and the higher mantling and other somaclonal variation risks due to the very high proliferation rate.

11.4.4 Cloning and Recloning Efficiencies

The results and experience reported here are updates from Soh et al. (2011).

In general, all primary ortets are able to produce callus; however, only 76% of them become embryogenic. Callusing rates of explants for primary ortets ranged from 1% to 74% with a mean of 23%, while, embryogenesis rates of callus cultures can range from 0.1% to 24.9% with a mean of 3.0% (Table 11.2).

As with primary ortets, all clonal ortets are able to produce callus with rates of 1%–60% and mean of 24% (Table 11.3) compared to primary ortets with rates of

TABLE 11.2
Callusing and Embryogenesis Rates of Primary Ortets of Various Genotypes

Genotype	Number of Ortets	Callusing Rate of Explants (%)		Embryogenesis Rate of Callus Cultures (%)	
		Mean	Range	Mean	Range
Deli D × Dumpy AVROS P	225	24	1–74	2.9	0.1–24.9
Deli D × Dumpy.Yangambi.AVROS P	177	26	1–59	2.2	0.1–23.3
Deli D × Yangambi.AVROS P	57	24	1–67	4.7	0.1–21.1
Deli D × Cameroon P	43	18	1–46	1.7	0.3–5.2
Deli D × AVROS P	33	22	2–54	2.5	0.3–14.5
Ulu Remis D × D	12	20	1–48	2.2	0.3–3.9
Others	121	23	1–70	2.4	0.1–15.2

1%–74% and mean of 23%. However, in terms of embryoids production, clonal ortets are more responsive with 96% of the clonal ortets producing embryoids with rates of 0.1%–24% and mean of 5%.

The callusing and embryogenesis rates for both primary and clonal ortets were found to differ widely between and within genotypes. The influence of genotypes on the rates of callus induction has also been reported by Ginting and Fatmawati (1995) and Choo et al. (2014).

In short, callus can be induced from both primary and clonal ortets. On a palm basis, the clonal ortets are more amenable to culture with an embryogenesis rate of 96% and shoot regeneration of 85% as compared to primary ortets with only 76% and 56%, respectively (Table 11.4). About 5% of callogenic explants of clonal ortets differentiated into embryoids, which is one and a half times more than that for the primary ortets. Shoot regeneration from embryogenic lines for both types of ortets appears similar at 88%.

TABLE 11.3
Callusing and Embryogenesis Rates of Various Genotypes of Clonal Ortets

Genotype	Number of Ortets	Callusing Rate of Explants (%)		Embryogenesis Rate of Callus Cultures (%)	
		Mean	Range	Mean	Range
AVROS	101	23	2–56	5.0	0.1–24.9
Yangambi AVROS	98	26	1–60	5.2	0.1–23.3
Cameroon	76	26	6–43	5.3	0.1–21.1
NIFOR	18	10	2–39	5.2	0.3–5.2
Yangambi	16	25	11–39	6.1	0.3–14.5
Others	9	26	2–55	2.1	0.1–15.2

TABLE 11.4
Comparison of Cloning (Primary) and Recloning (Clonal) Efficiencies of AAR Ortets

	Success Rate (%)			
	Palm Basis		Explant Basis	
Culture Stage	Primary Ortet	Clonal Ortet	Primary Ortet	Clonal Ortet
Callusing	100	100	23	24
Embryogenesis	76	96	3	5
Shoot regeneration	56	85	88	88

11.4.5 Acclimatization in the Ramet Nursery

Transplanting of well-rooted tissue culture plantlets is carried out directly into sand beds or suitable potting mixtures containing peat, soil, or sand. Ramets have to be acclimatized under high humidity in plastic chambers under shade to ensure optimum survival. Growth of ramets in the nursery is initially slower than that of normal seedlings, because the latter have kernel food reserve, but after about 12 weeks, growth rates of ramets and of seedlings are comparable (Wooi et al. 1981). The average survival rates after transplanting for acclimatization is about 98.6% (Tan et al. 1999). This extremely high survival rate has contributed to the success of AAR's commercial oil palm cloning program. Hardened ramets are despatched to the plantation's prenursery for transplanting into soil in polybags for three months, after which they are then transferred into large polybags in the main nursery for a further nine months before field planting as per normal seedling practice.

11.4.6 Commercial Clonal Propagation Issues

11.4.6.1 Fruit Mantling

The mantling rate in AAR's clonal plantings is summarized in Table 11.5. The initial high rates between 1986 and 1995 were from protocol treatment trials in the early years. However, with improved protocols and stringent culture management, AAR was able to consistently keep the mantling rates to less than 3% in the clonal plantings since 1996.

TABLE 11.5
Mantling Rates in AAR Clones

Year Planted	Explant Source	Mantled Palms (%)
1986–1991	Ortets (immature leaves)	6.2
1989–1993	Ortet (embryos)	1.1
1992–1995	Seedlings (immature leaves)	7.6
1996–2007	Ortets (immature leaves)	0.0–2.7

Mantling incidence is currently kept at a low level by stringent culture management such as reduced use of plant growth hormones in the culture media, reduced time in culture, and limited production per embryogenic line or clone. The mantling rate in the field can be further reduced to less than 5% with the adoption of a clonal planting package of a minimum 5–10 clones per planting. Recently, Ong-Abdullah et al. (2015) reported that the mantling phenomenon is due to the loss of *karma transposon* methylation during the tissue culture process. The availability of a diagnostic kit/marker for screening of mantled clones at the nursery stage (if not earlier) would lead to greater confidence in clonal plantings and consequently their increased production.

11.4.6.2 Cloning Inefficiency

11.4.6.2.1 Ortet Availability

About 3 (cloning) to 10 (recloning) leaf explants out of every thousand cultured becomes a plant. Owing to this inefficiency, especially in embryogenesis, a large number of elite ortets is constantly required to scale up production. AAR worked out that to achieve an annual ramet production of 500,000 from the fifth year onward (2 years to obtain the first plantlet and 3 years to build up the critical mass of proliferating cultures from a handful of amenable clones), about 100 selected ortets needed to be cultured per year. These ortet numbers derived from progeny-test trials were unlikely to be met in most existing breeding programs as only about 20%–30% of palms are available as ortets per trial. A continuous series of progeny-test trials must be available to meet this continuous ortet demand, which is a daunting task to the oil palm breeder. One alternative approach is to reproduce the best progeny-tested family and plant them as large ortet gardens (Soh 1986, 1987, 1998). The other alternative is to reproduce the best tested clones, which usually also exist in larger numbers or can be planted as clonal ortet gardens. This is contingent on confidence in recloning efficiency especially with respect to mantling risk.

11.4.6.2.2 Process Inefficiency

The inefficient OPTC process also poses a great hindrance resource-wise to large-scale ramet production. Large specialized clean and controlled environment rooms and a big team of skilled operators are needed for media preparation, inoculation, culture selection and transfer, and plantlet acclimation. This can be achieved by improving the efficiency of the process and better utilization of resources. One such approach is the use of the liquid suspension system that can increase the speed of multiplication and growth of cultures. A strictly liquid system whereby cultures are initiated in the liquid medium right from the start has not been found to be feasible. The liquid suspension system pioneered by AAR (Wong et al. 1999a) is a hybrid system whereby proembryoids are initiated on gel culture and proliferated in liquid medium, and the plantlet regenerated again on gel culture. The inability to pick out the proembryoids amenable to liquid culture remains a bottleneck. R&D on innovative technologies, for example, automation, to reduce labor and skills are in progress, a number of which led by Malaysian Palm Oil Board (MPOB) are based on the liquid system. Whichever successful innovations have been made, the ultimate test is in the fidelity of the clones in the field.

11.4.7 Field Planting Issues

The advantage of genetically and phenotypically uniform F_1 hybrids and clones is that they can be planted in large areas supported by the most uniform field operations or inputs to realize their maximum heterosis or hybrid vigor. This is not currently the case with oil palm clones. To mitigate against a high mantling risk, a package of 4–5 clones with 4 palm rows per clone has been suggested and is practiced. A 4-palm row of a low sex ratio D × P or clone to be interspersed with every 4 × 4 clone strip has also been suggested to ensure adequate pollen supply to sustain weevil populations and pollination (Tan et al. 2003). The availability of a molecular marker for mantling for nursery screening would greatly reduce the risk of mantled palms being field planted and would perhaps obviate the need for multiclone packaging to hedge against high mantling risk. Even with low mantling risk from field planting palms, planting large monoclonal blocks could still be inadvisable in terms of risks of vulnerability to pest and disease outbreaks, inadequate pollination, and reduced agrobiodiversity, and consequent reduced crop resilience in a climate-changing world (see Chapter 13).

11.4.8 Clonal Field Planting Results

AAR's clonal field trial and commercial field planting yield and mill extraction rate (OER) from the dedicated clonal crop trials that have been run, including updated results from the publication of Soh et al. (2011), are given and the salient points discussed in Appendices 11A.1 through 11C.1.

11.5 SEED CERTIFICATION AND PLANT VARIETY PROTECTION

Choo Kien Wong, Wei Chee Wong, Abdul Razak Purba, and Aik Chin Soh

11.5.1 Seed Certification versus Plant Variety Protection

As farming progresses from the subsistence level to larger farms and business entities, the need to protect their agribusiness investments through the use of high-quality seeds or planting materials becomes apparent. Similarly, as seed production evolved from saving seeds from one's own farm to purchasing varietal seeds from private seed producers who guarantee uniform high-yielding and high-quality crops, resulting from the seed company's high level of investments in well-trained plant breeders, the latest seed production technology, time, and consequently money, the need by the seed company to protect their proprietary variety from theft by others becomes necessary, hence, the establishment of seed certification and plant variety protection (PVP) regulatory schemes.

There appears to be a disconnect or overlap between seed certification and PVP as both involve defining the variety in terms of its pedigree and economic traits. However, these are not always accurate or fully informative. Seed certification aims to protect the growers and seed clients by ensuring that the seeds they purchase meet the prescribed standards of varietal identity, purity, and germination and are free from pests, diseases, and weed seeds. PVP gives the breeder the right to exclude others from selling the variety, offering it for sale, reproducing it, and importing or exporting it.

Seed certification was initiated in the early twentieth century by the Swedes, leading to the formation of the International Crop Scientist Association, which became the Association of Official Seed Certifying Agencies (AOCSA) in 1969. In Malaysia, oil palm seed certification began with the Standard and Industrial Research Institute of Malaysia's or SIRIM's Malaysian Standard MS 157;1973 Oil Palm Seeds for Commercial Planting Certification Scheme. In Indonesia, the commercial seed producer has to be certified by the TP2V (*Tim Penilai dan Pelepas Varietas*, Team for Assessment and Releasing of Variety) of the National Seed Board (Appendix 11D.1).

The UPOV (The International Union for the Protection of New Varieties of Plants) PVP system was formulated in 1961 to recognize and protect the intellectual property rights of plant breeders in their varieties on an international basis. PVP has not been given serious consideration by seed producers, including oil palm seed producers in Malaysia until 2004 when obliged by the World Trade Organization's Tripps Agreement Article 27(3)b "to provide protection of new varieties of plants either by means of patent or an effective *sui generis* (own kind) or a combination of both" and Malaysia passed The Malaysian Protection of New Plant Act 2004, modeled after the UPOV and its *sui generis* system implemented in 2008. Thailand and Indonesia have been obliged similarly and have their own PVP Acts (Appendix 11E.1).

11.5.2 Application of DNA Fingerprinting in Genotype Identification and PVP

The advent of Marker Assisted Selection (MAS) technologies (Chapter 9) has facilitated the application of genetic markers such as microsatellites (SSR), single-nucleotide polymorphisms (SNPs), and insertion–deletion polymorphisms (*indels*) for crop genetic improvement (Väli et al. 2008; Van Inghelandt et al. 2010; Simko et al. 2012). The utilization of MAS is becoming popular when genetic markers can be closely linked to phenotypically important traits that are difficult and/or costly to be scored phenotypically or become evident late in the development of the crop species. Another major application of genetic markers is in revealing the genetic variation among a set of individuals within a population or between populations (see Chapter 3), and this basic information can also be used for genotype authentication. A primary use of genetic markers to an oil palm breeder is to ensure the legitimacy of the genetic materials, whether for breeding or for commercial sale deployment (Corley 2005). Commercial oil palm ramets exhibiting high clonal fidelity (Soh et al. 2011; Wong et al. 2011) can eventually receive PVP as the genotype is essentially fixed and replicated at a large scale (Durand-Gasselin 2009). The last requires palm genetic differentiation at the individual plant level especially with oil palm materials that are genetically narrow and related (Rosenquist 1986).

Most of the early attempts in developing oil palm genetic markers have laid the foundation of using them for genetic diversity studies (Billotte et al. 2001; Singh et al. 2008; Cochard et al. 2009). Billotte et al. (2001) reported success in revealing *Elaeis* genetic diversity using 21 characterized SSR loci. Later, Singh et al. (2007) reported an effective method to develop SSR markers where a set of 12 informative

SSR markers was identified and found suitable for fingerprinting the OPTC clones. This was adopted following the guidelines for the conduct of tests for Distinctness, Uniformity and Stability: Oil Palm (DUS-OP Guidelines, Plant Varieties Board Malaysia 2009). The established oil palm DNA fingerprinting platform supplements the genetic analyses, because it allows the inclusion of geographic origin, crossing/ancestry relationships, and marker distances, that is, similarity/dissimilarity (Figure 11.1, Wong et al. 2015).

Currently, the DUS-OP Guidelines (Plant Varieties Board Malaysia 2009) for the protection of Plant Breeder's Rights (PBR) for a new variety of oil palm is limited to T clones. In this case, DNA fingerprints will only be used when there is a dispute and not used as the basis to differentiate a palm from other varieties when the phenotype appears the same. Although there is no emphasis on DNA fingerprinting in the examination of DUS in oil palm, it remains a reliable and useful tool for distinguishing oil palm genotypes at population levels (see Figure 11.2, Wong et al. 2015), as well as different types of oil palm "varieties" as illustrated by Feyt and Durand-Gasselin (2006).

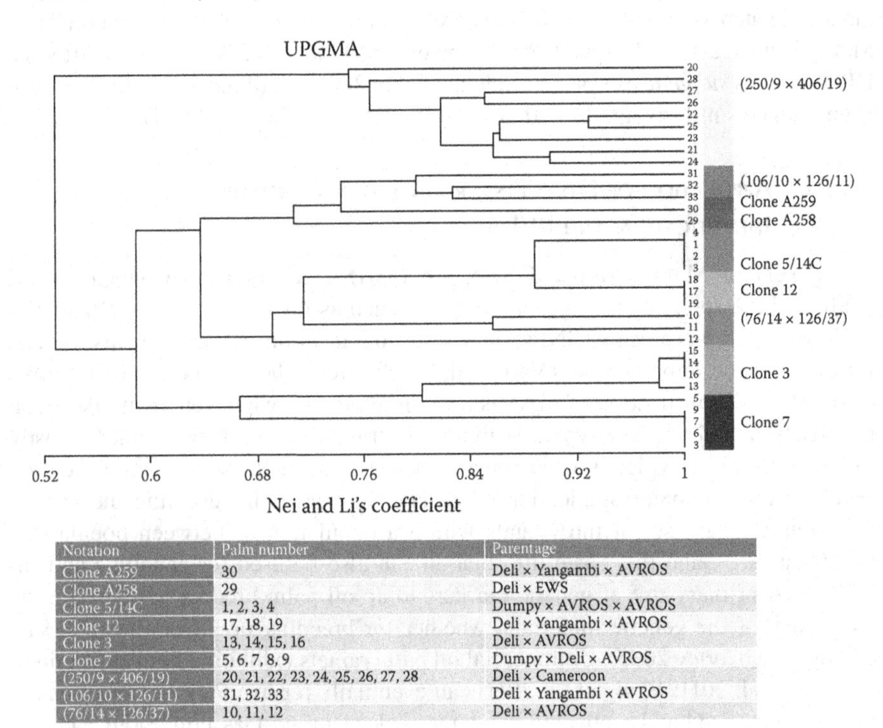

Notation	Palm number	Parentage
Clone A259	30	Deli × Yangambi × AVROS
Clone A258	29	Deli × EWS
Clone 5/14C	1, 2, 3, 4	Dumpy × AVROS × AVROS
Clone 12	17, 18, 19	Deli × Yangambi × AVROS
Clone 3	13, 14, 15, 16	Deli × AVROS
Clone 7	5, 6, 7, 8, 9	Dumpy × Deli × AVROS
(250/9 × 406/19)	20, 21, 22, 23, 24, 25, 26, 27, 28	Deli × Cameroon
(106/10 × 126/11)	31, 32, 33	Deli × Yangambi × AVROS
(76/14 × 126/37)	10, 11, 12	Deli × AVROS

FIGURE 11.1 A dendrogram based on UPGMA clustering of AAR's D × P materials and tissue culture clones using 17 polymorphic SSRs. Ramets from the same clone were clustered correctly under their respective clone origins (except for Ramet 13 in Clone 3 and Ramet 5 of Clone 7, which could be illegimate or due to inefficiency in the analytical technique). Seedling palms from crosses 250/9 × 406/19, 106/10 × 126/11, and 76/14 × 126/37 were distinguishable within and between crosses using SSR markers. (Reproduced from Wong, W.C. et al. 2015. *Journal of Oil Palm Research* 27: 113–127.)

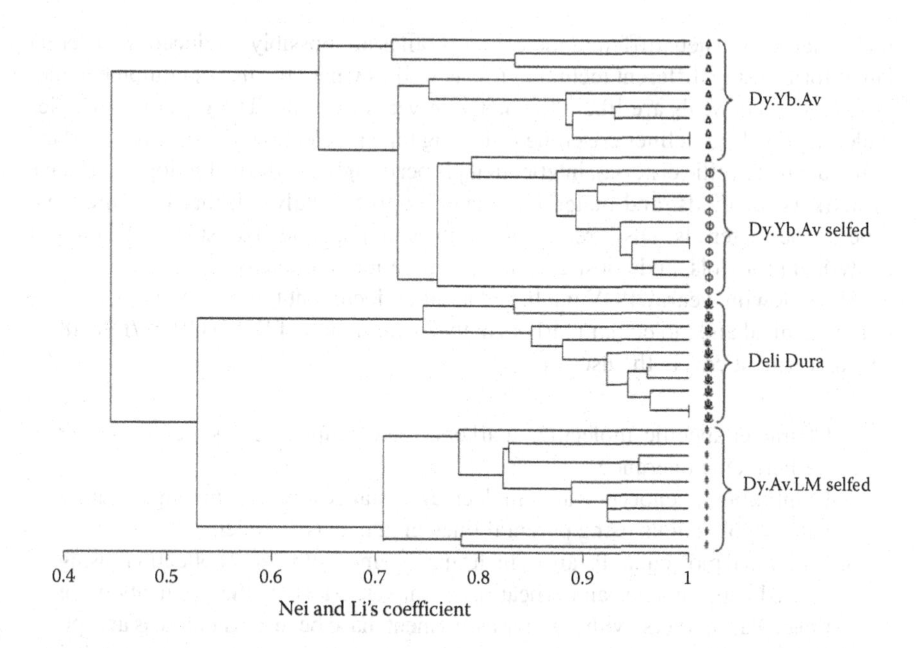

FIGURE 11.2 UPGMA clustering of four different segregating populations: Dumpy Yangambi AVROS (Dy.Yb.Av), (Dy.Yb.Av) selfed, Deli Dura, and Dumpy.AVROS.La Mé (Dy.Av.Lm) selfed reflects some identifiable genetic distance differences of the tested seedlings and confirms the usefulness of genetic markers to distinguish oil palm segregating materials at the population level, with individual palms uniquely identified. (Reproduced from Wong, W.C. et al. 2015. *Journal of Oil Palm Research* 27: 113–127.)

1. *Population hybrid varieties ("origin")*: The varieties classed by the major origin, for example, "Deli × Ekona" type. "Deli × Nigeria" type, "Deli × La Mé" type, etc.
2. *Full-sib family hybrid varieties*: The identity of the ancestors or grandparents from which the seeds were derived. Usually, these varieties can be considered as a subset of (1), as well as it being possible to classify the varieties derived from palms selected within parental families that underwent a generation of crossing within the family, either selfed or sib-crossed.
3. *Varieties reproducing families (sets of full sibs)*: The breeder has identified the best-performing family and has reproduced it by either cloning or selfing of the parents, offering the full advantage of a true-to-type reproduction of the selected cross. In the latter case, this will only be true if the palm to be self-pollinated is close to homozygosity.
4. *Clones:* Premium varieties. Superior genotypes by OPTC cloning.

In 2010, The International Union for the Protection of New Varieties of Plants (UPOV) in Geneva adopted and provided Biochemical and Molecular Techniques (BMT) guidelines for molecular marker selection and database construction (UPOV/INF/17/1). The purpose of these BMT guidelines is to provide guidance for developing harmonized methodologies in generating high-quality and reproducible molecular data, where different molecular profiles are possibly produced in different laboratories using different technologies, as well as where the use of equipment and/or reaction chemicals are likely to change in years to come. The general principles under the BMT guidelines cover the following topics: selection of molecular markers (general and specific criteria), intellectual property rights to the technology (inclusive of markers, methods, and materials), materials to be analyzed, that is, source and type of the materials, DNA reference samples and sample size, standardization of analytical protocols, and construction and validation of databases.

The following year, UPOV published another document to report the possible use of biochemical and molecular markers in the examination of DUS (UPOV/INF/18/1). The assessment covers the uses of

1. Character-specific molecular markers, for example, gene-specific marker for herbicide tolerance.
2. Combining phenotypic and molecular distances in the management of variety collections using parental lines in maize as a model.
3. Calibrated molecular distances in the management of variety collections using oilseed rape, maize, and wheat as examples. Most of the applications of molecular markers, with positive assessment, have been considered as acceptable for DUS. But the importance of examining the uniformity and stability on the same characteristics as used for distinctness must be emphasized.

As discussed above, the successive progression of applying molecular data in the examination of DUS for other crop species showed that it is increasingly necessary to review the practical limitations and identify the lessons that can be applied toward applying molecular markers in the examination of DUS for oil palm. More specifically, the evaluation of disease resistance characteristics such as basal stem rot disease (major fungal disease of oil palm in South East Asia) using visualized scoring of disease severity is not always straightforward because the assessments are often influenced by environmental factors. Aside from molecular markers, it may require additional biological assays, that is, pathogenicity testing to confirm/predict disease resistance traits. Currently, the two leading palm oil-producing countries, that is, Indonesia and Malaysia, are yet to become members of the UPOV system. However, the protection of PBR is becoming increasing important due to the high cost and competitiveness of developing new oil palm varieties through genome-assisted breeding (GAB), aiming to improve yield, quality, and other agronomically useful traits.

11.6 CONCLUSIONS

Aik Chin Soh, Sean Mayes, and Jeremy Roberts

The oil palm is seen as the crop for development through agriculture by developing countries in the tropics, especially West Africa and Latin America where the crop (*EG* and *EO*, respectively) originated, besides East Asian countries. It is likely that demand for planting materials will increase. Initially, this will be met by importation from the existing large plantation seed companies who are also likely to be the plantation investors as well; followed by franchised and locally developed seed production.

In short, oil palm planting material demand will rise and supply will become competitive due to the lucrative business opportunities this will generate. The demand for high-quality pedigree planting materials is likely to make seed certification and PVP mandatory by the suppliers. Seeds—whether as sexual or clonal hybrid seed—will continue to dominate the market. Clones will only assume a small premium market (for in-house use or buyers who are able to exploit them) until the remaining production inefficiency issues are resolved.

APPENDIX 11A.1 CLONAL FIELD TRIAL RESULTS

Many of the clones that were trial planted during 1997–1999 were better oil yielding than the AA D × P control available at that time (see Table 11A.1). With the availability of the AAHybrida1 control (which was 22% better in OY than AA D × P) from 1999 onward, the number of clones significantly better than the control in the year 2000 trial plantings were greatly reduced until the appearance of new clones in the 2004 trial plantings onward. Although there were definitely outstanding clones, by the time the commercial reclones are available, improved versions of AAHybrida would be available, reducing the original larger OY advantage of the outstanding clones.

TABLE 11A.1
Field Trial Oil Yield (OY) Performance of AAR Clones

Trial	Number of Clones	Mean OY of All Clones (% of Control)	Number of Clones >110% of Control	Number of Clones >105% of Control	Control Type	Recording Period Years (Y) and Months (M)
BCT13-97	12	120	10	11	AA D × P	7 Y 9 M
BCT14-97	11	118	8	11	AA D × P	6 Y 11 M
BCT15-98	8	138	8	8	AA D × P	7 Y 6 M
BCT16-98	24	119	20	22	AA D × P	8 Y
BCT17-99	14	124	11	11	AA D × P	8 Y 8 M
BCT18A-00	19	126	18	18	AA D × P	11 Y 2 M
BCT18B-00	7	119	6	7	AA D × P	10 Y 7 M
BCT19-00	10	94	2	2	AAHybrida1	10 Y 3 M
BCT20A-00	8	120	6	7	AA D × P	10 Y
BCT20B-00	8	125	6	6	AA D × P	10 Y 3 M
BCT21A-04	13	92	1	4	AAHybrida1	8 Y 9 M
BCT21B-04	13	112	7	12	AAHybrida1	8 Y 9 M
BCT22-07	16	110	9	10	AAHybrida1	5 Y 7 M

Note: AAHybrida1 control was 22% higher oil yielding than AA D × P control.

TABLE 11B.1
Commercial Fresh Fruit Bunch (FFB) Yield Performance of AAR Clones in Sabah, Malaysia

Estate	Area Ha	Planting Year	2005 FFB t/ha/ year	2006 FFB t/ha/ year	2007 FFB t/ha/ year	2008 FFB t/ha/ year	2009 FFB t/ha/ year	2010 FFB t/ha/ year	2011 FFB t/ha/ year
Pang Burong	35	1999	39.9	35.8	34.9	44.4	44.3	40.6	43.6
Sri Kunak	124	1999	32.0	33.0	30.5	32.8	31.6	33.6	33.8
Sigalong	87	1999	34.8	34.2	36.9	37.4	33.8	32.0	31.1
Average	246	1999	35.5	34.3	34.0	38.2	36.5	35.4	36.2

APPENDIX 11B.1 COMMERCIAL CLONAL FIELD FFB YIELD RESULTS

Table 11B.1 illustrates the consistently high commercial field FFB yield results of a group of estates in Sabah, East Malaysia, which has one of the earliest replantings with AAR clones. Unfortunately, there were no D × P replantings done on the estate at the same time in these estates for comparisons. It must be pointed out that FFB yields are very responsive to environmental influence and management inputs and thus variable. Valid or objective comparisons of clonal versus D × P hybrid commercial field yields will have to be observed over a larger number of comparable fields, estates, and locations and over a longer period.

APPENDIX 11C.1 RAMET CROP MILL OER OR OIL EXTRACTION RATE RESULTS

Better comparisons between the ramet crop and the normal D × P crop (as compared to yield results) can be made with dedicated ramet crop mill OER runs. Table 11C.1 shows

TABLE 11C.1
Mill Test Runs of Oil Extraction Rates (OERs) between Ramet and Normal Mill D × P Crops

Crop	Test 1	Test	Test 3	Test 4	Test 5	Test 6	Test 7	Mean
Ramet OER%	24.3	26.7	25.6	24.1	25.9	25.4	24.3	25.2
Normal mill (D × P) OER%	22.4	21.9	22.1	19.7	23.0	23.0	22.8	22.1
Difference OER %	1.9	4.8	3.5	4.4	2.9	2.4	1.5	3.1
Ramet ha.	464	423	251	1341	1367	1226	1857	
Ramet crop (t)	390	343	389	914	553	1094	1379	

Note: Test runs made in four different mills with young clonal plantings (with 20% D × P mixed into clonal block plantings to improve pollination).

the consistently better OERs of the ramet crop (24.1%–26.7%, mean = 25.2%) over the normal mill (mainly D × P) crop (19.7%–23.0%, mean = 22.1%), a mean advantage of 3.1% OER points.

APPENDIX 11D.1 SEED CERTIFICATION

11D.1.1 Malaysia

11D.1.1.1 Seeds

An oil palm seed producer needs to have its seeds certified by SIRIM by fulfilling a set of criteria. Malaysian Standard MS 157; Oil Palm Seeds for Commercial Planting (currently in its fourth revision) was prepared in consultation with oil palm breeders and other industry stakeholders. The certificate is endorsed by the Department of Standards Malaysia of the Malaysian government. With this, the producer can apply to MPOB for the license to sell oil palm seeds. MS 157, besides detailing the quality standards for seed processing and purity, seed sampling and test methods, packaging and labeling, and legal (definition) requirements, also spells out the minimum selection criteria of the D × P progeny-test performance and D mother palms for three categories of hybrid materials: parents of known pedigree and known progeny-test performance; parents of unknown pedigree but known progeny-test performance; parents of known pedigree but unknown performance as summarized below:

Minimum requirements for the *tenera* in the progeny test (for all three categories):

1. Fresh fruit bunch (FFB) yield, minimum: 170 kg/p/year
2. Oil to bunch (O/B) ratio, minimum: 25%
3. Kernel to bunch (K/B) ratio, minimum: 3%
4. Oil yield, minimum: 46.5 kg/p/year

The FFB yield values are derived from at least 4 consecutive years of recording. The bunch component values are derived from a minimum of 50 bunches involving at least 30 palms per D × P cross or 5 bunches per palm from at least 10 palms per D × T//T × D/T × T/T × P cross.

Minimum requirements for the *dura* parent palms (for first two categories):

1. Fresh fruit bunch (FFB) yield, minimum: 150 kg/p/year
2. Mesocarp to fruit (M/F) ratio, minimum: 57%
3. Shell to fruit (S/F) ratio, maximum: 33%
4. Kernel to fruit (K/F) ratio, minimum: 5%
5. Oil to dry mesocarp (O/DM) ratio, minimum: 75%
6. Oil to bunch (O/B) ratio, minimum: 19%

11D.1.1.2 Clones

SIRIM has drawn up an MS standard for commercial clone production, "Oil Palm Ortet Selection for Cloning—Specification."

Besides specifications on definitions, guidelines for production practices and facilities, packaging and transportation of plantlets, legal requirement, and certification mark, the selection criteria for ortets are given below.

11D.1.1.3 Cloning

Minimum standards required for the selection of an ortet (for both "materials of known pedigree and known performance of family and individual palm" and "materials of unknown pedigree and known performance from known seed producers" are

OY—50 kg/p/year and O/B—27%

With the proviso that the ortet is derived from a family size of at least 30 palms (10 palm per plot × 3 replicates); the OY based on at least 4 consecutive years of FFB yield recording and 5 analyses of O/B.

11D.1.1.4 Recloning

As above, except that O/B could be based on 5 bunches per clonal ortet or 30 analyses per clone, and the clonal ortet has no history of fruit or vegetative abnormalities, and the clone from which the clonal ortet is selected has not more than 5% abnormality based on at least 100 ramets field evaluated in the same year.

11D.1.2 Indonesia

The process to become an oil palm seed producer in Indonesia is detailed below.

The candidate must have at least an oil palm variety, either owned or franchised by other parties. The variety must pass the assessment conducted by the TP2V (*Tim Penilai dan Pelepas Varietas*, Team of Assessment and Releasing Variety), which is formed by the National Seed Board, whose members are breeders of various estate commodities as well as experts in various fields of science. TP2V members generally highlight more on progeny testing and unique features, which distinguish the variety candidate with other existing commercial varieties. Table 11D.1 describes a few criteria and procedures needed to be fulfilled by the potential plant variety.

After the TP2V passes the plant variety candidate, it is then proposed to the Minister of Agriculture to be approved as a new oil palm variety. The next step to become an oil palm seed producer is to obtain a seed production permit from the Ditjenbun (*Direktorat Jenderal Perkebunan*, Directorate General of Estate Plantations). Basing on the data used in releasing the variety, the Ditjenbun monitoring and evaluation team together with the local technical and operational unit (*Unit Pelaksana Teknis Daerah*, UPTD) or with the BBP2TP (*Balai Besar Proteksi dan Perbenihan Tanaman Perkebunan*) will inspect facilities and infrastructure owned by the proposing company. Several requirements during the process are described in Table 11D.2.

When the monitoring and evaluation team of Ditjenbun is satisfied that all the above requirements have been met, a designation letter will be issued to the new oil palm seed producer. With this letter, the seed producer will be able to obtain the seed production permit from the local provincial government.

11D.1.3 Other Countries

Seed certification schemes are being developed or implemented in some African and Latin American countries.

TABLE 11D.1
Requirements Needed to Fulfil for the Potential New Variety

No.	Criteria	Procedure
1	Observations on progeny trial	
	a. Production record	Mean of four consecutive harvesting years
	b. Bunch analysis	Palm oil analysis, which is synchronized with the soxhlet method, after the palm age is more than 5 years after planting
	c. Vegetative observation	Conducted at least twice during the trial
2	Crossings selection criteria	
	d. Fresh fruit bunch (FFB) production	≥175 kg/p/year
	e. Palm products (crude palm oil [CPO] + palm kernel oil [PKO])	≥6 t/ha/year
	f. Industrial extraction rate (laboratory soxhlet extraction rate × 0.855)	≥23%
	g. Height increment (measured at age 6 years after planting)	≤80 cm/year

TABLE 11D.2
Further Requirements to Be Fulfilled for the Potential New Variety

No.	Standard	Requirement
1	Mating design and seed production	Referring to the progeny trial of an existing oil palm variety
2	Technical aspects of seed garden development	Minimum 36 palms per D × D or T × T/P crossing
3	Palm physical condition	Healthy
4	System on parentage selection	
	Dura parentage (D × D) and *Pisifera* parentage (T × T/P)	Reproduction of parent palms, which have been progeny-tested
		Only palms confirmed as *Pisifera* from fertile fruit on observations and/or D × P progeny are utilized as pollen sources
5	Seed garden management	
	a. Selected female and male parent palms	Given permanent mark
	b. Pollination activities	Following the pollination standard operational procedure (SOP)
6	Usage of clone palms as female and male parent palms	Passed the test on flowering abnormalities on their progeny palms
7	Pollen preparation laboratory	
8	Seed preparation unit	
9	Seed dormancy unit	
10	Germination unit	
11	Packaging unit	
	Human resource qualification	

APPENDIX 11E.1 PLANT VARIETY PROTECTION

11E.1.1 Malaysia

The passing of the PVP Act in 2004 has motivated oil palm seed producers and breeders to be more systematic and targeted in characterizing the planting materials they produce. The planting materials need to be defined to be new, distinct, uniform, and stable; *new*—the propagating or harvested material of the candidate variety has not been sold commercially under the breeder's consent; *distinct*—on the filing date of application, the variety is clearly distinguishable from any other plant variety of common knowledge; *uniform*—subject to the variation that may be expected from the particular features of its propagation, it is sufficiently *uniform* in its relevant characteristics; *stable*—its relevant characteristics remain unchanged after repeated propagation or in the case of a particular cycle of propagation, at the end of each propagation cycle. "New," "distinct," "uniform," and "stable," are conditions needed to be met by breeders who are keen to register a plant variety. While the definition of "new" is straightforward and easily verifiable, "distinct," "uniform," and "stable" require test guidelines to confidently distinguish its credibility in meeting the definition. Some of the salient points of the guidelines for the conduct of test for DUS for oil palm are

1. Materials required
   a. Age of palm: at least 8 years after field planting with normal growth and not affected by any important pest or disease at the point of inspection
   b. Minimum 10 plants to be inspected if vegetatively propagated planting materials
   c. Minimum 30 plants to be inspected if seed-derived planting materials
2. Method of inspection
   a. Two inspection periods per year and one inspection per application
   b. Additional tests (optional): bunch analysis and DNA profiling
   c. Test location is normally one, but separate plots can be used
   d. Test design in selection of palms or parts of palms to be removed is done randomly
   e. Type of observation could be:
      i. Single measurement of a group of plants or parts of plants (MG)
      ii. Measurement of a number of individual plants or parts of plants (MS)
      iii. Visual assessment by a single observation of a group of plants or parts or plants (VG)
      iv. Visual assessment by observation of individual plants or part of plants (VS)
3. Assessment of distinctness, uniformity, and stability
   a. Distinctness—Examiners shall emphasize on consistent differences within the duration of test and clear differences on characteristics being qualitative, quantitative, or pseudo-qualitative
   b. Uniformity—Examiner shall emphasize on the number of off-types allowed
   c. Stability—Generally considered stable when proven uniform
4. Characteristics to be examined are 26
   a. Tree: 4 characteristics

   b. Frond: 8 characteristics
   c. Bunch: 3 characteristics
   d. Fruit: 8 characteristics
   e. Nut: 3 characteristics
5. Additional test (optional)
   a. Bunch analysis
   b. DNA profiling according to protocol using 10 SSR markers developed by Malaysia Palm Oil Board

The general flowchart for registration and grant breeder's right is as follows:

1. Application for registration
2. Preliminary examination (Denomination, Novelty)
3. Technical examination (Distinct, Uniform, Stable, Identifiable)
4. Approval by Plant Varieties Board
5. Publication
6. Deposit of samples
7. Grant of Plant Breeder's Right
8. Maintenance of Breeder's Right

11E.1.1.1 Other Countries

Likewise, other South East Asian countries have established similar systems for the protection of oil palm breeding material with test guidelines being fundamentally similar. The main difference among countries is the number of characteristics to be examined. There are 26 characteristics for Malaysia, 73 characteristics for Indonesia, and 30 characteristics for Thailand to be observed. Harmonization of oil palm test guidelines among Malaysia, Indonesia, and Thailand has been initiated, led by Malaysia. Such harmonized oil palm test guidelines provides options to breeders either to register under individual country or across countries. Apparently, African and Latin American countries (Brazil, Colombia, Ecuador) have also organized PVP for the oil palm under the UPOV system.

11E.1.1.2 Issues with PVP

There are also other issues that deter the urgency or ability for oil palm breeders to obtain PVP. First, oil palms seeds sold commercially are still essentially mixed hybrids derived directly or indirectly (introgressions) from a few known lineages, for example, Deli, AVROS, Yangambi, and La Mé, hence genetically and phenotypically variable and related, making it difficult to segregate the different planting materials phenotypically and genetically based on DUS criteria. Some of these materials have also been sold commercially for some time disqualifying them for PVP. Near-true F1 or single-cross hybrids, clonal hybrids (bi or semi) with DUS traits, have a better chance for PVP protection. It can be argued that in some annual crops, proprietary ownership of the inbred parents confers sufficient protection without resorting to PVP. However, the parents of oil palm that are near-true hybrids, clonal hybrids, and clones have undergone a long tedious expensive and rigorous scientific breeding process. It would be unacceptable for the elite parents to be stolen through their selfs or clones by illegitimate seed producers.

Clonal and transgenic varieties are covered by the UPOV system. There have been a couple of applications for PVP under The Malaysian Protection of New Plant Act 2004 but their outcomes are still as yet unknown.

Besides PVP, bilateral seller–buyer rights agreements/contract and plant patents are the other means of variety proprietary protection, although the latter is more commonly practiced with clonal, mutant, and transformed varieties in the United States.

REFERENCES

Billotte, N., Risterucci, A.M., Barcelos, E., Noyer, J.L., Amblard, P., and Baurens, F.C. 2001. Development, characterization and across-taxa utility of oil palm (*Elaeis guineensis* Jacq.) microsatellite markers. *Genome* 44: 413–425.

Chin, C.W. 1999. Oil palm breeding techniques. In: *Proceedings Seminar Science of Oil Palm Breeding* (Montpellier, 1992). pp. 49–64. Palm Oil Research Institute, Kuala Lumpur, Malaysia.

Choo, C.N., Wong, C.K., NurAkilla, M.R., Ee, C.C., Ilham, A.A., and Tan, C.C. 2014. Genotype effect on oil palm tissue culture callogenesis and embryogenesis. In: *Proceedings International Oil Palm Conference (IOPC).* International Oil Palm Research Institute, Bali, Indonesia.

Cochard, B. et al. 2009. Geographic and genetic structure of African oil palm diversity suggests new approaches to breeding. *Tree Genetics and Genomes* 5: 493–504.

Corley, R.H.V. 2005. Illegitimacy in oil palm breeding—A review. *Journal of Oil Palm Research* 17: 64–69.

Corley, R.H.V., Lee, C.H., Law, I.H., and Wong, C.Y. 1986. Abnormal flower development in oil palm clones. *Planter* 62: 233–240.

Donough, C.R., Ng, M., and Lai, C. 1993. Pamol's approach to quality control in controlled pollination for D × P seed production. *Planter* 69: 163–175.

Durand-Gasselin, T. 2009. ID checking by microsatellite-type markers (SSR) during the oil palm variety selection and production processes. Paper presented in Proceedings International Seminar on Oil Palm Genomics and Its Application to Oil Palm Breeding. ISOPB, Kuala Lumpur, Malaysia.

Escobar, C.R. 1980. An improved oil palm fruit depulperfor single bunch lots. *The Planter* 56: 540–542.

Feyt, H. and Durand-Gasselin, T. 2006. The UPOV sui generis system and its possible application to oil palm (*E. guineensis*). Paper presented at The International Seminar on Yield Potential in the Oil Palm. ISOPB, Phuket, Thailand.

Ginting, G. and Fatimwati. 1995. Propagation methodology of oil palm at Marihat. In: *Recent Developments in Oil Palm Tissue Culture and Biotechnology.* Eds. V. Rao, I.E. Henson, and N. Rajanaidu. pp. 33–37. Palm Oil Research Institute, Kuala Lumpur, Malaysia.

Hartley, C.W.S. 1988. *The Oil Palm*, 3rd Edition. Longman, London/New York.

Herrera, J., Alizaga, R., and Guevara, E. 1998. Use of Chemical Treatments to Induce Seed Germination in Oil Palm (*Elaeis guineensis* Jacq,). *ASD Oil Palm Papers* 18: 1–16.

Hussey, G. 1958. An analysis of the factors controlling the germination of the seed of the oil palm, *Elaeis guineensis* (Jacq.). *Annals of Botany* 22: 259–286.

Jacquemard, J.J.C., Meunier, J., and Bonnot, F. 1981. Genetic study of the reproduction of an *Elaeis guineensis* oil palm cross. Application to selected seed production and improvement. *Oléagineux* 36: 343–352.

Kushairi, A., Mohd Din, A., and Rajanaidu, N. 2011. Oil palm breeding and seed production. In: *Further Advances in Oil Palm Research (2000–2010)* Eds. B.W. Mohd, Y.M. Choo, and K.W. Chan. Malaysian Palm Oil Board, Kuala Lumpur, Malaysia.

Mok, C.K. 1982. Hear requirement for breaking dormancy of oil palm seeds after storage under different conditions. In: *The Oil Palm in Agriculture in the Eighties*. Vol 1. Eds. E. Pushparajah and P.S. Chew. pp. 197–206. Incorp. Soc. Planters, Kuala Lumpur.

Murashige, T. and Skoog, F. 1962. A revised medium for rapid growth and bioassays with tobacco tissue culture. *Physiologia Plantarum* 15: 473–496.

Noiret, J.M. and Ahizi Adiapa, P. 1970. Anomalies de L'Embryon Chez le Palmier a la Huile. Application a la production de semences. *Oleagineux* 25: 511.

Ong-Abdullah, M. et al. 2015. Loss of karma transposon methylation underlies the mantled somaclonal variant of oil palm. *Nature* doi: 10.1038/nature 15365.

Paranjothy, K. and Rohani, O. 1982. *In vitro* propagation of oil palm. In: *Proceedings of 5th International Congress. Plant Tissue and Cell Culture*. pp. 747–748. Japanese Association for Plant Tissue Culture, Tokyo, Japan.

Periasamy, A., Gopal, K., and Soh, A.C. 2002. Productivity improvements in seed processing techniques for commercial oil palm seed production. *Planter* 78: 420–441.

Plant Varieties Board Malaysia. 2009. *Guidelines for the Conduct of Tests for Distinctness, Uniformity and Stability: Oil Palm (Elaeis guineensis Jacq.). Protocols DNA Profiling of Oil Palm: Useful Explanations on Molecular Characteristics Using Simple Sequence Repeat*. Department of Agriculture, Malaysia.

Purba, A.R. 2010. Oil palm breeding: Indonesian experience. In: *Proceedings Indonesian Oil Palm Conference 2010*. AGR1.1, Yogjakarta, Indonesia.

Rao, V. and Kushairi, A. 1999. Quality of oil palm planting material. In: Proc. 1996 Seminar "Sourcing of Oil Palm Planting Materials for Local and Overseas Ventures". Eds. N. Rajanaidu and B.S. Jalani. pp. 189–197. Palm Oil Research Institute, Kuala Lumpur, Malaysia.

Rees, A.R. 1962. High temperature treatment and the germination of seed of oil palm, *Elaeis guineensis* (Jacq.). *Annals of Botany* 26: 569–581.

Rosenquist, E. 1986. The genetic base of oil palm breeding populations. In: *Proceedings of International Workshop "Oil Palm Germplasm and Utilisation."* pp. 27–56. Palm Oil Research Institute, Kuala Lumpur, Malaysia.

Simko, I., Eujayl, I., and van Hintum, T.J.L. 2012. Empirical evaluation of DArT, SNP and SSR marker-systems for genotyping, clustering and assigning sugar beet hybrid varieties into populations. *Plant Science* 184: 54–62.

Singh, R., Jayanthi, N., Tan, S.G., Panandam, J.M., and Cheah, S.C. 2007. Development of simple sequence repeat (SSR) markers for oil palm and their application in genetic mapping and fingerprinting of tissue culture clones. *Asia Pacific Journal of Molecular Biology and Biotechnology* 15: 121–131.

Singh, R., Noorhariza, M.Z., Ting, N.C., Rosli, R., Tan, S.G., Lim, L.E., Maizura, I., and Cheah, S.C. 2008. Exploiting an oil palm EST database for the development of gene-derived SSR markers and their exploitation for assessment of genetic diversity. *Biologia* 63: 227–235.

Smith, W.K. and Thomas, J.A. 1973. The isolation and *in vitro* cultivation of cells of *Elaeis guineensis*. *Oléagineux* 28: 128.

Soh, A.C. 1986. Expected yield increase with selected oil palm clones from current D × P seedling materials and its implications on clonal propagation, breeding and ortet selection. *Oléagineux* 41: 51–56.

Soh, A.C. 1987. Strategies in breeding and selection for oil palm clones. In: *Proceedings of Colloquium Breeding and Selection for Clonal Oil Palms*. pp. 52–62. Palm Oil Research Institute, Kuala Lumpur, Malaysia.

Soh, A.C. 1998. Review of ortet selection in oil palm. *Planter* 74: 217–226.

Soh, A.C., Wong, C.K., Ho, Y.W., and Choong, C.W. 2007. Oil palm. In: *Handbook on Plant Breeding. Oil Crops*. Eds. J. Vollmann and I. Rajcan. pp. 333–367. Springer, Dordrecht.

Soh, A.C., Wong, G., Tan, C.C., Chew, P.S., Chong, S.P., Ho, Y.W., Wong, C.K., Choo, C.N., Nor Azura, A., and Kumar, K. 2011. Commercial-scale propagation and planting of elite oil palm clones: Research and development towards realization. *Journal of Oil Palm Research* 23: 935–952.

Tailliez, B. 1970. Germination acceleree des grains de palmier a huile. Technique avec sub-srat. *Oleagineux* 25: 335–336.

Tan, C.C., Wong, G., and Soh, A.C. 1999. Acclimatization and handling of oil palm tissue cultured plantlets. *Paper Presented at the 1999 PORIM International Palm Oil Congress.* February 1–6, Kuala Lumpur, Bangi.

Tan, C.C., Soh, A.C., Wong, G., Hor, T.Y., Chong, S.P., and Gopal, K. 2003. Experiences and lessons from oil palm clonal evaluation trials and commercial test plantings. *Paper Presented at the International Palm Oil Congress on Palm Oil: The Power House for the Global Oil and Fats Economy.* Malaysian Palm Oil Board, Putrajaya, Malaysia, August 22, 2003.

UPOV/INF/17/1. 2010. Guidelines for DNA-profiling: Molecular marker selection and database construction ("BMT Guidelines"). Adopted by the council at its Forty-Fourth Ordinary Session, Geneva, October 21, 2010.

UPOV/INF/18/1. 2011. Possible use of molecular markers in the examination of distinctness, uniformity and stability (DUS). Adopted by the council at its Forty-Fifth Ordinary Session, Geneva, October 20, 2011.

Väli, U., Brandström, M., Johansson, M., and Ellegren, H. 2008. Insertion-deletion polymorphisms (Indels) as genetic markers in natural populations. *BMC Genetics* 9: 8. doi: 10.1186/1471-2156-9-8.

Van Inghelandt, D., Melchinger, A.E., Lebreton, C., and Stich, B. 2010. Population structure and genetic diversity in a commercial maize breeding program assessed with SSR and SNP markers. *Theoretical and Applied Genetics* 120: 1289–1299.

Wong, G., Chong, S.P., Tan, C.C., and Soh, A.C. 1999a. Liquid suspension culture—A potential technique for mass production of oil palm clones. In: Preprints, 1999 PORIM Int. Palm Oil Conf. pp. 3–11. Palm Oil Research Institute, Kuala Lumpur, Malaysia.

Wong, C.K., Choo, C.N., Ng, W.J., Kumar, K., Aida, N.N., Chin, S.Y., and Tan, C.C. 2011. Semi-clonal seeds: Development and performance, applied agricultural resources S.B.'s perspective. In: *Proceedings of International Palm Oil Congress (PIPOC).* Malaysian Palm Oil Board, Malaysia.

Wong, G., Tan, C.C., and Soh, A.C. 1997. Large scale propagation of oil palm clones—Experiences to date. *Acta Horticulturae* 447: 649–658.

Wong, G., Tan, C.C., Soh, A.C., and Chong, S.P. 1999b. Clonal propagation of oil palm through tissue culture. *Planter* 75: 221–230.

Wong, W.C., Teo, C.J., Wong, C.K., Mayes, S., Singh, R., and Soh, A.C. 2015. Development of an effective SSR-based fingerprinting system for commercial planting materials and breeding applications in oil palm. *Journal of Oil Palm Research* 27: 113–127.

Wooi, K.C. 1990. Oil palm (*Elaeis guineensis* Jacq.): Tissue culture and micropropagation. In: *Biotechnology in Agriculture and Forestry 10: Legumes and Oilseed Crops I.* Ed. Y.P.S. Bajaj. pp. 569–592. Springer-Verlag, Berlin.

Wooi, K.C. 1995. Oil palm tissue culture—Current practice and constraints. In: *Recent Developments in Oil Palm Tissue Culture and Biotechnology.* Eds. V. Rao, I.E. Henson, and N. Rajanaidu. pp. 21–32. Palm Oil Research Institute, Kuala Lumpur, Malaysia.

Wooi, K.C., Wong, C.Y., and Corley, R.H.V. 1981. Tissue culture of palms—A review. In: *Proceedings of COSTED Symposium. Tissue Culture of Economically Important Plants.* pp. 138–144. Singapore.

Zamzuri, I. 1999. Using double-layer technique in rooting of oil palm *in vitro* plantlets. In: Preprints, *1999 PORIM International Palm Oil Conference.* pp. 223–229. Palm Oil Research Institute, Kuala Lumpur, Malaysia.

12 Field Experimentation

Rob Verdooren, Aik Chin Soh, Sean Mayes, and Jeremy Roberts

CONTENTS

12.1 Introductory Overview (*Aik Chin Soh, Sean Mayes, and Jeremy Roberts*)....327
12.2 Experimental Designs and Analysis (*Rob Verdooren and Aik Chin Soh*)....328
12.2.1 Experiment....328
12.2.2 Experimental Design....329
12.2.3 Comparison of Treatment Effects....330
12.3 Statistical Models....331
12.3.1 Fixed Model....331
12.3.2 Random Model....331
12.3.3 Mixed Model....331
12.4 Applications in Oil Palm....333
12.4.1 Oil Palm Breeding Experiments....333
12.4.2 Experimental Designs....333
12.4.3 Plot Size and Replication....334
12.4.3.1 Power Analysis....335
12.5 Trial Conduct....336
12.5.1 Interplant and Interplot Competition....336
12.5.2 Control Treatments/Check Varieties....336
12.5.3 Missing/Dummy Plots....337
12.6 Data Collection and Management, Statistical Analyses, and Software....337
12.6.1 Data Collection and Management....337
12.6.2 Statistical Analysis and Software....338
12.7 Conclusions....338
Appendix 12A.1 Alpha (α) Incomplete Factorial Mating Design To Estimate Gca And Sca Effects Of *Dura* And P Parents....338
Appendix 12B.1 Field experimental design for oil palm breeding trials....345
References....350

12.1 INTRODUCTORY OVERVIEW

Aik Chin Soh, Sean Mayes, and Jeremy Roberts

Field experimentation is integral to the plant breeder's job function and the mastery of its techniques based soundly on statistical or biometrical principles as applied to the crop of interest. It is through these field experiments particularly those with a genetic analysis structure that it is possible to establish the various relevant information of plant breeding interest as discussed in Chapter 4 earlier. In addition the following could also be established:

- The breeding system of the germplasm and to devise a sampling system for future prospection in the various centers of crop diversity for genetic conservation or improvement of specific traits (see also Chapter 4).
- The adaptability of the germplasm and traits concerned in the various ecological (location) and agronomic (spacing, fertilizer, and irrigation) environments of interest to the plant breeder.
- The best varieties with broad adaptability and also those for specific adaptation.
- The minimal sample size (e.g., trees, bunches, fruits, and leaves) or number of measurements and the measurement periods (monthly, quarterly, and yearly) of observations to characterize a plant or crop mean for a trait.
- Molecular marker-assisted selection (MAS) procedures.

Genomic marker (molecular breeding) technology is progressing at a rapid pace with efficient low cost and high throughput genotyping systems available. The bottleneck in developing an efficient MAS procedure for economic quantitative traits, for example, yield, is the imprecision in measuring the phenotypic trait or the accuracy with which the phenotyping reflects the underlying genetic composition. Good genetic trial data are needed to develop the MAS technique and to validate the marker approach in further material after its development. This emphasizes the importance of good field experimental techniques and trials which have and will continue to provide a treasure trove of data for many useful subsequent scientific investigations. Currently, there are more biotechnologists ("molecular breeders") than plant breeders ("field breeders") joining the plant breeding scene. Besides training more plant breeders, biotechnologists should be trained to appreciate field experimental techniques and conversely field breeders are to be trained to appreciate biotechnological tools. The future of successful marker-assisted selection lies critically within the integration of the two disciplines.

12.2 EXPERIMENTAL DESIGNS AND ANALYSIS

Rob Verdooren and Aik Chin Soh

12.2.1 Experiment

An experiment is a planned enquiry usually in the form of a hypothesis to obtain new/deny/confirm facts. A field breeding experiment typically comprises

Treatments—the procedures, usually for plant breeding; variety/genotypes/genetic populations and sometimes also, location, fertilizer, irrigation, the effects of which are measured and compared.

Plot—a unit of experimental material, for example, one to many plants, an area of land, where a treatment is applied according to the randomization process. The treatment plots are replicated, randomized, and laid out in an appropriate experimental design.

A field experiment has to reckon with multiple sources of variation which affect the ability to pick out real treatment effects and differences. The typical causes of variation in the field are

Inherent (genetic) plant differences and soil heterogeneity, for example, fertility, terrain, moisture content, and physical structure

Abiotic (climate/weather) and biotic (pest and disease) factor induced differences

Human introduced variation, for example, differences induced by nonuniform handling of experimental material by the researcher or field workers

Experimental design and field experimentation techniques employ three basic principles to attempt to circumvent these and demonstrate that the treatment differences are not biased.

Randomization—different treatments are applied to different plots at random, that is, every plot has equal chance of being assigned with any of the treatments to ensure no bias, for example, "desirable" treatments assigned to good plots and less favored to poor plots.

Replication—repeated plots receiving the same treatment.

The combination of randomization and replication ensures a precise (closeness of repeated measurements) and accurate (closeness to the true mean) estimate of the treatment effect (mean) and provides a valid estimate of the experimental error variance of the plots having the same treatment, reflective of inherent field variation, which is the statistical basis or yardstick to judge whether the measured treatment differences are true or not.

Blocking—a full (complete blocking) or partial set (incomplete blocking) of a replicate set of treatments are handled more closely and uniformly together under the same soil and homogeneous growing conditions. The treatments must be randomized inside each block. This is usually done to remove unidirectional/gradient effect complications, for example, terrain, soil fertility, soil moisture which would otherwise inflate the experimental error (and bias or deviation from true mean) and prevent the detection of true treatment differences and effects. With experiments having a large set of treatments or with big plots, for example, in oil palm with 6×6 palms planted on the corners of equilateral triangles with edges of 9 m, complete blocking requiring larger land areas would also likely encounter intra-block soil heterogeneity and hence incomplete blocking is advisable.

12.2.2 Experimental Design

An experimental design is a set of rules to allocate the treatments to the experimental units. Each experimental design with its own statistical analysis according to the randomization procedure partitions the total variation (variance) into the known sources/components of variation which are individually tested against the residual (unaccounted) or experimental error variance using Fisher's variance ratio (or "F-test") based on the null hypothesis (H_0) of no difference between treatment effects and random error. If the treatment variance ratio (F) exceeds the expected value at the 5% error rate or $\alpha = 0.05$, that is, 95% confidence rate, the treatment effects are deemed significant, the H_0 rejected, and the H_A (alternative hypothesis) of true treatment effects accepted. The F test for $\alpha = 0.01$, that is, 99% confidence rate is sometimes used as well.

Experimental designs commonly used:

Completely randomized design (CRD)—the simplest design with no blocking, can handle a variable number of treatments and replicates commonly used for sites that are relatively uniform, for example, laboratories. The design provides maximum degrees of freedom (df) for estimating the error variance for efficient detection of treatment differences under uniform conditions.

Randomized complete block design (RCBD)—the most commonly used and robust field experimental design. As its name implies, each block comprises one complete replicate of all the treatments. Blocking is ideal for removing unidirectional effect (e.g., slope, plant age) and placing the blocks or replicates along the gradient of the variation and thus reducing the size of the error term. As indicated earlier with a large number of treatments and thus a large block size, intra-block variation due to heterogeneous growing conditions would negate the advantage of blocking to reduce the size of the experimental error.

Latin square design—this design is able to remove bidirectional (row and column blocking) effects (e.g., slope and plant age). It is a less commonly used design because of the cumbersome field layout especially with a larger number of treatments and also susceptibility to larger intra-block variation.

Split-plot design—this design is commonly adopted in genotype (variety/hybrid/clone) × environment (e.g., plant spacing, fertilizer, and irrigation) or G × E experiments whereby the E comprises the main plot treatments and G the subplot treatments. This is done for ease of treatment application, plot management, and statistical efficiency as the subplot treatment and interaction effects are tested more efficiently. Variety × location trials are in a sense split-plot experiments,

Incomplete block designs—variety testing trials commonly comprise large treatment/entry numbers (tens/hundreds/thousands). As such complete block designs are inefficient and incomplete blocks desirable if not mandatory. Lattices (balanced/unbalanced/square and Row and Column and α Designs are commonly used. In Appendixes 12A.1 and 12B.1 worked examples are provided to illustrate the use of α Design for both the mating/crossing design and field experimental layout for oil palm breeding experiments.

12.2.3 Comparison of Treatment Effects

Multiple range tests, for example, Fisher's least-significant difference (*LSD*), Tukey's honestly significant difference (*HSD*) tests are commonly used. Duncan's multiple range test although also commonly used is not recommended. These tests can only be used on balanced (equal replication) experimental designs. With unequal replications (missing plots, incomplete block designs), individual standard error of difference (*SED*) has to be generated for each paired comparison. The Bonferroni procedure can then be used to set the experimental error at the significance level α and to test each of the k paired comparisons with a significant level α/k. Statisticians also frown upon indiscriminate comparisons with multiple range tests as the errors rates committed are much higher than $\alpha = 0.05$ than envisaged. Orthogonal contrasts/comparisons based on one df in the numerator is preferable as it usually answers or tests each hypothesis or question in turn.

12.3 STATISTICAL MODELS

12.3.1 Fixed Model

This is most commonly used in variety testing type trials conducted in RCBD. Replicate effects are considered fixed. Treatment effects are also considered fixed as they have been selected. The inference population is the selected set of treatments where the results apply.

12.3.2 Random Model

If the treatments are a randomly chosen sample or subset from a population of treatments, the inference population is the population rather than the subset. A good example would be the testing of plants or their randomly mated progenies chosen at random from an open-pollinated/randomly mated population of plants in a genetic experiment. Random model is commonly used in genetic experiments where the objective is to estimate genetic and environmental components of variance and heritability.

12.3.3 Mixed Model

This arises in a two (or more) factorial experiment where one factor is considered fixed, namely the treatments and the other factor, for example, incomplete blocks are considered as random. Usually a set of incomplete blocks formed a replication of all treatments; such a design is called resolvable. The treatments and the replications are fixed factors and the incomplete blocks inside the replications are considered as random. This allows the recovery of inter-block information which gives a better estimate of the treatment differences.

Mixed model is useful in the analysis and interpretation of unbalanced (unequal replications, missing/unequal treatments) and "messy" data. The messy factors are treated as fixed effects and removed to provide a better estimate of the random (treatment) effects. The best linear unbiased prediction (BLUP) is such a procedure and has found use in animal and plant (tree) breeding to remove differential age, location, and nutrition complications to estimate the breeding values of the breeding parents. Further discussions and worked examples are given in Chapter 6.

The term general linear models are used in the analyses of these: for the *fixed model*

$$y = XB + e,$$

where y is the response vector, X the matrix of treatment effects and e the random error; and for the *mixed model*

$$y = XB + ZU + e,$$

where Z is the incidence matrix. U the vector of random effects.

In the *fixed model* $y = XB + e$, where the expectation vector of e, $E(e) = 0_n$, the vector of *n* zeroes, we have that the expectation vector of y, $E(y) = XB$. In this fixed model the covariance matrix or dispersion matrix of the normally distributed random error e is $\sigma^2 I_n$ with I_n the $n \times n$ identity matrix and n is the number of independent observations. In the fixed model the covariance matrix of y, D, is $D = \sigma^2 I_n$. The fixed parameters B are Best Linear Unbiased Estimate (BLUE) with the ordinary least-squares method by solving the normal equations $(X'X)\,B = X'y$. In the case that the error e is normally distributed, the solution of the ordinary normal equations gives the minimum variance unbiased estimate of B.

In the *mixed model* $y = XB + ZU + e$ we have two random components U and e. The random vector U has expectation vector $E(U) = 0_q$ when we have q normally distributed random parameters of **U** and covariance matrix $\sigma^2_U I_q$ and the normally distributed random vector e has expectation vector $E(e) = 0_n$ and covariance matrix $\sigma^2_e I_n$. The random vector U and e are independent from each other. In the mixed model the covariance matrix of y, D, is $D = \sigma^2_U ZZ' + \sigma^2_e I_n$ or with ratio $\varphi = \sigma^2_U/\sigma^2_e$, $D = \sigma^2_e [\varphi ZZ' + I_n]$. The fixed parameters B are BLUE with the generalized least-squares method by solving the generalized normal equations $(X' D^{-1} X)\,B = X' D^{-1} y$. Note that the ratio φ must be known. In practice estimates for σ^2_U and σ^2_e are obtained using the restricted maximum likelihood method (REML). The computer packages IBM-SPSS statistics and SAS gives these REML estimates for σ^2_U and σ^2_e and insert these estimates in the ratio φ and gives with the mixed model procedure the solution of $(X' D^{-1} X)\,B = X' D^{-1} y$. These estimates of B from the generalized normal equations are in the case of using an estimate for the ratio φ officially not BLUE any more but in practice they are better than the solution of the ordinary least-squares normal equations. In the case that the REML estimate for σ^2_U is zero the mixed model procedure then automatically gives the solution of the ordinary normal equations. The REML procedure to estimate the variance components σ^2_U and σ^2_e also gives the best estimates according to the iterative minimum quadratic unbiased estimates (MINQUE) procedure in the case that U and e are not normally distributed. Hence, the REML procedure is the preferred procedure to estimate the variance components.

12.4 APPLICATIONS IN OIL PALM

12.4.1 OIL PALM BREEDING EXPERIMENTS

Oil palm breeding experiments comprise the following types:

D × D trials—these are done for the selection of mother palms for D × P hybrid seed production and of D parents for further breeding (with or without subsequent D × P/T progeny testing).

T × T/P trials—these are done for the selection of female-sterile P pollen parents for D × P hybrid seed production (with or without subsequent D × P progeny testing) and for the selection of T parents for further breeding (with or without subsequent D × T progeny testing).

With the variable presence of female-sterile and consequently more vigorous P segregants outcompeting their T and D siblings, yields based on the T and D siblings would be biased.

D × P/T trials—these are done to select the D and P parents for D × P hybrid seed production and for further breeding. Superior families and individuals are also selected from these trials for clonal (using clonal parents) hybrid seed production and clone production respectively.

G × E trials—these are done to test the adaptability of different genotypes (G, i.e., hybrids, clones) under different environmental (E, i.e., location, moisture stress, spacing, and fertilizer) conditions.

12.4.2 EXPERIMENTAL DESIGNS

The early work on oil palm field experimentation in W. Africa by Webster (1938), Ollagnier (1951), and Chapas (1961) have had great influence on subsequent oil palm breeding trials which probably run into thousands in number in the rest of the world. A typical breeding trial comprises 10–50 treatments in plots of 12–30 palms laid down in 3–6 replicates in RBCD. As emphasized earlier, in some experiments due to irregularity (topography, soil heterogeneity) of the land, discrete (square, rectangular) complete blocks with contiguous plots are hard to achieve and RCBD an inappropriate layout. The most suitable designs are resolvable incomplete block designs where a set of incomplete blocks forms a complete set of the treatments or replication.

CRD with single palm plots with very high (26) replication planted in two blocks and in lattice designs have been attempted (Rao et al. 1990). Such a design is questionable due to interplant competition effects as the neighboring palms are from different crosses. Plots with more palms from the same cross would have reduced interplant competition and are advisable.

Soh et al. (1990) studied the coefficient of variation (CV%) and relative SED (RSE = $(2\ \mathrm{CV/r})^{1/2}$ in 30 D × P/T, D × D, T × T/P trials in coastal (more uniform soils) and inland (more variable soils) areas and found that the CVs and RSEs were generally lower in the former (7.3%, 4.6%) than the latter (9.7%, 6.4%). They also reported that in a relatively genetically uniform Deli × AVROS D × P trial planted in a relatively more uniform coastal soil in CRD with single palm plots replicated five times the CV was 25.4% and in another trial with prospected open-pollinated Nigerian accessions planted in a cubic lattice design on inland soil, the CV was 11.1% when analyzed as such and 12.0% when analyzed as RCBD.

12.4.3 Plot Size and Replication

Choice of suitable plot size and number of replications in oil palm has been influenced by the early work of Webster (1938), Ollagnier (1951), and Chapas (1961) in West Africa. Based on the CVs of yield data from D × D uniformity trials, Webster (1938) recommended a plot size of 12–32 palms for single year yields while Ollagnier (1951) suggested 6–12 palms with high replication. Smith (1938) and Goh and Alwi (1988) disliked the CV approach and favored the soil heterogeneity index (*b*) approach for studying the effect of changing plot size instead. The former showed that in the calculation of variance over the whole experimental area for each of several plot sizes, the variance per unit area for yield V_x, in plots of size *x*, is related to *x* by the formula:

$$V_x = \frac{V_1}{x^b} \text{ or } \log V_x = \log V_1 - b \log x$$

where V_1 is the variance on a standard unit sized plot. b ($0 < b < 1$) measures the similarity between plots: when neighboring plots are highly correlated *b* approaches 0 and 1 when uncorrelated. The *b* value can be established using V_x values obtained from several x values using a uniformity trial. Koch and Rigney (1951) showed that b values are obtainable from experiments where a hierarchal analysis can be made, for example, split-plot design, incomplete block design. Such designs are not so commonly used in oil palm breeding trials. Linn and Binns (1984a,b) later showed that b values are obtainable from intra-block correlations in RCBD experiments:

$$\sigma_m^2 = \frac{\sigma_B^2}{\sigma_B^2 + \sigma^2}$$

where $\sigma^2{}_B$ = block variance and σ^2 = error variance

$$b = \frac{1 - \log[m - (m - 1)(1 - \sigma^2{}_m)]}{\log m}$$

where m = number of treatments

Suitable plot sizes are then computed from:

$$x^b = \frac{2(t_1 + t_2)^2 (1 - \sigma_m^2) C^2}{rd^2}$$

where t_1 and t_2 are the $\alpha = 0.05$ and $2(1-P) = 0.4$ percentage points of the t distribution; x the unit plot size; C the CV; r the number of replicates; and *d* the true difference (d%) between the two treatments to be detected (Binns 1982). Using information from seven large uniformity trials involving D × P materials and Smith's (1938) variance law, Alwi and Chan (1990) computed b values (0.3–1.0) and constructed optimum plot sizes and replications for various specified *d*, *C*, and *b* values, using the formula of Cochran and Cox (1957):

$$x^b = \frac{2(t_1 + t_2)^2 C^2}{rd^2}.$$

For example, with a *b* value 0.6, *C* of 10% and to detect a 10% *d*, a 30 palm plot replicated four times or a 21 palm plot replicated five times would be needed. The trials were sited on inland soils.

Soh et al. (1989, 1990) essentially constructed a similar table but using the formula of Linn and Binns (1984a,b), which takes blocking into account, on the two sets of coastal and inland trials discussed earlier. On coastal soils with mean values of b = 0.6 and C = 7% to detect a 10% d, 5 replicates of 25 palm plots or 6 replicates of 18 palm plots would be required. For inland soils with mean b = 0.7 and C = 9% and d = 10%, 6 replicates of 29 palm plots or 8 replicates of 19 palm plots would be needed. Soh et al. (1990) concluded that for Peninsula Malaysia coastal areas, trials laid out in RCBD with plot size 12–16 palms replicated 5–4 times, respectively, could detect differences of 10%–15% whereas to detect such differences in inland areas plot sizes of 20–25 palms replicated 6–5 times would be needed. Bonnot (1990) also discussed the considerations in choice of plot size and replications including the cost factor following Smith (1938):

$$\frac{K_1 + K_2 x}{x_b}$$

where K_1 is the cost of an experimental plot and K_2 the cost per tree.

In most oil palm breeding trials cost however has seldom been an important consideration.

12.4.3.1 Power Analysis

All the above approaches to determine suitable plot size and number of replications were essentially based on relative SED or confidence limits to detect differences between two treatments under the null hypothesis where Treatment A mean is no different from Treatment B mean (H_0: $\mu_A - \mu_B = 0$) or the alternative hypothesis that Treatment A is better than Treatment B (H_A: $\mu_A - \mu_B > 0$) using the *t* or *z*-test with $\alpha = 0.05$ probability significance level. With the 95% confidence limits or testing zero difference at 5% significance level, the *power* of the test or the probability that the test correctly rejects a false H_0 of equality between two treatments is only 50%. A *power* of 80% would be desirable and this can be achieved by increasing the sample size, for example, number progenies. In his paper "Guidelines for comparing commercial oil palm varieties. Statistical procedure employed on an existing data base" Breure (2017) used this approach to determine the number of D × P progenies per P and the number of Ps per P source (constituting a variety) needed for testing between D × P varieties.

12.5 TRIAL CONDUCT

12.5.1 Interplant and Interplot Competition

Except for perennial fruit trees which tend to be planted widely spaced and thus a single tree plot is justifiable, most crops including oil palm are planted closely. In such a situation, interplant and interplot competition warrant consideration in the choice of plot size. Single or few palm plots are likely to encounter more competition from their neighbors. With genetically variable materials, a larger plot (sample) size would be needed to characterize its mean than a more uniform material (inbreds, single cross hybrids, and clones). But if the trial comprises a variable range of such materials, larger plots would be needed and the core palms (minus the border/guard rows) used in the statistical analysis. The previous computations on optimum/suitable plot sizes did not take this into consideration and in practice most oil palm breeding experiments do not have guard rows. If the plot sizes are 12–20 palms, that is, not particularly small, sufficiently replicated (more than three) with similar genetic material treatments and interests lie in treatment differences and not means, the bias is deemed negligible. If bias is suspected, the core palms (2 in 12 palm plot size, 6 in 20 palm plot size), should be analyzed instead with the caveat that the core palms look normal for the variety or progeny.

Despite understanding the soil and plant heterogeneity and measures (experimental design, blocking) taken to minimize them in terms of uniform routine trial upkeep if not across the whole trial at least on a block/replicate basis, intra-block variability does arise. Covariance analysis (e.g., different early growth/age, differential growth and variable number of P palms per plot, missing palms), moving mean/neighboring plot covariance adjustments are sometimes performed (Pearce and Moore 1976; Mak et al. 1978; Shorter and Butler 1986). Soh et al. (1990) attempted the latter but found the improvement in efficiency minimal.

Proponents of single palm plots planted in CRD cited the following advantages (Rao et al. 1990):

- Applicability to commercial fields as seeds planted is a mixture of hybrids
- Differential replication allows maximum treatments and use of genetic materials
- Multiple palm plots and discrete blocks difficult to achieve with trial sites on rolling land
- Missing plots are easily handled
- Control of intra-block error

12.5.2 Control Treatments/Check Varieties

One of the major setbacks of D × P trials in Malaysia earlier was the lack of a consistent check variety/standard cross across all experiments to allow tracking of breeding progress. The check variety can be a D × P cross or clone. Preferably the check variety should represent the current commercial variety. As the parent palms of a cross or clone have a limited life it is good to identify and include a younger check variety for linking or straddling across trials of different generations. Clonal parents of hybrids and clone checks may persist longer as they can be recloned but reclones may deteriorate due to somaclonal variation, for example, mantling affecting yield. Both varieties/genotypes with general/wide adaptability and specific/local adaptability should be included as controls.

12.5.3 Missing/Dummy Plots

It is not uncommon not to have the full complement of replicate plots due to insufficient seeds or accidental loss of plants, "filler" or "dummy" progenies or palms may be planted to keep the integrity of the trial layout. Earlier in the statistical analysis of such trials, missing-plot techniques, for example, Kuiper–Corsten iteration, Rubin's method (Pearce et al. 1988) may be applied to estimate the treatment effects, but these methods do not provide exact standard errors. Using the GLM or generalized least squares approach these obsolete techniques would not be necessary because in the GLM the normal equations are solved directly and the standard errors of treatment differences can be correctly estimated.

12.6 DATA COLLECTION AND MANAGEMENT, STATISTICAL ANALYSES, AND SOFTWARE

12.6.1 Data Collection and Management

The following data are usually collected routinely on an individual palm basis:

Vegetative growth (yearly)—leaf (frond) number, production, weight, length, and area; stem height, increment, and girth

Yield (usually each 10-day harvesting round)—number and weight of fruit bunches

Bunch quality (sampled across each year and over recording years)—percentage fruit to bunch, fruit weight, mesocarp content, shell content, kernel content, oil content, and sometimes oil quality (fatty acid and micronutrient composition)

With these measurements, the following parameters on per palm/plot/hectare per year basis can be derived: bunch and oil yields, bunch and harvest indices, leaf area index, leaf area ratio, dry matter production, and stem growth increment (see Chapters 2 and 5 for more detailed discussions).

Most breeding trials have a recording life of 7–9 years. Based on the correlation of early yield (starting at year 3) and subsequent mature years (years 10–16) the first 4–5 years mean yield was reckoned to be a good reflection of the later yielding ability of the palm (Blaak 1965; Corley et al. 1988). Cirad prefers to extend the yield recording period to 6 years with the first 3 years as immature or pre palm competition phase yield and the next 3 years as post competition phase yield with the latter presumably more reflective of the later year yields.

Large amounts of data are generated in oil palm breeding trials and have to be maintained for long periods. Efficient data collection and management systems are thus essential. Mostly the bunch (FFB) yields of the palms are recorded on paper at each harvesting round. Nowadays it is also possible to use data loggers and barcoding for this purpose. Later these data must be compiled to get the total FFB yield per palm in a year. The FFB yields per palm must then be summed up to the yield per plot, because the plot is the experimental unit on which the analysis is based. The PC program Excel is a good program to do this compilation. Recording errors can be quite easily spotted within the Excel worksheet.

12.6.2 Statistical Analysis and Software

Most of the statistical analyses for the various experimental designs are found in the standard software packages, for example, SAS. GENSTAT, IBM-SPSS statistics; the GLM approach can handle most situations. For a fixed model the GLM approach gives the correct estimates and the correct analysis. In a mixed model, for example, treatments and replications are fixed with incomplete blocks inside the replications. The MIXED module of SPSS or the PROC MIXED of SAS gives the correct estimates of the treatment differences with the use of the recovery of the inter-block information.

12.7 CONCLUSIONS

Field experimentation embraces the entire technical process of breeding toward plant variety development; at the start to identify the traits of economic interest, in the intervening stages to incorporate the traits into a superior yielding genotype, the validation of the stable incorporation of the desirable trait and genotype, and finally the testing of the candidate variety in adaptability trials across environments and agronomic treatments before the commercial release of the new variety. Field experimentation with perennial tree crops such as the oil palm is challenging with extended time and land requirements and the attendant plant and soil heterogeneities. However, new statistical tools are constantly being developed to handle such and other issues in other crops, for example, forestry, which could be adapted to the oil palm. Lastly, it is important that the oil palm breeder is competent in both the principles and practice of field experimentation and has access to consultation from a good biometrician. A well-conducted field experiment can be a gold mine of useful information (historical and new) not only to plant breeding and genetics but also to other disciplines, for example, plant and crop physiology, plant pathology, agronomy, and soil microbiology while with a poorly conducted one, much resources, time, and opportunities will be wasted.

APPENDIX 12A.1 ALPHA (α) INCOMPLETE FACTORIAL MATING DESIGN TO ESTIMATE GCA AND SCA EFFECTS OF *DURA* AND P PARENTS

As discussed in earlier chapters, *dura* × *pisifera*/*tenera* progeny testing is an essential trial in oil palm breeding. It is from this trial where the *dura* × *pisifera*/*tenera* parents for *dura* × *pisifera* hybrid seed production and for further breeding are selected based on their estimated general combining ability (GCA, additive genotype) and specific combining ability or specific combination (SCA, nonadditive genotype) effects. Usually a large number of potential *dura*, *pisifera*/*tenera* parents are involved. A complete factorial (North Carolina Mode 1) mating design to obtain full information of the GCA of the parents and the SCA of their crosses or progenies is too cumbersome. Estimating GCA of the parents is considered more useful as GCA effects are generally larger and that the selected parents will be used to cross to other parents (besides those progeny tested within the trial) for further breeding and seed production than SCA or specific combinations. SCA effects are also usually much smaller. Hence the use of a *connected* incomplete factorial mating design minimizing the number of cross combinations is needed to estimate the parental GCA at the expense of detecting complete SCA crosses.

12A.1.1 Statistical (Additive) Model

Expected yield of the T offspring ($D_i \times P_j$), $E(y_{ij}) = \mu + \alpha_i + \beta_j + \gamma_{ij}$

where μ (constant), α_i (the GCA effect of the D female parent) + β_j (the genotypic effect of the P male parent), and γ_{ij} (SCA of the cross $D_i \times P_j$)

In order to compare the entire set of the *dura* and the *pisifera* parents on the basis of the GCA values the parents must be crossed according to a so-called *connected* crossing scheme. A crossing scheme is called connected if for each *dura* pair (D_h, D_i), there is a chain of *duras* from D_h to D_i, in which each of the adjacent links of the chain occur together with the same *pisifera*. Otherwise the crossing scheme is called *disconnected*. In the same vein, the crossing scheme is connected if for each *pisifera* pair (P_k, P_j), there is a chain of *pisiferas* from P_k to P_j, in which each of the adjacent links of the chain occur together with the same *dura*. Another way to check whether the crossing scheme is connected, is to form a two-way table of the crosses with the A *duras* as rows and the B *pisiferas* as columns. The crossing scheme is connected if we cannot split the table in separate tables by interchanging rows and columns. Example 12A.1 illustrates this with C = 8 crosses made from A = 4 *duras* and *B* = 4 *pisiferas*.

EXAMPLE 12A.1

	Pisifera			
Dura	P_1	P_2	P_3	P_4
D_1	*		*	
D_2		*		*
D_3	*		*	
D_4		*		*

From the cross of D_1 with P_1, $D_1 \times P_1$, we can make a chain to the cross of D_3 with P_1, $D_3 \times P_1$; from $D_3 \times P_1$ we can go to the cross $D_3 \times P_3$, and from this cross $D_3 \times P_3$

we can go to the cross $D_1 \times P_3$, and then we come back to the cross $D_1 \times P_1$. In this chain we have missed D_2 and D_4. Hence this crossing scheme is disconnected.

When we have rearranged the two-way table as follows (interchange P_3 with P_2 and also interchange D_3 with D_2), we see directly that there are two disconnected sets of four crosses each. The first set contains the four connected crosses

		Pisifera			
		P_1	P_3	P_2	P_4
Dura	D_1	*	*		
	D_3	*	*		
	D_2			*	*
	D_4			*	*

$D_1 \times P_1$, $D_1 \times P_3$, $D_3 \times P_1$, and $D_3 \times P_3$; the second set contains the four connected crosses $D_2 \times P_2$, $D_2 \times P_4$, $D_4 \times P_2$, and $D_4 \times P_4$. In such a disconnected crossing scheme no unbiased estimate can be made for the difference in GCA effect between, for example, D_1 and D_2 or from the difference in GCA effect between P_3 and P_4.

A more practical method of checking whether a crossing scheme is connected is to draw a chain from one cross to another following a horizontal or vertical direction only. If all the crosses are connected by one continuous chain the crossing scheme is connected.

A necessary (but not sufficient) condition to have a connected design is that C must be at least equal to the sum of the degrees of freedom of intercept, A and B, that is, $C \geq 1 + (A - 1) + (B - 1) = A + B - 1$. In the example above we have $A = 4$ and $B = 4$, so $C \geq 4 + 4 - 1 = 7$ crosses are sufficient for a connected design.

In the above mentioned example the crossing scheme is connected if, for example, the following $C = 8$ crosses were made:

EXAMPLE 12A.2

		Pisifera			
		P_1	P_2	P_3	P_4
Dura	D_1	*	*		
	D_2		*	*	
	D_3			*	*
	D_4	*			*

Here, we have 8 crosses and the crossing scheme is still connected when $C = 7$, for example, if the cross $D_4 \times P_1$ is not made.

In the past many large crossing schemes were found to be not connected, because the oil palm breeder only included previous good *dura* and *pisifera* parents. Consequently only some of the *dura* parents could be compared with each other and likewise with the *pisifera* parents.

For large crossing schemes the simplest way is to make a data file with the *dura* and *pisifera* parents of the crosses and construct a hypothetical yield y by using for example random drawings of a normal distribution with mean 20 + index of *dura* + index of *pisifera* and standard deviation 1.

Run then with a statistical package such as IBM-SPSS-Statistics (previously SPSS) or SAS, the analysis of variance with the sum of squares (SS) Type III option (SS of *dura* after correction for *pisifera* and SS of *pisifera* after correction for *dura*). If the df of *dura* = A – 1 and the df of *pisifera* = B – 1, then the crossing scheme is connected, otherwise it is disconnected.

In our Example 12A.1 the IBM-SPSS-Statistics data file is

Dura	*Pisifera*
1	1
1	3
2	2
2	4
3	1
3	3
4	2
4	4

The IBM-SPSS-Statistics syntax is

```
COMPUTE y =RV.NORMAL(20 +dura +pisifera, 1).
EXECUTE.
UNIANOVA y BY dura pisifera
 /METHOD =SSTYPE(3)
 /INTERCEPT =INCLUDE
 /CRITERIA =ALPHA(.05)
 /DESIGN =dura pisifera.
```

In the output we find in the analysis of variance table with SS Type III for *dura* df = 2 and for *pisifera* df = 2, but we have 4 *duras* and 4 *pisiferas*; for a connected design we must have for *dura* df = 4–1 = 3 and for *pisifera* df = 4–1 = 3. Hence the crossing scheme is disconnected!
The SAS syntax is as follows:

```
data crossing;
input dura pisifera @@;
datalines;
1 1 1 3 2 2 2 4 3 1 3 3 4 2 4 4;
run;
data crossing_y;
set crossing;
y =RAND('NORMAL',20 +dura +pisifera,1);
run;
proc print data =crossing_y;
run;
proc glm data=crossing_y;
class dura pisifera;
model y = dura pisifera;
run;
```

In the SAS output we find in the analysis of variance table with SS Type III for *dura* df = 2 instead of 4 – 1 = 3 and for *pisifera* df = 2 instead of 4 – 1 = 3, hence the crossing scheme is disconnected!

Now we will discuss the construction of a good crossing design when we have A and B parents and we want to use C crosses in an incomplete factorial connected scheme where $A + B - 1 \leq C \leq A \times B$. The choice between several connected mating designs can best be tested on the standard error of the estimator for the difference in the GCA value of all the *dura* pairs and of the *pisifera* pairs. The standard error of the estimator for the difference in the GCA value between two *dura* parents D_i and D_j is $(S_{Dij}) \times \sigma$, or between two *pisifera* parents P_i and P_j is $(S_{Pij}) \times \sigma$, where σ is the residual standard deviation and the value of S_{Dij} and S_{Pij} depends solely on the mating scheme. The value of σ depends on the studied trait (e.g., yield), the variation between the plots in the experimental field, and the plot size.

For complete crossing schemes (as a complete factorial) with A *dura* and B *pisifera* parents, where each cross occurs on *r* plots, the standard error of the estimator of the difference between the GCA values of the *dura* parents is the same for all pairs of *dura* and S_{Dij} is $\sqrt{[2/(B \times r)]}$; also the standard error of the estimator of the difference between the GCA values of the *pisifera* parents is the same for all pairs of *pisiferas* and S_{Pij} is $\sqrt{[2/(A \times r)]}$. For incomplete mating designs the standard error of the estimator of the differences in GCA values varies across the parents. The quality of such mating designs can be measured by the average and range of the standard errors of the estimator of the differences between the GCA values of a pair of *dura* parents or of a pair of *pisifera* parents. As shown above, such quality evaluation can solely be based on the coefficients S_{Dij} and S_{Pij}.

We will demonstrate for an incomplete connected factorial how the coefficients S_{Dij} and S_{Pij} can be calculated. The IBM-SPSS-Statistics data file for Example 12A.2 after the generation of fancy yield ***y*** by using for example random drawings of a normal distribution with mean 20 + index of *dura* + index of *pisifera* and standard deviation 1 is

Dura	*Pisifera*	y
1	1	21.09
1	2	22.96
2	2	23.72
2	3	24.84
3	3	24.14
3	4	25.23
4	4	27.68
4	1	26.63

The IBM-SPSS-Statistics syntax is as follows:

```
COMPUTE y = RV.NORMAL(20 +dura +pisifera,1).
EXECUTE.
UNIANOVA y BY dura pisifera
 /METHOD +SSTYPE(3)
```

```
/INTERCEPT = INCLUDE
/EMMEANS = TABLES(dura) COMPARE ADJ(LSD)
/EMMEANS = TABLES(pisifera) COMPARE ADJ(LSD)
/PRINT = PARAMETER
/CRITERIA = ALPHA(.05)
/DESIGN = dura pisifera.
```

From the output we find the mean square error = SS(Error)/df(Error) = 0.99982; this MS(Error) is an estimate for σ^2. In the table of pairwise comparisons of the estimated marginal means (EMMEANS) we find the SED between *dura* 1 and 2 is 1.22464; hence $SD_{12}=1.22464/\sqrt{0.99982} = 1.225$. In the same way we find $S_{D13} = 1.414$; $S_{D14} = 1.225$; $S_{D23} = 1.225$; $S_{D24} = 1.414$; $S_{D34} = 1.225$; $S_{P12} = 1.225$; $S_{P13} = 1.414$; $S_{P14} = 1.225$; $S_{P23} = 1.225$; $S_{P24} = 1.414$; $S_{P34} = 1.225$.

With the following SAS syntax the *y*-values, after the random drawing from the normal distribution, are for $D_1P_1 = 22.5714$; $D_1P_2 = 21.9578$; $D_2P_2 = 22.3086$; $D_2P_3 = 26.2951$; $D_3P_3 = 24.1570$; $D_3P_4 = 26.5849$; $D_4P_4 = 26.7996$; $D_4P_1 = 24.0734$.

```
data crossing;
input dura pisifera @@;
datalines;
1 1 1 2 2 2 2 3 3 3 3 4 4 4 4 1
;
run;
data crossing_y;
set crossing;
y = RAND('NORMSL',20 + dura + pisifera, 1);
run;
proc print data = crossing_y;
run;
proc glm data = crossing_y;
class dura pisifera;
model y = dura pisifera;
LSMEANS dura / Pdiff = all;
LSMEANS pisifera/Pdiff = all;
estimate 'D1-D2' dura 1 -1 0 0;
estimate 'D1-D3' dura 1 0 -1 0;
estimate 'D1-D4' dura 1 0 0 -1;
estimate 'D2-D3' dura 0 1 -1 0;
estimate 'D2-D4' dura 0 1 0 -1;
estimate 'D3-D4' dura 0 0 1 -1;
estimate 'P1-P2' pisifera 1 -1 0 0;
estimate 'P1-P3' pisifera 1 0 -1 0;
estimate 'P1-P4' pisifera 1 0 0 -1;
estimate 'P2-P3' pisifera 0 1 -1 0;
estimate 'P2-P4' pisifera 0 1 0 -1;
estimate 'P3-P4' pisifera 0 0 1 -1;
run;
```

The estimate for σ^2 is MS(Error) = 1.18151552 and the standard error for D_1–D_2 is 1.33126754 hence $S_{D12} = 1.33126754 / \sqrt{1.18151552} = 1.225$. In the same way we can calculate the other S_{Dij} and S_{Pij} values. Of course we get the same S_{Dij} and S_{Pij} values as before with IBM-SPSS-Statistics.

To find a good mating design one can search for balanced or partially balanced incomplete mating designs. For such incomplete mating designs one can use the incomplete block designs. In such incomplete block designs there must be compared v varieties (or treatments) in blocks of sizes of k plots, where the block size k < v. Well-known incomplete block designs are lattices where $v = k \times k$ or rectangular lattices where $v = k \times (k + 1)$ (Cochran and Cox 1957). To extend the possibilities for v unequal to $k \times k$ or $k \times (k + 1)$ are the alpha-designs (see Patterson et al. 1978). A computer program *CycDesigN* is available to generate incomplete block designs such as *alpha*-designs and cyclic designs (http://www.vsni.co.uk/software/cycdesign/). To use such an incomplete block design the role of treatments is played by the *dura* and the role of the incomplete blocks by the *pisifera*. So we must look for incomplete block designs with A treatments and B blocks. The block size k is then chosen as C/B, where C is the number of crosses used. If there is no incomplete block design which fits the requirements, we can always start from a smaller design and add some extra treatments (= *dura*) to the blocks (= *pisifera*).

Illustrated here are some mating designs involving C = 40 crosses, A = 20 *duras*, and B = 10 *pisiferas*. In these designs each *dura* must be crossed with 2 *pisiferas*; furthermore each *pisifera* must be crossed with 4 *duras*. Two designs (I and II) were solely chosen intuitively on the basis of symmetry by two oil palm breeders and the last Design III was chosen by the statistician as an *alpha*-design, with $v = 20$ treatments (= *dura*), $k = 40/10 = 4$ as block size, $b = 10$ blocks (= *pisifera*), $r = 2$ replications, where the first replication consists of blocks 1–5 and the second replication consist of blocks 6–10.

Design I										
	Pisifera									
Dura	1	2	3	4	5	6	7	8	9	10
1	*									*
2	*									*
3	*	*								
4		*							*	
5		*	*							
6			*					*		
7			*	*						
8				*			*			
9				*	*					
10					*	*				
11					*	*				
12					*	*				
13						*	*			
14				*			*			
15							*	*		
16			*					*		
17								*	*	
18		*							*	
19									*	*
20	*									*

Design II										
	Pisifera									
Dura	1	2	3	4	5	6	7	8	9	10
1	*									*
2	*					*				
3	*	*								
4		*					*			
5		*	*							
6			*					*		
7			*	*						
8				*					*	
9				*	*					
10					*					*
11					*	*				
12	*					*				
13						*	*			
14		*					*			
15							*	*		
16			*					*		
17								*	*	
18				*					*	
19									*	*
20					*					*

Design III										
	Pisifera									
Dura	1	2	3	4	5	6	7	8	9	10
1	*					*				
2		*					*			
3			*					*		
4				*					*	
5					*					*
6	*									*
7		*				*				
8			*				*			
9				*				*		
10					*				*	
11	*								*	
12		*								*
13			*			*				
14				*			*			
15					*			*		
16	*							*		
17		*							*	
18			*							*
19				*		*				
20					*		*			

It can be directly seen that all these three mating designs are connected. In the following table the minimum, maximum, and average of the coefficients S_{Dij} and S_{Pij} for the standard errors of the estimators of the difference between GCA values of pairs of *dura* and *pisifera* parents, for the three designs, are given.

Mating Design	S_{Dij} for Pairs of *Dura*			S_{Pij} for Pairs of *Pisifera*		
	Min	Max	Average	Min	Max	Average
I	1.000	2.236	1.561	0.765	2.072	1.417
II	1.000	1.483	1.313	0.841	1.250	1.125
III	1.125	1.291	1.214	0.949	1.080	1.001

From the table it is clear that Design III (*alpha*-design) gives the smallest average values for S_{Dij} and S_{Pij} for the *dura* and the *pisifera* pairs, and the smallest ranges (max–min) for S_{Dij} and S_{Pij}, indicating that the *alpha*-design which always gives a *connected mating design* should be the design of choice.

APPENDIX 12B.1 FIELD EXPERIMENTAL DESIGN FOR OIL PALM BREEDING TRIALS

It is often difficult to find a sufficiently large homogeneous area to test C number of *tenera* progenies from *dura* and *pisifera* parents in a CRD. Even when we use RCBD where the C crosses occur only once, the experimental conditions for the plots in the block are quite different. This is due to the size of the plots; often comprising six rows of six palms, where the palms are laid down at the corners of an equilateral triangle with 9 m sides. For the yielding capacity of a cross the yield of the 16 (4 × 4) inner palms is recorded. Hence RCBD, with block size $k = C$, often cannot be used if $C > 10$. To circumvent the heterogeneous growing conditions in an experimental field one can use a randomized incomplete block design. With a smaller block size $k < C$ homogeneous parts of the experimental field can be found. The well-known incomplete block designs, such as balanced incomplete block designs often need too much replication; in most oil palm breeding trials the number of replications r is usually 3. Earlier incomplete block designs such as lattices (for the case $C = k \times k$) and rectangular lattices (for the case $C = k \times (k + 1)$) have been used (Cochran and Cox 1957). Often the number of tested progenies C does not fit into lattices or rectangular lattices. The *alpha*-designs introduced by Patterson et al. (1978) can handle many combinations of progenies C and block sizes k. The *alpha*-designs are available for many (r, k, s) combinations with r the number of replicates, k the block size, s the number of blocks per replicate; the number of blocks $b = rs$, and the number of treatments is $t = ks$. The *CycDesigN* computer program can generate incomplete block designs as *alpha*-designs and *cyclic* designs and is available as indicated earlier. The breeder often wants to use *resolvable* incomplete block designs where the design can be divided into r groups (= *replications*) such that each group contains each of the C crosses exactly once. The program CyCDesigN

(2014) can do this.All these above-mentioned designs are *connected*. A block design is called connected as explained for the mating designs where the role of *pisifera* is block and the role of *dura* is progeny. In a connected incomplete block design one can estimate all differences between the progenies. But later on we also want to estimate from the yield of the C *tenera* progenies, the general combining abilities (GCA) of the A *dura* and the B *pisifera* parents. Therefore, we must use a *connected* crossing design for the *dura* and the *pisifera*.

EXAMPLE 12B.1

Let us consider the case that we have made C = 10 connected *tenera* crosses T_1, T_2, ..., T_{10}, derived from A = 5 *dura* parents and B = 5 *pisifera* parents. In the following two-way table the crossing scheme is given; a dot (•) indicates a cross which has not been made.

The palm plot consists of six rows of six palms. For the yielding capacity of a cross the yield of the 16 inner palms is recorded. Suppose that the experimental field is very heterogeneous, and that we can only find homogeneous growing conditions (blocks) of five adjacent plots.

	Pisifera				
Dura	**P_1**	**P_2**	**P_3**	**P_4**	**P_5**
D_1	T_1	•	•	•	T_{10}
D_2	T_2	T_3	•	•	•
D_3	•	T_4	T_5	•	•
D_4	•	•	T_6	T_7	•
D_5	•	•	•	T_8	T_9

A resolvable *alpha*-design with block size $k = 5$ and with $r = 4$ replications is used. The index of the *tenera* crosses T_i are given by the program CycDesigN in the randomized resolvable *alpha*-design below.

Block			**Rep 1**		
1	9	4	1	6	2
2	5	7	8	10	3
			Rep 2		
1	5	4	3	8	1
2	6	2	7	9	10
			Rep 3		
1	10	4	9	6	3
2	7	8	1	5	2
			Rep 4		
1	9	1	3	5	7
2	10	4	8	6	2

12B.1.1 Statistical Analysis

After we have laid out the design of Example 12B.1 in the field, we gathered the yield **y** in ton/ha after a year. We made then the following data file of the results from Example 12B.1 for IBM-SPSS-Statistics, where we used consecutive block numbers 1–8 for the blocks in the replications.

Rep	Block	*Tenera*	*Dura*	*Pisifera*	y
1	1	9	5	5	20.10
1	1	4	3	2	17.50
1	1	1	1	1	15.70
1	1	6	4	3	18.10
1	1	2	2	1	14.60
1	2	5	3	3	16.20
1	2	7	4	4	18.70
1	2	8	5	4	21.10
1	2	10	1	5	18.80
1	2	3	2	2	16.70
2	3	5	3	3	17.20
2	3	4	3	2	15.80
2	3	3	2	2	17.90
2	3	8	5	4	18.10
2	3	1	1	1	15.10
2	4	6	4	3	16.70
2	4	2	2	1	15.00
2	4	7	4	4	20.00
2	4	9	5	5	21.70
2	4	10	1	5	18.60
3	5	10	1	5	16.00
3	5	4	3	2	15.60
3	5	9	5	5	21.50
3	5	6	4	3	17.20
3	5	3	2	2	16.00
3	6	7	4	4	18.40
3	6	8	5	4	17.30
3	6	1	1	1	14.70
3	6	5	3	3	16.70
3	6	2	2	1	13.50
4	7	9	5	5	20.00
4	7	1	1	1	16.00
4	7	3	2	2	14.60
4	7	5	3	3	16.70
4	7	7	4	4	17.20
4	8	10	1	5	17.20
4	8	4	3	2	14.90
4	8	8	5	4	19.10
4	8	6	4	3	18.30
4	8	2	2	1	13.80

The IBM-SPSS-Statistics syntax to find the GCA values for the *dura* and the *pisifera* and the adjusted *tenera* means (EMMEANS = estimated marginal means) is as follows:

```
Title 'Example 3, blocks random with module MIXED'.
MIXED y BY dura pisifera rep
 /CRITERIA =CIN(95) MXITER(100) MXSTEP(10) SCORING(1)
SINGULAR(0.000000000001) HCONVERGE(0, ABSOLUTE) LCONVERGE(0,
ABSOLUTE) PCONVERGE(0.000001, ABSOLUTE)
 /FIXED = rep dura pisifera | SSTYPE(3)
 /METHOD =REML
 /PRINT =SOLUTION TESTCOV
 /RANDOM =INTERCEPT | SUBJECT(block) COVTYPE(VC)
 /EMMEANS =TABLES(dura) COMPARE ADJ(LSD)
 /EMMEANS =TABLES(pisifera) COMPARE ADJ(LSD).
Title 'Example 3, tenera, blocks random with module MIXED'.
MIXED y BY tenera rep
 /CRITERIA =CIN(95) MXITER(100) MXSTEP(10) SCORING(1)
SINGULAR(0.000000000001) HCONVERGE(0, ABSOLUTE)
LCONVERGE(0, ABSOLUTE) PCONVERGE(0.000001, ABSOLUTE)
 /FIXED = rep tenera | SSTYPE(3)
 /METHOD =REML
 /PRINT =SOLUTION TESTCOV
 /RANDOM =INTERCEPT | SUBJECT(block) COVTYPE(VC)
 /EMMEANS =TABLES(tenera) COMPARE ADJ(LSD).
```

The SAS syntax to find the GCA values for the *dura* and the *pisifera* and the adjusted *tenera* means (LSMEANS = least square means; this LSMEANS is the same as what SPSS calls EMMEANS), is as follows:

```
data example3;
input rep block tenera dura pisifera y;
datalines;
```

1	1	9	5	5	20.10
1	1	4	3	2	17.50
1	1	1	1	1	15.70
1	1	6	4	3	18.10
1	1	2	2	1	14.60
1	2	5	3	3	16.20
1	2	7	4	4	18.70
1	2	8	5	4	21.10
1	2	10	1	5	18.80
1	2	3	2	2	16.70
2	3	5	3	3	17.20
2	3	4	3	2	15.80
2	3	3	2	2	17.90

(Continued)

2	3	8	5	4	18.10
2	3	1	1	1	15.10
2	4	6	4	3	16.70
2	4	2	2	1	15.00
2	4	7	4	4	20.00
2	4	9	5	5	21.70
2	4	10	1	5	18.60
3	5	10	1	5	16.00
3	5	4	3	2	15.60
3	5	9	5	5	21.50
3	5	6	4	3	17.20
3	5	3	2	2	16.00
3	6	7	4	4	18.40
3	6	8	5	4	17.30
3	6	1	1	1	14.70
3	6	5	3	3	16.70
3	6	2	2	1	13.50
4	7	9	5	5	20.00
4	7	1	1	1	16.00
4	7	3	2	2	14.60
4	7	5	3	3	16.70
4	7	7	4	4	17.20
4	8	10	1	5	17.20
4	8	4	3	2	14.90
4	8	8	5	4	19.10
4	8	6	4	3	18.30
4	8	2	2	1	13.80

```
;
run;

proc print data = example3;
run;
title 'Example 3;proc Mixed, blocks random';
proc mixed method =REML data = example3;
class dura pisifera rep block;
model y = rep dura pisifera/ddfm =satterthwaite;
random block/type =VC;
lsmeans dura/diff =all;
lsmeans pisifera / diff =all;
run;
title 'Example 3;proc Mixed, tenera, blocks random';
proc mixed method =REML data = example3;
class tenera rep block;
model y = rep tenera/ddfm=satterthwaite;
random block/type =VC;
lsmeans tenera/diff =all;
run;
```

From the output we find that the *dura* is significant (P-value = 0.005) and that the *pisifera* is significant (P-value = 0.001) and the *tenera* is significant (P-value = 0.000). The LSMEANS (=EMMEANS) are collected in the following EXCEL worksheet:

Example 3			**LS Mean**			**Additive**			**LS Mean**		**LS Mean**
Tenera	***Dura***	***Pisifera***	***Tenera***	***Dura***	***Pisifera***	**Mean D + P**	**SCA**	***Dura***	**GCA *Dura***	***Psifera***	**GCA *Pisifera***
1	1	1	15.375	16.533	15.703	15.0284	0.3466	1	16.533	1	15.703
2	2	1	14.225	16.077	15.703	14.5724	–0.3474	2	16.077	2	17.083
3	2	2	16.3	16.077	17.083	15.9524	0.3476	3	16.422	3	17.138
4	3	2	15.95	16.422	17.083	16.2974	–0.3474	4	17.993	4	17.442
5	3	3	16.7	16.422	17.138	16.3524	0.3476	5	19.013	5	18.672
6	4	3	17.575	17.993	17.138	17.9234	–0.3484				
7	4	4	18.575	17.993	17.442	18.2274	0.3476		17.2076		17.2076
8	5	4	18.9	19.013	17.442	19.2474	–0.3474		Mean		Mean
9	5	5	20.825	19.013	18.672	20.4774	0.3476				
10	1	5	17.65	16.533	18.672	17.9974	–0.3474				

The first *dura* mean in column 5 (for *dura* 1) is done with Excel-function VLOOKUP(B3,$I$3:$J$7;2;1).

The first *pisifera* mean in column 5 (for *pisifera* 1) is done with Excel-function VLOOKUP(C3,$K$3:$L$7;2;1). Then this has been done for all the other *tenera*. The additive mean of D + P (= *dura* + *pisifera*) for *tenera* = LSMEAN GCA *dura* + LSMEAN *pisifera* – mean GCA; for *tenera* 1 we get as additive mean = 16.533 + 15.703–17.2076 = 15.0284. The specific combining ability (SCA) of *tenera* is calculated as LSMEAN (*tenera*)—additive mean; the SCA for *tenera* 1 is 15.375–15.0284 = 0.13466. Then this has been done for all the other *teneras*. For the use of statistical selection procedures to select the best set of *dura* and *pisifera* parents with the indifference zone approach of selection of Bechhofer (1954) or the subset selection procedure of Gupta (1965) see Breure and Verdooren (1995) and Laan et al. (1989).

REFERENCES

Alwi, A. and Chan, K.W. 1990. Experimental plot size and shapes for oil palm experiments. In: Soh A.C., Rajanaidu N., and Basri, M.N.H. (eds). *Applications of Statistics to Perennial Tree Crops*. Palm Oil Research Institute of Malaysia. Kuala Lumpur. pp 51–61.

Bechhofer, R.E. 1954. A single-sample multiple decision procedure for ranking means of normal populations with known variances. *The Annals of Mathematical Statistics* 25: 16–39.

Binns, M.E. 1982. The choice of plot size in randomised block experiments. *Journal of the American Society of Horticultural Science* 107: 17–19.

Blaak, G. 1965. Breeding and inheritance in the oil palm (*Elaeis guineensis* Jacq.). Part III. Yield selection and inheritance. *Journal of the West African Institue of Oil Palm Research* 4: 262–283.

Bonnot, F. 1990. Thoughts on the size of experimental plots for oil palm. In: Soh, A.C., Rajanaidu, N., and Basri, M.N.H. (eds). *Applications of Statistics to Perennial Tree Crops*. Palm Oil Research Institute of Malaysia. Kuala Lumpur. pp 62–69.

Breure, C.J. 2017. Guidelines for comparing commercial oil palm varieties. *Journal of Oil Palm Research* 29: 23–34.

Breure, C.J. and Verdooren, L.R. 1995. *Guidelines for testing and selecting parent palms in oil palm, practical aspects and statistical methods*. ASD OIL PAPERS number 9, Occasional publication of ASD de Costa Rica S.A.

Chapas, L.C. 1961. Plot size and reduction of variability in oil palm experiments. *Empire Journal of Experimental Agriculture* 29: 212–224.

Cochran, W.G. and Cox, G.M. 1957. *Experimental Designs*. 2nd edn. John Wiley and Sons, New York, London, Sydney.

Corley, R.H.V., Lee, C.H., Law, I.H., and Cundall, E. 1988. Field testing of oil palm clones. In: Halim Hassan, A. et al. (ed). *Proceeding of 1987 International Oil Palm Conference on Progress and Prospects*. Palm Oil Research Insitute of Malaysia. Kuala Lumpur. pp 173–185.

CycDesigN. 2014. A Package for the Computer Generation of Experimental Designs, http://www.vsni.co.uk/software/cycdesign/.

Goh, K.H. and Alwi, A. 1988. Uniformity trials with oil palms in Malaysia. In: Hj, A.H. Chew, P.S. Wood, B.J., and Pushparajah, E. (eds). *Proceedings of 1987 International Oil Palm/Palm Oil Conference on Agriculture*. Palm Oil Research Institute of Malaysia. Kuala Lumpur. pp 677–684.

Gupta, S.S. 1965. On some multiple decision (selection and ranking) rules. *Technometrics* 7: 225–245.

Koch, E.J. and Rigney, J.A. 1951. A method of estimating optimum plot size from experimental data. *Agronomy Journal* 43: 17–21.

Laan, P. van der and Verdooren, L.R. 1989. Selection of populations. An overview and some recent results. *Biometrical Journal* 31: 383–420.

Linn, C.S. and Binns M.R. 1984a. The precision of cultivar trials within eastern cooperative tests. *Canadian Journal of Plant Science* 64: 587–591.

Linn, C.S. and Binns M.R. 1984b. Working rules for determining the plot size and numbers of plots per block in field experiments. *Journal of Agricultural Science* 103: 11–15.

Mak, C., Harvey, B.L., and Berdahl J.D. 1978. An evaluation of control plots and moving means for error control in barley nurseries. *Crop Science* 18: 870–873.

Ollagnier, M. 1951. Forme, Dimension des Parcelles et Nombre de Repetitions dans le Essai Culturaux sur Arachide et sur Palme et a Huile. *Oleagineux* 6: 707–710.

Patterson, H.D., Williams, E.R., and Hunter, E.A. 1978. Block designs for variety trials. *Journal of Agricultural Science* 90: 395–400.

Pearce, S.C., Clarke, G.M., Dyke, G.V., and Kempson, R.E. 1988. *A Manual of Crop Experimentation*. Charles Griffin and Company Ltd., London and Oxford University Press, New York.

Pearce, S.C. and Moore, C.S. 1976. Reduction of experimental error in perennial crops using adjustment by neighbouring plots. *Experimental Agriculture* 12: 267–272.

Rao, V., Donough, C., Rajanaidu, N., and Chow, C.S. 1990. Completely randomized design for oil palm breeding trials: The precision of oil palm breeding experiments in Malaysia. In: Soh, A.C., Rajanaidu, N., and Basri, M.N.H. (eds). *Applications of Statistics to Perennial Tree Crops*. Palm Oil Research Institute of Malaysia. Kuala Lumpur. pp 21–27.

Shorter, R. and Butler, D. 1986. Moving mean covariance adjustments on error and genetic variance estimates and selection of superior lines in peanuts (*Arachis hypogea* L.). *Euphytica* 35: 185–192.

Smith, H.E. 1938. An empirical law describing heterogeneity in yields of agricultural crops. *Journal of Agricultural Science* 28: 1–23.

Soh, A.C., Lee C.H., and Chin C.W. 1989. Suitable plot sizes and replications in oil palm breeding experiments. *SABRAO Journal* 21: 143–150.

Soh, A.C., Lee, C.H., Yong Y.Y., Chin, C.W., Tan Y.P., Rajanaidu, N., and Phua, P.K. 1990. The precision of oil palm breeding experiments in Malaysia. In: Soh, A.C., Rajanaidu, N., and Basri, M.N.H. (eds). *Applications of Statistics to Perennial Tree Crops.* Palm Oil Research Institute of Malaysia. Kuala Lumpur. pp 41–50.

Webster, C.C. 1938. A note on uniformity trial with oil palm. *Tropical Agriculture* 16: 15–19.

13 Future Prospects

Aik Chin Soh, Sean Mayes, Jeremy Roberts, Tasren Mahamooth, Denis J. Murphy, Sue Walker, Asha S. Karunaratne, Erik Murchie, John Foulkes, Marcel de Raissac, Raphael Perez, Denis Fabre, Kah Joo Goh, Chin Kooi Ong, and Hereward Corley

CONTENTS

13.1 Introductory Overview (*Aik Chin Soh, Sean Mayes, and Jeremy Roberts*)....354
13.2 Breeding for Sustainability and Climate Change Effects in Oil Palm (*Aik Chin Soh and Tasren Mahamooth*)....355
13.2.1 Sustainability and Climate Change Effects....355
13.2.2 Environmental Sustainability....356
13.2.3 Breeding for Resource-Use Efficiency....356
13.2.3.1 Nutrients....356
13.2.3.2 Water....357
13.2.3.3 Light....359
13.2.4 Breeding for Resilience or Adaptation to Environmental Stress....360
13.2.4.1 Extreme Weather....360
13.2.4.2 Temperature....360
13.2.4.3 Strong Winds....360
13.2.4.4 Poor Soils....361
13.2.5 Breeding for Biodiversity....361
13.2.6 Breeding Approaches....362
13.3 Transgenic Oil Palm (*Denis Murphy*)....363
13.3.1 Transgenic Crops....363
13.3.2 Why Develop Transgenic Oil Palm?....364
13.3.3 Target Traits: Fatty Acid Modification and Other Traits....365
13.3.4 Historical Approaches to Oil Palm Transformation....366
13.3.5 Current Technical Challenges....367
13.3.6 New Technologies for Genome Manipulation: CRISPR....369
13.3.7 Can Transgenic Oil Palm Varieties Achieve Commercial Success?....370
13.3.8 Conclusions and Future Prospects....371
13.4 Oil Palm Yield Modeling (*Sue Walker and Asha S. Karunaratne*)....372
13.4.1 Crop Models....372
13.4.2 Oil Palm Crop Models....373
13.4.3 Explanation about APSIM: Oil Palm Module....374
13.4.4 An Example: Model Results at Rompin, Malaysia (1997–2014)....375
13.4.5 Future Use of Crop Models to Formulate Breeding Strategies....377

13.5 New Methods for Crop Phenotyping (*Erik Murchie and John Foulkes*)...... 378
13.5.1 Definition of Phenotyping, Its Application and Current Use 378
13.5.2 Types of Phenotyping and the State of Play 379
13.5.3 Canopy (Shoot) Phenotyping Methods.. 380
13.5.4 Root Phenotyping Methods 383
13.5.5 Moving Forward: An Affordable Technological Cornucopia?....... 385
13.6 New Challenges in Oil Palm Phenotyping in Relation to Climate (*Marcel de Raissac, Raphael Perez, and Denis Fabre*) 386
13.6.1 Background.. 386
13.6.2 Plant Architecture and Light Interception.................................... 388
13.6.3 Photosynthesis Phenotyping for Drought Tolerance Using the Fluorescence Transient Technique... 391
13.7 Prospects of Integrating Beneficial Plant-Microbial Interactions in Oil Palm Improvement Strategies (*Tasren Mahamooth and Kah Joo Goh*) 394
13.8 Transitioning the Oil Palm Plantation to a More Diversified Cropping System (*Chin Kooi Ong*).. 396
13.8.1 Oil Palm Monoculture.. 396
13.8.2 Lessons from Agroforestry... 397
13.9 Summing Up: An Ideotype for Yield and Sustainability (*Hereward Corley*).. 397
13.9.1 Oil Palm Sustainability.. 397
13.9.2 Yield Potential 398
13.9.3 Dry Matter Production.. 399
13.9.4 Dry Matter Partitioning 400
13.9.5 Bunch Composition and FFB Yield... 401
13.9.6 Disease and Stress Tolerance.. 401
13.9.7 Other Characteristics 402
13.9.7.1 Long Bunch Stalks 402
13.9.7.2 Low Lipase 402
13.9.7.3 Suitability for Machine Harvesting...................................... 402
13.9.7.4 Fruit Color 402
13.9.7.5 Yield Stability.. 403
13.9.7.6 Fertilizer Efficiency... 404
13.9.7.7 Oil Composition 404
13.9.8 Crossing Programs.. 404
13.9.9 Time Scale 405
13.9.10 Conclusions.. 405
References... 406

13.1 INTRODUCTORY OVERVIEW

Aik Chin Soh, Sean Mayes, and Jeremy Roberts

For the next few decades, world agriculture has to grapple with the great challenge of not only feeding but also contributing to the societal well-being of the rising world population which is expected to reach 10 billion by 2050. To exacerbate the issue, at the December 2015 COP21 (http://www.c2es.org/international/negotiations/cop21-paris/summary), the world has accepted that climate change and global warming is here to stay: due to the rapid rise in atmospheric CO_2 attributed to largely an anthropogenic cause; and has agreed to implement mitigation measures to reduce the temperature rise to less than 2°C for the next few decades. This is unlikely to be met and a 3–7°C rise would be more likely. How would oil palm respond to this? In terms of meeting the increased per capita dietary fat requirement, the oil palm is poised to take center stage being the highest yielding oil crop (5–10 times more oil per hectare per year than others, e.g., soybean) and consequently requiring a far smaller land area to produce the same yield as other species. However, increased yield in existing production areas is challenging due to continuous mono-cropping related issues while expansion into new forest and marginal (e.g., peat) land faces increasing greenhouse gas (GHG) emission and biodiversity loss concerns. In this chapter, the various issues involved and the R&D efforts needed are discussed and the progress toward achieving the goals that have been set is evaluated.

13.2 BREEDING FOR SUSTAINABILITY AND CLIMATE CHANGE EFFECTS IN OIL PALM

Aik Chin Soh and Tasren Mahamooth

13.2.1 Sustainability and Climate Change Effects

Agricultural sustainability is supported by the three interlocking pillars of profit, planet, and people. The profit sustainability and its consequent society well-being benefits have been so good that oil palm growers from the largest corporate plantation to the smallest holding are continuously replanting the same piece of land (up to the third and fourth cycle) with monocrop oil palm, despite the attendant risks of reduced yield due to soil degradation and increased pest and disease damage. Likewise, other tropical developing countries are joining in the act especially with investments resulting from the expansion efforts of established corporate plantation companies. Such expanded oil palm plantings will inadvertently venture into agro-ecological zones suboptimal for oil palm.

Mitigation and adaptation to climate change effects is essentially environmental sustainability. The oil palm is a net sequester of carbon except when replanting forest and peat land and during forest fires (www.carbonstockstudy.com, 2016). However, excess applied nitrogen fertilizers escape from the oil palm plantation and are volatized to give nitrous oxide and ammonia. These components, together with ozone derived from the palm's emission of volatile organic compounds, are potent GHGs. The Green Revolution which eradicated famine and improved societal well-being in many developing countries was derived from the breeding of super yielding wheat, rice, and maize varieties that were responsive to high fertilizer, irrigation, and chemical pesticide inputs made possible by the prevailing cheap fossil fuel-based energy. These cultivars are planted intensively and extensively as monocultures for ease of mechanized harvesting and cultivation. Their impacts on the environment, biodiversity, and ecosystem services were not of prime concern then, as world hunger was the major focus of attention. Oil palm breeding essentially followed this lead. The current scenario is one of high energy and fertilizer prices and an expected increase in the frequency of extreme weather events due to climate change (IPCC Climate Change 2014 Synthesis Report). Oil palms that are high yielding, more efficient in resource use (e.g., nutrients, water, light, soil), resistant to *Ganoderma* and other potential pests and diseases, resilient to extreme weather stresses, and can be planted in alternative cropping systems that encourage biodiversity should be pursued.

This ensuing discussion focuses on breeding for environmental sustainability, as breeding for profit sustainability in terms of improved yield potential, ease of harvesting, and resistance to *Ganoderma* have been deliberated earlier (see Chapter 5).

13.2.2 Environmental Sustainability

During the course of crop domestication and subsequent intensification beginning with the development of high yielding, uniform, and high agricultural input responsive varieties, many of the adaptation and environmental stress resilience traits available in the wild crop progenitors may have been lost. Such traits could be useful with crop expansion into marginal areas and with climate change (Heslop-Harrison 2012; Kole et al. 2015).

13.2.3 Breeding for Resource-Use Efficiency

13.2.3.1 Nutrients

Current oil palm cultivars have been bred for good response to high fertilizer inputs, particularly N fertilizer. N fertilizer cost has increased tremendously due to rising fossil fuel energy cost. Hirel et al. (2011) reported that more than 50% of the N fertilizer applied to annual crops could have leaked out of the system. These losses are through volatilization (as ammonia and nitrous oxide, which are GHGs) and through leaching and runoff causing eutrophication of the rivers and contaminating drinking water, that is, polluting the environment. Studies on oil palm, however, indicated only around 10% N lost, mainly through surface runoff (Maene et al. 1979; Kee and Chew 1996; Bah et al. 2014), and minimal contamination of the waterways has been observed. This could be attributed to judicious applications of site-specific fertilizer recommendations, multi-split fertilizer doses, and maintenance of vegetation (cover crops, soft weeds) in the palm inter-rows. This, however, could change with the current and future scenarios of frequent extreme weather events and scarcity of skilled labor for efficient fertilizer application.

Nitrogen use efficiency (NUE) can be defined as the yield of crop per unit of available N in the soil (including residual soil N present and fertilizer N applied). This NUE comprises two separate processes: uptake efficiency (NupE; the ability of a plant to remove N from soil as nitrate and ammonium ions) and utilization efficiency (NutE; the ability to use N to produce the crop). Plant uptake efficiency is primarily a set of root characteristics, principally architectural (density and depth of roots), but is also related to function (uptake and translocation of resources). Prolific shallow roots are required to capture applied fertilizer, particularly immobile species such as phosphate, and deeper roots are likely to be important for accessing water and deeper N reserves.

The second key trait involving efficient production of useable biomass depends on canopy function (photosynthesis), architecture, longevity, and efficient remobilization of nutrients from discarded or non-harvested material to crop biomass (Hirel et al. 2007). Palms that make more efficient use of N fertilizer would be desirable. NutE would need to be good in marginal areas to make full use of residual nutrient available in the soil. NupE would be more desirable for good areas to improve uptake and reduce fertilizer usage and wastage. These traits were suggested by Moll et al. (1981) and Bertin and Gallais (2001) in their studies on genetic variability of NUE response under low and high N fertilization in maize. Breeding for NUE in oil palm has yet to take off although differences in leaf nutrient levels among clones and progenies planted under the same fertilizer application regime (Donough et al. 1996; Goh et al. 2009; Lee et al. 2011) have been observed. Soil microorganisms, for example, rhizobium N fixation bacteria, arbuscular mycorrhiza (AM) interacting with plants can also be involved in N uptake and immobilization. Attempts to engineer N-fixing ability from legumes to non-legumes using model plant and crop species have yet to produce tangible results, although studies are still persisting. Application of AM fungal inoculum into oil palm seedlings is a recommended practice (to promote nutrient uptake particularly P and some trace elements) in some companies. Differences in microbial diversity associated with different clonal plantings have been observed and may have bearing on the differences in the nutrient uptake (Tasren Mahamooth, *per comm.*). Breeding for more efficient microbial symbioses, for example, rhizobia, AM fungi could be an interesting alternative for increasing plant productivity using the same amount of synthetic N fertilizer (Reynolds et al. 2009).

13.2.3.2 Water

Water is a very important resource to the plant. It provides the H and O components together with atmospheric CO_2 to form carbohydrates (CHO_n) in photosynthesis. Water is essential for the roots to absorb nutrients and transport them through the vascular system for subsequent diffusion to various parts of the plant for its various essential metabolic processes; for transpiration to cool the plant while supporting metabolic processes; and to provide turgidity to maintain the cell, tissue structure, and form. The availability of water is the major constraint on world crop productivity (Parry and Hawkesford 2010) and thus the challenge is to increase crop production with less available water. Only about 10% of the water available to crops is used productively in transpiration which means that there are significant opportunities to improve water productivity both by increasing the water allocated to transpiration and the efficiency with which transpired water produces biomass. For oil palm being grown essentially as a rain-fed crop, availability of water has not been viewed as a serious issue except in certain regions, for example, South Thailand and parts of West Africa and Indonesia. Moreover, the nature of "drought" can mean different things in different environments, ranging from prolonged and regular periodic absence of rainfall, to erratic and short-term lack of rainfall leading to short-term stress. Both aspects are likely to be important in the future as global warming shifts annual rainfall patterns, but potentially also introduces greater weather volatility and more frequent extreme events. Both floods and droughts are expected to increase in frequency and intensity, with consequent reductions in yield in those regions most affected. Marginal (fringe latitude areas of the oil palm belt, sandy soils, podsols, and peat soil) areas, for example, in Colombia, Thailand, South Sumatra, West Africa, and Central and South America would be most prone to drought stresses.

The following are the physiological effects of drought stress on oil palm with consequential negative effects on photosynthesis, plant growth, and yield: stomata closure restricting diffusion of CO_2; delayed leaf opening reducing leaf area; conversion of starch to sugars within the trunk to maintain respiration and bunch development; premature desiccation of older leaves; and canopy snapping and palm death. Moisture stress during the lagged production period will have the following effects: floral initiation inhibition (33–36 months), male sex inflorescence differentiation (19–25 months), and immature bunch abortion (8–11 months) (Caliman and Southworth 1998).

Early drought tolerance breeding work in oil palm and other crops dealt with plant survival as the index rather than production or yield under severe (or even mild) moisture stress (Nouy et al. 1999; Corley and Tinker 2015). Irrigation experiments in low rainfall, seasonal, or periodic drought stress areas have shown differential genotypic responses to irrigation (Lamade et al. 1998; Méndez et al. 2012). In such situations, the concept of water-use efficiency (WUE) is perhaps more pertinent. WUE can be defined as a ratio of CO_2 assimilation or total biomass yield or crop yield to water consumed as transpiration or evapotranspiration or water input into the system (Sinclair et al. 1984). Based on dry matter (non-oil equivalent) yield per unit of water transpired, yet Henson (1995) estimated WUE of 2.5 g/kg water transpired for well-grown mature oil palm as compared to 30–37 g/kg for maize and 8–9 g/kg for soybean (www.fao.org/fileadmin/.../files/.../TR_07_web.pdf).

Physiological and molecular approaches to select for WUE in crops include improved hormonal physiology, increases in stomatal conductance, osmotic adjustments (total and reducing sugar content), and improved root depth and spread, reduced shoot:root growth ratio and increased harvest index or HI (reduced plant height and leaf area with increased yield). Increased abscisic acid (ABA) sensitivity or its production together with improved osmotic adjustment have been pursued as a selection trait because ABA is involved with guard cell deflation and stomatal closure to conserve moisture, while osmotic adjustment is a cellular adaptation that enhances dehydration tolerance and supports yield under water stress (Fess et al. 2011). Rapid osmotic adjustment helps to maintain high leaf water content and turgor, assisting the crop to continue moderate rates of transpiration and photosynthesis under reduced leaf water potential, promoting cell stability, and avoiding yield losses. However, the intrinsic conundrum with WUE is that closed stomata retain leaf water vapor, but do not allow the entry of CO_2 for carbon fixation, so it seems very likely that different levels of environmental water deficit will require different genotypes and different selection criteria. In wheat, carbon isotope discrimination (CID; carbon assimilation favors $^{12}CO_2$ when stomata are open, but will fix $^{13}CO_2$ when stomata are closed and the $^{13}CO_2$:$^{12}CO_2$ ratio rises, so CID is an integrative measure of how closed stomata are over a period of time, with stored carbon reflecting that integration and this ratio has been used to select for drought tolerant varieties in Australia. However, in UK environments, the relationship between CID and yield is often inverted, as under favorable conditions, open stomata are often preferable for yield production. A balance is needed between plant leaf water retention and CO_2 acquisition and the optimum is likely to be different for different levels of soil water deficit, vapor pressure deficit, and the duration of stress.

Oil palms in W. Africa experience reasonably long periods of drought and hence W. African commercial planting materials tend to be more drought tolerant because they have been bred under such conditions. There are also drought tolerance breeding programs (Houssou et al. 1987; Cornaire et al. 1989, 1994; Nouy et al. 1999; Okwuagu and Ataga 1999). The semi-wild collections obtained by MPOB from the drier regions of Nigeria would include drought tolerant genotypes. In their irrigation experiments on nursery seedlings of four O×G hybrids, Méndez et al. (2012) found that Hybrid U185 recorded the highest photosynthetic rate and the lowest respiratory rate under optimal moisture conditions. While under moderate and severe water deficits, Hybrid U1937 showed the highest photosynthetic rate with the lowest respiratory rate by moving assimilates mainly toward the roots, that is, adjusting its water potential (active accumulation of sugars). Based on these results, they proposed two environment–genotype tolerance relationships based on high photosynthetic rate, low leaf respiration rate, water potential adjustment, and WUE, under adequate moisture conditions and under water deficit.

13.2.3.3 Light

The oil palm requires about 5 hours of sunshine daily for good bunch production. This requirement is easily met for oil palm areas in the wet tropics particularly S.E. Asia except perhaps during the seasonal haze resulting from field burning (forest, farm, plantation). However, this should become less frequent and limited due to the ban on open burning for land clearing and rapid fire control measures. Many oil palm areas in W. Africa experience prolonged hazy skies during the harmattan wind season which has a negative effect on oil yields.

Henson (1995) estimated in well-grown mature palms, 70% of the incident light was absorbed by the canopy, 15% reflected, and 15% reaching the ground. In terms of photosynthetic active radiation (PAR), the corresponding figures were 90%, 3%, and 6%, respectively, suggesting limited opportunity for improvement in this aspect. However, in full sunlight, the upper leaves are oversaturated with light while the lower leaves are undersaturated despite the leaflets' ability to angle themselves to the incident light. There is thus possibility of further improvement in light interception by altering the canopy architecture of the palm, for example, short erect leaves, which would allow more light to filter through to the lower leaves. Large progeny differences exist in light interception (Lamade and Setiyo 1996a,b; Perez et al. 2016). Also during the immature (first 3 years) phase of a planting sizable amounts of light filter through to the inter-rows currently occupied by legume cover crops and weeds that could be replaced with economic intercrops. Likewise, the light space available under aging palms (last 3–5 years) could also be utilized by economic shade-loving intercrops.

The efficiency of conversion of light energy to dry matter or radiation-use efficiency (RUE) of oil palm at 1.3 or 1.6 g/MJ based on oil content was considered low for a C_3 plant species (Squire and Corley 1987; Gerritsma and Wessel 1997). High photorespiratory losses were attributed to be a primary cause. Genetic differences for RUE between progenies and clones have been reported (Gerritsma 1988; Smith

1993; Lamade and Setiyo 1996a,b). RUE traits sought in other crops include stay green phenotypes (leaves remain green longer), early vigor, minimizing stomatal and mesophyll resistance, increasing photosynthetic capacity by utilizing the best photosynthetic enzymes, minimizing downregulation under stress, and decreasing photorespiratory losses. Ribulose 1,5 bisphosphate carboxylase/oxygenase (Rubisco) is the enzyme involved in both photosynthesis and photorespiration, with the energy gained by the former reduced by latter. Natural variants with greater affinity for CO_2 or higher carboxylase capacity, relative to the competing oxygenase activity, are sought. There has been a long standing effort to install the more efficient C_4 photosynthetic pathway (e.g., maize, sugarcane, tropical grasses) into C_3 species (e.g., rice). IRRI has embarked on such a program funded by the Gates Foundation since 2012. It is a challenging task because of the need to engineer not only the biochemical pathway but also the anatomical architecture (Kranz anatomy; mesophyll cells fix atmospheric CO_2 into C_4 acids, these are then transported into oxygen depleted bundle sheath cells to be decarboxylated, the CO_2 released is then recaptured by Rubisco).

13.2.4 Breeding for Resilience or Adaptation to Environmental Stress

13.2.4.1 Extreme Weather

Extreme weather events (temperature, wind storm, floods) and their frequency of occurrence are likely to be experienced in the fringe areas of the oil palm belt (high and low latitudes, altitude, coastal, and insular areas) and more generally with climate change.

13.2.4.2 Temperature

The effects of extreme high and low temperatures on oil palm are still little known. Palms exposed to 4 months of 8°C night temperature (22°C day temperature) became chlorotic and had arrested growth but resumed development when normal growing conditions were resumed. With prolonged low temperature of less than 21°C, floral abortion increases and fruit ripening is delayed. Putative cold tolerant varieties developed by ASD de Costa Rica from wild collections made from Bamenda Highlands of Cameroon and in Kigoma District, Tanzania at 1000–2000 m asl are now available (Blaak and Sterling 1996; Chapman et al. 2003). With a goal to gain a molecular understanding of the cold stress response and to expand African oil palm cultivation in subtropical regions in southern China, the cold stress response in oil palm was analyzed by deep RNA sequencing and transcriptome data analysis (Lei et al. 2014; Xu et al. 2014).

Lim et al. (2011) reported that temperatures reaching 45°C on tin tailing soil could inhibit palm growth probably through depressed nutrient and water uptake by the affected roots. Paterson et al. (2015) have also carried out initial suitability mapping in S.E. Asia for oil palm based on two climate change scenarios, which suggest that heat may become a limiting factor for oil palm in this region in the future.

13.2.4.3 Strong Winds

Oil palms growing in coastal and insular areas of the tropical wind storm belt (typhoon, hurricane, cyclone) of the Pacific Ocean and Caribbean Sea are most prone. Short sturdy palms with small canopies and strong deep and spreading roots would withstand strong winds better.

13.2.4.4 Poor Soils

Manifestations of Mg deficiency occur on oil palms planted in sandy areas in some parts of Papua New Guinea and Indonesia. Tolerance to Mg deficiency is an objective of the breeding program in Papua New Guinea (Breure et al. 1986). Palms planted on deep peat tend to lodge. Dwarf or smaller palms would circumvent this problem. Peat plantings are also prone to micronutrient deficiencies, particularly copper, zinc, and boron. As these deficiencies can be corrected easily with micronutrient applications, a breeding approach is unnecessary. Aluminum, iron, and manganese toxicity can occur in acid sulfate soils. Again these can be ameliorated through agro-management practices, for example, water table management. Salinity may pose a problem on coastal plantings especially with the predicted sea level rise and would require an agro-management and/or breeding solution.

13.2.5 Breeding for Biodiversity

Biodiversity in the context of the oil palm plantation is agricultural biodiversity or agrobiodiversity which "encompasses the variety and variability of animals, plants, and microorganisms which are necessary to sustain key functions of the agroecosystem, its structure and processes for, and in support of, food production and food security" (FAO: SD Dimensions: Environment: Environmental conventions and agreements). Agrobiodiversity consists of the genetic diversity within the species, the species diversity, and the ecosystem diversity, which comprises the variation between agroecosystems within a region. As proposed by Southwood and Way (1970), the level of agrobiodiversity in an agroecosystem is dependent on the diversity of vegetation within and around it, the permanence of its various crops, the intensity of management, and the extent of its isolation from natural vegetation.

Intensive cultivation will undoubtedly lead to a reduction in agrobiodiversity, attributed primarily to mono-cropping and the major reduction in plant species compared to that associated with a natural ecosystem. Other factors such as agro-management practices, that is, fertilizers and agrochemical applications, have been reported to exert further stresses on agrobiodiversity. As reviewed by Haichar et al. (2014), various studies have reported similar observations in which plants are capable of modifying their soil environment through the release of carbon compounds (rhizodeposits), giving rise to the rhizosphere effect. The release of carbon compounds influences microorganisms within (endorhizosphere) or on the root surface (rhizoplane) and outside the roots (ectorhizosphere) (Alami et al. 2000; Jones et al. 2004). With all plants, root exudates represent an important component of rhizosphere communications between plants and rhizosphere-inhabiting microorganisms. Communications involve a broad range of substrates and signaling molecules produced by plants. Bais et al. (2004) reported that there are over 100,000 different low-molecular mass natural products or secondary metabolites which are produced by plants. Taking these findings into context with the transformation from a natural to a mono-cropping ecosystem, the major drivers of below ground activity are root exudates of varying composition (from multiple plant species) and quantity and quality of leaf matter will gradually diminish (in context to composition and possibly abundance). All these macro processes and their changes above and below ground will gradually alter ecosystem processes (i.e., nutrient cycling) and the food web (i.e., distribution and abundance of primary consumers [fungi and bacteria involved in organic matter decomposition] and their successive consumers [secondary and higher-level consumers]). These changes have been well documented with various monocrops and their effects from land-use conversion studies. They thus provide a good case for a multispecies oil palm cropping system, for example, mixed cropping, intercropping (Suboh et al. 2009; Ponniah 2013). Breeding programs tailored for such multi-crop systems would be needed.

13.2.6 Breeding Approaches

Agro-management measures are more expedient in ameliorating and adapting to the effects of climate change but could be limited by the palm genotype. This is where plant breeding comes in but is usually a longer term effort. The molecular breeder with the help of the plant physiologist seeks to identify the key limiting steps in the physiological processes to manipulate via gene discovery, genetic transformation, for example, mutagenesis, TargetIng Local Lesions IN Genomes (TILLING), transgenics, clustered, regularly interspaced, short palindromic repeats (CRISPR) (see Section 13.3.6), and marker-assisted selection (MAS). Whereas, the field breeder (with the help of the crop physiologist/modeler and agronomist) based on the results of his multi-genotype × multi-environment or G×E trials seeks to identify the best adapted genotypes and the relative contributions of the resource use efficiency and stress adaptation traits both generally and in specific environments. The molecular breeding approach usually deals with major genes or quantitative trait loci (QTL). The transformed trait and plant, however, may not achieve the desired effect when planted as a crop. The traits sought for resource use efficiency and adaptation to environmental stress are complex and interlinked with crop ecophysiology via the crop photosynthesis and respiration and transpiration processes interacting with environmental factors (Smith 1989, 1993) which are fine-tuned in adapted species. A transformed trait may upset this balance through negative feedback mechanisms. The field breeding approach although somewhat empirical is more pragmatic as it selects on the sum total or combination of traits for adaptation. Current high yielding oil palm near true F_1 hybrid, clonal hybrid, and clonal cultivars are narrow genetically, uniform, and have been bred for response to high fertilizer, nutrient, and moisture inputs. Consequently, the resource use efficiency and stress tolerance traits that allow its progenitors to adapt to the variable conditions in the wild may have been lost. New breeding programs to face the challenges of climate change should focus on: selection for resource use efficiency and stress adaptation traits rather than maximum yield per se; introgression of these genes into elite and wider breeding populations; planting of diverse genotype and less genetically uniform cultivars; integration of underutilized crops with better resource use efficiency and stress adaptation traits in mixed or intercropped oil palm; and breeding for synergistic or complementary crop interactions. As the effects of climate change are likely to be different in different regions or even within individual countries, the development of "local" varieties is desirable.

Many breeding programs have made genetic collections from the centers of diversity in W. Africa (*Elaeis guineensis*) and S. America (*Elaeis oleifera*), some in marginal areas. An immediate pragmatic field approach would be to collaboratively assemble the diverse data and genotypes available and undertake international collaborative G × E trials to test diverse genotypes (germplasm, landraces, and breeding lines) in the different agro-ecological conditions and cropping systems of interest. Crop models (Kraalingan et al. 1989; Huth et al. 2014) can help in the preliminary selection of the genotypes to match with the sites' agro-ecological conditions. Subsequently, breeding programs *in situ* to achieve the ideotypes for the different agro-ecological zones can be developed (Donald 1968; Corley 2006a). Parallel physiological and molecular studies on the genotypes can be undertaken in controlled environment conditions to give a better understanding of the underlying processes and bottlenecks. Recombinant inbred lines (RILs) and near isogenic lines are ideal for such studies besides expediting cultivar release in response to market needs. Rapid methods of developing such genotypes in oil palm involving genomics-assisted breeding (combination of conventional breeding and genomic tools) are available, for example, single seed descent (Bah Lias Research Station Annual Report, 2010), doubled haploids (Dunwell et al. 2010; Iswandar et al. 2010; Nelson et al. 2009), MAS, high-throughput DNA sequencing and genotyping, and genome-wide selection (Wong and Bernado 2008; Cros et al. 2013; Singh et al. 2013a,b, 2014).

More accurate field phenotyping tools for measuring and subsequent development of selection aids or markers for such physiological traits can also be developed (*Proceedings of 3rd International Phenotyping Symposium*, 2014, MSSRF Chennai; see Sections 13.5 and 13.6). High oil yielding, high harvest-indexed palms with small erect canopies borne on short sturdy trunks and well-anchored with strong roots spreading wider and deeper within the soil would form the ideotype traits for most agro-ecological conditions and cropping systems.

The synergistic efforts of agronomists, crop physiologists and modelers, and biotechnologists and breeders are needed to expedite the development of sustainable high yielding, resource use efficient, and stress resilient oil palm cultivars for the future.

13.3 TRANSGENIC OIL PALM

Denis Murphy

13.3.1 Transgenic Crops

The goal of any transgenic crop modification is to use recombinant DNA methods to alter gene expression in order to create new varieties for breeders that may be difficult or impossible to produce using conventional approaches. The earliest commercial transgenic oilseed crop was a variety of rapeseed (canola) produced by Calgene and first grown under license by farmers in the United States in 1995 (Murphy 2006). Interestingly, this transgenic rapeseed, termed Laurical™, had been modified to produce a high-lauric oil with the aim of replacing imported palm kernel oil in

US markets. Laurical was produced by transferring a gene encoding a C12-specific thioesterase from the California Bay plant, *Umbelluria californica*, into rapeseed. Although yields of lauric acid were relatively low in the early versions of Laurical, the addition of further transgenes eventually enabled Calgene breeders to produce a seed oil that contained in excess of 60% w/w lauric (Voelker et al. 1996).

However, even this relatively lauric-rich oil crop was unable to compete commercially with the standard commodity lauric oil, which was (and still is) obtained from the far cheaper and more plentiful (and non-genetically modified [GM]) palm kernel oil. Therefore, the Laurical brand of transgenic rapeseed was a commercial failure and was only grown for a few seasons during the mid-1990s in the southern USA. This demonstrates some of the major challenges involved in producing transgenic crop varieties that also apply to oil palm. In the case of Laurical, the technical challenges to produce a high-lauric oil were eventually resolved but the commercial realities prevented the crop from succeeding.

Since the mid-1990s, there have been no further releases of GM crops specifically modified to produce novel oils and none of the major "first generation" GM crops released between 1996 and 2010 were modified for output traits such as oil composition. The reasons for this include the technical challenges in producing GM oil crops with acceptable fatty acid content, plus the lack of markets for such modified oils. Moreover, even if these technical and commercial hurdles can be overcome, new transgenic varieties still face problems due to public opposition to GM crops that has existed in some parts of the world since 1999. Such opposition has been particularly intense in Europe right up to the present day. Hence, in late 2015, many local regions in the EU took advantage of a change in regulations and instituted formal bans on the cultivation of all GM crops, even varieties that had been approved by the European Commission (Nelson 2015). In the case of a long-lived perennial tree crop like oil palm, all three of these challenges, namely technical feasibility, commercial prospects, and anti-GM sentiment, are considerably more severe when compared with annual oil crops such as rapeseed and soybean.

13.3.2 Why Develop Transgenic Oil Palm?

Production of the first experimental transgenic plants was reported in 1983 and over the next two decades, many hundreds of transgenes were inserted into dozens of plant species. During the 1980s and 1990s, the oil palm industry in Malaysia was expanding rapidly with a trebling of oil output between 1980 and 1995 (Abdullah 2005). Another aspect of this era was the aspiration of Malaysia to develop into a leading high-tech economy as outlined by then Prime Minister, Dr. Mahathir, as part of the "2020 vision" for the country (Murphy 2003). At the same time, several biotech companies, including Calgene, Dupont, and Monsanto were developing transgenic methods for oil modification in annual crops. However, oil modification would result in new output traits with a different crop product that would require segregation during all stages of growth, harvesting, transportation, and processing in order to avoid cross contamination between the different types of oil. This means that output traits are more expensive to manage in commercial farming and processing systems and therefore need to command a price premium. Another issue with many output traits is that they can entail the extensive manipulation of metabolic pathways and the insertion of several, sometimes many, transgenes.

Despite these caveats, the late 1990s were a relatively optimistic time for GM technology with the first large-scale crop releases occurring in 1998 and with the backlash from anti-GM groups still in the future. At this time, there was considerable pressure against imported palm oil from activist groups, such as the US soybean industry. The major criticism of what was referred to as "tropical oils" was based on the perceived nutritional disadvantages of relatively highly saturated oil. Existing accessions of oil palm were from a relatively restricted genetic base and did not display much variation in fatty acid profile, which ruled out a facile conventional breeding program. A mutagenesis approach would also be complicated by the need to manipulate several independent genes. Given the success of Calgene in using GM methods to modify high-oleic rapeseed into a high-lauric variety after a few years of R&D, the prospect of creating a transgenic oil palm with a lower amount of saturated fatty acids was very enticing (see next section). Moreover, if it was possible to manipulate acyl content at will, other more niche novel-oil varieties could be created, such as high stearate for solid fat markets or ultrahigh laurate for cosmetic/cleaning products (Sambanthamurthi et al. 2009; Murphy 2014a; Barcelos et al. 2015).

13.3.3 Target Traits: Fatty Acid Modification and Other Traits

Research and development work on transgenic oil palm started in earnest in the mid-1990s in PORIM (now MPOB) (Cheah et al. 1995; Sambanthamurthi et al. 2009). Even at this early stage, it was recognized that the choice of target traits was important and that disruption of the existing highly successful commodity palm oil products should be avoided. It was also recognized that oil palm transformation was a multi-decade endeavor that was unlikely to bear fruit before the 2020s. This is illustrated by the following quote about the goals of the GM oil palm program which is from the leader of the PORIM/MPOB transformation group, GKA Parveez:

> Among these targets are high-oleate and high-stearate oils, and the production of industrial feedstock such as biodegradable plastics. The efforts in oil palm genetic engineering are thus not targeted as commodity palm oil. Due to the long life cycle of the palm and the time taken to regenerate plants in tissue culture, it is envisaged that commercial planting of transgenic palms will not occur any earlier than the year 2020 (Parveez et al. 2000).

By far the most important target trait for the transgenic oil palm is the high-oleic content in the mesocarp oil. Existing varieties of oil palm produce a mesocarp oil that typically contains only about 40%–45% oleic acid whereas competitor crops such as rapeseed, olive, soybean, and sunflower have varieties containing well above 60% and in some cases as high as 85% oleate. The main reason for the low oleate content of palm oil is that the mesocarp has an active palmitoyl thioesterase that results in C16 palmitate being channeled toward triacylglyceride (TAG) accumulation instead of being elongated to stearate and desaturated to oleate. In order to change the mesocarp oil into a high-oleate profile, it is therefore necessary to manipulate at least three, and possibly more, genes. Firstly, palmitoyl thioesterase needs to be downregulated to stop palmitate being diverted to TAG. Secondly, the key elongation gene, KASII needs to be upregulated so that the palmitate is efficiently converted to C18 stearate. Thirdly, the stearoyl-acyl carrier protein (ACP) desaturase needs to be upregulated so that stearate is efficiently desaturated to oleate. Experience with other crops, such as the transgenic Laurical rapeseed described above, has shown that it may also be necessary to transfer additional acyltransferase genes to ensure that the new fatty acids are efficiently assembled onto triacylglycerols. For example, evidence from several plant species suggests that the type-2 acyl-CoA:diacylglycerol acyltransferase (DGAT2) can stimulate accumulation of exotic fatty acids in storage TAG (Kroon et al. 2006; Shockey et al. 2006).

Due to these complications, until very recently, the levels of the novel fatty acids in most transgenic plants have been relatively modest and were far from achieving commercial viability (see reviews by McKeon 2005; Murphy 2006; Cahoon et al. 2007; Dyer and Mullen 2008). This means that it is likely to be some time before an ultrahigh oleate trait can be created in oil palm, but this goal is still important due to the usefulness of oleate-rich feedstocks for both food and nonfood applications (Murphy 2014a). It also means that the engineering of oil palm for more minor fatty acid traits, such as high palmitate or palmitoleate, or for acyl derivatives such as bioplastics (Yunus et al. 2008b; Parveez et al. 2015) is going to be even more problematic. Given that these will be niche products with much more limited markets, it may be appropriate to question whether large R&D investments in a transgenic strategy aimed at such targets are justified at this stage.

In addition to oil modification, transgenic approaches can, in principle, be applied to any other trait where suitable genes have been identified. This means that there is a long list of possible GM varieties of oil palm, but high R&D costs are likely to limit their development to traits with a guaranteed agronomic and commercial potential. For example, expression of Bt insecticidal proteins has been proven to be an effective strategy to control certain insect pests in a range of crops and is being investigated in oil palm. A more important threat is the fungal pathogen, *Ganoderma boninense*, and the possibility of modifying lignin to make oil palm plants more resistant is being examined (Paterson et al. 2009). As of 2015, progress in addressing any of these traits has been limited and the high costs of existing transgene technology mean that scarce resources would be better targeted to the single key trait of high-oleate mesocarp oil where the biochemical basis of the trait is reasonably well understood and there is a guaranteed global market for the resulting oil (Murphy 2014a).

13.3.4 Historical Approaches to Oil Palm Transformation

Early work on oil palm transformation was initially focused on the assessment of suitable tissue culture systems for transformation and suitable vectors for transgene delivery. This work was based largely at PORIM/MPOB with additional input from several university-based groups in Malaysia, as outlined in Abdullah (2005). During the 1990s, *Agrobacterium*-mediated gene transfer was still largely restricted to dicot plant species. At this time, a few monocots such as rice and maize had been transformed, but oil palm was well outside the host range of *Agrobacterium* so initial strategies for oil palm transformation used the newer method of biolistics (Parveez 1998; Parveez and Christou 1998; Parveez et al. 2000). A disadvantage of biolistics is that it often results in the integration of multiple transgene copies, whereas *Agrobacterium*-mediated transformation is more likely to introduce either single-copy or low-copy-number transgenes. Nevertheless, it was possible to bombard oil palm cultures with marker genes such as beta-glucuronidase (GUS) and achieve transient expression of the transgene (Parveez et al. 1997). It was also possible to experiment with different gene promoters in order to investigate levels of gene expression and tissue specificity (Chowdhury et al. 1997).

In order to improve transformation efficiency, new tissue culture systems were developed and in particular, immature oil palm embryos were used as the target tissue using both biolistics and *Agrobacterium*-mediated transformation (Ruslan et al. 2005; Bhore and Shah 2012). This resulted in transformed tissues that expressed marker transgenes such as those encoding *Bacillus thuringiensis* (Bt) insecticidal proteins and the cowpea trypsin inhibitor (CpTI) (Ismail et al. 2010). In terms of logistics, these experiments required the transformation of thousands of tissue culture samples. Even after several stages of selection and regeneration, there were many hundreds of plants that required housing in containment facilities and subsequent analysis. It took 3–5 years to generate transgenic oil palm plants by either particle bombardment or *Agrobacterium*-mediated transformation and both processes were highly inefficient. The frequency of escapes and chimeric plants was high because of the long selection process during callus formation and the fact that somatic embryogenesis encourages the growth of non-transformed cells (Masani et al. 2014).

13.3.5 Current Technical Challenges

Despite several decades of intensive research, the development of transgenic oil palm is still a relatively young science and is fraught with many technical challenges. For example, it is still unclear whether biolistics or *Agrobacterium*-mediated gene transfer will be the gene delivery method of choice (Izawati et al. 2012; Parveez and Bahariah 2012). The choice of plant material is also crucial with some of the best options including various types of callus and embryo culture. Then, there is the issue of which gene promoters to use. Unlike most existing commercial transgenic crops where constitutive viral promoters are used, the manipulation of palm oil composition will require deployment of strong mesocarp- or kernel-specific gene promoters, ideally sourced from the oil palm genome itself.

Also, in order to achieve a high-oleic oil, it will probably be necessary to down-regulate some genes while adding or up-regulating other genes. Even when the primary transgenic plantlets have been produced, they will still need to be grown on for 3–5 years to obtain fruits that can be screened for oil content. Finally, these primary transformants will need to be taken through several sexual generations, backcrossed with existing elite lines, and then multiplied via micropropagation before any new commercial transgenic varieties can be released. Among some of the recent research directions for transgenic oil palm are the following:

- Improved transformation vectors for oil palm (Yunus and Parveez 2008a)
- Development of positive selectable marker systems such as phosphomannose isomerase (Bahariaha et al. 2013) and 2-deoxyglucose-6-phosphate phosphatase (Izawati et al. 2015)
- Transforming oil palm protoplasts by polyethylene glycol (PEG)-mediated transfection and microinjection (Masani et al. 2014)
- Using more rigorous molecular analytical methods to detect transgene presence (Nurfahisza et al. 2014)
- Improved tissue-specific gene promoters for mesocarp expression (Taha et al. 2012)

It seems likely that the development of routine methods for transforming oil palm with gene constructs optimized for commercial phenotypes will take several more decades. The initial target of having commercial transgenic oil palm varieties under cultivation in the 2020s is now looking rather optimistic. However, a target date in the late 2020s is not necessarily totally unrealistic if new genome editing technologies prove useful, as will be discussed below. It is, therefore, still important that R&D programs continue for the long-term future of crop improvement. This is because there are likely to be several important genetic traits that cannot be created using non-transgenic approaches. For example, although it may be possible to create a medium–high (55%–65%) oleic oil phenotype via conventional methods, for example, by using identified accessions obtained from Africa (Murphy 2014a), a more desirable ultrahigh (80%–90%) oleic trait might only be possible via transgenic technology. Another example where the use of transgenic methods is essential is the production of completely novel compounds, such as biopolymers like polyhydroxyalkanoates, in oil palm. In this case, it is necessary to transfer several bacterial genes to the crop. Meanwhile, as outlined below, newer technologies such as CRISPR, RNAi, trait stacking, chromosome engineering, pathway engineering, and more efficient gene cassettes will play important roles in expanding the scope of transgenic crops in the future.

There are several ways in which existing transgene technology can be improved to make it technically easier, more efficient, and broader in its scope, and better able to address concerns expressed by certain sections of the public. Some technical issues and areas of public concern are listed below:

- In the future, it will be desirable to generate transgenic crops that do not contain selection markers, such as genes for antibiotic or herbicide tolerance.
- First-generation transgenic plants were created using random insertion of transgenes, which can lead to variations in transgene behavior and other unpredictable pleiotropic effects. Therefore, targeted insertion of transgenes is required in the future.
- The spread of transgenes into wild populations via cross-pollination can be prevented using genetic use restriction technologies (GURTs).

- In order to prevent risk of contamination of human or animal food chains, biocontainment strategies should be incorporated into transgenic plants expressing certain bioactive products such as vaccines or pharmaceuticals.

13.3.6 New Technologies for Genome Manipulation: CRISPR

In the last few years, biological advances have started to make it much easier to manipulate genomes in more radical and precise ways than were possible with traditional late twentieth century GM technologies (Murphy 2016). For example, several new and potentially revolutionary forms of plant and animal (including humans) GM technologies known as "genome editing" have recently been developed. Probably, the most powerful is the CRISPR system (Bhaya et al. 2011; Mao et al. 2013; Hsu et al. 2014). In 2015, the CRISPR/Cas9 system was described in a Nature article as "*the biggest game changer to hit biology since PCR*" (Ledford 2015a). Applying this method to crop improvement has opened up many new possibilities for radical genome modifications (Zhang et al. 2014; Belhaj et al. 2015). Genome editing will greatly accelerate crop breeding by enabling precise and predictable genetic modifications directly in elite individuals as well as simultaneous modification of multiple traits.

This means that breeders will progress well beyond the random insertion of single or small numbers of genes into a genome (as in traditional GM) to the highly precise insertion into a defined location of large numbers of genes, chromosome segments, or pseudo-segments encoding entire metabolic pathways into virtually any plant species. Methods such as transcription activator-like effector nucleases (TALEN) and CRISPR/Cas9 can be used for gene knockouts, for example, to eliminate unwanted genes that adversely affect food quality, or confer susceptibility to pathogens, or that divert metabolism away from valuable end products. An example reported in 2014 was the use of both TALEN and CRISPR/Cas9 to target the genes of the mildew-resistance locus in wheat. This resulted in the production of plants resistant to powdery mildew disease, which is a serious crop disease (Wang et al. 2014). In late 2015, a much improved version of CRISPR was reported where a different endonuclease system called Cpf1 was used instead of Cas9 (Ledford 2015b; Zetsche et al. 2015). The new CRISPR/Cpf1 system is much easier to deploy in cells requiring genome editing and will enable the technology to be used more widely and effectively to create new crop (and animal) phenotypes.

Precise nucleotide exchanges using oligonucleotide donor sequences can also be used to modify the regulatory sequences upstream of genes that determine agricultural performance leading to improved crop yields. In some cases, it will be possible to reprogram genes so that they are expressed in different organs at much higher rates. For example, seed-specific genes regulating storage lipid accumulation could be reprogramed to function in leaves or roots in order to generate high levels of oil in such vegetative tissues (Vanhercke et al. 2013; Murphy 2014b). The oil could be used for food or fodder, as a source of renewable oleochemicals for industry, or even for biofuels. By 2050, it is almost certain that entire metabolic pathways will be transferred to the plant and tissue of choice so that some crops could become low-cost production systems for industrial green chemicals including high-value specialty chemicals (Murphy 2014b), nutraceuticals such as omega-3 oils (Napier et al. 2015), and pharmaceuticals such as vaccines, antibodies, and drugs (Murphy 2007).

In addition to transgenic approaches, modern breeding is able to profit from a host of advanced technologies such as association genetics (Rafalski 2010), molecular mutagenesis including TILLING (Shu 2009), and MAS (Ribaut et al. 2010). Deployment of these technologies is already enabling breeders to begin to address complex traits such as drought tolerance and disease resistance (Tuberosa and Salvi 2006; Xu 2010). Such technologies could be used together with GM methods to develop additional useful traits such as dwarf trees, increased oil content in mesocarp and kernel tissues, and wider resistance/tolerance profiles to a range of pests and pathogens (Murphy 2014a).

13.3.7 Can Transgenic Oil Palm Varieties Achieve Commercial Success?

One of the major lessons in GM technology from the last 30 years is that the commercial success of a new crop variety bears little relation to the ingenuity of the technology that has enabled its production. Particularly, in the 1990s, there were numerous attempts to engineer transgenic crops with modified oil compositions. However, even where the required oil profile was achieved, lack of attention by researchers to signals from commercial markets ultimately resulted in failure. In the future, more attention should be paid to understanding market demands and consumer expectations before embarking on the costly and often risky process of developing a transgenic crop variety, especially in the case of oil palm.

In the transgenic Laurical variety developed by Calgene (see above), levels of lauric acid in excess of 60% were produced in transgenic rapeseed oil, which is even higher than in palm kernel oil and a very impressive technical achievement. However, rapeseed is a relatively low yielding and land-hungry crop compared with oil palm and its oil is much more expensive to produce. This situation was exacerbated by the additional costs of segregating the new transgenic rapeseed plants and their oil from other oils in a supply chain that was dominated by low-cost C18-rich oil crops that were produced mainly for food and feed markets. The end result was that Laurical rapeseed was only ever grown on a small scale by a few contract farmers and the oil did not come close to displacing palm kernel oil in US markets.

There are many similar examples of oil crops being engineered to produce novel fatty acids, or even high-value proteins or biodegradable plastics, where the basic GM technology was sound but the market conditions were simply not suitable for commercial production of the new materials. In the case of oil palm, over the past 20 years, a large number of potential target traits have been pursued in various research programs, mostly involving oil modification but, apart from the high-oleate trait, the commercial potential of such varieties remains questionable. The high-oleate trait could be a game changer for palm oil as this trait would enable it to be used directly in premium food markets such as salad and cooking oils and as a non-hydrogenated feedstock for margarine manufacture. This would enable palm oil to compete more effectively with the major temperate oilseeds, which have much lower yields and higher land requirements, and may result in it displacing these less efficient crops in many global markets (Murphy 2014a).

13.3.8 Conclusions and Future Prospects

A major issue for transgenic oil palm is consumer acceptability of products from GM crops. Despite a lack of clear scientific evidence to support adverse claims from anti-GM activists, many politicians in Europe remain wary about allowing GM crops or products to be grown or imported into the region (Nelson 2015). Given that oil palm already suffers from an image problem (especially in Europe) due to its perceived environmental impacts, the industry is rightly cautious about possible future GM developments. Indeed, in many cases, palm oil producers actually stress that the crop is entirely non-GM at the present time.

However, the current status of GM crops may change radically over the next decade as the technology becomes ever more refined and globalized and is applied to more crops with traits that have better consumer appeal than current GM varieties (Murphy 2016). By 2015, over half of all GM crops were being grown in developing countries, including several countries in Africa. There has been an increasing tendency for transgenic crops to be developed by public-sector institutions in developing countries and then distributed at low cost to poorer farmers. An example is golden rice developed by IRRI in the Philippines and scheduled for release to farmers in 2016–2017 (Kowalski 2015; Moghissi et al. 2016). As developing countries produce new varieties of GM crops with traits, such as improved nutritional quality, drought tolerance, and disease resistance, it will become increasingly difficult to defend the 1990s-era argument that GM crops represent a tool for the imposition of global food hegemony by Western governments and multinationals (Engdahl 2007).

The new genome editing systems, such as CRISPR (as discussed above), will widen the scope of traits that can be modified while genomics and other omic technologies will make it much easier to identify key genes that regulate even relatively complex traits such as salt or drought tolerance. Another significant aspect of the new genome editing systems is that they can make it virtually impossible to detect any resultant modifications in a genome. This is in contrast to conventional late-twentieth century GM technologies, where the presence of novel DNA can be readily detected (Ledford 2015a,b; Lunshof 2015). Therefore, some new GM plants may not carry any proof whether they were produced either via a GM-type technology or via one of several non-GM technologies that could have been used instead. This development has the potential to undermine the entire framework that is currently used for the regulation of GM crops because, for example, it would become impossible to distinguish between a new plant variety that has arisen via spontaneous mutation from a plant produced via deliberate mutagenesis in the lab or a GM plant that has been modified by genome editing.

These factors of improved genome editing technologies, ability to alter many more traits, more public-sector and public-good R&D, and increased attractiveness to consumers around the world, mean that the coming decades are more likely to see an expansion in the number and range of GM crops and traits. For these reasons, one can be cautiously optimistic about the future of transgenic oil palm and the possibility that commercial releases might be carried out before the end of the 2020s.

13.4 OIL PALM YIELD MODELING

Sue Walker and Asha S. Karunaratne

13.4.1 Crop Models

Crop models will allow one to test the effects of a range of desired prospective traits proposed to be incorporated into the new oil palm lines. This is possible as the crop models use specific variables to describe the outcome or effect of certain characteristics on the final yield of fruit bunches over the growth period of oil palm trees. Crop-climate models are formulated using several alternative logically frameworks or approaches, and so different models have been built to be sensitive to different parameters. The two most common limiting factors that are used as the driving force for crop-climate models are based on physiological and developmental process that can be radiation-limiting (Monteith 1977; Tsubo et al. 2001) or water-limiting (Hanks 1983; Monteith 1986). Usually, the climate data are used as input information which allows the models to be linked to a specific place and time and thus yield is simulated based on these variables. The crop models use equations to generate potential growth and development trajectories which can then be modified according to the factors which will limit this based on the current and cumulative environmental conditions.

The radiation-limiting equations are constructed around the principle that photosynthetically active radiation (PAR) is absorbed by the leaf and then converted into chemical energy. This energy is converted into carbohydrates and made available for the growth and development of the crop plant and is partitioned according to the growth stages and age of the plant. Therefore, the cumulative PAR over a growing season is then related to the sum of the increase of dry matter or the cumulated biomass of the plant (Tsubo et al. 2001). So one can formulate the rate of growth equations to be driven by the daily PAR based on the length of daylight and latitude and time of year. Different types of plants have different growth rates and the slopes of such curves can be based on historical field measurements from past experiments. Different species also have different conversion rates for PAR to carbohydrates as well as the ratios of energy used for respiration, growth, and storage to harvestable organs. Thus, when there is less PAR available, there will be slower growth and lower yields. The way in which the crop-climate models can be used to assist in breeding and selection is by characterizing these growth and partitioning equations according to the genotype and characteristics of different varieties or cultivars.

The water-limiting type of models have equations based on the fact that plants are composed of a high percentage of water, and also transpire high volumes of water each day. This means that the amount of water available to the plants can be used as a driving force to determine the growth of the plant organs. As water transpired exits the leaf via the stomata while carbon dioxide enters the leaf at the same time, if the amount of transpiration can be determined, then it can be related to the amount of carbohydrates accumulated (Hanks 1983). Therefore, potential evapotranspiration, as calculated from the climate data, can be used to simulate the potential cumulative dry matter or biomass (Ritchie and Basso 2008). Then, as the water available for transpiration decreases, the plant becomes stressed and produces less growth (Hsiao and Xu 2000). Therefore, if these potential water-growth relationships have been characterized for certain plants, they can be related to the traits of specific varieties and cultivars. Then, the crop models can be used to optimize such traits under different prevailing conditions.

13.4.2 Oil Palm Crop Models

Not all crop models include perennial plantation trees or have been written specifically or been adapted for oil palm plantations. However, several have been developed and tested to help with plantation management. They include the following:

a. Agricultural Production Systems Simulator (APSIM www.apsim.info) is an agricultural systems model that has an oil palm module that simulates growth of oil palm vegetative parts and fruit bunches based on data from Papua New Guinea plantations (Huth et al. 2014).
b. ECOPALM (Combres et al. 2013) is a simple model that simulates variation in a number of harvested bunches. It uses photoperiod during sensitive phases, to form the seasonal production peaks together with a plant-scale index that optimizes inflorescence development and sensitive stages. It was formulated from observed data from Ivory Coast and validated with those from Benin and Indonesia.
c. Oil Palm Simulator (OPSIM) (Van Kraalingen et al. 1989) is a dynamic model that simulates growth and yield formation of oil palm dependence on weather data and plant characteristics, via photosynthesis to production. It has been tested using long-term data from two planting density trials.
d. Oil Palm Production Simulator (OPRODSIM) (Henson 2007) uses daily climatic data to predict growth and yield responses to climatic conditions starting at planting. It is a mechanistic model based on soil water status and crop water use, and could effectively simulate annual bunch yields and annual changes in some vegetative parameters. Bunch yields were well simulated, although it improved with a 2-year lag with respect to the weather data.
e. PALMSIM (Hoffmann et al. 2014, 2015) used solar radiation to simulate potential growth of oil palm and has been evaluated against measured oil palm yields from optimal water and nutrient management at a monthly time step for a range of sites across Indonesia and Malaysia. A potential yield map for Indonesia and Malaysia was generated with PALMSIM yield as a first step toward a decision support tool to identify degraded sites with potential.
f. Soil Nitrogen Overview—Oil Palm (SNOOP version) (de Barros et al. 2012) is a model that estimates the dynamics of water and nitrogen in soils under oil palm cultivation, and was developed from the environmental policy integrated climate (EPIC) model modified for Brazilian conditions. It can be used as a decision support system for N management, as it simulated many water and nitrogen parameters.

g. WaNuLCAS (Van Noordwijk et al. 2011) was developed to simulate tree–soil–crop dynamics in agroforestry systems and has an oil palm module, using water, nitrogen, and radiation interactions. The outputs can be used to help farmers make choices between tree species for a specific site for agroforestry systems.

13.4.3 Explanation about APSIM: Oil Palm Module

As oil palm is a perennial tree crop, the crop-climate modeling is complicated. It involves many variables that influence the growth and development of both vegetative and reproductive organs from year to year. The APSIM team led by Neil Huth has developed an APSIM-Oil Palm module based on data from Papua New Guinea plantations (Huth et al. 2014). This module simulates the growth of vegetative parts; namely fronds stem and a root; as well as the fruit bunches. Both the frond and attendant fruit bunch growth are based on age as a cohort and are initiated and developed successively according to the rank. The oldest (lowest) fronds are cut together with their respective oldest, lowest bunches as a cohort at harvesting (Huth et al. 2014). An age-dependent plastochron (assumed to be the same as the phyllochron) is used as the shortest time interval for frond initiation until the palm is 10 years old. This can, however, be influenced by temperature via a daily relative development rate (Combres et al. 2013). The frond appears as a "spear leaf" and expands for five phyllochrons following a logistic equation to a maximum size of 14 m^2 at age 14 years (Huth et al. 2014).

RUE (1.22 g MJ^{-1} for oil palm) is used to generate photosynthate, but it is also dependent on frond nitrogen content. It is then partitioned into stem, root and most importantly oil palm fruit growth. The fruits are borne on bunches in cohorts, but also undergo phases in gender determination, inflorescence abortion, and bunch growth or failure (Jones 1997; Combres et al. 2013). Fruit production is via bunch growth rate which is dependent on a ratio of maximum bunch size (according to age of palm) and duration of the growth phase (Huth et al. 2014).

This oil palm model can now be used to explore the opportunities and direction for adapting various building blocks that determine fruit bunch production according to genetic traits. As the model was built using information from the Papua New Guinea plantations, it can be linked to known genetic traits of those oil palm lines. The next steps needed would be parameterization of the model for a range of these lines with known and contrasting characteristics, especially those sensitive to environmental factors of temperature and water availability. Then, the model parameters related to growth, development, and switches for bunch production or failure, whether from temperature, abortion, or gender can also be assessed according to known characteristics of lines currently growing in S.E. Asian or African countries. Once this is done, there can be some model experimentation with the sensitivity of each of the parameters over a prescribed range within each parameter to determine its effect on fruit bunch production. This will help to determine which of the traits need to receive attention during a focused modeling-breeding team effort. In addition, the performance of a range of calibrated lines under various climate change scenarios or for stepped temperature and rainfall intervals, will then help to identify the traits that are likely to enhance oil production under climate change conditions.

13.4.4 An Example: Model Results at Rompin, Malaysia (1997–2014)

In a preliminary modeling exercise, the Papua New Guinea calibration was used for a potential plantation site near Rompin, in Pahang state, along the east coast of Peninsular Malaysia. It is a low-lying area (<70 m above average sea level) presently covered with secondary regrowth tropical rainforest and near a designated wildlife reserve (Endau-Rompin National Park) (Gopalan et al. 2015). The APSIM-oil palm model was run using publically available data for both climatic and soil information, as a preliminary estimation of the production for that area. Long-term reanalysis climate data obtained from NASA website (NASA-POWER 2015) was used at 1° latitude by 1° longitude grid, corresponding to approximately 110 km. Climate data for 17 years from January 1997 until December 2014 acquired at 2.87N and 103.257E with an elevation of 69 m included daily values of precipitation (mmd^{-1}), minimum air temperature (°C), maximum air temperature (°C), relative humidity (at 2 m, %), wind speed (at 10 m, ms^{-1}), and radiation (MJm^{-2} day^{-1}). Soil information from ISRIC—World Soil Information at 1 km^2 grid was used for six different soil horizons (0–200 cm) with values for organic carbon (fine earth fraction) (gkg^{-1}); sand%, clay%, silt% content mass fraction; coarse fragments >2 mm, pH in H_2O; cation exchange capacity ($cmolkg^{-1}$); and bulk density (kgm^{-3}) (ISRIC 2015). Soil types used were SapricHistosols (saHS, high organic matter), HisticFluvisols (hiFL, high sediments), GleyicAcrisols (glAC, clay rich), and Haplic Gleysols (haGL, saturated clay). As soil water holding capacity and hydraulic conductivity are needed for the model, they were determined using USDA-ARC Hydraulic Properties Calculator USDA (2015) to derive texture class; saturated hydraulic conductivity; % volume for wilting point, field capacity and saturation.

As the APSIM-Oil Palm model was not yet calibrated with Malaysian genetic lines, the assumption is made that the parameters give a reasonable representation of an average oil palm plantation. Therefore, results obtained serve as an indication of possible yields, which require further validation. In addition, as soil and climate data used were from publicly available large-scale databases, these should be considered as preliminary results. For this APSIM model exercise, oil palm is rainfed and planted once, then was run for 22 years giving predicted fruit bunch mass for a 17-year period, starting in 1997 with 6-year-old palms. In order to visually represent yield variation with time, results from each year were imported into geographic information system (GIS) with mapping applications. Oil palm production varied according to the climate throughout the years, with 1998 (7y), 1999 (8y), and 2005 (14y) giving the lowest yields (2–3 t ha^{-1}), while only 1 year produced the maximum yield (Figure 13.1) (Walker et al. 2015). These values need to be verified with appropriate historic field data as the model is not fully calibrated. More accurate predictions would also be obtained with a ground truthing validation exercise and detailed field analysis to provide good input values for crop models.

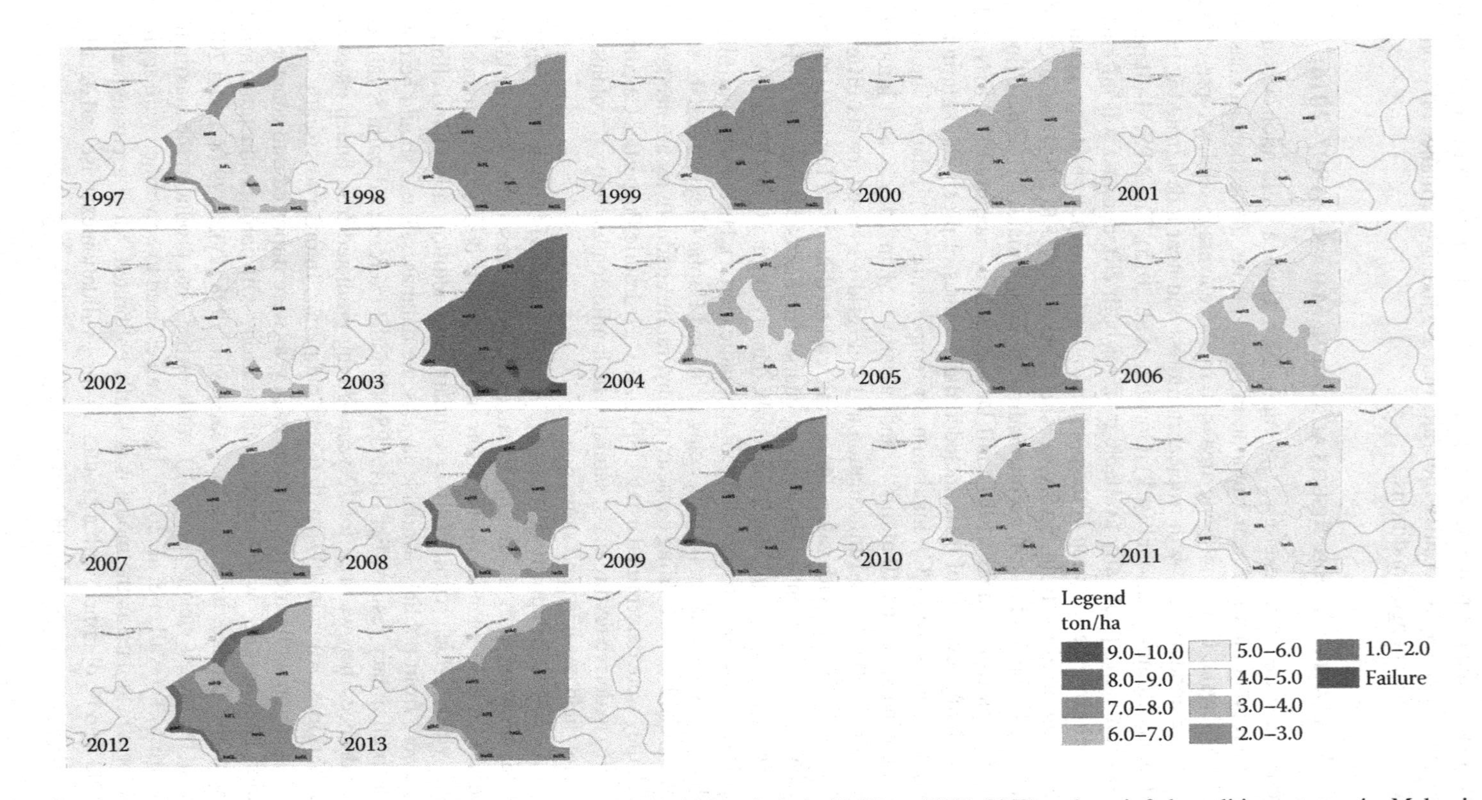

FIGURE 13.1 (See color insert.) Predicted oil palm fruit bunch yield variation with Time (1997–2013) under rainfed conditions at rompin, Malaysia on four soils. (Adapted from Walker, S. et al. 2015. Crop diversification for Malaysian low-lying marginal land. In: Poster presented at *ISUPS*, India.)

13.4.5 Future Use of Crop Models to Formulate Breeding Strategies

There are many opportunities for breeders and crop modelers to work together, and make a significant input into guiding the direction of the oil palm breeding strategies and take a major step ahead in increasing the productivity and sustainability of oil palm plantations.

Firstly, one should identify the major limiting factors from an environmental physiological, and genetic potential point of view, then use the crop models to predict the opportunities available to optimize these using focused breeding and site selection. This would then effectively be using a Genotype × Management × Environment (G × M × E) framework (Messina et al. 2009) to be sure to enhance the prospects for success with future developments. This is dependent on the intrinsic relationships within the crop model between the physiological parameters and the genetic traits. As these aspects have been part of the APSIM policy and intent, there is good reason to expect that one can use its crop models with a built-in physiological basis of genetic variation to explore adaptive traits within a species. Then, by assessing the model outcomes for a combination of a range of G, M, and E, the crop model can be used to predict and thus understand these interactions, and therefore develop a focused search strategy for the breeding program (Hammer et al. 2001). Innovative adaptive management strategies, such as intercropping or double row planting, can also be included to this added layer of complexity, and models in a landscape type model (e.g., Soil and Water Assessment Tool [SWAT] http://swat.tamu.edu/) that considers the overall resource use across the extended area.

A second alternative is to use specific mutants or populations of RILs to investigate the relationship between physiological and genetic information and origin of lines (Hammer et al. 2001). This will assist in the design and selection of multilocation × genotype lines and also suggest weightings during the analysis of the data. Together with detailed climate analysis, the outcomes of such trials can be weighted according to the frequency of the specific occurrence of the combination of climate parameters, in order to overcome the bias from such field trials (Podlich et al. 1999). Under the current increased climate variability, it is important that these aspects are addressed in order to obtain trustworthy results. Hammer et al. (2001) stated that even if a linkage seems to be apparent, an issue of credibility of the model remains as to whether it can truly properly predict the consequences as some subtle interactions may not be represented in the model equations.

Thirdly, as DNA markers can also be linked to specific physiological characteristics and environmental responses, modeling can be used to identify important molecular markers to enhance crop improvement by providing an effective bridge between genetic architecture and phenotypic expression (Hammer et al. 2001). In this way, crop modeling has been used to evaluate alternative breeding strategies using gene to phenotype bridging, and these methods can also be applied to oil palm breeding (Podlich et al. 1999). This then provides an integrated platform for the evaluation of the potential options for a way forward whether the genes effect is additive or interactive, as the model can help with the explanation over a wider range of possible climates than traditional field trials. In addition, the effects can be accumulated or segregated when using different growth and development rates over different cycles of selection depending on the breeding strategy simulated by the model. Hammer et al. (2001) indicated that insights gained would not have been possible without an integrating approach with an explanatory role provided by the crop model. Therefore, this approach can offer some opportunities for exploring how selection strategies using physiologically based molecular markers might enhance the effectiveness of oil palm breeding.

13.5 NEW METHODS FOR CROP PHENOTYPING

Erik Murchie and John Foulkes

13.5.1 Definition of Phenotyping, Its Application and Current Use

There is no doubt that we need to produce substantial and sustainable genetic yield gains in major and minor crops over the coming decades in order to ensure food security and safeguard environmental concerns (Parry et al. 2005; Lobell et al. 2008; Pingali 2012). Moreover, climate change, availability of resources, and changing patterns of land use and degradation will make this problem difficult (Ray et al. 2012, 2013; Challinor et al. 2014).

Rapid exploitation and exploration of crop genetic resources is of paramount importance. There is a need for rapid and high resolution physiological screening of genetically well-characterized populations such as mapping populations, landraces, and inbred varieties in order to identify novel and important traits. For example, wheat production has plateaued in much of the world at least partly due to a lack of variation for key traits (Reynolds et al. 2009; Charmet 2011) caused by genetic bottlenecks during its evolution. However, important and substantial agronomic variation still remains in wild relatives, landraces, and ancestors. Until recently, this could not be exploited due to the difficulty of characterizing genetic material derived from crosses (introgressions). However, technological advances in genotyping (Winfield et al. 2015) mean that this is no longer the case and there is a possibility that the limitation for substantial yield improvement will rest with the high-throughput and frequent measurement of plant physiological properties, linked with the appropriate genotyping. The increasing numbers of mapping populations, tagged mutant collections, and crop species with genomes sequenced makes the need for matching high resolution physiology with high resolution genotyping all the more urgent (Furbank and Tester 2011; Pieruschka and Poorter 2012; Araus and Cairns 2014; Großkinsky et al. 2015).

A phenotype is often defined as the set of an organism's observable characteristics or traits including morphology, development, biochemical or physical properties, and as defined by both genetic alteration and environment. A phenotype is also used to define a smaller subset of traits that are relevant to the subject organism and area of study. As it applies to agriculture "phenotyping" is normally taken to apply to measurements of phenology, morphology, biochemistry, physiology, and disease status of plants during growth. In some cases, this represents the automation of existing manual assessments. Phenomics is, therefore, the measure of the entire "phenome" and as such can be limited by the ability to handle large quantities of data accurately and with biological meaning. The advent of new technologies to rapidly measure more subtle traits vastly expands the possibilities. Furbank and Tester (2011) point out that the commonly used annotation "no visible phenotype" when screening mutants is simply a lack of ability to measure "invisible" traits such as photosynthesis that may be very important. In this chapter, the intention is to provide an overview of common crop phenotyping techniques that may be useful for the nonexpert.

13.5.2 Types of Phenotyping and the State of Play

It may be possible to return to the origins of plant breeding and claim with some credibility that field walking, scoring, and selection for beneficial traits is phenotyping. Many traditional plant physiological techniques such as spectral reflectance measurements, chlorophyll fluorescence (CF), and gas exchange have now been rebranded as phenotyping (Pask et al. 2012). However, the image of phenotyping is now usually one of high-throughput automated techniques, sometimes on an "industrial" scale where vast numbers of plants can be regularly measured for key traits.

This has partly been driven by the reduction in cost of sensors for imaging plant structure and function, such as spectral reflectance and fluorescence. The need for automation has led to the creation of a new plant phenotyping industry (see companies such as Lemnatech, Photon Systems Instruments) which has been driven by the requirements of large seed companies and large public research institutions in Europe, Australia, and elsewhere (e.g., European Plant Phenotyping Network, UK Plant Phenotyping Network, USDA, Australian Plant Phenomics Facility). The plant phenotyping movement is rapidly expanding. Figure 13.2 demonstrates how the use of this term in peer-reviewed literature is increasing at an exponential rate, almost threefold in the last 5 years.

The highest level of automation is seen in glasshouses and growth rooms where plants are moved toward sensors via conveyor belt or robot or where the plants are static and instruments are delivered to the plant via robot (Fiorani and Schurr 2013; Flood 2015). This type of system is suitable for studies where growth in an artificial environment is sufficient for exposing the required phenotypic variation. It will generate highly reproducible data due to the control over the growth environment and the conditions during measurement.

However, there are multiple reasons why the application of phenomics to the field will be critical to support crop breeding. For example, many of the required traits will be induced or expressed in different field environments that cannot be accurately replicated (e.g., Poorter et al. 2012) and $G \times E$ interactions must be accounted for. Highest economic yields of cereal plants are usually obtained in field conditions. The cost of growth and analysis per hectare is substantially lower in the field meaning that field screening will be the default option for many breeding and pre-breeding programs. The development of sophisticated field phenotyping techniques will therefore be important (Araus and Cairns 2014). However, there is an argument to be made that the cost of sensors has plummeted and so low cost but high-tech field phenotyping techniques should be soon available to farmers, breeders, and scientists in poor and rich countries alike (Araus and Cairns 2014; Deery et al. 2014). This is amply demonstrated by the emergence of companies that produce next generation multi parameter tools and focus on crowd sourcing data from growers (see photosynQ.org).

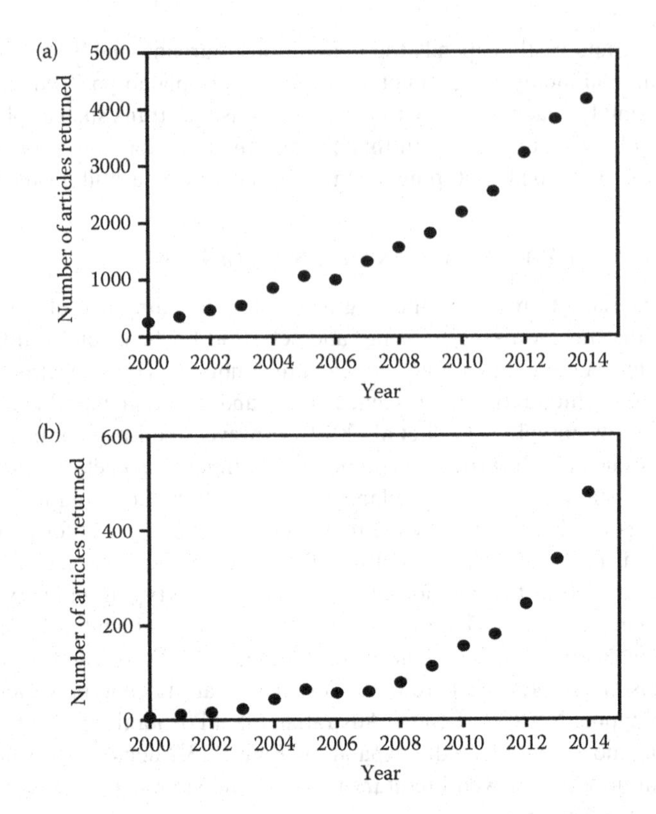

FIGURE 13.2 Exponential growth in the number of articles returned when searching for the term "Plant Phenotyping" in (a) Google Scholar articles with all of the words anywhere in the article and (b) Web of Science (searching within "topic"). These two searches were used to demonstrate how the use of the term has increased, in addition to the inclusion of plant phenotyping as a distinct area of research.

This chapter takes the view that phenotyping consists of a series of nondestructive techniques for which a high degree of commonality between field and controlled environments exists, although the challenge for crops must exist in the field. Therefore, we will cover a selection of these techniques with examples from both environments. It is fair to say that above-ground phenotyping has advanced further than below ground due to the practical difficulties of sensing in soil. The importance of roots should not be underestimated and here we discuss both (Ahmadi et al. 2014).

13.5.3 Canopy (Shoot) Phenotyping Methods

Most high-throughput canopy phenotyping involves cameras and sensors that make automated nondestructive measurements of plants or canopies on a frequent basis over the plant life span. The means of positioning depend on the growth scenario and can include robot and XY arm in a growth room (in which case measurements can be very frequent indeed [hours] [Flood 2015] or via hand [Pask et al. 2012], land vehicle [Deery et al. 2014], aerial vehicle [blimp or unmanned aerial vehicle (UAV)] [Furbank and Tester 2011], or even satellite [Coops et al. 2010; Joiner et al. 2014]).

Commonly, canopy phenotyping utilizes red green blue (RGB) (morphology, three-dimensional [3D] analysis), spectral reflectance (size, pigment content, and function), CF (photosynthesis and stress), thermal imaging (plant temperature, water loss, and photosynthesis), and LIDAR (3D analysis of canopies).

The most straightforward canopy technique can be referred to as "digital growth analysis" (Tester and Furbank 2011) which is an image of projected leaf area, taken in the visible regions (RGB, or longer wavelengths see below) using a standard digital camera at regular points during growth. The correlation between green area and dry weight of the plant can be extremely high (e.g., Rajendran et al. 2009) making it a useful low-tech technique with freely available tools for analysis (see plant-image-analysis.org). With sufficiently frequent measurements, the impact of factors that inhibit growth can be detected. Visible regions can give information on pigmentation and the spatial distribution of color as affected by disease and senescence (Fiorani and Schurr 2013). Occlusion and overlapping prevent accurate measurements of leaf area in mature canopies. RGB can also be used for 3D reconstruction of plant canopies where multiple views of plants are used to generate a point cloud, most easily in a controlled lighting environment. The point cloud, if sufficiently accurate, can be used to generate a two-dimensional (2D) surface reconstruction of the plant for light ray tracing and photosynthesis modeling (Pound et al. 2014; Burgess et al. 2015).

The spectra of light reflected from vegetation (spectral reflectance) can be used to conveniently and remotely assess vegetation properties (Babar et al. 2006; Malenovský et al. 2009). Normalized wavelength ratios can be used to isolate signals that relate to specific signals (such as chlorophyll) and correlate with physiological properties. This gives information on pigment composition, nutrient content, water content, and photosynthetic functioning with the advantage that measurements can be made remotely, even from satellite (Malenovský et al. 2009). When large numbers of wavelengths are considered together, this is termed hyperspectral analysis and, at expense, can be imaged at high resolution (Fiorani and Schurr 2013). Perhaps, the most common measurement is normalized difference vegetation index (NDVI) which is used to detect and quantify living vegetation across a large range of conditions (Govaerts and Verhulst 2010; Cabrera-Bosquet et al. 2012). Commercial NDVI instruments are commonly used in crop research and in agricultural management for measurements of canopy leaf area. Reflectance measurements are affected greatly by ambient light and canopy arrangement. Accuracy in handheld devices can be increased by the use of light emitting diodes (LEDs) as a modulated "excitation" source. Contact devices may provide the greatest accuracy because they are not affected by ambient light but cannot be used for remote measurements. Indices have also been used to detect biomass, relative leaf water content, and other leaf constituents such as nitrogen (Babar et al. 2006; Fiorani and Schurr 2013).

Some indices can provide sophisticated information: the photochemical reflectance index (PRI) can be derived from wavelengths corresponding to changes in the levels of key carotenoids (Peñuelas et al. 1995; Evain 2004). PRI has been shown to correlate with a variety of leaf photosynthetic parameters including light saturation of photosynthesis and photo-protective mechanisms (Murchie and Niyogi 2011;

Murchie et al. 2015). One of the most interesting conclusions from the PRI work thus far is the correlation with plant RUE (Filella et al. 2009; Garbulsky et al. 2011). It can be used to accurately detect the onset of photosynthetic activity in boreal tree species (Wong and Gamon 2015). However, detailed mechanistic links are currently confounded by the action of several physiological factors or "drivers" of the PRI signal (Garbulsky et al. 2011).

Canopy temperature, including point measurements and thermal imaging, is an inexpensive technique for assessing transpiration rates, since water evaporation from the leaves will have a cooling effect. The canopy temperature depression (difference between canopy temperature and ambient air temperature) has been used as an indicator of photosynthetic rates and even yield (Amani et al. 1996; Fischer et al. 1998; Pask et al. 2012). A recent significant advancement was the combination of CF with thermal imaging to provide a way of imaging WUE (McAusland et al. 2013).

CF has been used for many decades as a direct measurement of photosynthetic events in the thylakoid membrane and by comparison provides a much greater degree of insight into the functioning of the leaf. It is one of the commonest measurements in plant physiology. The principle is as follows: light energy absorbed by chlorophyll has three major fates: it can be utilized in photosynthesis (photochemistry), it can be re-emitted as heat in a controlled process called non photochemical quenching (NPQ), or it can be re-emitted as light at a longer wavelength (fluorescence). Fluorescence is typically around 1% of the amount of light absorbed and can be easily measured either in the dark (so-called continuous excitation fluorescence) or in the light (pulse-modulated fluorescence). The three fates for absorbed light energy can be considered to compete with each other and therefore, the yield of fluorescence can be used to indicate the efficiency of the other two processes (photosynthesis and NPQ). There is no scope for a detailed explanation of CF quenching analysis here, for this see reviews (Maxwell and Johnson 2000; Baker 2008; Murchie and Lawson 2013).

The upshot is that at its simplest level a user can point a handheld fluorometer at a leaf and within 1–2 seconds achieve a measurement of the operating efficiency of photosystem II (PSII) (Baker 2008). This parameter, called ϕPSII, yield, $\Delta F/Fm'$ or Fq'/Fm' has been used to correlate well with leaf-level gas exchange, not just in the laboratory (Sharkey et al. 2007) but also the field (Gaju et al. 2016). Therefore, ϕPSII can be used as a proxy for photosynthetic rate although caution must be applied in suboptimal conditions such as stomatal closure (Baker 2008; Murchie and Lawson 2013). CF is used to measure "stress" as it applies to inhibition or downregulation of PSII (Murchie and Niyogi 2011) and after a brief period of dark adaptation, the parameter Fv/Fm has proved useful in detecting tolerance to specific stresses (Baker and Rosenqvist 2004), even in screening wheat cultivars for drought tolerance (Flagella et al. 1996). Continuous CF has been used to accurately detect disease onset (Ajigboye et al. 2016). It is also useful when combined with leaf gas exchange for the measurement of important parameters such as photorespiration (Sharkey et al. 2007).

The usefulness of CF comes at a price because it is difficult to achieve accurate measurements remotely, largely due to the requirement for a saturating bright pulse of light to close PSII reaction centers and for dark adaptation to measure NPQ accurately. In addition, the 3D arrangement of leaves makes even illumination difficult. In this way, the measurement of PRI may prove useful, if the problems with interpretation can be overcome (see above). Therefore, most common CF measurements are made with inexpensive handheld instruments where measurements are extremely quick (Ajigboye et al. 2016). At expense, large heavy panels with highlight output can be mobilized in the field to overcome some of these problems and make measurements of whole canopies such as the recent "scanalyzer" construction at Rothamsted (UK).

Technological variations in CF may help to overcome the limitations for remote sensing, for example, long-term monitoring devices (Porcar-Castell 2011) and the laser-induced fluorescence transient (LIFT). LIFT is a means of providing spatial distribution data from up to 50 m from the target but currently remains a bespoke device (Rascher and Pieruschka 2008; Pieruschka et al. 2010). Another technique for remote sensing of CF is through the "telluric" lines in the daylight spectrum (absorption lines resulting from the presence of specific gases in the atmosphere in the sun's atmosphere). This type of remote fluorescence measurement is made from aircraft and even satellites, but as it uses ambient light to excite fluorescence it is rather limited, but this has not prevented its use in collecting data for vegetation models and attempts to measure crop photosynthesis (Guanter et al. 2014; Joiner et al. 2014).

There has been a lot of recent interest in the use of new phenotyping techniques for the measurement of 3D architecture of plants. 3D analysis can provide information on spatial and temporal light distribution and photosynthesis models. Light detection and ranging (LIDAR) is perhaps the most well-known technique for the measurement of 3D canopy properties commonly applied to trees (Omasa et al. 2007; Wang et al. 2008). Laser scanning of crop canopies, although costly, is now frequently included in automated phenotyping setups because measurements are not affected by ambient light (Paulus et al. 2014). RGB from multiple images using standard single lens reflex (SLR) is substantially cheaper but requires post-processing (Pound et al. 2016). All techniques suffer from the problem of occlusion which can be overcome by removing and scanning individual plants (Burgess et al. 2015).

13.5.4 Root Phenotyping Methods

While advances in genomic approaches to tackle complex traits have been made, the lack of high-throughput and large-scale phenotyping methods for root traits remains a bottleneck to gene discovery and selection for such traits in breeding programs. Recent progress in root measurement methodology has enhanced our ability to visualize, quantify and conceptualize root system architecture traits (RSAT) and their relationship to plant productivity. In controlled environment conditions, the use of soil-filled root-observation chambers (rhizotrons) and nondestructive digital imaging techniques offers some promise (Manschadi et al. 2006, 2010), but may be less suitable for screening of root traits that are expressed at later stages of crop development. Laboratory screens have focused mainly on seedlings, with seedlings growing on germination paper (Hund et al. 2009; Wojciechowski et al. 2009; Bai et al. 2013; Atkinson et al. 2015) or in growth pouches (Bonser et al. 1996; Liao et al. 2004; Hund et al. 2009). In addition, seedling roots grown in clear pots clearly distinguished from the dark soil have been used to image for seminal root number and angle in a large-scale, high-throughput method (Richard et al. 2015). Several non soil-filled methods for evaluating roots are available, including hydroponic, aeroponic, and agar-plate systems (Gregory et al. 2009). Root measurements in laboratory screens tend to be more precise and more reproducible because the plants are grown in a more homogeneous environment compared to the field. Although several screening tests have been designed to generate accurate and robust data from seedlings grown under artificial conditions, these phenotypes have only rarely been extrapolated to field conditions partly because of the pronounced plasticity of root growth and development processes. Laboratory-based methods can be limited in their ability to reproduce field-like conditions (Passioura 2006, 2010; Poorter et al. 2012). For example, soil-environment × genotype interactions significantly affect the root length of wheat cultivars grown in sandy soil compared to agar plates (Gregory et al. 2009). Encouragingly, seedling root traits based on a paper-based germination screen were shown to be linked to mature plant traits such as height and grain yield in 94 doubled-haploid lines derived from a cross between Savannah and Rialto (Atkinson et al. 2015). At an intermediate scale, x-ray computed tomography (Gregory et al. 2003; Lontoc-Roy et al. 2006; Hargreaves et al. 2009; Mooney et al. 2012; Mairhofer et al. 2013), magnetic resonance imaging (Metzner et al. 2015), mini-rhizotrons (Lontoc-Roy et al. 2006; Poorter et al. 2012; Vamerali et al. 2012), and rhizotrons (Nagel et al. 2012; Lobet and Draye 2013) are promising techniques.

Field phenotyping methods for roots in cereals were reviewed by Manske et al. (2001) and Polomski and Kuhn (2002), including the use of rhizotrons, mini-rhizotrons, and assessments of root parameters from soil cores (root washing and root counts/image analysis). However, most of these approaches are low-throughput. There are two relatively high-throughput field phenotyping techniques: the core-break method (Köpke 1979) and shovelomics (Trachsel et al. 2011). In the core-break method, a root auger is used to take soil-root cores from the field, the cores are then broken transversely and the roots on the exposed cross sections counted (Manske et al. 2001). The number of roots visible is then used to estimate root length density and mass from established calibrations. A field study in Australia on a range of genotypes (cultivars, near isogenic lines [NILs], and RILs) by Wasson et al. (2014) indicated that the core-break method can directly identify variation in deep root traits to speed up selection of genotypes. Shovelomics involves the excavation and visual scoring of crown roots extracted from the field, and results in maize have been shown to be well correlated with total root depth and root system total length (Trachsel et al. 2011). In this technique, the researcher excavates the root at a radius of 20 cm around the hypocotyl and 20 cm below soil surface. The root is excavated and washed in water containing mild detergent to remove soil, and the washed root placed on a phenotyping board consisting of a large protractor to measure root angles and score length and density classes of lateral roots and to measure root stem diameters with a caliper. Since the quantification of mature root systems is highly dependent on the researcher reducing the repeatability of measured quantities, an image-based estimation of root traits from extracted crown roots has been developed (Bucksch et al. 2014). An algorithm is then used to automatically calibrate the root trait measurements. Finally, soil core, root washing, and scanning have been successful in describing RSAT of adult plants in the field and controlled environment scale, and have been widely used as a standard technique to compare new methods against (Metzner et al. 2015). This method involves taking soil cores from the field, extracting the roots by washing, and digitizing the root image using a flatbed scanner. The measurement of the RSAT from the digitized image is carried out using appropriate software. The most commonly used software for the analysis of RSAT and root growth kinetics are the commercial WinRHIZO (Regent Instruments, Canada) (Arsenault et al. 1995), the OPTIMAS analysis software (Media Cybernetics, Bethesda, Maryland), and the public domain ImageJ (Schneider et al. 2012). However, fine secondary or tertiary roots can be lost, cut off, or adhere to each other during the cleaning process, making it difficult to analyze the entire root system.

The development of methods that measure changes in the root DNA concentration in soil to allow comparison of soil DNA concentrations from different wheat genotypes could eliminate the need for separation of roots from soil and permit large-scale phenotyping of root genotypes and responses to environmental stresses in the field (Huang et al. 2013). Field phenotyping for root traits in breeding programs is not currently feasible, so genetic progress will depend on the development of high-throughput controlled environment screens or molecular markers for root traits for MAS. The use of nondestructive digital imaging techniques offers some promise, but may be less suitable for screening of root traits that are expressed at later stages of crop development.

13.5.5 Moving Forward: An Affordable Technological Cornucopia?

Commercial interest in developing and selling technology for phenotyping is expanding with some companies producing completed systems (Lemnatech, Photon Systems Instruments) and others selling individual components or sensors that can be incorporated into custom builds. It is important that such developments meet the need for exploration of the increasing level of available genetic resources (Tester and Furbank 2011; Araus and Cairns 2013) and are linked to sophisticated modern analyses of plant function and composition, for example, as provided by metabolomics, transcriptomics, proteomics, and ionomics (Salt et al. 2008; Colmsee et al. 2012).

For research and breeding, individual institutes are likely to have specific needs, focusing on photosynthesis, WUE, root analysis, and so on. Therefore, the technology will need to remain modular and is likely to always remain a set of flexible and changing technologies. The cost of sensors is decreasing and there is a hope that sophisticated measurements will be available to researchers and farmers regardless of financial status. The emergence of low-cost multifunctional devices from companies such as PhotosynQ may be the future and replace sole use instruments such as chlorophyll sensors for nitrogen management (Peng et al. 1996) and CF for screening photosynthesis (Murchie and Lawson 2013).

In the case of field phenotyping, the sensors must by necessity be brought to the plants, or scanned from a distance. In the lab or glasshouse, the infrastructure for moving plants and sensors is costly but more straightforward. In field conditions, the numbers of plants, distances involved, and the potentially harsh conditions demand a separate subset of engineering approaches involving development of appropriate sensors, physical support, and 3D mobility. In some cases, immobile physical support, similar to that seen in the lab, has been deployed in the field, for example, the "scanalyser" at Rothamsted, UK.

Sensor mobility can be achieved using specially adapted wheeled vehicles (so-called "phenomobiles") with attached sensors that are suspended above the crop (Tester and Furbank 2011; Deery et al. 2014). Free airborne sensors capable of close proximity to the plants have attracted a lot of interest and these include aeroplanes, inflatable devices, and unmanned aerial vehicles (so-called pheno-copter (Chapman et al. 2014)). The latter are now cheap and easily available and where the payload is sufficient they are capable of making sophisticated hyperspectral measurements and can be accurately positioned by global positioning system (GPS) (Chapman et al. 2014).

Most phenotyping publications and approaches are developed for smaller model species and crop species, less than 1 m in height. There is no reason why seedlings of tree species (such as oil palm) are not amenable to such methods, even in controlled condition platforms. Larger plants can be accommodated with field equipment that is designed to scan from a distance using aerial platforms. LIDAR is used for 3D structural measurements of trees and forests and estimating biomass (Drake et al. 2003). Models and techniques for estimating oil palm productivity via remote sensing (satellite) are reviewed in Tan et al. (2012). Waring et al. (2010) discussed the roles for ground LIDAR measurements in validating remote sensing techniques for predicting forest growth. The rapid expansion of phenotyping techniques should lend itself to the improvement of tree crops such as oil palm and automation should mean that measurements over the long timescales required are simple to achieve.

13.6 NEW CHALLENGES IN OIL PALM PHENOTYPING IN RELATION TO CLIMATE

Marcel de Raissac, Raphael Perez, and Denis Fabre

13.6.1 Background

The development scheme of oil palm gives some specificity in relation to phenotyping: oil palm architecture is characterized by a mono-axial shoot (the stem), with a unique apical meristem producing phytomers in a regular succession (Henry 1958; Halle and Oldman 1970). These phytomers consist of one leaf, one bound male or female inflorescence, and a section of stem (also called an internode). The number of phytomers produced annually varies from 30–40 in young palms and declines to 20–25 leaves in older palms. The crown presents from 30 to 50 expanded leaves, while more than 40–50 hidden leaves and inflorescences are differentiating and growing under the apical meristem. Consequently, a palm presents at a given time phytomers set in a continuous gradient of developmental stages: from leaf or inflorescence initiation in the apical dome to mature leaf bearing male or female inflorescence and finally to mature bunch and senescent leaf.

This perennial undetermined scheme of development has to be considered in breeding strategies for climate adaptation. Indeed, climate conditions can vary within and between years in an unpredictable way, causing climatic constraints affecting more or less intensively and temporally all phytomers, each of them being at a different developmental stage. In contrast, a soil physical or chemical constraint is likely to affect phytomers in a constant way during their whole life cycle. It is thus much more complex to deal with phenotyping for climate adaptation than nutrient use efficiency.

Commonly, oil palm is grown in areas where climate conditions are constant all over the year, with low-temperature amplitude, regular rainfall, and even solar radiation distributions. Nevertheless, extension of the oil palm growing areas has been done in regions where climate conditions are suboptimal, that is, seasons with low temperatures (Central America), inadequate water supply and drought (West Africa), or low radiation (South America). At all of these sites, oil palm has to face intra-annual variations, but also inter-annual ones. The consequence for breeding and phenotyping is the need for materials with high phenotypic plasticity, able to react to climate variation by different adaptive mechanisms.

Moreover, with predicted climate change, oil palm will have to be adapted to new conditions to maintain its productivity. If the rise in temperature of 1–2°C is not expected to affect oil palm growth and development (with the hypothesis that there is no drastic increase in maximum daily temperatures), the predicted change in some regions for the total annual rainfall and its seasonal distribution requires materials which are more drought tolerant. In parallel, regions with naturally low radiation could suffer from more cloudy weather and adaptation for better radiation interception would be required, while the increase in CO_2 concentration in the air would undoubtedly improve the carbon assimilation processes.

Finally, the improvement of oil palm performance to cope with current and future climate variation will require new phenotyping methods allowing:

- Selection of material with high phenotypic plasticity to improve adaptation to local and transitory constraints, mainly drought; this means inductive responses to drought that help the plant to maintain yield production or protect yield morphogenesis processes during stress periods.
- Continuing selection for constitutive traits for high yield, drought tolerance, and performance under low radiation levels.

Regarding phenotypic plasticity to drought adaptation, a tentative approach for long-term phenotyping is currently carried out through a G × E experiment conducted by a group of private and public partners (Smart-RI, SIAT, PalmElit, Inrab, and Cirad). In this experiment, a set of progenies has been planted in Indonesia, Nigeria, and Benin, the three sites displaying mostly an increasing drought gradient. Long-term monitoring of palms is carried out, with observations of development and growth of each phytomer according to a unique protocol. The data are saved in a database, as well as daily meteorological data, in order to quantify the effects of temporary climate variations (and particularly drought spells characterized by intensity and duration) on the phytomer morphogenesis, the bunch organogenesis, and the final production. The results would allow the quantification of adaptive mechanisms and selection of progenies adapted to local climate constraints, in addition to the global high yielding breeding objectives of ongoing programs. In addition, it would lead to a focus on the basic phenotypic traits to be observed, opening the way for routine phenotyping on larger sets of progenies and promising genetic studies.

Breeding strategies would rely on innovative approaches to better comprehend oil palm productivity. Considering the Monteith equation (Monteith 1977), it is possible to dissect yield as a succession of three basic physiological processes:

$$Y = \varepsilon a \times \varepsilon b \times \varepsilon c \times PARi, \quad (13.1)$$

where:

Y is the bunch yield
εa is the light interception efficiency (capability to intercept solar radiation)
εB is the light conversion efficiency (capability to assimilate the CO_2)
εc is the allocation efficiency (capability to mobilize assimilates to yield organs)
PARi is the intercepted photosynthetically active radiation

Hence, besides working on conventional and integrative phenotypic traits such as the number of bunches and their weight—which contribute to εc-, investigations on maximizing both εa and εb could be a promising option. Such prospects could allow not only enhanced productivity under optimal conditions but also preservation of growth, carbohydrate storage, and potential yield under stress conditions.

13.6.2 Plant Architecture and Light Interception

The study of plant architecture has raised specific questions with regard to the interaction between structures and functions in plants, specifically the influence of photosynthetic organ geometry on light absorption. Several studies investigated plant architecture to highlight the main traits involved in light interception (Chazdon 1985; Takenaka 1994; Valladares and Brites 2004; Pearcy et al. 2005). Leaf area density and the spatial distribution of leaves has consequences on the self-shading of canopy components and therefore on light interception at plant scale (Chazdon 1985; Parveaud et al. 2008). Likewise, internode length and phyllotaxis are likely to be involved in maximizing light interception, acting on sun and shade gradients within the plant. Specific leaf geometry, such as leaf length, petiole length, and leaf angles (inclination and azimuth), are also directly implicated in light capture efficiency (Takenaka 1994; Falster and Westoby 2003; Valladares and Brites 2004).

But studying the influence of structural traits (separately or in combination) on plant performance is not straightforward. Modeling approaches on virtual plants have thus been developed to test hypotheses and set up virtual experiments concerning processes that could otherwise take years in field conditions (Fourcaud et al. 2008), especially for perennial plants like oil palm. Such models are called functional–structural plant models (FSPM) and enable researchers to virtually explore functioning processes on explicit representations of plant architecture (Vos et al. 2010; Dejong et al. 2011). FSPM may thus be interesting tools prior to developing phenotyping strategies as they allow focus on the main structural variables involved in functioning processes. This overall procedure could be seen as model-assisted phenotyping.

In oil palm, a modeling approach was recently developed to reconstruct 3D representations of five progenies (Perez 2016; Perez et al. 2016). This study aimed at developing the basis of a FSPM for oil palm, with the aim to identify aerial architectural traits maximizing light capture. The development of a computational reconstruction of oil palm required the identification and phenotyping of the set of architectural traits needed to explicitly represent the overall plant architecture. Consequently, variables related to the topology and the geometry of plant organs (stem, leaf, and leaflets) were defined. A phenotyping method, based on direct measurements, was developed and allowed to describe the architecture of 60 plants (see Figure 13.3). An allometric-based approach, coupled with a dedicated plant simulator (Vpalm), was then calibrated from field measurements to generate virtual oil palms through progeny-specific parameters. Results pointed out significant architectural differences between the progenies, such as petiole length, number of leaflets, and leaf angles.

It is likely that there is an optimal plant architecture for a specific environment, defined either by agronomic practices (planting density) and/or climatic conditions. To test the influence of architectural traits on light interception, the 3D mocks-ups of oil palm were used to estimate light interception by the MMR model (Dauzat and Eroy 1997; Dauzat et al. 2008) and compare light capture efficiency of the studied progenies (Figure 13.4). First results pointed out contrasting light capture between

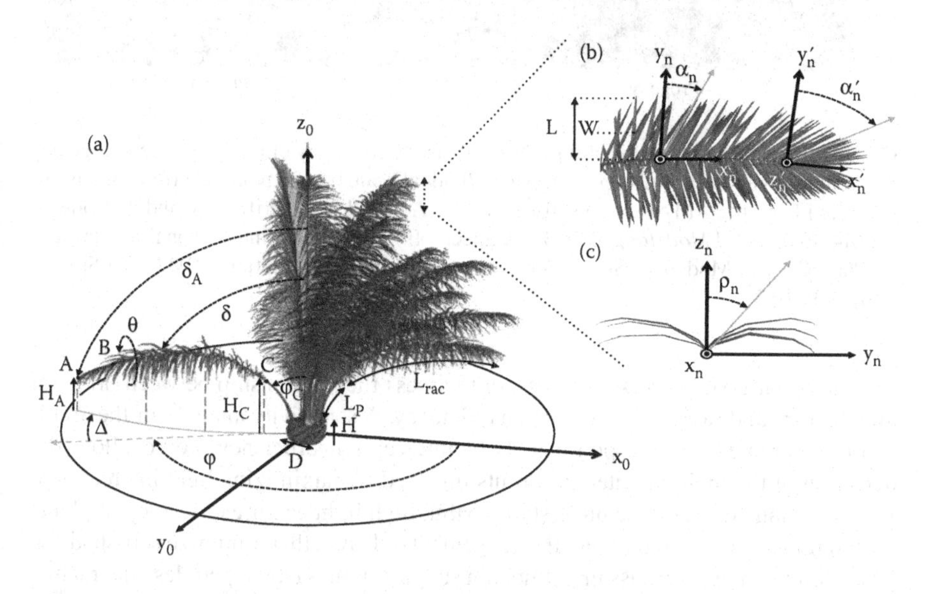

FIGURE 13.3 Geometric variables for assessing and generating 3D oil palm architecture. (a) Variables at plant and leaf scale. H: stem height; D: stem basis diameter; elevation angles (δ) are measured from the vertical reference z_0. Rachis azimuth (Δ) is measured through projected points along the rachis on the plane (x_C, y_C). Phyllotaxis (φ) is measured as the azimuth angle from one leaf insertion relatively to the following one. Rachis twist (θ) is measured as the rotation angle of the rachis local plane from a vertical plane. A, B, and C are reference points on rachis. (b) A detailed top view of a leaf in a horizontal plan. Leaflet length (L) and width (W) are measured in a sample of 10 leaflets per leaf. Axial insertion (azimuth angles αn) is measured with reference to local rachis planes (x_n, y_n). (c) Detailed front view of a leaf in a plane transversal to rachis axis. Radial insertion (elevation angle [ρ_n]) is measured with reference to local rachis plane (z_n, y_n). (Adapted from Perez, R.P.A. et al. 2016. *Journal of Experimental Botany*, erw203.).

FIGURE 13.4 (See color insert.) Light interception efficiency of two contrasted progenies at a young stage (4 years after planting) at individual scale (a) and plot scale (b). Outputs are generated from the AMAPstudio software suite. (Adapted from Griffon, S. and de Coligny, F. 2014. *Ecological Modelling*, 290: 3–10. Special Issue of the 4th International Symposium on Plant Growth Modeling, Simulation, Visualization and Applications (PMA'12) Special Issue of PMA'12.)

progenies but further work is necessary to assess the relationship between oil palm architecture and radiation interception efficiency. A sensitivity analysis of the architectural parameters in response to light interception efficiency would allow the detection of the main architectural traits involved in maximizing light interception. But as demonstrated on the understorey palm, high light capture efficiency of plants is maintained thanks to the spatial arrangement of leaves that minimize self-shading. This requires more biomass investment in support tissues (stem, petioles, and rachis) (Takenaka et al. 2001), and highlights the importance of studying light capture in combination with the conversion and the allocation of carbon within the plant (see Section 13.6.3). This study highlighted the existence of phenotypic variability of oil palm for architecture and light capture, which raises the interesting point that architectural traits have recently been considered as possible criteria for plant breeding (Costes et al. 2004; Sakamoto and Matsuoka 2004). Indeed some studies draw out the existence of genetical determinism of architectural traits for annual crops (Sakamoto and Matsuoka 2004; Wang and Li 2005; Li et al. 2015) and perennial crops (Segura et al. 2007, 2008).

This first step initiated on studying oil palm architecture by Perez et al. (2016) revealed the first indication of the potential for understanding genotype-to-phenotype relationships on oil palm architecture and paves the way for further studies to discuss the possible advantage of considering architecture explicitly in breeding programs. Model-assisted phenotyping may contribute to detect the main architecture traits involved in maximizing plant performance. By limiting the number of architectural traits measured, a higher number of progenies could be phenotyped providing consistent data for genetic studies. Indeed, genetic studies require reproducible and simple measurements easily conducted on a large number of individuals. Consequently, the complementary development of intensive phenotyping methodologies for plant architecture would be necessary. Besides, new elaborated methods for phenotyping oil palm architecture would allow the study of fine geometric traits such as leaf and leaflets angle, which are still labor intensive. The use of remote sensing tools like LiDAR, in combination with the development of software to analyze 3D data, provides promise for high-throughput phenotyping of plant architecture (Côté et al. 2009; Hackenberg et al. 2014). Such technology would present multiple benefits: (i) phenotyping methods for tall plants like oil palms, (ii) nondestructive methods allowing replicated measurements over time to better understand the developmental processes of plant growth and structure, and (iii) collecting rapidly complete and detailed structures of several plants in the same time.

13.6.3 Photosynthesis Phenotyping for Drought Tolerance Using the Fluorescence Transient Technique

Oil palm yield is known to be particularly affected by water availability, often considered as the main limiting factor. Maintenance of photosynthesis with water stress has been shown to be the main adaptive mechanism, that allows carbohydrate storage and its further remobilization when growth of vegetative and reproductive organs is inhibited (Legros et al. 2009a,b; Jazayeri et al. 2015). As oil palm is a perennial crop, efficient, rapid, and early screening techniques for drought tolerance and photosynthesis performance are needed to support breeding programs.

From recent years, chlorophyll *a* fluorescence techniques were shown to be a nonintrusive method, which can be used to give insights into the ability of a plant to tolerate environmental stresses and provide powerful and simple tools for assessment of photosynthetic electron transport and related photosynthetic processes (Kalaji et al. 2012).

They are based on the principle that many abiotic stresses like drought can, directly or indirectly, affect the photosynthetic activity of leaves and as a consequence alter the chlorophyll *a* fluorescence kinetics. The analysis of changes in chlorophyll *a* fluorescence kinetics provides detailed information on the structure and function of the photosynthetic apparatus, especially PSII. These existing methods are generally based on high-frequency monitoring of CF emitted by a dark adapted leaf, as in the OJIP test (Strasser and Strasser 1995; Oukarroum et al. 2007). From the data set collected, some useful parameters can be extracted in relation to objectives: when phenotyping for drought tolerance, performance index (PI) arises as a most valuable parameter to discriminate oil palm progenies for their sensitivity. The PI is a function of three components of photosynthesis, which provides quantitative information about the plant physiological status, and has the advantage of being linearly correlated to photosynthesis in a C3 plant. A derivation of this parameter has been proposed by Strauss et al. (2006) to calculate a new parameter called the drought factor index (DFI). This index expresses the relative drought-induced reduction of the PI in response to a water constraint. It is calculated by the formula

$$DFI = \log A + 2\log B$$

where A is the relative PI measured at intermediate water stress levels and B is the relative PI at a severe stress level.

The stress level is defined by the fraction of transpirable soil water (FTSW), widely used for the evaluation of plant responses to water deficit (Muchow and Sinclair 1991; Ray and Sinclair 1997; Bindi et al. 2005; Davatgar et al. 2009). The FTSW threshold indicates the timing of stomatal closure in response to soil water deficit. This approach assumes that the soil available water used for transpiration varies between field capacity (when transpiration is at the maximum rate, and so FTSW = 1) and wilting point (when transpiration of plants under stress equals 10% of maximum transpiration FTSW = 0).

One illustration of the method is given by a screen carried out in the Smart-RI research station. FTSW values were computed by the procedure previously described by Ray and Sinclair (1997); FTSW was respectively 0.4 for intermediate stress level and 0.15 for severe stress level. The relative PI was calculated as $PI_{drought}/PI_{control}$ using a handy-PEA fluorimeter (Hansatech).

The physiological effects of drought on four oil palm progenies at young stage (6 months old in nursery, grown in pots) were analyzed. The photosynthetic PI parameter measured by CF was used to calculate the DFI, revealing differences between progenies as a function of plant water stress sensitivity. Drought-sensitive genotypes that exhibit the largest reduction in PI during the latter stages of drought stress also have the lowest DFI values (Figure 13.5).

Moreover, the DFI ranking of the tested progenies was in line with the photosynthetic potential maintenance observed in nursery (Figure 13.6) meaning that leaf photosynthetic potential of sensitive progenies was more affected by water stress

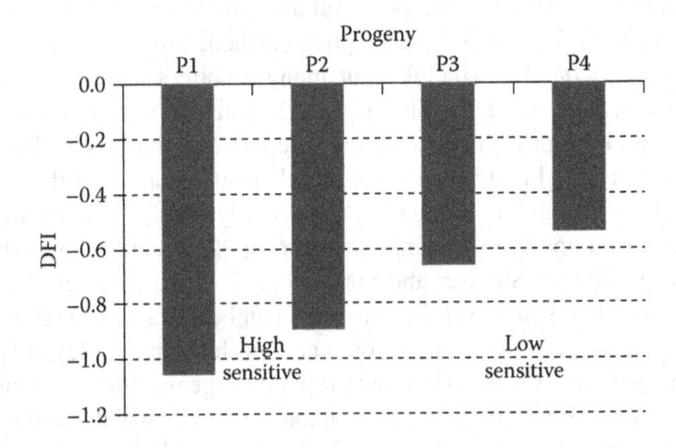

FIGURE 13.5 DFI ranking variability for four progenies in the nursery (2015).

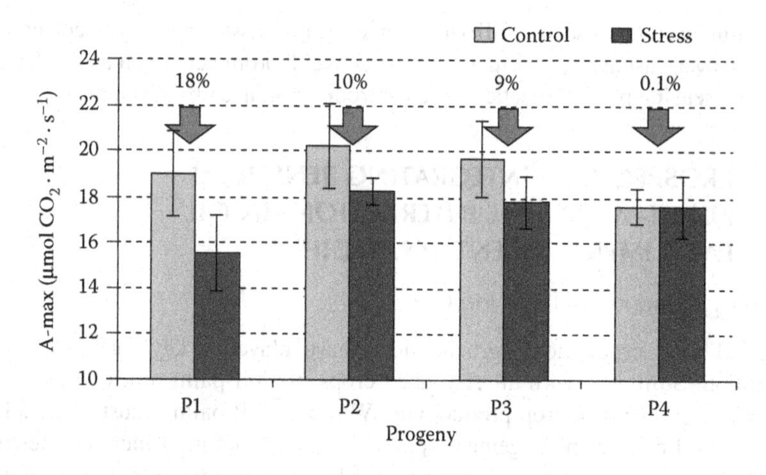

FIGURE 13.6 Photosynthetic potential of the four progenies in the nursery.

(dark bars), compare to controls (clear bars) and better maintenance was observed for low sensitive progenies.

To analyze the test relevancy of this approach in relation to oil palm growth and production, a field evaluation was conducted with the same progenies (5 years old) during 2 years in the dry season and photosynthesis measurements were performed using a Walz GFS-3000 portable photosynthesis system.

The results confirm the sensitivity ranking observed in the nursery (see Figure 13.7). Furthermore, a correlation between net carbon assimilation maintenance under water stress and yield has been found.

Finally, CF has become an important tool to assess photosynthetic performance at different levels, and portable devices taking measurements at the leaf level have proven to be efficient tools. The challenge is now to measure photosynthesis performance at the canopy level and further research is needed to improve and adapt new methodologies

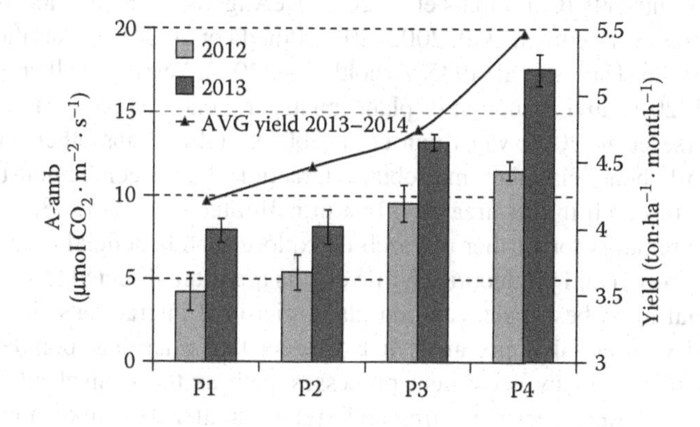

FIGURE 13.7 Net photosynthesis of the same four progenies in the field, in dry season (2012–2013).

like sun-induced fluorescence (SIF) to provide a way to assess photosynthetic processes passively over vast areas, or remote sensing active fluorometers, such as LIFT capable of remote sensing photosynthetic efficiency from a distance up to 50 m.

13.7 PROSPECTS OF INTEGRATING BENEFICIAL PLANT-MICROBIAL INTERACTIONS IN OIL PALM IMPROVEMENT STRATEGIES

Tasren Mahamooth and Kah Joo Goh

In global food production, agrochemicals have played a key role in the Green Revolution. Similarly, with all cultivated crops, the oil palm relies on high inputs of fertilizers to sustain crop productivity. While the oil palm industry has adopted an integrated nutrient management approach by incorporating inorganic fertilizers and organic bio-waste or necromasses, further improvements are still attainable (Corley and Tinker 2015; Gillbanks 2003; Goh and Härdter 2003). No doubt, the cultivation of any crop and agro-management of the land will inadvertently alter soil properties, that is, abiotic and biotic properties (Ding et al. 2013; Araujo et al. 2014). Contemporary research has demonstrated the profound effects of agriculture on microbial activity and diversity, which would include shifts in microbial species and their abundance, and even changes in microbial-associated ecosystem processes (Zhao et al. 2005; Munson et al. 2011; Saygin et al. 2011). It is through such concerns over the perturbations on microbial diversity, abundance, and activity, and consequently soil fertility that has spurred fundamental and applied research to look for new technologies and agro-practices that can contribute to sustainable crop cultivation and productivity. One of the promising technologies relies on harnessing beneficial microbial-plant interactions. These positive interactions between host plants acquiring beneficial microbial services have been demonstrated to increase agriculture productivity and furthermore spur the application of biofertilizers. Microbial services may include production of plant growth regulators or phytohormones (Fallik et al. 1989; Fuentes-Ramírez et al. 1993; Bastián et al. 1998; Son et al. 2014), provision of nutrients (Carvalhais et al. 2013; DeAngelis 2013), enhanced tolerance to abiotic stresses (Redman et al. 2002, 2011; Pineda et al. 2013), alteration of plant functional traits (Harris et al. 1985; Arnold et al. 2003; Perrine-Walker et al. 2007; Davitt et al. 2010) and induction of plant innate immune responses (Leeman et al. 1995; Pieterse et al. 2001; van Loon et al. 2006), and perhaps other mechanisms as well. While beneficial plant-microbial interactions have been demonstrated with many crops research in this area on oil palm is limited (see Table 13.1). Hence, an opportunity remains for further research to explore such beneficial conferred traits to oil palm growth and production. While efforts are often directed toward identifying beneficial microbes, understanding plant-microbial interactions should also be investigated with equal importance. It is evident that microbial populations have an integral role in many ecosystem processes such as those involved in the biogeochemical cycling (nutrient cycling and organic matter decomposition) and even biological control of plant pathogens, yet our current understanding and ability to harness their conferred benefits in agroecosystems is still lacking. Therefore, a better

TABLE 13.1
Effect of Beneficial Microorganisms on Oil Palm Growth

Conferred Benefit to Oil Palm Growth	Beneficial Microorganisms	Palm Age	Growth Responses[a]	References
		N_2-fixing and Phytohormone Production		
Increase in plant growth and nutrient acquisition	*Azospirillum brasilense ATCC* 29729, *Azospirillum lipoferum* CCM 3863 *Bacillus* sp. UPMB 10 *Bacillus* sp. UPMB 13	Nursery seedlings	Increase in root biomass and chlorophyll content	Amir et al. (2003, 2005)
	Azospirillum brasilense ATCC 29729, *Acetobacter diazotrophicus R12*	*In vitro* plantlets	Increase in root and shoot biomass, chlorophyll content	Azlin et al. (2007, 2009)
	Herbaspirillum seropedicae, *Microbacterium* sp., *Acetobacter* sp.	Nursery seedlings	Increase in root and shoot biomass attributed to IAA production	Noor et al. (2009, 2012)
		Arbuscular Mycorrhizae		
	Glomus intraradices	Nursery seedlings	Increase in shoot biomass	Galindo-Castaneda and Romero (2013)
	Glomus etunicatum *Gigaspora rosea* *Acaulospora morrowiae* *Scutellaspora heterogama*	Nursery seedlings	Increase in plant biomass but response limited to compatibility of AMF species	Sundram (2010)
	Glomus sp.	Nursery seedlings	Increase in root and shoot biomass. Increase in P uptake	Blal et al. (1990)
Induction of plant innate immune responses	*Trichoderma harzianum* T32 AMF consortium	Nursery seedlings	Induction of oil palm defence-related genes	Tan et al. (2015)

[a] Growth responses with significant differences between inoculated and control treatments as reported in the selected studies.

understanding of how different factors including plant genotypes, agro-management practices, and agrichemicals influence soil microbial communities is required. Identifying oil palm lineages with better host-symbiotic interrelationships may also serve as one of the many criteria toward oil palm improvement. Inferring from plant-microbial interactions with different genotypes of sugarcane (de Oliveira et al. 2003; Lopes et al. 2012) and even maize (Picard and Bosco 2006; Miyauchi et al. 2008), it remains plausible that different oil palm lineages may exert differential symbiotic associations, attributed to genotype and the presence of symbiotic-associated genes.

13.8 TRANSITIONING THE OIL PALM PLANTATION TO A MORE DIVERSIFIED CROPPING SYSTEM

Chin Kooi Ong

13.8.1 Oil Palm Monoculture

Can we avoid the widespread deforestation which has occurred in S.E. Asia and develop an oil palm agroforestry that is both productive and environmentally friendly? Currently, the production of biofuels presents a new economic opportunity for many developing countries, as well as developed countries to reduce their greenhouse emissions and enhance energy security. Their societal value depends on the extent to which they can address these needs, while at the same time minimizing social and environmental costs. In Latin America, where soybean and sugar cane are being grown for biofuels, massive land-use changes are taking place at the expense of tropical rainforests. In Africa, much of the tropical rainforests could be changed to monoculture commercial oil palm plantation as practiced in S.E. Asia. Recent analysis of carbon payback times for crop-based biofuels in S.E. Asian humid tropics suggests that the current use of S.E. Asian palm oil is far from climate neutral. A wide range of payback time scales ranging from a few years to 1628 years exists for several biofuel crops (Phalan 2009). But for even the best perennial oil crops like oil palm, the carbon payback time ranges from decades to centuries when forested or peat soils are cultivated (www.carbonstocks.org). Using oil palm and palm oil processing wastes for the generation of energy and preventing the conversion of tropical forest into oil palm plantations by establishing further plantings on non-peaty degraded soil can, however, lead to large cuts in the emission of carbon-based GHGs currently associated with the palm oil lifecycle.

The oil palm industry has been fully aware of its societal and environmental responsibility and has taken a very active role in reducing its large carbon footprint as a way to reduce contributions to climate change (www.carbonstocks.org). At the same time, the industry is also very reluctant to change its monoculture management practice. Part of this hesitation is explained by the lack of profitable alternatives to the large-scale industrial plantation offered by the intercropping or agroforestry approaches. The only documentation of a serious attempt to diversify an oil palm plantation was reported recently in the Brazilian Amazon by Miccolis (2015), which described an attempt at using ecologically diverse agroforestry systems in the state of Para northern Brazil, suggesting that it may be possible. The treatments consisted of oil palm planted between wide rows of agroforestry systems on 6 ha plots using slash and mulch leguminous species and organic fertilizer to build up soil fertility. The sites were either highly degraded orchards or pasture lands. The species includes *Theobroma cacao*, *Euterpa oleracea*, *Oenocarpus* sp., *Calophyllum brasiliensis*, *Cajanus cajan*, *Carnavalia* sp., *Tithonia diversifolia*, *Crotolaria spectrabilis*, *Inga edulis*, and *Gliricidia septum*. Results from the 4th year's oil palm bunch yields are, on average, higher than those of monocrop oil palm in the trial. Furthermore, carbon stocks in the multispecies plots were better than in secondary forests or conventional agroforestry systems, and biodiversity is flourishing under oil palm agroforestry. These bunch yields are lower than the highest yields reported in Malaysia but reasonable for marginal lands. Thus, it appears this is the first evidence that oil palm agroforestry can provide society with a socially, economically, and environmentally feasible alternative to monoculture oil palm plantation.

13.8.2 Lessons from Agroforestry

So what are the key lessons from the Brazilian Amazonian oil palm agroforestry experience? In 2010, the Brazil produced only 1% of the world's palm oil but large areas of the country were considered suitable for oil palm and it is expected that there is high potential for employment especially on existing degraded land instead of virgin rainforests. Farmers and large companies are nervous of the monoculture model from S.E. Asia and they fear the risks of plant diseases like Ganoderma white rot. Farmers still want to retain their own production of local staples and cash crops like cassava, beans, cacao, black pepper, and cupucu (*Theobroma grandiflorum*). The aim of the project is to avoid over-shading of oil palm and to determine what species will grow well in local conditions, increase soil fertility, and provide mulch for oil palm and other species. At the 4th year oil palm agroforestry treatments oil palms planted at 81 and 99 palms/ha in a double palm avenue system (Suboh et al. 2009) intercropped with 20 crop species (trees, annuals, creepers) produced 7.7 t/ha per year fresh fruit bunches compared to 5 t/ha per year by the monocrop oil palm planted at the conventional 143 palms/ha. In addition, the carbon stock increased from 60 to 75 Mg carbon/ha, which was attributed to improved soil health. It was hypothesized that the improved soil health was due to the Mexican sunflower (*T. diversifolia*) which has a higher nutrient content and biomass production. This preliminary finding is similar to many agroforestry reports from Africa where Mexican sunflower was shown to improve soil fertility on acid infertile soils (Jama et al. 2000). A longer term study is needed to show that a diversified oil palm agroforestry can indeed be a viable alternative to the monoculture approach common throughout S.E. Asia.

13.9 SUMMING UP: AN IDEOTYPE FOR YIELD AND SUSTAINABILITY

Hereward Corley

13.9.1 Oil Palm Sustainability

The oil palm is highly sustainable in agricultural terms: the oldest plantations have been in existence for a century, with regular replanting and steadily increasing yields. More broadly, though, economic, social, and environmental sustainability are also needed (see Chapter 2). Corley (2006b) considered that the greatest problem confronting the crop in the long term would be economic: as labor costs rise inexorably, it will be increasingly difficult for the oil palm industry to remain competitive with highly mechanized annual oil crops. By 2050, demand for palm oil for edible purposes may exceed 150 Mt, 2.5 times world production in 2014 (Corley 2009a). At today's yield level (world average 3.8 t/ha in 2012), an additional 24 Mha of palms might be required. This figure takes no account of possible demand for biodiesel, which could be effectively unlimited (to replace 10% of world diesel consumption would need over 30 Mha of palms; Corley and Tinker 2015). Clearly, environmental damage from oil palm expansion will be minimized if average yields can be increased significantly. Increased yields will lead to lower costs per ton of oil, but breeders also need to consider ways of improving labor productivity. Both yield and labor productivity are discussed in this chapter; social aspects of sustainability are largely outside the breeder's remit.

13.9.2 Yield Potential

The best yields in breeder's trials are about 11.5 t oil/ha; Fischer et al. (2014) found that in many crops realistically attainable best "farm" (plantation) yields are about 23% below the yield potential, so attainable estate-scale yields should be over 9 t/ha. However, for a perennial crop, we must allow for the replanting cycle and the time required for the introduction of new material. In the past, breeders have increased yields by an average of about 1% per year (Corley and Tinker 2015), so with a 25-year cycle today's plantings should yield about 25% more than those due for replanting, and 12.5% more than the average of all plantings. Thus, if an estate yield of 9 t/ha is attainable with today's material, the average attainable from existing estates is probably about 8 t/ha. Mathews and Foong (2010) quoted yields of over 7.5 t/ha, indicating that the yield gap can be closed with good management.

The oil palm has much higher yields than other oil crops; production costs are correspondingly low, but may be expected to increase. Gan and Ho (1994) estimated that the palm oil industry could absorb a four-fold increase in cost per man-day between then and 2020, but the discrepancy in labor requirements between palm oil and other oils is much greater than a factor of four. Stringfellow (2000) calculated that it takes 0.07 man-days to produce 1 t of soya bean oil, compared with 2 man-days for 1 t of palm oil, 30 times more. Fry (2002) predicted that, unless labor productivity could be improved, Malaysian palm oil would soon have a higher production cost than Brazilian and Argentinian soya bean oil. That has not happened yet, but labor productivity, especially harvester productivity, must be increased if the crop is to remain competitive.

While agronomists work to close the yield gap, the objective for breeders must be to increase the potential yield beyond 12 t/ha. Phenotypic selection for oil yield has achieved much in the past, but definition of the components of growth and yield for an "ideotype," a model plant with characteristics intended to give maximum yield, gives a target for breeders (see Chapters 5 and 9). Combined with new methods of phenotyping (see Chapter 9), MAS (see Chapter 6) and perhaps transformation (see Chapter

9), this could greatly speed up progress. Various authors have described or discussed oil palm ideotypes (Breure and Corley 1983; Squire 1984; Squire and Corley 1987; Corley 2006a; Henson 2007; Corley 2013). A common feature is the ability to yield well at high planting density; this must be combined with high oil/bunch.

13.9.3 Dry Matter Production

For maximum dry matter production (DMP) almost complete light interception is needed, together with a high photosynthetic rate of both exposed and shaded leaves. Light interception, and hence DMP, reaches a plateau at a leaf area index (LAI) of about 7 (see Figure 13.8). For a mature palm with average leaf area of 10 m^2 and 40 leaves (400 m^2/palm), a density of 175 palms/ha is needed to give an LAI of 7. Smaller palms such as crosses with the Pobe dwarf may have a mean leaf area of less than 8 m^2, so a planting density of over 200 palms/ha would be needed with such material.

For maximum life-time yield, this average leaf area should be reached as early as possible. Typically, leaf area does not reach a maximum until about 10 years after planting, but Breure (1985, 2010) showed that in some palms, the maximum is reached much earlier (Calabar in Figure 13.9). Squire and Corley (1987) estimated that for Breure's palms with rapid leaf expansion (Breure 1985), light interception would reach 90% by about the 4th year after planting, 5 years earlier than for those with slow expansion.

Lamade and Setiyo (1996a,b) found a difference in light distribution through the canopy between two contrasting families in Indonesia (extinction coefficient of 0.46 for one, 0.39 for the other). With a lower extinction coefficient, more light will reach the lower leaves in the canopy, thus increasing average rate of photosynthesis. To maintain total light interception, LAI must be increased if the extinction coefficient is lower.

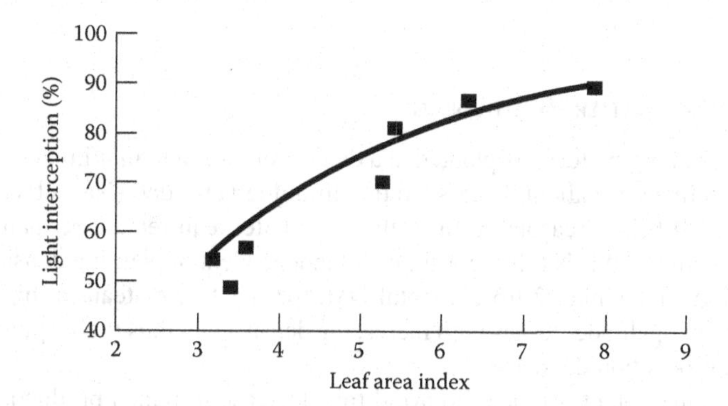

FIGURE 13.8 Light interception in relation to leaf area index: Showing plateau at LAI of about 7. (Adapted from Corley, R.H.V. and Tinker, P.B. 2015 *The Oil Palm*, 5th edn. World Agriculture Series, Wiley, New York; based on data from Squire, G.R. and Corley, R.H.V. 1987. Oil palm. In: *Tree Crop Physiology*. Eds. Sethuraj, M.R. and Raghavendra, A.S., Elsevier Science Publishers, Amsterdam, The Netherlands, pp. 141–167.)

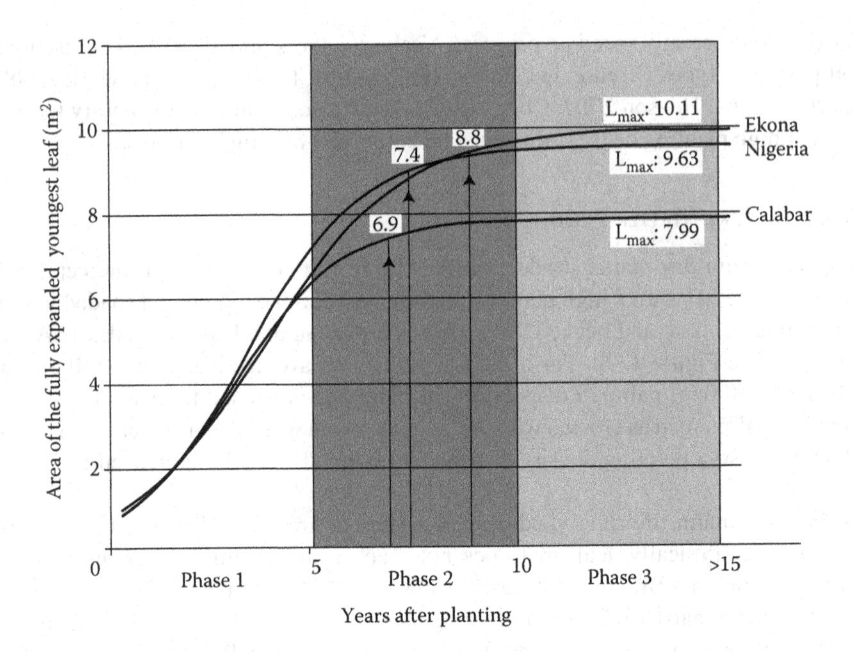

FIGURE 13.9 Increase in mean area per leaf with palm age for D × P progenies from Ekona, Nigeria, and Calabar *Pisiferas*. Phase 1 is the period up to canopy closure; at the end of phase 2, leaf area reaches a stable value, which is maintained throughout phase 3. Vertical arrows indicate time to reach 95% of maximum area. (Adapted from Breure, C.J. 2010. *NJAS—Wageningen Journal of Life Science* 57: 141–147.)

Total DMP of close to 40 t/ha.yr has already been recorded in planting density trials, so this appears a reasonable target for the ideotype. Rapid leaf expansion is needed so that this level can be reached as early as possible in the life of the planting.

13.9.4 Dry Matter Partitioning

If current planting material is planted at a density of 175/ha to maximize DMP, yield tends to decline after about 8 years, so that thinning is needed (Nazeeb et al. 2000; Palat et al. 2012). The reason for this is that dry matter requirement per palm for vegetative growth (VDM) is fairly constant and independent of planting density (Corley 1973; Corley and Tinker 2015). As total DMP per hectare plateaus at high density, production per palm decreases with increasing density, and thus the surplus available for bunch production decreases.

To maximize yield, we need to avoid this decrease in bunch production at high density. Palms with high bunch index (BI, bunch dry matter [BDM] as proportion of total dry matter) will have a higher optimum LAI (Corley 1973). To give a high BI, trunk and leaf dry matter must be reduced. Selection for short trunks will decrease trunk dry matter. Lower leaf dry matter can be obtained either from reduced individual leaf weight, or reduced leaf production rate. To reduce leaf weight without also decreasing light interception, a high leaf area ratio (LAR, ratio of leaf area to leaf dry weight) is needed. Leaf number can be reduced without reducing bunch number, because not all leaves carry bunches. The ideal palm might produce no more than 18 leaves per year at maturity, with 5 male flowers, 13 bunches and no aborted inflorescences. Leaf longevity must also be increased if total photosynthetic production is to be maintained.

In some palms, VDM does not remain constant, but declines as density increases (Corley and Donough 1992). A possible method of screening progenies for this characteristic was described by Corley (1976), based on the fact that effects of severe pruning are generally similar to the effects of high density, but this has not been followed up.

A reasonable target is a BI of 60% (above-ground dry matter only). Data from trials suggest that this should be achievable, but to maintain high BI at high density, it must be combined with a high LAR, and probably with some flexibility of VDM in response to density.

13.9.5 Bunch Composition and FFB Yield

A feasible target appears to be about 35% oil/bunch. This could be achieved with 70% fruit/bunch (F/B), 90% mesocarp/fruit, and 56% oil/fresh mesocarp (70% dry matter in mesocarp, 79% oil/dry mesocarp). All these ratios can be found in current breeding material; indeed, figures of 35% O/B have already been published, but given the known biases in bunch analysis methods (Rao et al. 1983) such laboratory figures may not translate into comparably high factory oil extraction ratios.

BDM content depends on F/B, because fruit has a higher dry matter content than stalk and spikelets (Corley et al. 1971). The relationship is

$$\text{BDM} = 0.37\ \text{F/B} + 0.29$$

Thus, with F/B = 0.7, dry matter content will be 0.55 (55%). From this, we can estimate the expected fresh fruit bunch (FFB) yield: DMP of 40 t/ha.yr and BI = 60% gives 24 t/ha.yr bunch biomass. With 55% dry matter, FFB yield will be 24/0.55 = 43.6 t/ha.yr. At a planting density of 175/ha, with 13 bunches per palm/yr (see above), mean bunch weight will be 19.2 kg.

13.9.6 Disease and Stress Tolerance

With an FFB yield of 43 t/ha.yr and 35% O/B, an oil yield of 15 t/ha should be achieved, but this would be under optimal conditions, implying no stress. However, in S.E. Asia, *Ganoderma* is likely to be a problem even in the best environments, so tolerance to this disease is clearly desirable. For Africa resistance to *Fusarium* wilt is needed. Nursery or *in vitro* selection methods for tolerance or resistance to both these diseases already exist (Mepsted et al. (1995) described an *in vitro* test for *Fusarium* resistance). Another important disease is fatal yellowing in Latin America, probably caused by *Phytophthora palmivora*; although there appears to be variation in tolerance to this, no nursery selection method has yet been developed. Breeding for resistance to these diseases is discussed in Chapter 5E.

Drought tolerance is also important in many areas (Chapter 5); even in Peninsular Malaysia occasional droughts can lead to significant loss of yield. By comparing droughted and irrigated plots, Corley and Palat (2013) showed that there was significant genetic variation in drought tolerance. No satisfactory method has yet been described for selecting drought tolerant material in the absence of drought, but if root density is important (see Chapter 5), then root phenotyping should be investigated (see Chapter 9).

13.9.7 Other Characteristics

To increase labor productivity, harvesting must be partially or completely mechanized. There are several possible ways in which this might be approached (see Chapter 5).

13.9.7.1 Long Bunch Stalks

Such a trait would make harvesting easier, whether by hand or machine. Priwiratama et al. (2010) indicated that long stalks may be a dominant character.

13.9.7.2 Low Lipase

Low lipase activity will reduce free fatty acid (FFA) buildup after harvesting, and could allow longer harvesting intervals without loss of quality. Ngando-Ebongue et al. (2008, 2011) found variation in lipase activity which was correlated with FFA buildup. Morcillo et al. (2013) identified a recessive gene controlling lipase activity, and developed markers for selection.

13.9.7.3 Suitability for Machine Harvesting

There are at least two options for this. With non-abscinding fruit (Donough et al. 1995), whole bunches could be harvested on a long harvesting interval. Oil/bunch would be maximized, as oil synthesis continues in attached fruit after the time when abscission normally starts (Mohanaraj and Donough 2013). However, the inheritance of the non-abscinding character is not at all clear (Corley et al. 2006; Priwiratama et al. 2010), and the best option might be to knock out one or more of the cell wall-degrading enzymes in the abscission zone (Henderson 1998) either by DNA transformation or genome editing.

As an alternative to non-abscinding fruit, loose fruit only could be collected from the base of the palm (Ching and Jasni 2003); machines for loose fruit collection already exist (e.g., Abd Rahim et al. 2011), but detached inner fruit are held tightly within the bunch and do not fall. Thus, a change in bunch structure would be required, with spikelets more widely spaced along the bunch stalk. This characteristic is not normally measured, so it is not known whether useful variation exists.

13.9.7.4 Fruit Color

An obvious color change such as seen in *Virescens* fruit at ripening could be a useful ripeness signal to the harvester, particularly if combined with non-abscinding fruit, where the usual loose fruit signal would be missing. There is some evidence that the color change occurs before the optimum time for harvest (Rao 1998), though Priwiratama et al. (2010) showed that this was not so for all palms. However, the color change might still be used with a "flag-harvesting" system (León Queruz 2003; Mosquera et al. 2009), in which ripe bunches are identified by a separate team before harvesting. Flag harvesting should also increase harvesters productivity, as they need not waste time looking for bunches.

13.9.7.5 Yield Stability

Yield stability from month to month is important from a practical viewpoint (planning of harvesting and milling). One way to achieve this is by judicious mixing of genotypes; Nouy et al. (1999) showed that, in the relatively uniform environment of North Sumatra, different families had quite different yield patterns, with peaks as much as 6 months apart in some years.

With a more severe dry season, the yield peak tends to be more uniform (see Chapter 5). In Benin, with a 5-month dry season, Cros et al. (2013) showed that, although all families peaked at the same time, there were differences in height of the yield peak. Thus, it should be feasible to select genotypes with greater yield stability even in such environments. This might also have a physiological benefit, because photosynthetic rate may decrease when the load of bunches on the palm is low (see Figure 13.10). If that is so, then maintaining a constant bunch load would help to maximize photosynthesis. The data in Figure 13.10 came from palms about 2 years after planting. At that age, there is negligible trunk storage capacity, so lack of sink demand has an immediate effect on photosynthetic rate. In older palms, there may be no decrease in photosynthesis while trunk starch reserves accumulate (Legros

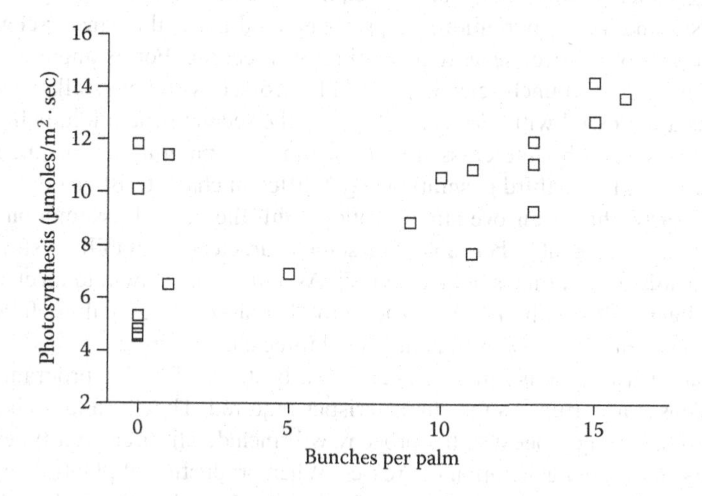

FIGURE 13.10 Relation between the number of bunches and the rate of photosynthesis. (Adapted from Corley, R.H.V. and Tinker, P.B. 2015. *The Oil Palm*, 5th edn. World Agriculture Series, Wiley, New York; based on data from Henson, I.E. 1990. Photosynthesis and source-sink relationships in oil palm (*Elaeis guineensis*). *Transactions of the Malaysian Society of Plant Physiology* 1: 165–171.)

et al. 2009a,b), but eventually this accumulation stops, and in the absence of bunches photosynthesis must decline, as in younger palms (Pallas et al. 2013).

13.9.7.6 Fertilizer Efficiency

For high palm oil yields, high fertilizer inputs (nitrogen, phosphate, and potassium) are usually required, yet there is almost no N, P, or K in palm oil. Thus, there may be considerable scope for improving fertilizer efficiency (see Chapter 2). Much can probably be achieved by agronomic changes (Corley 2009b), but breeding might also contribute. Several authors have demonstrated progeny or clone × fertilizer interactions (Donough et al. 1996; Azman et al. 1998; Kushairi et al. 2001), so it appears likely that genetic variation exists in either uptake or use of nutrients.

13.9.7.7 Oil Composition

There is much interest in producing a more liquid oil (high iodine value = higher unsaturation) (see Chapter 9), mainly because olein fetches a higher price than stearin. However, annual crop breeders are able to change oil composition much more rapidly than oil palm breeders, and farmers can switch to new varieties on an annual basis. Also, it is worth remembering that much liquid vegetable oil is actually hardened before use. In my view, therefore, palm oil breeders should not try to compete with annual crop breeders in terms of oil composition, but should concentrate on increasing yields and improving labor productivity, so that palm oil remains the cheapest general-purpose oil.

13.9.8 Crossing Programs

Breeding material with many of the required characteristics already exists (see Chapters 3 and 4). A "pyramiding" approach could allow the various characteristics to be combined over several generations of breeding. For example, in the first generation, high oil/bunch selections could be crossed with long stalks, while high BI palms are crossed with *virescens*. Then, in the second generation, palms within these crosses could be intercrossed with the aim of combining four of the required characteristics. In the third generation, eight different characteristics might be combined. Clearly, this is an oversimplification, with the crucial assumption that the characters are reasonably heritable. For some characters, heritability estimates are already available, but this is not true for all. As a first step, Priwiratama et al. (2010) crossed long stalks with *virescens* and with "late abscinding" palms, followed by crosses between these groups to combine all three characteristics.

Choice of fruit type is immaterial at the early stages of such a program; parents can be chosen according to the characteristics required. There is a lot to be said for T × T crosses, simply because the progeny will include all three fruit types, giving flexibility in the next generation of crosses. When production of planting material is planned, T × T crosses give the option of either D × P seed (with numbers scaled up by cloning or selfing selected *duras*, and avoiding inbreeding), or *tenera* clones. In a long-term program such as proposed there is no reason to maintain the Deli as a separate population. It is likely that Deli *dura* can be improved for some characteristics by introgressing material of other origins.

Year 1	1st generation, 2-way crosses
Year 2/3	Marker selection of desired combinations, field planting
Year 4/5	Flowering
Year 5	2nd generation, 4-way crosses
Year 6/7	Marker selection and planting 2nd generation
Years 5–8	Standard phenotypic recording of 1st generation crosses
Year 8/9	2nd generation flowering
	Selection based on markers, plus phenotypic data from 1st generation
Year 9	Crosses for 3rd generation, bringing in other characteristics
Year 10/11	Marker selection and planting 3rd generation.
	Possible marker selection of ortets for clone trials

FIGURE 13.11 Possible time scale for marker-assisted "Pyramiding" of characteristics.

13.9.9 Time Scale

With a generation time of 8–10 years, it would take several decades to get anywhere near the ideotype. However, markers for some of the required characteristics already exist, and for others are currently being developed (see Chapter 6). With markers, selection of the most promising individuals within a cross can be done before field planting, thus maximizing the gain at each generation. Such selected individuals can then be crossed for the next generation as soon as they start flowering, giving a generation time of 4–5 years. When it comes to selection for the third generation, phenotypic data will be available from the first generation. Thus, selection at that stage can be concentrated on descendants of palms with the required phenotype, and perhaps with improved predictions from marker genotype (see Figure 13.11).

Commercial planting material could be produced at any stage from such a program by cloning individuals with the desired characteristics (see Chapters 6 and 8). Ortet selection could be done using markers with germinated seeds, in the first stage of a two-stage selection process (Corley 2013).

13.9.10 Conclusions

For oil palm to remain competitive with other oil crops, both yield and labor productivity must be improved. In addition to high yield, requirements include ease of harvesting, preferably through mechanization, and drought and disease tolerance. Individually, most of the necessary characteristics have been identified, and the required values have been achieved; the challenge for breeders is to combine them in commercial planting material.

REFERENCES

Abd Rahim, S., Mohd Ramdhan, K., and Mohd Solah, D. 2011. Innovation and technologies for oil palm mechanization. In: *Further Advances in Oil Palm Research (2000–2010)*, Vol. 1. Eds. Basri Wahid, M., Choo, Y.M., and Chan, K.W., Malaysian Palm Oil Board, Kuala Lumpur, Malaysia, pp. 569–597.

Abdullah, R. 2005. A decade of oil palm gene manipulation. Where Are We Now? In: *9th Int. Conf. Agricultural Biotechnology: Ten Years After*, CEIS-University, Rome, Italy.

Ahmadi, N., Audebert, A., Bennett, M.J. et al. 2014. The roots of future rice harvests. *Rice* 7: 29.

Ajigboye, O.O., Bousquet, L., Murchie, E.H., and Ray, R.V. 2016. Chlorophyll fluorescence parameters allow the rapid detection and differentiation of plant responses in three different wheat pathosystems. *Functional Plant Biology* 43: 356–369.

Alami, Y., Achouak, W., Marol, C., and Heulin, T. 2000. Rhizosphere soil aggregation and plant growth promotion of sunflowers by an exopolysaccharide-producing *Rhizobium* sp. strain isolated from sunflower roots. *Applied Environmental Microbiology* 66: 3393–3398.

Amani, I., Fischer, R.A., and Reynolds, M.P. 1996. Canopy temperature depression association with yield of irrigated spring wheat cultivars in a hot climate. *Journal of Agronomy and Crop Science* 176: 119–129.

Amir, H.G., Shamsuddin, Z.H., Halimi, M.S., Marziah, M., and Ramlan, M.F. 2005. Enhancement in nutrient accumulation and growth of oil palm seedlings caused by PGPR under field nursery conditions. *Communications in Soil Science and Plant Analysis* 35: 15–16.

Amir, H.G., Shamsuddin, Z.H., Halimi, M.S., Ramlan, M.F., and Marziah, M. 2003. N_2 fixation, nutrient accumulation and plant growth promotion by rhizobacteria in association with oil palm seedlings. *Pakistan Journal of Biological Sciences* 6: 1269–1272.

Araujo, A.S.F., Borges, C.D., Siu, M.T., Cesarz, S., and Eisenhauer, N. 2014. Soil bacterial diversity in degraded and restored lands of Northeast Brazil. *Antonie van Leeuwenhoek Journal of Microbiology* 106: 891–899.

Araus, J.L. and Cairns, J.E. 2014. Field high-throughput phenotyping: The new crop breeding frontier. *Trends in Plant Science* 19: 52–61.

Arnold, A., Mejia, L., Kyllo, D., Rojas, E., Maynard, Z., Robbins, N., and Herre, E.A. 2003. Fungal endophytes limit pathogen damage in a tropical tree. In: *Proceedings of the National Academy of Sciences of the United States of America* 100: 15649–15654.

Arsenault, J.-L., Pouleur, S., Messier, C., and Guay, R. 1995. WinRHIZO™, a root-measuring system with a unique overlap correction method. *Horticultural Science* 30: 906.

Atkinson, J.A., Wingen, L.U., Griffiths, M. et al. 2015. Phenotyping pipeline reveals major seedling root growth QTL in hexaploid wheat. *Journal of Experimental Botany* 66: 2283–2292.

Azlin, C.O., Amir, H.G., Chan, L.K., and Zamzuri, I. 2007. Effect of plant growth-promoting rhizobacteria on root formation and growth of tissue cultured oil palm (*Elaeis guineensis* Jacq.). *Biotechnology* 6: 549–554.

Azlin, C.O., Amir, H.G., Chan, L.K., and Zamzuri, I. 2009. Microbial inoculation improves growth of oil palm plants (*Elaeis guineensis* Jacq.). *Tropical Life Sciences Research* 20: 71–77.

Azman, H.M.Y., Lee, C.H., and Razak, I.A. 1998. Oil and kernel yields of oil palm in relation to planting densities, materials and fertiliser rates. In: *Proc. 1996 Int. Conf. Oil and*

Kernel Production in Oil Palm—A Global Perspective. Eds. Rajanaidu, N., Henson, I.E. and Jalani, B.S., Palm Oil Research Institute of Malaysia, Kuala Lumpur, Malaysia, pp. 109–119.

Babar, M.A., Reynolds, M.P., van Ginkel, M., Klatt, A.R., Raun, W.R., and Stone, M.L. 2006. Spectral reflectance to estimate genetic variation for in-season biomass, leaf chlorophyll, and canopy temperature in wheat. *Crop Science* 46: 1046.

Bah, A., Husni, M.H.A., Teh, C.B.S., and Rafii, M.Y. 2014. Runoff loss of nutrients as impacted by conventional and controlled release fertilizers application. *Advances in Agriculture* 2014: 1–9, January 2014, doi: 10.1155/2014/285387.

Bahariaha, B., Parveez, G.K.A., Masani, M.Y.A. et al. 2013. Biolistic transformation of oil palm using the phosphomannose isomerase (pmi) gene as a positive selectable marker. *Biocatalysis and Agricultural Biotechnology* 2: 295–304.

Bai, C., Liang, Y., and Hawkesford, M.J. 2013. Identification of QTLs associated with seedling root traits and their correlation with plant height in wheat. *Journal of Experimental Botany* 64: 1745–1753.

Bais, H.P., Fall, R., and Vivanco, J.M. 2004. The role of root exudates in rhizosphere interactions with plants and other organisms. *Annual Review of Plant Biology* 57: 233–266.

Baker, N.R. 2008. Chlorophyll fluorescence: A probe of photosynthesis in vivo. *Annual Reviews of Plant Biology* 59: 89–113.

Baker, N. R., and Rosenqvist, E. 2004. Applications of chlorophyll fluorescence can improve crop production strategies: An examination of future possibilities. Journal of Experimental Botany. 55: 1607–1621.

Barcelos, E., de Almeida Rios, S., Cunha, R.N.V. et al. 2015. Oil palm natural biodiversity and the potential for yield improvement. *Frontiers in Plant Science* 6: 190.

Bastián, F., Cohen, A., Piccoli, P., Luna, V., Baraldi, R., and Bottini, R. 1998. Production of indole-3-acetic acid and gibberellins A1 and A3 by *Acetobacter diazotrophicus* and *Herbaspirillum seropedicae* in chemically-defined culture media. *Plant Growth Regulation* 24: 7–11.

Belhaj, K., Chaparro-Garcia, A., Kamoun, S., Patron, N.J., and Nekrasov, V. 2015. Editing plant genomes with CRISPR/Cas9. *Current Opinions in Biotechnology* 32: 76–84.

Bertin, P. and Gallais, A. 2001. Physiological and genetic basis of nitrogen use efficiency in maize. II. QTL detection and coincidences. *Maydica* 46: 53–68.

Bhaya, D., Davison, M., and Barrangou, R. 2011. CRISPR-Cas systems in bacteria and archaea: Versatile small RNAs for adaptive defense and regulation. *Annual Review of Genetics* 45: 273–297.

Bhore, S.J. and Shah, F.H. 2012. Genetic transformation of the American oil palm (*Elaeis oleifera*) immature zygotic embryos with antisense palmitoyl-acyl carrier protein thioesterase (PATE) gene. *World Applied Sciences Journal* 16: 362–369.

Bindi, M., Bellesi, S., Orlandini, S., Fibbi, F., Moriondo, M., and Sinclair, T.R. 2005. Influence of water deficit stress on leaf area development and transpiration of Sangiovese grapevines grown in pots. *American Journal of Enology and Viticulture* 56: 68–72.

Blaak, G. and Sterling, F. 1996. The prospects of extending oil palm cultivation to higher elevations through cold tolerant planting material. *The Planter* 72: 645–652.

Blal, B., Morel, C., Gianinazzi-Pearson, V., Fardeau, J.C., and Gianinazzi, S. 1990. Influence of vesicular-arbuscular mycorrhizae on phosphate fertilizer efficiency in two tropical acid soils planted with micropropagated oil palm (*Elaeis guineensis* Jacq.). *Biology and Fertility of Soils* 9: 43–48.

Bonser, A.M., Lynch, H., and Snapp, S. 1996. Effect of phosphorus deficiency on growth angle of basal roots in *Phaseolus vulgaris*. *New Phytologist* 132: 281–288.

Breure, C.J. 1985. Relevant factors associated with crown expansion in oil palm (*Elaeis guineensis* Jacq.). *Euphytica* 34: 161–175.

Breure, C.J. 2010. Rate of leaf expansion: A criterion for identifying oil palm (*Elaeis guineensis* Jacq.) types suitable for planting at high densities. *NJAS—Wageningen Journal of Life Science* 57: 141–147.

Breure, C.J. and Corley, R.H.V. 1983. Selection of oil palms for high density planting. *Euphytica* 32: 177–186.

Breure, C.J., Rosenquist, E.A., Konimor, J., and Powell, M.S. 1986. Oil palm introductions to Papua New Guinea and the formulation of selection methods at Dami Oil Palm Research Station. In: *Int. Workshop on Oil Palm Germplasm and Utilization*. Eds. Soh, A.C., Rajanaidu, N., and Nasir, M., Palm Oil Research Institute of Malaysia, Kuala Lumpur, Malaysia, pp. 189–193.

Bucksch, A, Burridge, J., York, L.M., Das, A., Nord, E., Weitz, J.S., and Lynch, J.P. 2014. Image-based high-throughput field phenotyping of crop roots. *Plant Physiology* 166: 470–486.

Burgess, A.J., Retkute, R., Pound, M.P., Preston, S.P., Pridmore, T.P., Foulkes, J., Jensen, O., and Murchie, E.H. 2015. High-resolution 3D structural data quantifies the impact of photoinhibition on long term carbon gain in wheat canopies in the field. *Plant Physiology* 169: 1192–1204.

Cabrera-Bosquet, L., Crossa, J., von Zitzewitz, J., Serret, M.D., and Araus, J.L. 2012. High-throughput phenotyping and genomic selection: The frontiers of crop breeding converge. *Journal of Integrative Plant Biology* 54: 312–320.

Cahoon, E.B., Shockey, J.M., Dietrich, C.R. et al. 2007. Engineering oilseeds for sustainable production of industrial and nutritional feedstocks: Solving bottlenecks in fatty acid flux. *Current Biology* 10: 236–244.

Caliman, J.P. and Southworth, A. 1998. Effect of drought and haze on the performance of oil palm. In: *1998 Int. Oil Palm Conf. "Commodity of the Past, Today and the Future"*. Eds. Jamika, A. et al., Indonesian Oil Palm Research Institute, Medan, Indonesia, pp. 250–274.

Challinor, A.J., Watson, J., Lobell, D.B., Howden, S.M., Smith, D.R., and Chhetri, N. 2014. A meta-analysis of crop yield under climate change and adaptation. *Nature Climate Change* 4: 287–291.

Chapman, K., Escobar, R., and Griffee, P. 2003. Cold tolerant or altitude adapted oil palm hybrid development initiatives in Asia/Pacific region. In: *Proc. 2001 Int. Palm Oil Congress—Agriculture*. Malaysian Palm Oil Board, Kuala Lumpur, Malaysia, pp. 282–288.

Chapman, S., Merz, T., Chan, A., Jackway, P., Hrabar, S., Dreccer, M., Holland, E., Zheng, B., Ling, T., and Jimenez-Berni, J. 2014. Pheno-copter: A low-altitude, autonomous remote-sensing robotic helicopter for high-throughput field-based phenotyping. *Agronomy* 4: 279–301.

Charmet, G. 2011. Wheat domestication: Lessons for the future. *Comptes Rendus Biologies* 334: 212–220.

Carvalhais, L.C., Dennis, P.G., Fan, B., Fedoseyenko, D., Kierul, K., Becker, A., von Wiren, N., and Borriss, R. 2013. Linking plant nutritional status to plant-microbe interactions. *PLoS One* 8: e68555.

Chazdon, R. 1985. Leaf display, canopy structure, and light interception of two understory palm species. *American Journal of Botany* 72: 1493–1502.

Cheah, S.C., Sambanthamurthi, R., Siti Nor Akmar, A. et al. 1995. Towards genetic engineering oil palm (*Elaeis guineensis* Jacq.). In: *Plant Lipid Metabolism*. Eds. Kader, J.C., and Mazliak, P., Kluwer, Dordrecht, The Netherlands, pp. 570–572.

Ching, P. and Jasni, K. 2003. Mechanising estate operations: The Pamol experience. *The Planter* 79: 95–108.

Chowdhury, M.K.U., Parveez, G.K.A., and Saleh, N.M. 1997. Evaluation of five promoters for use in transformation of oil palm (*Elaeis guineensis* Jacq.). *Plant Cell Reports* 5: 277–281.

Colmsee, C., Mascher, M., Czauderna, T. et al. 2012. OPTIMAS-DW: A comprehensive transcriptomics, metabolomics, ionomics, proteomics and phenomics data resource for maize. *BMC Plant Biology* 12: 245.

Combres, J.-C., Pallas, B., Rouan, L., Mialet-Serra, I., Caliman, J.-P., Braconnier, S., Soulie, J.-C., and Dingkuhn, M. 2013. Simulation of inflorescence dynamics in oil palm and estimation of environment-sensitive phenological phases: A model based analysis. *Functional Plant Biology* 40: 263–279.

Coops, N.C., Hilker, T., Hall, F.G., Nichol, C.J., and Drolet, G.G. 2010. Estimation of light-use efficiency of terrestrial ecosystem from space: A status report. *Bioscience* 60: 788–797.

Corley, R.H.V. 1973. Effects of plant density on growth and yield of oil palm. *Experimental Agriculture* 9: 169–180.

Corley, R.H.V. 1976. Effects of severe leaf pruning on oil palm, and its possible use for selection purposes. *Malaysian Agricultural Research Development Institute Research Bulletin* 4: 23–28.

Corley, R.H.V. 2006a. Potential yield of oil palm—An update. In: Int. Soc. Oil Palm Breeders Symp. "Yield Potential in Oil Palm II", Phuket, Thailand, November 27–28.

Corley, R.H.V., 2006b. Is the oil palm industry sustainable? *The Planter* 82: 121–135.

Corley, R.H.V. 2009a. How much palm oil do we need? *Environmental Science and Policy* 12: 134–139.

Corley, R.H.V. 2009b. Where do all the nutrients go? *The Planter* 85: 133–147.

Corley, R.H.V. 2013. Oil palm breeding—Yesterday, today and tomorrow. In: Int. Society Oil Palm Breeders. Int. Seminar "Oil Palm Breeding—Yesterday, Today and Tomorrow", Kuala Lumpur, Malaysia, November 18.

Corley, R.H.V. and Donough, C.R. 1992. Potential yield of oil palm clones—The importance of planting density. In: *Proc. Workshop Yield Potential in the Oil Palm*. Eds. Rao, V., Henson, I.E., and Rajanaidu, N., International Society Oil Palm Breeder, Kuala Lumpur, Malaysia, pp. 58–70.

Corley, R.H.V., Donough, C.R., Teo, K.W., and Rao, V. 2006. Progress with non-shedding oil palms. *The Planter* 82: 799–807.

Corley, R.H.V., Hardon, J.J., and Tan, G.Y. 1971. Analysis of growth of the oil palm (*Elaeis guineensis* Jacq.). 1. Estimation of growth parameters and application in breeding. *Euphytica* 20: 307–315.

Corley, R.H.V. and Palat, T. 2013. Maximising lifetime yield for greater economic sustainability. In: *Int. Palm Oil Congress "Green Opportunities from the Golden Crop"*, Kuala Lumpur, Malaysia, November 19–21.

Corley, R.H.V. and Tinker, P.B. 2015. *The Oil Palm*, 5th edn. World Agriculture Series, Wiley, New York.

Cornaire, B., Daniel, C., Zuily-Fodil, Y., and Lamade, E. 1994. Oil palm performance under water stress. Background to the problem, first results and research approaches. *Oleagineux* 49: 1–11.

Cornaire, B., Houssou, M., and Muenier, J. 1989. Breeding for drought resistance in the oil palm 2. Kinetics of stomata opening and protoplasmic resistance. In: Paper presented at Int. Conf. 'Palms and Palm Products', November 21–25, Nigerian Institute for Oil Palm Research, Benin City, Nigeria.

Costes, E., Lauri, P., Laurens, F., Moutier, N., Belouin, A., Delort, F., Legave, J., and Regnard, J. 2004. Morphological and architectural traits on fruit trees which could be relevant for genetic studies: A review. *Acta Horticulturae* 663: 349–355.

Côté, J.-F., Widlowski, J.-L., Fournier, R.A., and Verstraete, M.M. 2009. The structural and radiative consistency of three-dimensional tree reconstructions from terrestrial lidar. *Remote Sensing and Environment* 113: 1067–1081.

Cros, D. et al. 2013. Practical aspects of genomic selection in oil palm (*Elaeis guineensis*). In: Paper presented in 2013 ISOPB Symp., Bali, Indonesia.

Dauzat, J., Clouvel, P., Luquet, D., and Martin, P. 2008. Using virtual plants to analyse the light-foraging efficiency of a low-density cotton crop. *Annals of Botany* 101: 1153–1166.

Dauzat, J. and Eroy, M. 1997. Simulating light regime and inter-crop yields in coconut based farming systems. *European Journal of Agronomy* 7: 63–74.

Davatgar, N., Neishabouri, M.R., Sepaskhah, A.R., and Soltani, A. 2009. Physiological and morphological responses of rice (*Oryza sativa* L.) to varying water stress management strategies. *International Journal of Plant Production* 3: 19–32.

Davitt, A.J., Stansberry, M., and Rudgers, J.A. 2010. Do the costs and benefits of fungal endophyte symbiosis vary with light availability? *New Phytologist* 188: 824–834.

DeAngelis, K.M. 2013. Rhizosphere microbial communication in soil nutrient acquisition. In: *Molecular Microbial Ecology of the Rhizosphere*. Ed. de Bruijin, F.J., John Wiley & Sons, Hoboken, New Jersey, pp 823–833.

de Barros, I. et al. 2012. SNOOP: A simulation model for the dynamics of water and nitrogen in oil palm. In: Congresso Brasileiro De Mamona, Simpósio Internacional De Oleaginosas Energéticas, and Fórum Capixaba De Pinhão Manso, Guarapari, Brazil, 2012. Desafios e Oportunidades: Anais. Campina grande: Embrapa Algodão, p. 102.

Deery, D., Jimenez-Berni, J., Jones, H., Sirault, X., and Furbank, R. 2014. Proximal remote sensing buggies and potential applications for field-based phenotyping. *Agronomy* 4: 349–379.

DeJong, T.M., Silva, D.D., Vos, J., and Escobar-Guitiérrez, A.J. 2011. Using functional-structural plant models to study, understand and integrate plant development and ecophysiology. *Annals of Botany* 108: 987–989.

de Oliveira, A.L., de Lima Canuto, E., Reis, V.M., and Baldani, J.I. 2003. Response of micropropagated sugarcane varieties to inoculation with endophytic diazotrophic bacteria. *Brazilian Journal of Microbiology* 34: 59–61.

Ding, C.G., Piceno, Y.M., Heuer, H., Weinert, N., Dohrmann, A.B., Carrillo, A., Andersen, G.L., Castellanos, T., Tebbe, C.C., and Smalla, K. 2013. Changes in soil bacterial diversity as a consequence of agricultural land use in a semi-arid ecosystem. *PLoS One* 8: e59497.

Donald, C.M. 1968. The breeding of crop ideotypes. *Euphytica* 17: 385–403.

Donough, C.R., Corley, R.H.V., and Law, I.H. 1995. "Non-shedding" mutants of the oil palm. *International Society Oil Palm Breeders Newsletter* 11: 10–11.

Donough, C.R., Corley, R.H.V., Law, I.H., and Ng, M. 1996. First results from an oil palm clone × fertiliser trial. *The Planter* 72: 69–87.

Drake, J.B., Knox, R.G., Dubayah, R.O., Clark, D.B., Condit, R., Blair, J.B., and Hofton, M. 2003. Above-ground biomass estimation in closed canopy Neotropical forests using lidar remote sensing: Factors affecting the generality of relationships. *Global Ecology and Biogeography* 12: 147–159.

Dunwell, C.M., Wilkinson, M.J., Nelson, S. et al. 2010. Production of haploids and double haploids in oil palm. *BMC Plant Biology* 10: 218.

Dyer, J.M. and Mullen, R.T. 2008. Engineering plant oils as high-value industrial feedstocks for biorefining: The need for underpinning cell biology research. *Physiologia Plantarum* 132: 11–22.

Engdahl, F.W. 2007. *Seeds of Destruction: The Hidden Agenda of Genetic Manipulation*, Global Research.

Evain, S. 2004. A new instrument for passive remote sensing: 2. Measurement of leaf and canopy reflectance changes at 531 nm and their relationship with photosynthesis and chlorophyll fluorescence. *Remote Sensing of Environment* 91: 175–185.

Fallik, E., Okon, Y., Epstein, Y.E., Goldman, A., and Fischer, M. 1989. Identification and quantification of IAA and IBA in *Azospirillum brasilense* inoculated maize roots. *Soil Biology and Biochemistry* 21: 147–153.

Falster, D.S. and Westoby, M. 2003. Leaf size and angle vary widely across species: What consequences for light interception? *New Phytologist* 158: 509–525.

Fess, T.L., Kotcon, J. B., and Benedito, V. A. 2011. Crop breeding for low input agriculture: A sustainable response to feed a growing world population. *Fess, T.L., Kotcon, J. B., and Benedito, V. A. 2011. Crop breeding for low input agriculture: A sustainable response to feed a growing world population. Sustainability* 3: 1742–1772.

Filella, I., Porcar-Castell, A., Munné-Bosch, S., Bäck, J., Garbulsky, M.F., and Peñuelas, J. 2009. PRI assessment of long-term changes in carotenoids/chlorophyll ratio and short-term changes in de-epoxidation state of the xanthophyll cycle. *International Journal of Remote Sensing* 30: 4443–4455.

Fiorani, F. and Schurr, U. 2013. Future scenarios for plant phenotyping. *Annual Review of Plant Biology* 64: 267–291.

Fischer, R.A., Byerlee, D., and Edmeades, G.O. 2014. *Crop Yields and Global Food Security: Will Yield Increase Continue to Feed the World?* ACIAR Monograph No. 158. Australian Centre for International Agricultural Research, Canberra, Australia, 634 pp.

Fischer, R.A., Rees, D., Sayre, K.D., Lu, Z.-M., Condon, A.G., and Saavedra, A.L. 1998. Wheat yield progress associated with higher stomatal conductance and photosynthetic rate, and cooler canopies. *Crop Science* 38: 1467.

Flagella, Z., Campanile, R.G., Ronga, G., Stoppelli, M.C., Pastore, D., De Caro, A., and Di Fonzo, N. 1996. The maintenance of photosynthetic electron transport in relation to osmotic adjustment in durum wheat cultivars differing in drought resistance. *Plant Science* 118: 127–133.

Flood, P.J. 2015. Natural genetic variation in *Arabidopsis thaliana* photosynthesis. *PhD thesis*, Wageningen University, Wageningen, The Netherlands.

Fourcaud, T., Zhang, X., Stokes, A., Lambers, H., and Körner, C. 2008. Plant growth modelling and applications: The increasing importance of plant architecture in growth models. *Annals of Botany* 101: 1053–1063.

Fry, J. 2002. The competitive position of palm oil in the global oil market. In: *Indonesian Oil Palm Res. Inst. 2002 Int. Oil Palm Conf.*, Bali, Indonesia, July 8–12.

Fuentes-Ramírez, L.E., Jiménez Salgado, T., Abarca Ocampo, I.R., and Caballero-Mellado, J. 1993. *Acetobacter diazotrophicus*, an indoleacetic acid producing bacterium isolated from sugarcane cultivars of México. *Plant and Soil* 154: 145–150.

Furbank, R.T. and Tester, M. 2011. Phenomics—Technologies to relieve the phenotyping bottleneck. *Trends in Plant Science* 16: 635–644.

Gaju, O., DeSilva, J., Carvalho, P., Hawkesford, M. J., Griffiths, S., Greenland, A., et al. 2016. Leaf photosynthesis and associations with grain yield, biomass and nitrogen-use efficiency in landraces, synthetic-derived lines and cultivars in wheat. *Field Crop Research* 193: 1–15.

Galindo-Castaneda, T. and Romero, H.M. 2013. Mycorrhization in oil palm (*Elaeis guineensis* and *E. oleifera* × *E. guineensis*) in the pre-nursery stage. *Agronomia Colombiana* 31: 95–102.

Gan, L.T. and Ho, C.Y. 1994. Impact of rising labour cost on choice of plantation crops for the estate sector. In: *Proc. 1994 Int. Planters Conf., Management for Enhanced Profitability in Plantations.* Ed. Chee, K.H., Incorporated Society of Planters, Kuala Lumpur, Malaysia, pp. 499–511.

Garbulsky, M.F., Peñuelas, J., Gamon, J., Inoue, Y., and Filella, I. 2011. The photochemical reflectance index (PRI) and the remote sensing of leaf, canopy and ecosystem radiation use efficiencies: A review and meta-analysis. *Remote Sensing of Environment* 115: 281–297.

Gerritsma, W. 1988. *Light Interception, Leaf Photosynthesis and Sink Source Relations in Oil Palm.* Wageningen Agricultural University, Wageningen, The Netherlands.

Gerritsma, W. and Wessel, M. 1997. Oil palm: Domestication achieved? *Netherlands Journal of Agricultural Science* 45: 463–475.

Gillbanks, R.A. 2003. Standard agronomic procedures and practices. In: *Oil Palm—Management for Large and Sustainable Yields*. Eds. Fairhurst, T. and Härdter, R., International Potash Institute, Basel, Switzerland, pp. 115–150.

Goh, K.J. and Härdter, R. 2003. General oil palm nutrition. In: *Oil Palm—Management for Large and Sustainable Yields*. Eds. Fairhurst, T. and Härdter, R., International Potash Institute, Basel, Switzerland, pp. 191–230.

Goh, K.J., Ng, P.H.C., and Lee, C.T. 2009. Fertilizer management and productivity of oil palm in Malaysia. In: *Proc. Int. Planters Conf. Plantation Agriculture and Environment*. Ed. Pushparajah, E., Incorporated Society of Planters, Kuala Lumpur, Malaysia, pp. 49–88.

Gopalan, Y., Walker, S., Jahanshiri1, E., and Virdis, S.G.P. 2015. Predicting paddy rice production under an integrated cropping system on Malaysian east coast. In: PAWEES-INWEPF, Crops For The Future, Semenyih, Malaysia, 2015.

Govaerts, B. and Verhulst, N. 2010. *The Normalized Difference Vegetation Index (NDVI) Greenseeker™ Handheld Sensor: Toward the Integrated Evaluation of Crop Management. Part A: Concepts and Case Studies*. CIMMYT, Mexico.

Gregory, P., Bengough, A.G., Grinev, D., Schmidt, S., Thomas, W.(B.)T.B., Wojciechowski, T., and Young, I.M. 2009. Root phenomics of crops: Opportunities and challenges. *Functional Plant Biology* 36: 922–929.

Gregory, P., Hutchison, D., Read, D., Jenneson, P., Gilboy, W., and Morton, E. 2003. Non-invasive imaging of roots with high resolution X-ray micro-tomography. *Plant and Soil* 255: 351–359.

Griffon, S. and de Coligny, F. 2014. Amapstudio: An editing and simulation software suite for plants architecture modelling. *Ecological Modelling*, 290: 3–10. Special Issue of the 4th International Symposium on Plant Growth Modeling, Simulation, Visualization and Applications (PMA'12) Special Issue of PMA'12.

Großkinsky, D.K., Svensgaard, J., Christensen, S., and Roitsch, T. 2015. Plant phenomics and the need for physiological phenotyping across scales to narrow the genotype-to-phenotype knowledge gap. *Journal of Experimental Botany* 66: 5429–5440.

Guanter, L., Zhang, Y., Jung, M. et al. 2014. Global and time-resolved monitoring of crop photosynthesis with chlorophyll fluorescence. In: *Proceedings of the National Academy of Sciences of the United States of America* 111: 1327–1333.

Hackenberg, J., Morhart, C., Sheppard, J., Spiecker, H., and Disney, M. 2014. Highly accurate tree models derived from terrestrial laser scan data: A method description. *Forests* 5: 1069–1105.

Haichar, F.Z., Santaella, C., Heulin, T., and Achouak, W. 2014. Root exudates mediated interactions below ground. *Soil Biology and Biochemistry* 77: 69–80.

Hammer, G.L., Kropff, M.J., Sinclair, T.C., and Porter, J.R. 2001. Future contributions of crop modelling—From heuristics and supporting decision making to understanding genetic regulation and aiding crop improvement. *European Journal of Agronomy* 18: 15–31.

Hanks, R.J. 1983. Yield and water use relationships: An overview. In: *Limitations to Efficient Water Use in Crop Production*. Eds. Taylor, H.M. et al., ASA CSSA SSSA, Madison, Wisconsin, pp. 393–411.

Hallé, F. and Oldeman, R. 1970. *Essai sur l'architecture et la dynamique de croissance des arbres tropicaux*. Masson et Cie, Paris, France

Hargreaves, C.E., Gregory, P.J., and Bengough, A.G. 2009. Measuring root traits in barley (*Hordeum vulgare* ssp. *vulgare* and ssp. *spontaneum*) seedlings using gel chambers, soil sacs and X-ray microtomography. *Plant and Soil* 316: 285–297.

Harris, D., Pacovsky, R., and Paul, E. 1985. Carbon economy of soybean rhizobium-*glomus* associations. *New Phytologist* 101: 427–440.

Henderson, J. 1998. Ripening and abscission in fruit of the oil palm (*Elaeis guineensis* Jacq.): A biochemical investigation. *PhD thesis*, Open University, Oxford Research Unit, Oxford, 228 pp.

Henry, P. 1958. Croissance et développement chez Elaeis guineensis Jacq. de la germination à la première floraison. *Revue générale de botanique* 66: 5–34.

Henson, I.E. 1990. Photosynthesis and source-sink relationships in oil palm (*Elaeis guineensis*). *Transactions of the Malaysian Society of Plant Physiology* 1: 165–171.

Henson, I.E. 1995. Carbon assimilation, water use, and energy balance of an oil palm plantation using micrometeorological techniques. In: *Proc. 1993 PORIM Int. Palm Oil Congress—Agriculture.* Eds. Jalani, B.S. et al., Palm Oil Research Institute of Malaysia, Kuala Lumpur, Malaysia, pp. 137–158.

Henson, I.E. 2007. Modelling the effects of physiological and morphological characters on oil palm growth and productivity. *Oil Palm Bulletin* 54: 1–26.

Heslop-Harrison, J.S.P. 2012. Traits with ecological functions. *Annals of Botany* 110: 139–140.

Hirel, B., Le Gouis, J., Ney, B., and Gallais, A. 2007. The challenge of improving nitrogen use efficiency in crop plants: Towards a more central role for genetic variability and quantitative genetics within integrated approaches. *Journal of Experimental Botany* 58: 2369–2387.

Hirel, B., Tétu, T., Peter, J., Lea, P.J., and Dubois, F. 2011. Improving nitrogen use efficiency in crops for sustainable agriculture. *Sustainability* 3: 1452–1485.

Hoffmann, M.P., Castaneda Vera, A., van Wijk, M.T., Giller, K.E., Oberthür, T., Donough, C., and Whitbread, A.M. 2014. Simulating potential growth and yield of oil palm (*Elaeis guineensis*) with PALMSIM: Model description, evaluation and application. *Agricultural Systems* 131: 1–10.

Hoffmann, M.P., Donough, C., Oberthür, T. et al. 2015. Benchmarking yield for sustainable intensification of oil palm production in Indonesia using PALMSIM. *The Planter* 91: 81–96.

Houssou, M., Meunier, J., and Daniel, C. 1987. Breeding oil palm (*Elaeis guineensis* Jacq.) for drought tolerance. Preliminary results in Benin. In: *Proc. 1987 Int. Oil Palm/Palm Oil Conf. Progress and Prospects. Conf. I—Agriculture.* Eds. Hassan, H.A.H., Chew, P.S., Wood, B.J., and Pushparajah, E., Palm Oil Research Institute of Malaysia, Kuala Lumpur, Malaysia, pp. 23–26.

Huang, C.Y., Kuchel, H., Edwards, J., Hall, S., Parent, B., Herdina, P.E., Hartley, D.M., Langridge, P. and McKay, A.C. 2013. A DNA-based method for studying root responses to drought in field-grown wheat genotypes. *Scientific Reports* 3: 3194.

Hsiao, T.C. and Xu, L.-K. 2000. Sensitivity of growth of roots versus leaves to water stress: Biophysical analysis and relation to water transport. *Journal of Experimental Botany* 51: 1595–1616.

Hsu, P.D., Lander, E.S., and Zhang, F. 2014. Development and applications of CRISPR-Cas9 for genome engineering. *Cell* 157: 1262–1278.

Huang, C.Y., Kuchel, H., Edwards, J., Hall, S., Parent, B., Herdina, P.E., Hartley, D.M., Langridge, P., and McKay, A.C. 2013. A DNA-based method for studying root responses to drought in field-grown wheat genotypes. *Scientific Reports* 3: 3194.

Hund, A., Trachsel, S., and Stamp, P. 2009. Growth of axile and lateral roots of maize: I development of a phenotying platform. *Plant and Soil* 325: 335–349.

Huth, N.I., Banabas, M., Nelson, P.N., and Webb, M. 2014. Development of an oil palm cropping systems model: Lessons learned and future directions. *Environmental Modelling & Software* 62: 411–419.

Ismail, I., Lee, F.S., Abdullah, R. et al. 2010. Molecular and expression analysis of cowpea trypsin inhibitor (CpTI) gene in transgenic *Elaeis guineensis* Jacq leaves. *Australian Journal of Crop Science* 4: 37–48.

ISRIC 2015. ISRIC World Soil Information. ISRIC SoilGrids 1 km visualisation and distribution website. [online] SoilGrids 1 km. Available at: http://soilgrids1km.isric.org/ (Accessed March–May 2015).

Iswandar, H.E.I., Dunwell, J.M., Forster, B.P., Nelson, S.P.C., and Caligari, P.D.S. 2010. Doubled haploid ramets via embryogenesis of haploid tissue cultures. In: *Paper Presented in ISOPB Seminar 2010*, Bali, Indonesia.

Izawati, A.M.D., Masani, M.Y.A., Ismanizan, I., and Parveez, G.K.A. 2015. Evaluation on the effectiveness of 2-deoxyglucose-6-phosphate phosphatase (*DOGR1*) gene as a selectable marker for oil palm (*Elaeis guineensis* Jacq.) embryogenic calli transformation mediated by *Agrobacterium tumefaciens*. *Frontiers in Plant Science* 6: 727.

Izawati, A.M.D., Parveez, G.K., and Masani, M.Y. 2012. Transformation of oil palm using *Agrobacterium tumefaciens*. *Methods in Molecular Biology* 847: 177–188.

Jama, B., Palm, C.A., Buresh, R.J., Niang, A., Gachengo, C., Nzizuheba, G., Armadalo, B. 2000. Tithonia diversifolia as a green manure for soil fertility improvement in western Kenya: A review. Agroforestry Systems 49: 201–221.

Jazayeri, S.M., Rivera, Y.D., Camperos-Reyes, J.E., and Romero, H.M. 2015. Physiological effects of water deficit on two oil palm (Elaeis guineensis Jacq) genotypes. *Agronomia Colombiana* 33: 164–173.

Joiner, J., Yoshida, Y., Vasilkov, A.P. et al. 2014. The seasonal cycle of satellite chlorophyll fluorescence observations and its relationship to vegetation phenology and ecosystem atmosphere carbon exchange. *Remote Sensing of Environment* 152: 375–391.

Jones, D.L., Hodge, A., and Kuzyakov, Y. 2004. Plant and mycorrhizal regulation of rhizodeposition. *New Phytologist* 163: 459–480.

Jones, L.H. 1997. The effects of leaf pruning and other stresses on sex determination in the oil palm and their representation by a computer model. *Journal of Theoretical Biology* 187: 241–260.

Kalaji, H.M., Carpentier, R., Allakhverdiev, S.I., and Bosa, K. 2012. Fluorescence parameters as early indicators of light stress in barley. *Journal of Photochemistry and Photobiology B: Biology* 112: 1–6.

Kee, K.K. and Chew, P.S. 1996. Nutrient loss through surface runoff and soil erosion—Implications for improved fertilizer efficiency in mature oil palms. In: *Proc. PORIM Int. Oil Palm Congress*. Palm Oil Research Institute of Malaysia, Kuala Lumpur, Malaysia, pp. 153–169.

Kole, C. et al. 2015. Application of genomics-assisted breeding for generation of climate resilient crops: Progress and prospects. *Review Frontiers in Plant Science* August 11. 6: 563.

Köpke, U. 1979. Ein Vergleich von Feldmethoden zur Bestimmung des Wurzelwachstums landwirtschaftlicher Kulturpflanzen. *Doctoral thesis*, University of Göttingen, Göttingen, Germany.

Kowalski, S.P. 2015. Golden rice, open innovation, and sustainable global food security. *Industrial Biotechnology* 11: 84–90.

Kraalingan, D.W.C., Breure, K.J., and Spitters, C.S.T. 1989. Simulation of oil palm growth and yield. *Agricultural and Forest Meteorology* 46: 227–244.

Kroon, J.T., Wei, W., Simon, W.J., and Slabas, A.R. 2006. Identification and functional expression of a type 2 acyl-CoA: Diacylglycerol acyltransferase (DGAT2) in developing castor bean seeds which has high homology to the major triglyceride biosynthetic enzyme of fungi and animals. *Phytochemistry* 67: 2541–2549.

Kushairi, A., Rajanaidu, N., and Jalani, B.S. 2001. Response of oil palm progenies to different fertiliser rates. *Journal of Oil Palm Research* 13: 84–96.

Lamade, E. and Setiyo, I.E. 1996a. Test of Dufrêne's production model on two contrasting families of oil palm in North Sumatra. In: *Proc. 1996 PORIM Int. Palm Oil Congress—Competitiveness for the 21st Century*. Eds. Ariffin, D., Basri Wahid, M., Rajanaidu, N., Tayeb Dolmat, M., Paranjothy, K., Cheah, S.C., Chang, K.C., and Ravigadevi, S., Palm Oil Research Institute of Malaysia, Kuala Lumpur, Malaysia, pp. 427–435.

Lamade, E. and Setiyo, I.E. 1996b. Variation of in maximum photosynthesis of oil palm in Indonesia: Comparison of three morphologically different clones. *Plantations, Recherche, Developpment* 3: 429–435.

Lamade, E., Setiyo, I.E., Muluck, C., Hakim, M. et al. 1998. Physiological studies of three contrasting clones in Lampung (Indonesia) under drought in 1997. In: Paper presented at Int. Conf. 'Developments in Oil Palm Plantation Industry for the 21st Century', Bali, Indonesia, September 21–22.

Leeman, M., Van Pelt, J. A., Den Ouden, F. M., Heinsbroek, M., Bakker, P. A. H. M., and Schippers, B., 1995. Induction of systemic resistance by *Pseudomonas fluorescens* in radish cultivars differing in susceptibility to fusarium wilt, using a novel bioassay. *European Journal of Plant Pathology* 101: 655–664.

Ledford, H. 2015a. CRISPR, the disruptor. *Nature* 522: 20–24.

Ledford, H. 2015b. Alternative CRISPR system could improve genome editing. Smaller enzyme may make process simpler and more exact. *Nature* 526: 17.

Lee, C.T., Zaharah, A.R., Chin, C.W., Mohamed, M.H., Mohd Shah, N., Tan, C.C., and Wong, M.K. 2011. Variation of leaf nutrient concentrations in oil palm genotypes and their implication on oil yield. *Paper presented in ISOPB/MPOB. International Seminar on Breeding for Sustainability in Oil Palm*, , Kuala Lumpur, Malaysia, November 18, p. 86.

Legros, S., Mialet-Serra, I., Caliman, J.P., Siregar, F.A., Clément-Vidal, A., Fabre, D., and Dingkuhn, M. 2009a. Phenology and growth adjustments of oil palm (*Elaeis guineensis*) to sink limitation induced by fruit pruning. *Annals of Botany* 104: 1183–1194.

Legros, S., Mialet-Serra, I., Clement-Vidal, A., Caliman, J.P., Siregar, F.A., Fabre, D., and Dingkuhn, M. 2009b. Role of transitory carbon reserves during adjustment to climate variability and source-sink imbalances in oil palm (*Elaeis guineensis*). *Tree Physiology* 29: 1199–1211.

Lei, Y., Xia, W., Mason, A.S. et al. 2014. RNA-sequence analysis of oil palm under cold stress reveals a different C-repeat binding factor (CBF) mediated gene expression pattern in *Elaeis guineensis* compared to other species. *PLoS One* 9: e114482.

León Queruz, A. 2003. La identificación de palmas con racimos a cosechar: una estrategia para incrementar la productividad de la agroindustria de la palma de aceite. In: *Int. Oil Palm Conf.*, Cartagena, Colombia, September 23–26.

Li, C., Li, Y., Shi, Y., Song, Y., Zhang, D., Buckler, E.S., Zhang, Z., Wang, T., and Li, Y. 2015. Genetic control of the leaf angle and leaf orientation value as revealed by ultra-high density maps in three connected maize populations. *PloS One* 10: e0121624.

Lim, K.H., Goh, K.J., Kee, K.K., and Henson, I.E. 2011. Climatic requirements of oil palm. In: *Agronomic Principles & Practices of Oil Palm Cultivation*, Eds. Goh, K.J., Chiu, S.B., Paramanathan, S., Agricultural Crop Trust, Malaysia, pp. 1–37.

Liao, M., Fillery, I., and Palta, J. 2004. Early vigorous growth is a major factor influencing nitrogen uptake in wheat. *Functional Plant Biology* 31: 121–129.

Lobell, D.B., Burke, M.B., Tebaldi, C., Mastrandrea, M.D., Falcon, W.P., and Naylor, R.L. 2008. Prioritizing climate change adaptation needs for food security in 2030. *Science* 319: 607–610.

Lobet, G. and Draye, X. 2013. Novel scanning procedure enabling the vectorization of entire rhizotron-grown root systems. Plant Methods 9: 1–11.

Lontoc-Roy, M., Dutilleul, P., Prasher, S.O., Han, L., Brouillet, T., and Smith, D.L. 2006. Advances in the acquisition and analysis of CT scan data to isolate a crop root system from the soil medium and quantify root system complexity in 3-D space. *Geoderma* 137: 231–241.

Lopes, V.R., Bespalhok-Filho, J.C., de Araujo, L.M., Rodrigues, F.B., Daros, E., and Oliveira, R.A. 2012. The selection of sugarcane families that display better associations with plant growth promoting rhizobacteria. *Journal of Agronomy* 11: 42–52.

Lunshof, J. 2015. Regulate gene editing in wild animals. Nature 521: 127.

Maene, L.M., Thong, K.C., Ong, T.S., and Mokhtarudin, A.M. 1979. Surface wash under mature oil palm. In: *Proc. Symp. Water in Malaysian Agriculture*. MSSS, Kuala Lumpur, Malaysia, pp. 203–216.

Mairhofer, S.S., Zappala, S., Tracy, S., Sturrock, C., Bennett, M.J., Mooney, S.J., and Pridmore, T.P. 2013. Recovering complete plant root system architectures from soil via X-ray mu-Computed Tomography. *Plant Methods* 9: 1–7.

Malenovský, Z., Mishra, K.B., Zemek, F., Rascher, U., and Nedbal, L. 2009. Scientific and technical challenges in remote sensing of plant canopy reflectance and fluorescence. *Journal of Experimental Botany* 60: 2987–3004.

Manschadi, A.M., Christopher, J., de Voil, P., and Hammer, G.L. 2006. The role of root architectural traits in adaptation of wheat to water-limited environments. *Functional Plant Biology* 33: 823–837.

Manschadi, A.M., Christopher, J.T., Hammer, G.L., and de Voil, P. 2010. Experimental and modelling studies of drought-adaptive root architectural traits in wheat (*Triticum aestivum* L.). *Plant Biosystems* 144: 458–462.

Manske, G.G.B., Ortiz-Monasterio, J.I., and Vlek, P.L.G. 2001. Techniques for measuring genetic diversity in roots. In: *Application of Physiology in Wheat Breeding*. Eds. Reynolds, M.P., Ortiz-Monasterio, J.I., and McNab, A., CIMMIYT, Mexico, pp 208–218.

Mao, Y., Zhang, H., Xu, N. et al. 2013. Application of the CRISPR-Cas system for efficient genome engineering in plants. *Molecular Plant* 6: 2008–2011.

Masani, M.Y.A., Noll, G.A., Parveez, G.K.A. et al. 2014. Efficient transformation of oil palm protoplasts by PEG-mediated transfection and DNA microinjection. *Plos One* 9: e98631.

Mathews, J. and Foong, L.C. 2010. Yield and harvesting potentials. *The Planter* 86: 699–709.

Maxwell, K. and Johnson, G.N. 2000. Chlorophyll fluorescence: A practical guide. *Journal of Experimental Botany* 51: 659–668.

McAusland, L., Davey, P.A., Kanwal, N., Baker, N.R., and Lawson, T. 2013. A novel system for spatial and temporal imaging of intrinsic plant water use efficiency. *Journal of Experimental Botany* 64: 4993–5007.

McKeon, T.A. 2005. Genetic modification of seed oils for industrial applications. In: *Industrial Uses of Vegetable Oils*. Ed. Erhan, S.Z., AOCS Press, Urbana, Illinois, pp 1–13.

Méndez, Y.D.R., Chacón, L.M., Bayona, C.J., and Romero, H.M. 2012. Physiological response of oil palm interspecific hybrids (*Elaeis oleifera* H.B.K. Cortes *versus Elaeis guineensis* Jacq.) to water deficit. *Brazilian Journal of Plant Physiology* 24(4).

Mepsted, R., Flood, J., Paul, T., Airede, C., and Cooper, R.M. 1995. A model system for rapid selection for resistance and investigation of resistance mechanisms in *Fusarium* wilt of oil palm. *Plant Pathology* 44: 749–755.

Messina, C., Hammer, G., Dong, Z., Podlich, D., and Cooper, M. 2009. Modelling crop improvement in a G × E × M framework via gene–trait–phenotype relationships. In: *Crop Physiology: Applications for Genetic Improvement and Agronomy*. Eds. Sadras, V., and Calderini, D., Academic Press, Burlington, Massachusetts, pp. 235–266.

Metzner, R., Eggert, A., van Dusschoten, D., Pflugfelder, D., Gerth, S., Schurr, U., Uhlmann, N., and Jahnke, S. 2015. Direct comparison of MRI and X-ray CT technologies for 3D imaging of root systems in soil: Potential and challenges for root trait quantification. *Plant Methods* 11: 1–11.

Miccolis, A. 2015. *The Sustainability of Biofuels Production in Brazil: The Role of Agroforestry Systems*. SCOPE Report, April 14, 2015. World Agroforestry Centre.

Miyauchi, M.Y.H., Lima, D.S., Noguira, M.A., Lovato, G.M., Murate, L.S., Cruz, M.F., Ferrira, J.S., Zangaro, W., and Andrade, G. 2008. Interactions between diazotrophic bacteria and mycorrhizal fungus in maize genotypes. *Science and Agriculture* 65: 525–531.

Moghissi, A.A., Pei, S., and Liu, Y. 2016. Golden rice: Scientific, regulatory and public information processes of a genetically modified organism. *Critical Reviews in Biotechnology* 36(3): 535–541.
Mohanaraj, S.N. and Donough, C.R. 2013. Harvesting practices for maximum yield in oil palm: Results from a reassessment at IJM plantations, Sabah. In: Poster presented at Int. Palm Oil Congress. "*Green Opportunities from the Golden Crop*", Palm Oil Board, Kuala Lumpur, Malaysia, November 19–21, pp. 23–27.
Moll, R.H., Kamprath, E.J., and Jackson, W.A. 1982. Analysis and interpretation of factors which contribute to efficiency of nitrogen utilization. *Agronomy Journal* 74: 562–564.
Monteith, J.L. 1977. Climate and the efficiency of crop production in Britain. *Philosophical Transaction of the Royal Society of London, B* 281: 277–294.
Monteith, J.L. 1986. How do crops manipulate water supply and demand? *Philosophical Transaction of the Royal Society of London, A* 316: 245–289.
Mooney, S.J., Pridmore, T.P., Helliwell, J., and Bennett, M.J. 2012. Developing X-ray computed tomography to non-invasively image 3-D root systems architecture in soil. *Plant and Soil* 352: 1–22.
Morcillo, F., Cros, D., Billotte, N. et al. 2013. Improving palm oil quality through identification and mapping of the lipase gene causing oil deterioration. *Nature Communications* 4: 2160.
Mosquera, M., Fontanilla, C.A., Martínez, R., Sánchez, A.C., and Alarcón, W. 2009. Identifying oil palms with ripe bunches before harvesting, IRBBH: A strategy for increasing labor productivity. In: *Proc. Int. Palm Oil Conf., Agriculture, Biotechnology and Sustainability, Vol. III*, Malaysian Palm Oil Board, Kuala Lumpur, Malaysia, pp. 993–1006.
Muchow, R.C. and Sinclair, T.R. 1991. Water deficits effects on maize yields modeled under current and "greenhouse" climates. *Agronomy Journal* 83: 1052–1059.
Munson, S.M., Belnap, J., and Okin, G.S. 2011. Responses of wind erosion to climate-induced vegetation changes on the Colorado Plateau. In: *Proceedings of the National Academy of Sciences of the United States of America* 108: 3854–3859.
Murchie, E.H., Ali, A., and Herman, T. 2015. Photoprotection as a trait for rice yield improvement: Status and prospects. *Rice* 8: 31.
Murchie, E.H. and Lawson, T. 2013. Chlorophyll fluorescence analysis: A guide to good practice and understanding some new applications. *Journal of Experimental Botany* 64: 3983–3998.
Murchie, E.H. and Niyogi, K.K. 2011. Manipulation of photoprotection to improve plant photosynthesis. *Plant Physiology* 155: 86–92.
Murphy, D.J. 2003. Working to improve the oil palm crop. *Inform* 14: 670–671.
Murphy, D.J. 2006. Molecular breeding strategies for the modification of lipid composition. *In Vitro Cellular and Developmental Biology—Plant* 42: 89–99.
Murphy, D.J. 2007. Future prospects for oil palm in the 21st century: Biological and related challenges. *European Journal of Lipid Science and Technology* 109: 296–306.
Murphy, D.J. 2014a. The future of oil palm as a major global crop: Opportunities and challenges. *Journal of Oil Palm Research* 26: 1–24.
Murphy, D.J. 2014b. Using modern plant breeding to improve the nutritional and technological qualities of oil crops. *Oilseeds & Fats Crops and Lipids* 21: D607, doi: 10.1051/ocl/2014038.
Murphy, D.J. 2016. The potential of biosciences for agricultural improvement: Looking forward to 2050. In: *The Bioscience Revolution in Europe and Africa*. Eds. Ademola, A. and Morris, J., Routledge, Abingdon, pp 145–170.
Nagel, K.A., Putz, A., Gilmer, F. et al. 2012. GROWSCREEN-Rhizo is a novel phenotyping robot enabling simultaneous measurements of root and shoot growth for plants grown in soil-filled rhizotrons. *Functional Plant Biology* 39: 891–904.

Napier, J.A., Usher, S., Haslam, R. et al. 2015. Transgenic plants as a sustainable, terrestrial source of fish oils. *European Journal of Lipid Science Technology*, 117(9): 1317–1324.

NASA-POWER 2015. Prediction of Worldwide Energy Resource. POWER: Agroclimatology Daily Averaged Data [online]. Available at: http://power.larc.nasa.gov/ (Accessed March–May 2015).

Nazeeb, M., Barakabah, S.S., and Loong, S.G. 2000. Potential of high density oil palm plantings in diseased environments. *The Planter* 76: 699–710.

Nelson, A. 2015. Half of Europe opts out of new GM crop scheme, Guardian, October 1, 2015. Available at: www.guardian.com.

Nelson, S.P.C., Wilkinson, M.J., Dunwell, J.M., Forster, B.P., Wening, S., Sitorus, A., Croxford, A., Ford, C., and Caligari, P.D.S. 2009. Breeding for high productivity lines via haploid technology. In: (Unedit.) *Proc. Agriculture, Biotechnology, & Sustainability Conf. PIPOC 2009, Vol. 1.* Malaysian Palm Oil Board, Kuala Lumpur, Malaysia, pp. 203–225.

Ngando-Ebongue, G.F., Koona, P., Nouy, B., Zok, S., Carriere, F., Zollo, P.H.A., and Arondel, V. 2008. Identification of oil palm breeding lines producing oils with low acid values. *European Journal of Lipid Science and Technology* 110: 505–509.

Ngando-Ebongue, G.F., Nouy, B., Koona, P., and Arondel, V. 2011. Assessment of lipase activity in the mesocarp of the fruit of the oil palm: A useful tool for the identification of low acidity lines. In: Int. Palm Oil Congress "Palm Oil: Fortifying and Energising the World", Malaysian Palm Oil Board, Kuala Lumpur, Malaysia, November 15–17.

Noor, A.O., Amir, H.G., Chan, L.K., and Othman, A.R. 2009. Influence of various combinations of diazotrophs and chemical N fertilizer of plant growth and N_2 fixation capacity of oil palm seedlings (*Elaeis guineensis* Jacq.). *Thai Journal of Agricultural Sciences* 42: 139–149.

Noor, A.O., Tharek, M., Keyeo, F., Chan, L.K., Zamzuri, I., Ahmad Ramli, M.Y., and Amir, H.G. 2012. Influence of indole-3-acetice acid (IAA) produced by diazotrophic bacteria on root development and growth of in vitro oil palm shoots (*Elaeis guineensis* Jacq.). *Journal of Oil Palm Research* 25: 100–107.

Nouy, B., Baudouin, L., Djégui, N., and Omoré, A. 1999. Oil palm under limiting water supply conditions. *Plantations, Recherche, Développement* 6: 31–45.

Nurfahisza, A.R., Rafiqah, M.A., Masani, M.Y.A. et al. 2014. Molecular analysis of transgenic oil palm to detect the presence of transgenes. *Journal of Oil Palm Research* 26: 96–103.

Okwuagu, C.O. and Ataga, C.D. 1999. Oil palm breeding programme in Nigeria. In: *Proc. Symp. "The Science of Oil Palm Breeding"*. Eds. Rajanaidu, N., and Jalani, B.S., Palm Oil Research Institute of Malaysia, Kuala Lumpur, Malaysia, pp. 131–138.

Omasa, K., Hosoi, F., and Konishi, A. 2007. 3D lidar imaging for detecting and understanding plant responses and canopy structure. *Journal of Experimental Botany* 58: 881–898.

Oukarroum, A., El Madidi, S., Schansker, G., and Strasser, R.J. 2007. Probing the responses of barley cultivars (*Hordeum vulgare* L.) by chlorophyll *a* fluorescence OLKJIP under drought stress and re-watering. *Environmental Experimental Botany* 60: 438–446.

Palat, T., Chayawat, N., and Corley, R.H.V. 2012. Maximising oil palm yield by high density planting and thinning. *The Planter* 88: 241–256.

Pallas, B., Mialet-Serra, I., Rouan, L., Clement-Vidal, A., Caliman, J.P., and Dingkuhn, M. 2013. Effect of source/sink ratios on yield components, growth dynamics and structural characteristics of oil palm (*Elaeis guineensis*) bunches. *Tree Physiology* 33: 409–424.

Parry, M.A.J. and Hawkesford, M.J. 2010. Food security: Increasing yield and improving resource use efficiency. Symposium on 'Food supply and quality in a climate-changed world'. In: *Proceedings of the Nutrition Society (2010)* 69: 592–600.

Parry, M.A.J., Rosenzweig, C., and Livermore, M. 2005. Climate change, and risk global food supply of hunger. *Philosophical Transactions of the Royal Society B: Biological Sciences* 360: 2125–2138.

Parveaud, C.-E., Chopard, J., Dauzat, J., Courbaud, B., and Auclair, D. 2008. Modelling foliage characteristics in 3D tree crowns: Influence on light interception and leaf irradiance. *Trees* 22: 87–104.

Parveez, G.K.A. 1998. Optimization of parameters involved in the transformation of oil palm using the biolistics method. *PhD Thesis*, Universiti Putra Malaysia, Serdang, Malaysia.

Parveez, G.K.A. and Bahariah, B. 2012. Biolistic-mediated production of transgenic oil palm. *Methods in Molecular Biology* 847: 163–175.

Parveez, G.K.A., Bahariah, B., Ayub, N.H. et al. 2015. Production of polyhydroxybutyrate in oil palm (*Elaeis guineensis* Jacq.) mediated by microprojectile bombardment of PHB biosynthesis genes into embryogenic calli. *Frontiers in Plant Science* 6: 598.

Parveez, G.K.A., Chowdhury, M.K.U., and Saleh, N.M. 1997. Physical parameters affecting transient GUS gene expression in oil palm (*Elaeis guineensis* Jacq.) using the biolistic device. *Industrial Crops and Products* 6: 41–50.

Parveez, G.K.A. and Christou, P. 1998. Biolistic-mediated DNA delivery and isolation of transgenic oil palm (*Elaeis guineensis* Jacq.) embryogenic callus cultures. *Journal of Oil Palm Research* 10: 29–38.

Parveez, G.K.A., Masri, M.M., Zainal, A. et al. 2000. Transgenic oil palm: Production and projection. *Biochemical Society Transactions* 28: 969–972.

Pask, A., Pietragalla, J., Mullan, D., and Reynolds, M.P. 2012. *Physiological Breeding II: A Field Guide to Wheat Phenotyping*. CIMMYT, Mexico, DF.

Passioura, J.B. 2006. The perils of pot experiments. *Functional Plant Biology* 33: 1075–1079.

Passioura, J.B. 2010. Scaling up: The essence of effective agricultural research. *Functional Plant Biology* 37: 585–591.

Paterson, R.R.M., Moen, S., and Lima, N. 2009. The feasibility of producing oil palm with altered lignin content to control Ganoderma disease. *Journal of Phytopathology* 157: 649–656.

Paterson, R.R.M., Kumar, L.,Taylor S., and Lima, N. 2015. Future climate effects on suitability for growth of oil palms in Malaysia and Indonesia. Scientific Reports 5: Article number 14457.

Paulus, S., Behmann, J., Mahlein, A.-K., Plümer, L., and Kuhlmann, H. 2014. Low-cost 3D systems: Suitable tools for plant phenotyping. *Sensors* 14: 3001–3018.

Pearcy, R., Muraoka, H., and Valladares, F. 2005. Crown architecture in sun and shade environments: Assessing function and trade-offs with a three-dimensional simulation model. *New Phytologist* 166: 791–800.

Peng, S., Garcia, F.V., Laza, R.C., Sanico, A.L., Visperas, R.M., and Cassman, K.G. 1996. Increased N-use efficiency using a chlorophyll meter on high-yielding irrigated rice. *Field Crops Research* 47: 243–252.

Peñuelas, J., Filella, I., and Gamon, J.A. 1995. Assessment of photosynthetic radiation-use efficiency with spectral reflectance. *New Phytologist* 131: 291–296.

Perez, R.P.A. 2016., Toward a functional-structural model of oil palm: Evaluation of genetic differences between progenies for architecture and radiation interception efficiency. In: Paper presented at ICOPE 2016, Bali, Indonesia.

Perez, R. P., Pallas, B., Moguédec, G. L., Rey, H., Griffon, S., Caliman, J.-P., Costes, E., and Dauzat, J. 2016. Integrating mixed-effect models into an architectural plant model to simulate inter- and intra-progeny variability: A case study on oil palm (*Elaeis guineensis Jacq.*). *Journal of Experimental Botany*, erw203. doi:10.1093/jxb/erw203

Perrine-Walker, F.M., Gartner, E., Hocart, C.H., Becker, A., and Rolfe, B.G. 2007. *Rhizobium*-initiated rice growth inhibition caused by nitric oxide accumulation. *Molecular Plant Microbe Interactions* 20: 283.

Phalan, B. 2009. The societal and environmental impacts of biofuels in Asia: An overview. *Applied Energy* 86: 521–529.

Picard, C. and Bosco, M. 2006. Heterozygosis drives maize hybrids to select elite 2,4-diacethylphloroglucinol-producing *Pseudomonas* strains among resident soil populations. *FEMS Microbiology Ecology* 58: 193–204.

Pieruschka, R., Klimov, D., Kolber, Z.S., and Berry, J.A. 2010. Monitoring of cold and light stress impact on photosynthesis by using the laser induced fluorescence transient (LIFT) approach. *Functional Plant Biology* 37: 395.

Pieruschka, R. and Poorter, H. 2012. Phenotyping plants: Genes, phenes and machines. *Functional Plant Biology* 39: 813–820.

Pieterse, C.M.J., Ton, J., and van Loon, L.C. 2001. Cross-talk between plant defence signaling pathways: Boost or burden? *Agriculture Biotechnology Network* 3: 1–18.

Pineda, A., Dicke, M., Pieterse, C.M.J., and Pozo, M.J. 2013. Beneficial microbes in a changing environment: Are they always helping plants to deal with insects? *Functional Ecology* 27: 574–586.

Pingali, P.L. 2012. Green revolution: Impacts, limits, and the path ahead. In: *Proceedings of the National Academy of Sciences of the United States of America* 109: 12302–12308.

Podlich, D.W., Cooper, M., and Basford, K.E. 1999. Computer simulation of a selection strategy to accommodate genotype-by-environment interaction in a wheat recurrent selection program. *Plant Breeding* 118: 17–28.

Polomski, J. and Kuhn, N. 2002. Root research methods. In: *Plant Roots: The Hidden Half*, 3rd edn. Eds. Waisel, Y., Eshel, A., and Kafkafi, U., Marcel Dekker, Inc., New York, pp. 300–306.

Ponniah, R. 2013. Oil palm in India with difference: In various agroclimatic conditions and with inter/mixed cropping under irrigated condition. In: *Proc. 2013 PIPOC*, Kuala Lumpur, paper AP5, pp. 18–25.

Poorter, H., Bühler, J., van Dusschoten, D., Climent, J., and Postma, J.A. 2012. Pot size matters: A meta-analysis of the effects of rooting volume on plant growth. *Functional Plant Biology* 39: 839.

Porcar-Castell, A. 2011. A high-resolution portrait of the annual dynamics of photochemical and non-photochemical quenching in needles of *Pinus sylvestris*. *Physiologia Plantarum* 143: 139–153.

Pound, M., French, A., Murchie, E., and Pridmore, T. 2014. Automated recovery of 3D models of plant shoots from multiple colour images. *Plant Physiology* 144: 1688–1698.

Pound, M. P., French, A. P., Fozard, J. A., Murchie, E. H., and Pridmore, T. P. 2016. A patch-based approach to 3D plant shoot phenotyping. *Machine Vision and Application*. doi:10.1007/s00138-016-0756-8.

Priwiratama, H., Djuhjana, J., Nelson, S.P.C., and Caligari, P.D.S. 2010. Progress of oil palm breeding for novel traits: Late abscission, virescence and long bunch stalks. *Int. Oil Palm Conf.*, Yogyakarta, Indonesia, June 1–3.

Rafalski, J.A. 2010. Association genetics in crop improvement. *Current Opinion in Plant Biology* 13: 174–180.

Rajendran, K., Tester, M., and Roy, S.J. 2009. Quantifying the three main components of salinity tolerance in cereals. *Plant, Cell and Environment* 32: 237–249.

Rao, V. 1998. Ripening in the virescens oil palm. In: *Proc. 1996 Int. Conf. Oil and Kernel Production in Oil Palm—A Global Perspective*. Eds. Rajanaidu, N., Henson, I.E., and Jalani, B.S., Palm Oil Research Institute of Malaysia, Kuala Lumpur, Malaysia, p. 226.

Rao, V., Soh, A.C., Corley, R.H.V. et al. 1983. A critical reexamination of the method of bunch quality analysis in oil palm breeding. *Palm Oil Research Institute of Malaysia, Occ. Paper* 9: 1–28.

Rascher, U. and Pieruschka, R. 2008. Spatio-temporal variations of photosynthesis: The potential of optical remote sensing to better understand and scale light use efficiency and stresses of plant ecosystems. *Precision Agriculture* 9: 355–366.

Ray, D.K., Mueller, N.D., West, P.C., and Foley, J.A. 2013. Yield trends are insufficient to double global crop production by 2050. *PloS One* 8: e66428.

Ray, D.K., Ramankutty, N., Mueller, N.D., West, P.C., and Foley, J.A. 2012. Recent patterns of crop yield growth and stagnation. *Nature Communications* 3: 1293.

Ray, J.D. and Sinclair, T.R. 1997. Stomatal closure of maize hybrids in response to soil drying. *Crop Science* 37: 803–807.

Redman, R.S., Kim, Y.O., Woodward, C.J.D.A., Greer, C., Espino, L., Doty, S.L., and Rodriguez, R.J. 2011. Increased fitness of rice plants to abiotic stress via habitat adapted symbiosis: A strategy for mitigating impacts of climate change. *PLoS One* 6: e14823.

Redman, R.S., Sheekan, K.B., Stout, R.G., Rodriguez, R.J., and Henson, J.M. 2002. Thermotolerance generated by plant/fungal symbiosis. *Science* 298: 1581.

Reynolds, M., Foulkes, M.J., Slafer, G.A., Berry, P., Parry, M.A.J., Snape, J.W., and Angus, W.J. 2009. Raising yield potential in wheat. *Journal of Experimental Botany* 60: 1899–1918.

Ribaut, J.M., deVicente, M.C., and Delannay, X. 2010. Molecular breeding in developing countries: Challenges and perspectives. *Current Opinion in Plant Biology* 13: 213–218.

Richard, C.A., Hickey, L.T., Fletcher, S., Jennings, R., Chenu, K., and Christopher, J.T. 2015. High-throughput phenotyping of seminal root traits in wheat. *Plant Methods* 11: 13.

Ritchie, J.T. and Basso, B. 2008. Water use efficiency is not constant when crop water supply is adequate or fixed: The role of agronomic management. *European Journal of Agronomy* 28: 273–281.

Ruslan, A., Zainal, A., Heng, W.Y. et al. 2005. Immature embryo: A useful tool for oil palm (*Elaeis guineensis* Jacq.) genetic transformation studies. *Electronic Journal of Biotechnology* 8: 25–34.

Sakamoto, T. and Matsuoka, M. 2004. Generating high-yielding varieties by genetic manipulation of plant architecture. *Current Opinion in Biotechnology* 15: 144–147.

Salt, D.E., Baxter, I., and Lahner, B. 2008. Ionomics and the study of the plant ionome. *Annual Reviews of Plant Biology* 9: 709–733.

Sambanthamurthi, R., Singh, R., Parveez, G.K.A. et al. 2009. Opportunities for the oil palm via breeding and biotechnology. In: *Breeding Plantation Crops*. Eds. Mohan Jain, M., Priyadarshan, P.M., Springer, New York, 654pp.

Saygin, S.D., Basaran, M., Ozcan, A.U., Dolarslan, M., and Timur, O.B. 2011. Land degradation assessment by geo-spatially modeling different soil erodibility equations in a semi-arid catchment. *Environmental Monitoring Assessments* 180: 201–215.

Schneider, C.A., Rasband, W.S., and Eliceiri, K.W. 2012. NIH Image to ImageJ: 25 years of image analysis. *Nature Methods* 9: 671–675.

Segura, V., Cilas, C., and Costes, E. 2008. Dissecting apple tree architecture into genetic, ontogenetic and environmental effects: Mixed linear modelling of repeated spatial and temporal measures. *New Phytologist* 178: 302–314.

Segura, V., Denancé, C., Durel, C.-E., and Costes, E. 2007. Wide range QTL analysis for complex architectural traits in a 1-year-old apple progeny. *Genome* 50: 159–171.

Sharkey, T.D., Bernacchi, C.J., Farquhar, G.D., and Singsaas, E.L. 2007. Fitting photosynthetic carbon dioxide response curves for C(3) leaves. *Plant, Cell and Environment* 30: 1035–1040.

Shockey, J.M., Gidda, S.K., Chapital, D.C. et al. 2006. Tung tree DGAT1 and DGAT2 have non-redundant functions in triacylglycerol biosynthesis and are localized to different subdomains of the endoplasmic reticulum. *Plant Cell* 18: 2294–2313.

Shu, Q.Y. Ed. 2009. *Induced Plant Mutations in the Genomics Era*. FAO, Rome. Available at: http://www.fao.org/docrep/012/i0956e/i0956e00.htm.

Sinclair, T.R., Tanner, C.B., and Bennett, J.M. 1984. Water-use efficiency in crop production. *BioScience* 34: 36–40.

Singh, R., Low, E.-T.L., Ooi, L.C.-L. et al. 2013a. The oil palm SHELL gene controls oil yield and encodes a homologue of SEEDSTICK. *Nature* 500: 340–344.

Singh, R., Ong-Abdullah, M., Low, E-T.L. et al. 2013b. Oil palm genome sequences reveals divergence of interfertile species in old and new worlds. *Nature* 500: 335–339.

Singh, R., Low, E-T.L., Ooi, L.C-L. et al. 2014. The oil palm VIRESCENS gene controls fruit colour and encodes a R2R3-MYB. *Nature Communications* 5: Article No. 4106. doi:10.1038/ncomms5106

Smith, B.G. 1989. The effects of soil water and atmospheric vapour pressure deficit on stomata behavior and photosynthesis in the oil palm. *Journal of Experimental Botany* 40: 647–651.

Smith, B.G. 1993. Correlations between vegetative and yield characteristics in and photosynthetic rate and stomata conductance in the oil palm (*Elaeis guineensis* Jacq.). *Elaeis* 5: 12–26.

Son, J.S., Smayo, M., Hwang, Y.J., Kim, B.S., and Ghim, S.Y. 2014. Screening of plant growth-promoting rhizobacteria as elicitor of systemic resistance against grey leaf spot disease in pepper. *Applied Soil Ecology* 73: 1–8.

Southwood, T.R.E. and Way, M.J. 1970. Ecological background to pest management. In: *Concepts of Pest Management.* Eds. Rabb, R.L., and Guthrie, F.E., North Carolina State University, Raleigh, North Carolina, pp. 6–29.

Squire, G.R. 1984. *Light Interception, Productivity and Yield of Oil Palm.* Internal Report, Palm Oil Research Institute of Malaysia, Kuala Lumpur, Malaysia.

Squire, G.R. and Corley, R.H.V. 1987. Oil palm. In: *Tree Crop Physiology.* Eds. Sethuraj, M.R. and Raghavendra, A.S., Elsevier Science Publishers, Amsterdam, The Netherlands, pp. 141–167.

Strasser, B.J. and Strasser, R.J. 1995. Measuring fast fluorescence transients to address environmental questions: The JIP-test. In: *Photosynthesis: From Light to Biosphere.* Ed. Mathis, P., Kluwer Academic Publishers, Dordrecht, The Netherlands, pp 977–980.

Strauss, A.J., Krüger, G.H.J., Strasser, R.J., and Van Heerden, P.D.R. 2006. Ranking of dark chilling tolerance in soybean genotypes probed by the chlorophyll *a* fluorescence transient O-J-I-P. *Environmental and Experimental Botany* 56: 147–157.

Stringfellow, R. 2000. The competitiveness of the palm oil industry now and in the future. In: *Proc. Int. Planters Conf. Plantation Tree Crops in the New Millennium: The Way Ahead.* Ed. Pushparajah, E., Incorporated Society of Planters, Kuala Lumpur, Malaysia, pp. 199–214.

Suboh, I., Norkaspi, K., and Raja, Z.R.O. 2009. Double Row Avenue System for Crop Integration with Oil Palm. MPOB TT424. Malaysian Palm Oil Board, Kuala Lumpur, Malaysia.

Sundram, S. 2010. Growth effects of arbuscular mycorrhiza fungi on oil palm (*Elaeis guineensis* Jacq.) seedlings. *Journal of Oil Palm Research* 22: 796–802.

Taha, R.S., Ismail, I., Zainal, Z., and Abdullah, S.N.A. 2012. The stearoyl-acyl-carrier-protein desaturase promoter (Des) from oil palm confers fruit-specific GUS expression in transgenic tomato. *Journal of Plant Physiology* 169: 1290–1300.

Takenaka, A. 1994. Effects of leaf blade narrowness and petiole length on the light capture efficiency of a shoot. *Ecological Research* 9: 109–114.

Takenaka, A., Takahashi, K., and Kohyamas, T. 2001. Optimal leaf display and biomass partitioning for efficient light capture in an understorey palm, licuala arbuscula. *Functional Ecology* 15: 660–668.

Tan, K.P., Kanniah, K.D., and Cracknell, A.P. 2012. A review of remote sensing based productivity models and their suitability for studying oil palm productivity in tropical regions. *Progress in Physical Geography* 36: 655–679.

Trachsel, S., Kaeppler, S., Brown, K., and Lynch, J. 2011. Shovelomics: High throughput phenotyping of maize (*Zea mays* L.) root architecture in the field. *Plant and Soil* 341: 75–87.

Tsubo, M., Walker, S., and Mukhala, E. 2001. Comparisons of radiation use efficiency of mono-/inter-cropping systems with different row orientations. *Field Crops Research* 71: 17–29.

Tuberosa, R. and Salvi, S. 2006. Genomics-based approaches to improve drought tolerance of crops. *Trends in Plant Science* 11: 405–412.

USDA 2015. Soil Water Characteristics: Hydraulic Properties Calculator [online]. Available at: http://hydrolab.arsusda.gov/soilwater/Index.htm (Accessed March–May 2015).

Valladares, F. and Brites, D. 2004. Leaf phyllotaxis: Does it really affect light capture? *Plant Ecology* 174: 11–17.

Vamerali, T., Bandiera, M., and Mosca, G. 2012. Minirhizotrons in modern root studies. In: *Measuring Roots*. Ed. Mancuso, S., Springer-Verlag, Berlin, Germany, pp. 341–362.

Vanhercke, T., Wood, C.C., Stymne, S. et al. 2013. Metabolic engineering of plant oils and waxes for use as industrial feedstocks. *Plant Biotechnology Journal* 11: 196–210.

Van Kraalingen, D.W.G., Breure, C.J., and Spitters, C.J.T. 1989. Simulation of oil palm growth and yield. *Agricultural Forest Meteorology* 46: 227–244.

van Loon, L.C., Rep, M., and Pieterse, C.M.J. 2006. Significance of inducible defense related proteins in infected plants. *Annual Review of Phytopathology* 44: 135–162.

Van Noordwijk, M., Lusiana, B., Khasanah, N., and Mulia, R. 2011. *WaNuLCAS Version 4.0: Background on a Model of Water, Nutrient and Light Capture in Agroforestry Systems*. World Agroforestry Centre (ICRAF), Bogor, Indonesia, p. 224.

Voelker, T.A., Hayes, T.R., Cranmer, A.M. et al. 1996. Genetic engineering of a quantitative trait: Metabolic and genetic parameters influencing the accumulation of laurate in rapeseed. *Plant Journal* 9: 229–241.

Vos, J., Evers, J.B., Buck-Sorlin, G.H., Andrieu, B., Chelle, M., and de Visser, P.H.B. 2010. Functional–structural plant modelling: A new versatile tool in crop science. *Journal of Experimental Botany* 61: 2101–2115.

Walker, S., Virdis, S.G.P., Gopalan, Y., Karunaratne, A.S., and Jahanshiri, E. 2015. Crop diversification for Malaysian low-lying marginal land. In: Poster presented at ISUPS, Madurai, India.

Wang, Y., Cheng, X., Shan, Q. et al. 2014. Simultaneous editing of three homoeoalleles in hexaploid bread wheat confers heritable resistance to powdery mildew. *Nature Biotechnology* 32: 947–951.

Wang, Y. and Li, J. 2005. The plant architecture of rice (*Oryza sativa*). *Plant Molecular Biology* 59: 75–84.

Wang, Y., Weinacker, H., and Koch, B. 2008. A lidar point cloud based procedure for vertical canopy structure analysis and 3D single tree modelling in forest. *Sensors* 8: 3938–3951.

Waring, R.H., Coops, N.C., and Landsberg, J.J. 2010. Improving predictions of forest growth using the 3-PGS model with observations made by remote sensing. *Forest Ecology and Management* 259: 1722–1729.

Wasson, A.P., Rebetzke, G.J., Kirkegaard, J.A., Christopher, J., Richards, R.A., and Watt, M. 2014. Soil coring at multiple field environments can directly quantify variation in deep root traits to select wheat genotypes for breeding. *Journal of Experimental Botany* 65: 6231–6249.

Winfield, M.O., Allen, A.M., Burridge, A.J. et al. 2015. High-density SNP genotyping array for hexaploid wheat and its secondary and tertiary gene pool. *Plant Biotechnology Journal* 14: 1195–1206.

Wojciechowski, T., Gooding, M.J., Ramsay, L., and Gregory, P.J. 2009. The effects of dwarfing genes on seedling root growth of wheat. *Journal of Experimental Botany* 60: 2565–2573.

Wong, C.K. and Bernado, R. 2008. Genome-wide selection in oil palm: Increasing selection gain per unit time and cost with small populations. *Theoretical and Applied Genetics* 116: 815–824.

Wong, C.Y.S. and Gamon, J.A. 2015. Three causes of variation in the photochemical reflectance index (PRI) in evergreen conifers. *The New Phytologist* 206: 187–195.

Xu, L., Xia, W., Mason, A.S. et al. 2014. Exploiting transcriptome data for the development and characterization of gene-based SSR markers related to cold tolerance in oil palm (*Elaeis guineensis*). *BMC Plant Biology* 14: 384.

Xu, Y. 2010. *Molecular Plant Breeding*. CABI, Oxford, UK.

Yunus, A.M.M., Chai-Ling, H., and Parveez, G.K.A. 2008b. Construction of PHB and PHVB transformation vectors for bioplastics production in oil palm. *Journal of Oil Palm Research* 2: 37–55.

Yunus, A.M.M. and Parveez, G.K.A. 2008a. Development of transformation vectors for the production of potentially high oleate transgenic oil palm. *Electronic Journal of Biotechnology* 11(3).

Zetsche, B., Gootenberg, J.G., Abudayyeh, O.O. et al. 2015. Cpf1 is a single RNA-guided endonuclease of a Class 2 CRISPR-cas system. *Cell.* 163, 1–13.

Zhang, H., Zhang, J., Wei, P. et al. 2014. The CRISPR/Cas9 system produces specific and homozygous targeted gene editing in rice in one generation. *Plant Biotechnology Journal* 12: 797–807.

Zhao, W.Z., Xiao, H.L., Liu, Z.M., and Li, J. 2005. Soil degradation and restoration as affected by land use change in the semiarid Bashang area, northern China. *Catena* 59: 173–186.

Index

A

AAR, *see* Applied agricultural resources
ABA, *see* Abscisic acid
ABCE model, 204, 205
Abiotic stress, 15
  G × E interaction, 99–100
  oil palm resistance to drought, 101–104
  and resilience, 100
  tolerance, 99
Abscisic acid (ABA), 358
ABW, *see* Average bunch weight
ACC, *see* 1-Aminocyclopropane-1-carboxylic acid
Acclimatization in ramet nursery, 310
ACC oxidase (ACO), 97
Acetylated histone H4 (AcH4), 208
ACGT, *see* Asiatic Centre for Genome Technology
ACO, *see* ACC oxidase
ACS, *see* 1-Aminocyclopropane-1-carboxylic acid synthase
Actual yield (AY), 14
Acute wilt, 109
Adaptability
  abiotic stress tolerance and resilience, 100
  G × E interaction, 99–100
  oil palm resistance to drought, 101–104
  trial tests, 197
Additive model, *see* Statistical model
AF, *see* Amerelecimento Fatal
AFLPs, *see* Amplified fragment-length polymorphisms
African oil palm, 21
Africa, oil palm diseases, 119–120
Agricultural Production Systems Simulator (APSIM), 373
  APSIM-oil palm module, 374–375
Agricultural sustainability, 355
Agriculture "phenotyping", 378–379
Agro-management measures, 362
Agrobacterium, 367
  Agrobacterium-mediated gene transfer, 366–367
  PVX-*Agrobacterium* infection, 117
Agrobiodiversity, 212, 312, 361
Agrochemicals, 394
Agroforestry, 374
  lessons from, 397
  oil palm, 396
Agronomic requirements of crop, 13; *see also* Ecological requirements of crop
  fertilizer, 13–14
  light, 13
  water, 14
Algemeene Vereniging van Rubberplanters ter Oostkust van Sumatra (AVROS), 2, 23, 25–26, 147
Allelic frequencies, 42, 249
Alpha-designs, 345–346
Alpha (α) incomplete factorial mating design, 338–345
α-naphthalene acetic acid (NAA), 307, 308
AM, *see* Arbuscular mycorrhiza
Amerelecimento Fatal (AF), 112–115
American oil palm (*Elaeis oleifera*), 2, 21, 45
1-Aminocyclopropane-1-carboxylic acid synthase (ACS), 97
1-Aminocyclopropane-1-carboxylic acid (ACC), 97
Amplified fragment-length polymorphisms (AFLPs), 42, 49, 173, 238
ANACOVA and estimation of genetic correlation, 76, 77
Aneuploids, 230
Angola, 21, 31, 35, 44, 101, 171
  Angolan palms, 39
  E. guineensis, 35
Annual frond production, 94
Annual height increment (HInc), 91, 93, 94
ANOVA
  classical 2-way, 69–70
  for estimation of repeatability, 73
  and joint regression analysis, 78
  for NCM1 Design, 74
  for NCM2 design, 75
AOCSA, *see* Association of Official Seed Certifying Agencies
Apical meristem(s), 7, 12, 193, 230, 307, 386
Applications in oil palm, 333
  control treatments/check varieties, 336–337
  experimental designs, 333–334
  missing/dummy plots, 337
  oil palm breeding experiments, 333
  plot size and replication, 334–336
Applied agricultural resources (AAR), 2, 147, 175, 303
  AA Hybrida Deli *Dura* lineage, 177

Applied agricultural resources (*Continued*)
breeding for resistance, 176
breeding improvement, 178
commercial FFB yield performance of AAR clones, 318
comparison of cloning and recloning efficiencies of AAR ortets, 310
field trial OY performance of AAR clones, 317
Hybrida I Pisifera lineage, 177
mantling rates in AAR clones, 310
APSIM, *see* Agricultural Production Systems Simulator
Arabidopsis, 116, 205, 209, 262
class A–E genes, 205
PAP1 gene, 98
RPS2 in tomato, 116
seed oil transcription factor WRINKLED1, 94
Arachis hypogaea L, *see* Peanut
Arbuscular mycorrhiza (AM), 357
Area under disease progression curve (AUDPC), 128
"Arm ratio", 231, 232
"Artificial nose" tool, 305
Artificial seed, *see* Synthetic seed
ASD, 88, 167, 287, 304
chronology of ASD backcrossing program, 285
de Costa Rica, 2
ASD'S O × G interspecific hybrid breeding and compact clone program, 285–286
BC program, 286–287
chronology, 285
FFB and oil yield and growth characteristics, 287, 288
genetic diversity, 286
"high density" types, 285
using of *E. oleifera* to breed for OG varieties, 288–289
tissue culture and clonal performance, 288
Asiatic Centre for Genome Technology (ACGT), 241
Association of Official Seed Certifying Agencies (AOCSA), 313
Asymmetric DNA exchange, 209
AUDPC, *see* Area under disease progression curve
Auxiliary breeding program, 175, 177
Auxiliary traits, selection for, 89
Average bunch weight (ABW), 38
AVROS, *see* Algemeene Vereniging van Rubberplanters ter Oostkust van Sumatra
AY, *see* Actual yield

B

Bacillus thuringiensis (Bt), 367
Backcross program (BC program), 284, 285
BC_2F_3 and BC_3F_2 populations, 287
breeding, 146
Bacterial Artificial Chromosome library (BAC library), 241
Bad Karma, 207
Balai Besar Proteksi dan Perbenihan Tanaman Perkebunan (BBP2TP), 320
Barcode
labeling, 254–255
reading system, 254–255
Basal stem rot disease (BSR disease), 105, 110, 316
Base pairs (bp), 249
Base peak chromatogram (BPC), 245
Base SI, 149, 153, 155
Basidiocarps, 110
Bayesian random regression (BRR), 156–157
BBP2TP, *see Balai Besar Proteksi dan Perbenihan Tanaman Perkebunan*
BC program, *see* Backcross program
Beneficial plant-microbial interactions, integrating, 394
beneficial microorganisms on oil palm growth, 395
oil palm lineages, 396
Bénin—Presco—Colé, 172
Best linear unbiased estimates (BLUE), 153
Best linear unbiased prediction (BLUP), 59, 149, 153, 331–332
condition, 269
single-crosses model, 155–156
sire model, 154–155
Beta-glucuronidase (Beta-GUS), 367
Beta-GUS, *see* Beta-glucuronidase
β-1,3-Glucanase, 244
BI, *see* Bunch index
Biallelic QTLs, 266
Biclonal seed, 196
Binga, 26, 166, 175, 179
Bingerville, 168, 169
Biochemical and Molecular Techniques (BMT), 315, 316
Biodiversity, breeding for, 361–362
Biomarkers, 200
Biotic stress, 15
breeding for resistance to oil palm diseases, 119–125
diseases, 105
host–pathogen interactions in breeding for disease resistance, 106–119
pests, 105
physiological disorders, 105
resistance, 105

screening for resistance to *Ganoderma* sp. in Malaysia, 125–130
Biotrophic fungal pathogens, 107
BIS, *see* Breeding information system
"Block 500", genetic setup, 170
Blocking, 329, 330
Blowing technique, 302
BLUE, *see* Best linear unbiased estimates
BLUP, *see* Best linear unbiased prediction
BMT, *see* Biochemical and Molecular Techniques
BNO, *see* Bunch number
Bogor palms, 1, 25
Bonferroni procedure, 331
bp, *see* Base pairs
BPC, *see* Base peak chromatogram
BPROs, *see* Breeding populations of restricted origins
Bread wheat (*Triticum aestivum* L.), 21, 24
Breeders, 227
equation, 254
oil palm, 65
plant, 10, 47, 58
Breeding
approaches, 362–363
for biodiversity, 361–362
conventional oil palm breeding limitations, 237
conventional to molecular guided, 236
discovery of genes influencing agronomic traits, 238–241
environmental sustainability, 356
genomic tools application in oil palm breeding, 246–256
germplasm management, characterization, and conservation, 237–238
marker-assisted selection to improving breeding efficiency, 241–243
oil palm, 236
post-genomics era, 243–245
progress in oil palm, 246
for resilience, 360–361
for resource-use efficiency, 356–360
selection, 246
sustainability and climate change effects, 355–356
for sustainability and climate change effects in oil palm, 355
unraveling genome, 238–241
Breeding drought-resistant crosses methods
distribution of production, 103
offset production peak, 103
Breeding for resistance
Africa, 119–120
bud rot diseases, 124–125
fusarium vascular wilt, 120–121
FW, 109
Ganoderma BSR, 121–124
Ganoderma stem rots, 112
Latin America, 120
to oil palm diseases, 119
and resistance expression, 109
South East Asia, 120
spear rot, 113
Breeding information system (BIS), 39
Breeding plans and selection methods
BC breeding, 146
best linear unbiased prediction, 153–156
breeding for clonal propagation, 147–148
clonal hybrid seeds, 148
clonal variety, 148
DH breeding, 146
genomic selection, 156–160
IL lines, 146
inbred-hybrid variety breeding, 145
mass selection, 144–145
methods and variety types in oil palm, 144–145
NIL lines, 147
parent palm selection methods, 148–149
recurrent selection, 145
RIL breeding, 146
RIV breeding, 146
SI, 150–151
SIS for multiple traits in oil palm, 151–153
SSD, 146
top-cross test breeding, 145
Breeding populations, 24–25, 27, 362
commercial oil palm planting materials and, 29
diversity and improvement, 171
La Mé F_0 and F_1, 169
Malaysian, 61
NIFOR, 27
Breeding populations of restricted origins (BPROs), 25, 27, 168, 196
Breeding programs, 168
in AAR, SAIN, and SUMBIO, 175
AVROS, 25–26
Bénin, 168
Bénin—Presco—Colé, 172
Binga, 26
breeding strategy, 169–171
breeding tools, 172–173
Calabar, 27
Côte d'Ivoire, 168
current program, 171–172
Deli, 25
derived and recombinant BPROs, 27
Ekona, 26–27
generic MRRS scheme, 166
generic MRS scheme, 166
genetic base of, 24
genetic progress, 173–175
La Me, 26

Breeding programs (*Continued*)
palmelit/cirad's breeding program, 168–169
pre-RRS strategy, 169
second-cycle trials, 171–172
Socfin Indonesia, 168–169
Yangambi, 26
Breeding strategy, 169, 195, 377–378
diversity and improvement of breeding populations, 171
RRS scheme, 170–171
selection criteria, 169–170
Breeding tools
identity checking, 172
linkage mapping, 173
in vitro culture, 172–173
Breeding values (BV), 59, 67–68, 155
Broad-sense heritability, 63
BRR, *see* Bayesian random regression
BSR disease, *see* Basal stem rot disease
Bt, *see Bacillus thuringiensis*
B-type genes, 205
Bud rot, 112–115, 124–125
Bud rot diseases, 86, 124
breeding for resistance to PC complex, 124–125
Bunch
composition, 404
quality, 337
quality characters, 38–39
Bunch dry matter, 93
Bunch index (BI), 9, 32, 38, 87, 91, 93–94, 152
Bunch number (BNO), 38, 65
Bunch weight (BW), 32, 65
mean, 154
plasticity of, 70
BV, *see* Breeding values
BW, *see* Bunch weight

C

Calabar, 27
Callogenesis, 199, 200, 308
Callus, 172, 231, 308, 309, 367
bulk, 201
callusing rates of primary and clonal ortets, 308, 309
differentiation process, 308
embryogenic, 308
formation, 199
phase, 192
primary, 172
Cameroon, *E. guineensis*, 34
Cameroonian, introgression programs, 40
Canopy
efficiency, 88
phenotyping methods, 380–383
temperature, 382
Capillary electrophoresis mass spectrometry (CE-MS), 210–211
Carbohydrates (CHO_n), 87, 357
Carbon isotope discrimination (CID), 358
CBD, *see* Convention on biological diversity
CBS, *see* Combined Breeding Scheme
cc score, *see* Cover crop score
CD, *see* Crown disease
cDNA microarrays, 258
CE-MS, *see* Capillary electrophoresis mass spectrometry
CEBiP, 116
Cell-organism differentiation, 208
Central America and Northern region of S. America, 46
"Centromere index", 231
"Centromeric index", 231
Cercospora elaeidis (*C. elaeidis*), 37–38
CF, *see* Chlorophyll fluorescence
CGIAR, *see* UN Consultative Group on International Agricultural Research
Chlorophyll *a* fluorescence techniques, 391
Chlorophyll fluorescence (CF), 379
Chloroplast, 208–209
Chromatin, 236
Chromosomes
analysis of oil palm, 229–231
inheritance, 228
Chronic wilt, 109
CID, *see* Carbon isotope discrimination
Cirad's process, 303
Circular canopy, 13
Clonal
D and P seed, 301
hybrid seeds, 148, 192
industry, 197
ortets, 308, 309
performance, 288
variety, 148
Clonal field
planting results, 312
trial results, 317
Clonal propagation, 192, 268
breeding for, 147–148, 195
clonal industry, 197
culture stages and durations, 198
cutting-edge molecular tools, 199
development of oil palm artificial seeds, 215–216
flameless sterilizer, 214
genetic improvement with clones, 194–195
historical development, 193–194
innovations, 214
liquid culture system, 201–203
oil palm liquid culture, 212–213
OPTRACKS, 215

Ortet selection and cloning strategies, 195–197
post-genomics for in vitro cultures, 209–211
potential of organelle research in tissue culture, 208–209
refining in vitro protocols, 201
revolutionizing OPTC research, 197
somaclonal variation associated with micro-propagation, 203–208
understanding embryogenesis process, 199–201
Clones, 315
Cloning
efficiencies, 308–309
R genes, 116–117
strategies, 195–197
Cloning inefficiency, 311
ortet availability, 311
process inefficiency, 311
Cluster analysis, 44, 48, 49
Clustered, regularly interspaced, short palindromic repeats (CRISPR), 362, 369–370
CRISPR/Cas9 system, 369
CRISPR/Cpf1 system, 369
CO_2 assimilation, 87
Coelaenomenodera minuta pest damage, 284
Colletotrichum species, 107
Combined Breeding Scheme (CBS), 289
Combined selection, 149
Combining ability, 67–68
Commercial clonal field FFB yield results, 318
Commercial clonal plant production, 306–307
acclimatization in ramet nursery, 310
callusing and embryogenesis rates of primary and clonal ortets, 309
clonal field planting results, 312
cloning and recloning efficiencies, 308–309, 310
commercial clonal propagation issues, 310–311
commercial oil palm tissue culture process, 307–308
field planting issues, 312
principles and procedures applied in Ortet selection, 307
Commercial clonal propagation issues, 310
cloning inefficiency, 311
fruit mantling, 310–311
Commercial oil palm tissue culture process, 307–308
Commercial planting material production, 297
commercial clonal plant production, 306–312
commercial seed production, 301–306
DNA fingerprinting application, 313–316
oil palm seed companies seed and Ramet market, 299–301
private seed companies, 298
PVP, 312–313
seed certification, 312–313
Commercial seed production, 301
controlled pollination, 301–303
innovations and improvements, 304–306
oil palm breeding programs, 301
seed establishment and pollen gardens, 301
seed processing and germination, 303–304
Completely randomized design (CRD), 38, 123, 330
Constraints, 247–248
Contemporary research, 394
Continuous excitation fluorescence, 382
Continuous ortet demand, 311
Controlled pollination, 301–303, 305
Conventional breeding programs, 28, 246
breeding selection, 246–247
constraints, 247–248
Conventional oil palm breeding, limitations of, 237
Convention on biological diversity (CBD), 22, 32
"Core" effectors, 117
Correlated
response to selection estimation, 76
trait, 149
Correlation estimation, 76
Côte d'Ivoire
breeding program, 168
RRS scheme, 170
Covariance analysis, 336
Cover crop score (cc score), 90
Cowpea trypsin inhibitor (CpTI), 367
Cpf1, 369
CPO, *see* Crude palm oil
CpTI, *see* Cowpea trypsin inhibitor
CRD, *see* Completely randomized design
CRISPR, *see* Clustered, regularly interspaced, short palindromic repeats
Crop; *see also* Plant
adoption and development, 21
agronomic requirements, 13–14
crop-climate models, 372
ecological requirements, 10–13
physiological traits, 9–10
Crop models, 372–373
future use of, 377–378
oil palm, 373–374
Crop phenotyping, 378–379
affordable technological cornucopia, 385–386
canopy phenotyping methods, 380–383
new methods for, 378
root phenotyping methods, 383–385
types, 379–380
Crossing scheme, 339, 342

Crown disease (CD), 61, 88, 91, 105, 227
markers, 62–63
severity, 94
Crude palm oil (CPO), 3, 9, 298
Cryopreservation technique, 40
Cupucu (*Theobroma grandiflorum*), 397
Cutting-edge molecular tools, 199
CycDesigN computer program, 344, 345–346
Cytogenetic analysis, 229–230
Cytological landmarks, 230
Cytology of oil palm, 229–236

D

Dabou (Dab), 25, 170
DamID, *see* DNA adenine methyltransferase identification
DAP, *see* Days after pollination
DAPI, *see* 4′,6-Diamidino-2-phenylindole
DArT, *see* Diversity arrays technology
Data collection and management, 337–338
Days after pollination (DAP), 97
$D \times D$ trials, 333
Defense-related genes, 118, 130
degrees of freedom (df), 330, 340
Deli, 25
Deli dura (D), 1, 25, 123, 170–171, 179, 247, 289, 404
AAR's AA Hybrida Deli *Dura* lineage, 177
BPRO, 23
control seedlings, 130
Dumpy, 109
germinated, 253
Demand, 14, 298
continuous ortet, 311
energy, 260
food, 256
high-quality pedigree planting materials, 317
for palm oil, 3, 5, 398
plantation, 166
seed, 304
Depericarper functions, 303
Desired gains index, 153
df, *see* degrees of freedom
DFI, *see* Drought factor index
DGAT2, *see* Type-2 acyl-CoA:diacylglycerol acyltransferase
DH, *see* Doubled haploid
4′,6-Diamidino-2-phenylindole (DAPI), 231
2,4-Dichlorophenoxyacetic acid (2,4-D), 307
Diffused bud rot, 124
Digital growth analysis, 381
Disease resistance, 175
characteristics, 316
FW, 113–114
Ganoderma stem rots, 114–115
innate immunity genes as candidates, 115–118
screening for, 113
spear rot(s), 115
Disease(s)
assessment, 126, 128–129
biotic stress resistance, 105
Disease severity index (DSI), 128
Distinctness, 314, 322
Distinctness, Uniformity and Stability: Oil Palm Guidelines (DUS-OP Guidelines), 314
Ditjenbun, 320
Diversified cropping system
agroforestry, 397
oil palm monoculture, 396–397
transitioning oil palm plantation to, 396
Diversity arrays technology (DArT), 250
"Djongo" palm, 23
DMP, *see* Dry matter production
DNA
markers, 250, 377–378
methylation, 206–207, 235–236
profiling, 323
in situ hybridization, 231
DNA adenine methyltransferase identification (DamID), 206
DNA fingerprinting, 305
application in genotype identification and PVP, 313
dendrogram based on UPGMA clustering, 314
oil palm varieties, 314–315
successive progression of applying molecular data, 316
UPGMA clustering of segregating populations, 315
DNA sequencing, 231
boom in, 248–249
Domestication, 20, 58, 96
Double-layer technique, 308
Doubled haploid (DH), 146
breeding, 146
lines, 384
production, 179
D×P, *see Dura*×*Pisifera*
$D \times P/T$ trials, 333
Drought factor index (DFI), 392
Drought, oil palm resistance to, 101–104
Drought tolerance, 24, 359, 371, 402
photosynthesis phenotyping for, 391–394
Dry heat method, 304
Dry matter partitioning, 400–401
Dry matter production (DMP), 93, 399–400
DSI, *see* Disease severity index
Dumpy (Dy), 25, 95, 147
Dura (D), 60, 88

fruit types, 1
palms, 227
traits of D palms in dura lines, 91
Dura×Pisifera (D×P), 298
alpha (α) incomplete factorial mating design, 338–345
Dura × pisifera/tenera parents, 338
DUS-OP Guidelines, *see* Distinctness, Uniformity and Stability: Oil Palm Guidelines
Dwarfing
genes, 96
markers, 62
traits, 25

E

EC, *see* Embryogenic suspension cultures
Ecological requirements of crop, 10; *see also* Agronomic requirements of crop
flood, 12
global warming effects and relationships, 12
rainfall, 10
RH, 11
salinity, 12–13
soil, 11
solar radiation, 11
temperature, 11
VPD, 11
wind, 11
ECOPALM crop model, 373
Ecuador, *E. oleifera* seeds, 47
EFB, *see* Empty fruit bunch
Effectoromics, 117
Effectors
core, 117
pathogen, 117–118
protein, 107
TAL, 117
EFR, *see* Elongation factor
Eg707 gene, 200
EgBrRK, see Putative brassinosteroid leucine-rich repeat receptor kinase
EgCKX, see Putative cytokinin dehydrogenase
EgLIP1 gene, 97
Ekona (EK), 26–27, 166, 175
Elaeidobius kamerunicus, see Weevil
Elaeis genus, 45
Elaeis guineensis (*E. guineensis*), 31, 33, 98–99, 105, 120, 230, 234, 283; *see also* Oil palm
Angola, 35
Cameroon, 34
E. oleifera introgression into, 284–285
Gambia, 35
Ghana, 36
Guinea, 35–36
Jacq, 101
Madagascar, 34–35
Nigeria, 34
Senegal, 35
Sierra Leone, 35
Tanzania, 34–35
Zaire, 34
Elaeis oleifera (*E. oleifera*), 33, 36–37, 95, 230, 232, 234, 283
to breed for OG varieties, 288–289
Central America and Northern region of S. America, 46
characteristics, 45
characteristics and genetic diversity of each origin, 47
Ecuador, 47
F_1 Hybrids, 284
French Guyana, 47
genetic diversity, 48–50
genetic resources, 45, 284–285
germplasm in UPB, 290
introgression into *E. guineensis*, 284–285
Peru, 47
phenotypic diversity, 47–48
prospections and collections, 46
South America, 46–47
Elaeis oleifera × Elaeis guineensis interspecific hybrid improvement, 283
ASD'S O × G interspecific hybrid breeding and compact clone program, 285–289
UPB, 289–293
utilization of *E. oleifera* genetic resources, 284–285
Elmina (E), 25
Elongation factor (EFR), 116
EMBRAPA, *see* Empresa Brasileira de Pesquisa Agropecuária
Embryogenesis, 311
influence of hormones on SE, 200–201
molecular aspects of SE, 199–200
process, 199
rates of primary and clonal ortets, 308, 309
Embryogenic suspension cultures (EC), 308
Embryoids, 308
induction, 308
primary, 201
true, 192
EMMEANS, *see* Estimated marginal means
Empirical data, GS studies based on, 265
Empresa Brasileira de Pesquisa Agropecuária (EMBRAPA), 47
Empty fruit bunch (EFB), 5, 14
Encoding nuclear transcription factors, 96
Enfermedad no identificada, see La ENI
Environmental
stress, 360–361
sustainability, 356

Environmental policy integrated climate model (EPIC model), 373
EO, *see Elaeis oleifera* (*E. oleifera*)
E. oleifera × *E. guineensis* programs (*OG* programs), 196
EPIC model, *see* Environmental policy integrated climate model
Epigenetic
  mechanism, 203
  modulation of genomes, 235–236
Epigenome, 205, 243
  DNA methylation, 206–207
  histone modification, 207–208
eQTL analysis, *see* Expression QTL analysis
ERF genes, *see* Ethylene response factor genes
Estimated marginal means (EMMEANS), 343, 348
ESTs, *see* Extended sequence tags
Ethylene response factor genes (ERF genes), 97
Evapotranspiration, 358, 372–373
Exotic germplasm, 37–38, 95
Expected yield (EY), 14, 339
Experimental design, 38, 173
  Alpha incomplete block, 179
  and analysis, 328–331
  field experimental design for oil palm breeding trials, 345–350
  field experimentation, 329–330
Explants, 199, 288, 307
  callusing rates, 308
  inflorescence and root, 305
  leaf, 311
  manipulating, 201
Expression QTL analysis (eQTL analysis), 229, 257, 262
Ex situ, 23
  conservation, 40
  field genebank, 40
  seed storage, 40
  in vitro storage, 40–41
Extended sequence tags (ESTs), 200, 238
Extreme weather, 360
EY, *see* Expected yield

F

FAC, *see* Factorial analysis of correspondences; Fatty-acid composition
Factorial analysis of correspondences (FAC), 48
Factorial design, 74, 123, 154
Family and individual palm selection (FIPS), 24, 145, 169, 247–248, 301
Fatal Yellowing AF, 112–115
Fatty-acid composition (FAC), 285
Fatty acid modification, 365–366
F/B, *see* Fruit/bunch
F/B%, *see* Percent fruit to bunch
Fertilizer, 13–14, 27
  efficiency, 404
  experiments, 88
  N fertilizer, 357
  nutrients, 14
Feulgen, 230
FFA, *see* Free fatty acid
FFB, *see* Fresh fruit bunch
F_1 Hybrids, 184, 284
Field breeders, 328
Field breeding experiment, 328–329
Field experimentation, 327–328
  alpha (α) incomplete factorial mating design, 338–345
  applications in oil palm, 333–336
  comparison of treatment effects, 330–331
  data collection and management, 337–338
  experiment, 328–329
  experimental design, 329–330
  field experimental design for oil palm breeding trials, 345–350
  genomic marker technology, 328
  statistical analysis and software, 338
  statistical models, 331–332
  trial conduct, 336–337
Field genebank, 40
Field phenotyping methods, 384, 385–386
Field planting issues, 312
Field resistance, 129
Field trial OY performance of AAR clones, 317
Field waste, 5
FIPS, *see* Family and individual palm selection
FISH, *see* Fluorescent *in situ* hybridization
Fisher's variance ratio, 329
Fixed model, 331, 332
"Flag-harvesting" system, 402–403
Flagellin, 116
Flameless sterilizer, 214
Flood, 12
Floral organ identity, 204–205
Floral tissues, 230
Flower development, 204–205
Fluorescence transient technique, 391–394
Fluorescent *in situ* hybridization (FISH), 228, 231
Foe, see F. oxysporum f. sp. *elaeidis*
Food demands, 257–258
Fourier transform ion cyclotron resonance mass spectrometry (FT-ICR-MS), 210–211
F. oxysporum f. sp. *elaeidis* (*Foe*), 108
FP, *see* Frond production
Fraction of transpirable soil water (FTSW), 392
Free fatty acid (FFA), 9, 62, 95, 402
Freezer-stored pollen, 303
French Guyana, *E. oleifera* seeds, 47
Fresh fruit bunch (FFB), 38, 65, 248, 287
  yield, 319, 401

Fresh seeds, 303
Frond production (FP), 38
Fructose-1,6-biphosphate aldolase, 259
Fruit
abscission, 98–99
color, 97–98, 402–403
mantling, 310–311
ripening, 97–98
shedding process, 98–99
Fruit/bunch (F/B), 38, 401
FS, *see* Full-sib
FSPM, *see* Functional–structural plant models
"F-test", 330
FT-ICR-MS, *see* Fourier transform ion cyclotron resonance mass spectrometry
FTSW, *see* Fraction of transpirable soil water
Full-sib (FS), 89
family hybrid varieties, 315
Functional DNA, 227
Functional–structural plant models (FSPM), 388
Fungal isolate–host genotype interactions, 110
Fusarium moniliforme (*F. moniliforme*), 113
Fusarium oxysporum (*F. oxysporum*), 23, 37–38, 105, 118
Fusarium vascular wilt, 120–121
Fusarium wilt (FW) 105, 107, 113–114
biology and impact, 108–109
breeding for resistance and resistance expression, 109
Fusarium wilt-resistant varieties, 175
pathogen variability, 109–110

G

GA, *see* Gibberellic acid
GAB, *see* Genome-assisted breeding
Gambia, *E. guineensis*, 35
Ganoderma BSR, 121
breeding for disease resistance against, 106–119
breeding for resistance to, 176
nursery screening of planting material, 122–124
Ganoderma sp., 87, 125, 244, 356, 401–402
alternative screening methods, 129
disease assessment, 128–129
field resistance and nursery–field screening relationships, 129
Ganoderma disease, 15, 125, 175, 257
G. boninense, 105, 121, 245, 366
G. lucidum, 118
in Malaysia, screening for resistance to, 125
molecular of oil palm defense response, 129–130
nursery screening methods, 126–127
physiological age, 127–128
resistant cultivar, 125–126
Ganoderma stem rots, 114–115
biology and impact, 110–112
breeding for resistance and resistance expression, 112
Gas chromatography mass spectrometry (GCMS), 210–211
GATA, 233
GBLUP, *see* Genomic best linear unbiased prediction
GBS, *see* Genotyping by sequencing
GCA, *see* General combining ability
GCMS, *see* Gas chromatography mass spectrometry
GEBV, *see* Genomic estimated breeding value
G × E interaction, *see* Genotype × environment interaction
Gelled culture system, 201–202
Gene(s), 238
ESTs projects, 238–239
gene-function data, 210
hubs, 99
influencing agronomic traits, 238
MAS, 238
sequencing hypomethylated genome of oil palm, 240–241
whole-genome sequencing, 241
General combining ability (GCA), 67, 88, 90, 145, 174, 247, 301, 339, 346
alpha (α) incomplete factorial mating design, 338–345
General linear model (GLM), 75
GeneThresher™ technology, 240
Genetically modified crops (GM crops), 364
technology, 370
Genetic covariance
and correlation, 68
estimation, 76
Genetic dissection, 250–252
Genetic diversity, 48, 286
amplified fragment-length polymorphisms, 49
isozymes, 48
markers, 49–50
RAPD, 48
restriction fragment-linked polymorphisms, 48–49
simple sequence repeat, 49
Genetic fingerprinting, 23, 59, 227, 307
Genetic improvement with clones, 194–195
Genetic markers application, 313
Genetic modification, 59
Genetic progress
disease resistance, 175
generic MRS scheme, 166
HINC, 174–175
oil yield progress, 173–174
progress of true second cycle of selection, 174

Genetic resources
centers of distribution of natural oil palm populations, 33
conservation, 40
development of core collection in oil palm, 42
E. guineensis, 33–36
E. oleifera, 36–37, 45–50
evaluation of oil palm genetic material, 38
exotic germplasm and quarantine procedures, 37–38
exploiting interesting traits from germplasm, 27–31
ex situ conservation, 40–41
future needs and traits, 27
general principles, 20–22
genetic base broadening, 23–24
genetic base in oil palm, 27
genetic base of breeding programs, 24–27
germplasm collection and sampling, 32
germplasm management, characterization, and conservation, 42–45
germplasm utilization, 39
introgression programs, 39–40
methods to oil palm genetic material evaluation, 38–39
MPOB's genetic resources program, 32–33
prospection and conservation, 22–23
in situ conservation, 41
Genetic setup "Block 500", 170
Genetic stratification, 251
Genetic use restriction technologies (GURTs), 368
Genetic variance and heritability estimation, 63, 71
applications, 65–66
genetic assumptions for, 65
mating designs and genetic analyses, 71–75
methods, 71
offspring–male parent regression data structure and analysis, 72
use of heritability estimation methods, 66–67
Genome-assisted breeding (GAB), 316
Genome-wide association study (GWAS), 251–252
Genome-wide selection (GWS), 149
Genome, 208
chloroplast genome, 209
DNA quantity, 49
editing technology, 369
editing systems, 371–372
organellar genomes code, 209
plant mitochondrial genomes, 209
of species, 227
technologies for genome manipulation, 369–370
"Genome Select" program, 254
Genomic best linear unbiased prediction (GBLUP), 157
Genomic breeding, 173
Genomic estimated breeding value (GEBV), 156, 252
Genomic *in situ* hybridization (GISH), 228
Genomic(s), 238, 256
maps, 230
marker technology, 328
NGS for, 229
oil palm, 243
technology, 228–229
Genomic selection (GS), 59, 149, 156, 228, 251–252
accuracy of, 157–158
extending GS approach to other traits, 268–269
for hybrid performances, 158–159
implementation of, 265
increasing clonal values using GS, 268
increasing genetic gain rate in oil palm yield, 265–268
indeed, 252–254
integrating multiple omics and genetic data in GS, 269
interactions with environment, 269
for oil palm, 159–160, 263
for perennial crops, 159
principles, 156–157
studies on empirical data, 265
studies on simulations, 263–264
Genomic tools application in oil palm breeding, 246
boom in DNA sequencing, 248–249
conventional breeding programs, 246–248
DNA markers development, 250
genetic dissection, 250–252
marker application to oil palm breeding, 252–254
molecular technology, 254
Genotype × environment interaction (G × E interaction), 68–70, 99–100
effect estimation, 76
trials, 333, 363
Genotype × Management × Environment framework (G × M × E framework), 377
Genotypes, 148
Genotyping, 264
Genotyping by sequencing (GBS), 158, 243
Geographic information system (GIS), 375
Germination, 303–304
Germplasm
characteristics of germplasm collections, 30–31
collection and sampling, 32
commercial oil palm planting materials and breeding populations, 29

exploiting interesting traits from, 27
management, characterization, and conservation, 42–45, 237–238
MPOB oil palm germplasm, 28
priority traits in oil palm, 29
utilization, 39
GHA851 progenitor, 27
Ghana, *E. guineensis*, 36
GHG emission, *see* Greenhouse gas emission
Gibberellic acid (GA), 62
Gibberellins, 96
GIS, *see* Geographic information system
GISH, *see* Genomic *in situ* hybridization
GleyicAcrisols (glAC), 375
GLM, *see* General linear model
Global positioning system (GPS), 386
Glyceraldehyde-3-phosphate dehydrogenase, 259
Glycerol-3-phosphate, 259
Glycerol-3-phosphate dehydrogenase, 259
Glycolysis, 259
GM crops, *see* Genetically modified crops
G × M × E framework, *see* Genotype × Management × Environment framework
Good Karma, 207
GPS, *see* Global positioning system
Greenhouse gas emission (GHG emission), 355
Green revolution, 10, 96, 355, 394
Gross CO_2 assimilation, 88
GS, *see* Genomic selection
Guinea, *E. guineensis*, 35–36
Gunung Malayu (GM), 25
GURTs, *see* Genetic use restriction technologies
GWAS, *see* Genome-wide association study
GWS, *see* Genome-wide selection

H

Haber–Bosch process, 13
Hansatech, 392
Haploid plants, 230
Harrisons and Crosfield (H&C), 2
Harvestable yield, 94
perennial tree crop, 94
phenotypic traits, 96–99
Harvest index (HI), 9, 88
H&C, *see* Harrisons and Crosfield
Heat-shock proteins (HSP), 210
Height (HT), 38, 89, 91, 152, 170
Height increment (HINC), 174–175
Hemibiotrophic fungi, 107
Heritability values, 155
Hermaphrodite inflorescences, 8
Heterobasidion, 118
Heterochromatin formation, 208
Heterozygous palms, 89
HGP, *see* Human Genome Project
HI, *see* Harvest index
High-resolution cytogenetic maps, 230, 231, 234
High-yielding palms (HY palms), 258
"High density" types, 285
Highlands Research Unit (HRU), 2
High VPD, 11
HINC, *see* Height increment
Histone modification, 207–208
H3K4^{Me}, *see* Methylation of histone H3 at lysine 4
HMG, *see* Hydroxymethylglutaric
Homeotic transformations, 206–207
Homozygous *dura* (DD), 1
Homozygous female sterile *pisifera*, 1
Honestly significant difference test (HSD test), 330–331
Hormones on SE, 200–201
Host–parasite relationship principles, 105
Host–pathogen interactions in breeding for disease resistance, 106
FW, 108–110
Ganoderma stem rots, 110–112
innate immunity genes as candidates for genetic modification, 115–118
microbial pathogens, 106
molecular interactions, 119
resistance durability as influencing by host and pathogen genetics, 107–108
spear rot, 112–115
Host, resistance durability as influencing by, 107–108
HR, *see* Hypersensitive reaction
HRU, *see* Highlands Research Unit
HSD test, *see* Honestly significant difference test
HSP, *see* Heat-shock proteins
HT, *see* Height
Human Genome Project (HGP), 249
Hybridization, 111, 125, 195
DNA *in situ* hybridization, 231
interspecific hybridization with native species, 113
Hybrid performances, GS for, 158–159
Hybrid U1937, 359
Hydroxymethylglutaric (HMG), 244
HY palms, *see* High-yielding palms
Hypersensitive reaction (HR), 107

I

IAN, *see* Instituto Agronômico do Norte
IBA, *see* Indole-3-butyric acid
IBPGR, *see* International Board for Plant Genetic Resource
Ideogram, 234
Ideotype breeding, 10

Ideotype for yield and sustainability
bunch composition and FFB yield, 401
crossing programs, 404–405
disease and stress tolerance, 401–402
DMP, 399–400
dry matter partitioning, 400–401
fertilizer efficiency, 404
fruit color, 402–403
long bunch stalks, 402
low lipase, 402
oil composition, 404
oil palm sustainability, 397–398
suitability for machine harvesting, 402
time scale, 405
time scale for marker-assisted "pyramiding" of characteristics, 405
yield potential, 398–399
yield stability, 403–404
Illegitimate seeds, 305
IL lines, *see* Isogenic lines
Incomplete block designs, 330, 331, 344, 345
Indels polymorphisms, *see* Insertion–deletion polymorphisms
Independent culling-level selection, 149–150
Individual/mass selection, 148
Individual palms, 148
Indole-3-butyric acid (IBA), 308
Indonesia
breeding strategy, 170–171
seed certification, 320
Socfin Indonesia, 168, 172
Indonesian Oil Palm Research Institute (IOPRI), 2
INEAC, *see* Institut National pour l'Etude Agronomique du Congo
Inflorescence, 8
androgynous, 203
female, 204
hermaphrodite, 8
isolation, 302
male, 302
tissue, 288
Innate immunity genes
as candidates for genetic modification for disease resistance, 115–118
defense-related genes, 118
innate immunity, 116–117
pathogen effectors and toxins, 117–118
Inoculation, 308
artificial, 121
germinated seed for, 115
method, 122
root, 128
Insecticide-infiltrated cotton wad collar, 302
Insertion–deletion polymorphisms (*indels* polymorphisms), 313
In situ
collections, 23
conservation, 41
hybridization, 232
Institut de Recherches en droit Des Affaires (IRDA), 98–99
Institut de Researches pour les Huiles Oleagineux (IRHO), 2, 26, 168, 172–173
Institut National pour l'Etude Agronomique du Congo (INEAC), 2, 26
Instituto Agronômico do Norte (IAN), 47
Integrated pest management (IPM), 15, 105
Inter-origin effect, 167
Intermediate quarantine station, 37
International Board for Plant Genetic Resource (IBPGR), 34
International Crop Scientist Association, 313
International experiment, 169
International Panel on Climate Change (IPCC), 100
International Rice Genome Sequencing Project (IRGSP), 249
International Union for Protection of New Varieties of Plants (UPOV), 313, 315
Internode, 386
length and phyllotaxis, 388
tissues, 96
Interplant and interplot competition, 336
Interspecific hybrids, 125, 284
Introgression programs, 39
Cameroonian, 40
Nigerian, 39
Zairean, 40
In vitro
manipulating explants, 201
plantlets, 148
post-genomics for in vitro cultures, 209–211
refining in vitro protocols, 201
storage, 40–41
In vitro culture, 172–173, 193, 285
oil palm, 215
plant material, 118
post-genomics for in, 209–211
Iodine value (IV), 38
Ionomics, 228, 256, 257, 385
IOPRI, *see* Indonesian Oil Palm Research Institute
IPCC, *see* International Panel on Climate Change
IPM, *see* Integrated pest management
IPTGRFA, 22
IRDA, *see* Institut de Recherches en droit Des Affaires
IRGSP, *see* International Rice Genome Sequencing Project
IRHO, *see* Institut de Researches pour les Huiles Oleagineux
Isobaric tags, 259, 261

Isogenic lines (IL lines), 147
Isozymes, 48
IV, *see* Iodine value

K

Karma retrotransposon, 207
K/B ratio, *see* Kernel to bunch ratio
Kernel/fruit (K/F), 38
Kernel oil yield (KOY), 65
Kernel to bunch ratio (K/B ratio), 319
Kernel to fruit ratio (K/F ratio), 319
Kernel yield (KY), 38
K/F, *see* Kernel/fruit
K/F%, *see* Percent kernel to fruit
K/F ratio, *see* Kernel to fruit ratio
KLM, *see* Kuala Lumpur Melanococca
Kosambi's distances, 251
KOY, *see* Kernel oil yield
Kuala Lumpur Melanococca (KLM), 37, 289
Kuiper–Corsten iteration, 337
KY, *see* Kernel yield

L

LA, *see* Leaf area
Laboratory-based methods, 384
Laboratory screens, 383–384
La ENI, 124
LAI, *see* Leaf area index
La Me, 26
La Mé F_0 and F_1 breeding populations, 169
La Me Ps (LM Ps), 25
LAR, *see* Leaf area ratio
Laser-induced fluorescence transient (LIFT), 383
LASSO, *see* Least absolute shrinkage selection operator
Latin America, oil palm diseases, 120
Latin square design, 330
Lauric acid, 364
Laurical™, 363, 364, 370
Lauric-rich oil crop, 364
LC-MS, *see* Liquid chromatography mass spectrometry
LD, *see* Linkage disequilibrium
LDM, *see* Leaf dry matter
Leaf area (LA), 91, 92, 170
Leaf area index (LAI), 9, 38, 88, 170, 399
Leaf area ratio (LAR), 9, 90, 91, 152, 401
Leaf dry matter (LDM), 93
Leaf length (LL), 287
Leaflet number (LN), 38
Leaf measurements, 92–93
Leaf weight (LW), 91
Least-significant difference test (LSD test), 330–331
Least absolute shrinkage selection operator (LASSO), 157
Least square means (LSMEANS), 348
LEDs, *see* Light emitting diodes
Lemnatech, 379
LemnaTec phenomics platforms, 229
Lethal bud rot, 112–115
Leucine-rich repeat (LRR), 200–201
LIDAR, *see* Light detection and ranging
LIFT, *see* Laser-induced fluorescence transient
Light, 13, 359–360
  interception, 388–391
  winds, 11
Light detection and ranging (LIDAR), 383
Light emitting diodes (LEDs), 381
Linear mixed model, 155
Linkage disequilibrium (LD), 157, 228
Linkage mapping, 173, 250–251
Lipase activity, 97
Liquid chromatography mass spectrometry (LC-MS), 210–211
  LC-MS BPC, 245
  LC-MS/MS, 244
Liquid culture system, 201–203
Liquid media, 308
Liquid suspension
  culture technology, 202–203
  pathway, 308
  system, 311
LL, *see* Leaf length
LN, *see* Leaflet number
Logistic growth functions, 92
Long bunch stalk markers, 62
Low lipase, 44, 402
  markers, 62
Low-yielding palms (LY palms), 258
LRR, *see* Leucine-rich repeat
LSD test, *see* Least-significant difference test
LSMEANS, *see* Least square means
L'Union Tropicale de Plantation (UTP), 168
LW, *see* Leaf weight
LY palms, *see* Low-yielding palms

M

Machine harvesting, suitability for, 402
Madagascar, *E. guineensis*, 34–35
Malaysia
  clones, 319
  cloning, 320
  other countries, 323
  PVP, 322–323
  recloning, 320
  screening for resistance to *Ganoderma* sp. in, 125–130
  seed certification, 319–320

Malaysian Agricultural Research and Development Institute (MARDI), 2
Malaysian Palm Oil Board (MPOB), 2, 31, 87, 96, 203, 215, 237, 311
  genetic resources program, 32–33
  oil palm genetic materials, 42
  oil palm germplasm, 28
Malaysian Protection of New Plant Act (2004), 313
MAMPs, *see* Microbial associated molecular patterns
Mantled abnormality, 205, 243
Mantling abnormality, 193
Mantling incidence, 306, 311
Manuring block, 13
MARDI, *see* Malaysian Agricultural Research and Development Institute
Marker-assisted recurrent selection (MARS), 263
Marker-assisted selection (MAS), 61–62, 147, 149, 238, 252–254, 313, 362
  to improving breeding efficiency, 241–243
  in oil palm selection schemes, 253
  procedures, 328
  on seedling populations, 197
Marker(s), 49–50
  application to oil palm breeding, 252
  density, 228
  "Genome Select" program, 254
  kits, 62
  MAS, 252–254
  systems, 228
MARS, *see* Marker-assisted recurrent selection
MAS, *see* Marker-assisted selection
Mass selection, 301
Mass spectroscopy (MS), 210
Mating designs and genetic analyses, 71–75
Maximum leaf area (L_{max}), 88
Maximum likelihood (ML), 75
Mean fruit weight (MFW), 37, 38
Mean nut weight (MNW), 38
Meiotic chromosomes, 230
Meristem culture, 192
Mesocarp to fruit ratio (M/F ratio), 38, 319
Messy factors, 331–332
Metabolic QTL (mQTL), 262
Metabolome, 210
Metabolomics, 210, 256
  applications in oil palm improvement, 243–245
  studies, 246
Methylation of histone H3 at lysine 4 ($H3K4^{Me}$), 208
Mexican sunflower (*T. diversifolia*), 397
M/F%, *see* Percent mesocarp to fruit
M/F ratio, *see* Mesocarp to fruit ratio
MFW, *see* Mean fruit weight
Mg deficiency, 361
Microarray analysis, 258
Microbial associated molecular patterns (MAMPs), 116
Microbial pathogens, 107
Microcyclus, 107
Micro-propagation, 203
  epigenome, 205–208
  understanding flower development, 204–205
MicroRNA (miRNA), 240
Mill waste, 5
Minimum quadratic unbiased estimates procedure (MINQUE procedure), 332
miRNA, *see* MicroRNA
Mitochondria, 208–209
Mitochondrial DNA markers, 111
Mixed D × P oil palm seed, 301
Mixed model, 331–332
ML, *see* Maximum likelihood
MNW, *see* Mean nut weight
Model-assisted phenotyping, 391
Modified reciprocal recurrent selection (MRRS), 74, 145, 166–167, 301
Modified recurrent selection (MRS), 145, 166–167, 301
MoFaTT, *see* MPOB Fast Transfer Technique
Moisture stress, 358
Molecular
  approaches, 358
  aspects of SE, 199–200
  breeders, 328
  breeding, 59
  diagnostic assay, 242
  marker, 21, 95–96
  MAS procedures, 328
  molecular-based assays, 42–45, 237–238
  of oil palm defense response toward *Ganoderma* sp, 129–130
  technology, 254
Molecular breeding technology, *see* Genomic marker technology
Molecular cytogenetics of oil palm, 232–235
  ideogram, 234
  locations of abundant repetitive DNA sequences, 233
  metaphase chromosomes of oil palm, 232
Molecular genetics, 59, 227
  causative mutation, 227–228
  conventional to molecular guided breeding, 236–246
  crop introgression program, 227
Monteith equation, 388
Month-old seedlings (MO seedlings), 126–127

Motorized vessel with fast media transfer (Movefast media transfer), 213
MoVess, *see* MPOB modified vessel
MPOB-Motovess, *see* MPOB motorized vessel
MPOB, *see* Malaysian Palm Oil Board
MPOB Fast Transfer Technique (MoFaTT), 212, 216
MPOB modified vessel (MoVess), 212
MPOB motorized vessel (MPOB-Motovess), 212
mQTL, *see* Metabolic QTL
MRRS, *see* Modified reciprocal recurrent selection
MRS, *see* Modified recurrent selection
MS, *see* Mass spectroscopy
Multi-genotype × multi-environment (G×E trials), 362, 363
Multiple-gene heterosis, 21–22
Multiple range test, 330–331
Multiple trait selection, 149–150
Mutagenesis approach, 365

N

NAA, *see* Napthaleneacetic acid; α-naphthalene acetic acid
Nagoya Protocol of CBD, 22
Napthaleneacetic acid (NAA), 200
Narrow sense heritability, 63
National Key Economic Area (NKEA), 28–29
NBS-LRR, *see* Nucleotide binding, leucine-rich repeat
NCM1 mating design, *see* North Carolina Model 1 mating design
NDVI, *see* Normalized difference vegetation index
Near-single cross D × P oil palm seed, 301
Near isogenic lines (NIL), 147, 384
Necrotrophs, 107
Nerica rice types, 24
Nested/Hierarchal Design, 73
Next-generation sequencing (NGS), 229
N fertilizer, 356, 357
NGS, *see* Next-generation sequencing
Nicotiana benthamiana (*N. benthamiana*), 116
NIFOR, *see* Nigerian Institute for Oil Palm Research
Nigeria, *E. guineensis*, 34
Nigerian Institute for Oil Palm Research (NIFOR), 2, 27, 31
Nigerian, introgression programs, 39
Nigrescens, 98, 242
NIL, *see* Near isogenic lines
Nitrogen use efficiency (NUE), 356, 357
NKEA, *see* National Key Economic Area
NMR, *see* Nuclear magnetic resonance
Non-abscinding fruit, 402
Non-*Elaeis* species, 24
Non-Mendelian diseases, 250
"Nonfunctional" DNA, 227
Non photochemical quenching (NPQ), 382
NOR, *see* Nucleolar organizing region
Normalized difference vegetation index (NDVI), 381
North Carolina Model 1 mating design (NCM1 mating design), 65, 73, 145
North Carolina Model 2 mating design (NCM2 mating design), 74
NPQ, *see* Non photochemical quenching
Nuclear magnetic resonance (NMR), 211
Nucleolar organizing region (NOR), 232
Nucleosome, 208
Nucleotide binding, leucine-rich repeat (NBS-LRR), 116
Nucleotides, 261
NUE, *see* Nitrogen use efficiency
NUE uptake efficiency (NupE), 356
Null hypothesis, 330, 335
NupE, *see* NUE uptake efficiency
Nursery–field screening relationships, 129
Nursery progeny seedlings, 197
Nursery screening
  early screening tests, 122–123
  interactions between *Ganoderma* isolates and oil palm progenies, 123
  methodology of testers, 123
  methods, 126–127
  of planting material, 122
  relationship between field and nursery tests, 123–124
Nutraceuticals, 370
Nutrient(s), 5, 14, 356–357
  balance approach, 14
  cycling, 362, 394

O

Objective traits
  adaptability and abiotic stress tolerance, 99–104
  biotic stress resistance, 105–130
  environmental sustainability, 87
  harvestable yield, 94–99
  selection for physiological traits in oil palm, 87–94
O/B ratio, *see* Oil to bunch ratio
OCP, *see* Original compact palm
O/DM ratio, *see* Oil to dry mesocarp ratio
OER, *see* Oil extraction rate
Oil crop(s), 20, 256
  Lauric-rich, 364
  twenty-first century omics platforms in, 256–257
Oil extraction rate (OER), 169–170, 284, 318

Oil palm, 7, 21, 22, 24, 27–28, 86–87, 175, 193, 227, 231, 236; *see also Elaeis guineensis* (*E. guineensis*); Transgenic oil palm
artificial seeds development, 215–216
BI, 93–94
breeding approaches, 362–363
breeding efforts in West Africa, 2
breeding experiments, 333
breeding for biodiversity, 361–362
breeding for resilience, 360–361
breeding for resource-use efficiency, 356–360
breeding for sustainability and climate change effects in, 355–356
breeding goals, 27
breeding programs, 301
breeding progress in, 246
breeding trials, field experimental design for, 345–350
canopy efficiency, 88
chromosomes, 229–231, 235–236
conventional oil palm breeding limitations, 237
crop, 1, 373–374
cytology of, 229–236
Deli D, 1
discovery of genes influencing agronomic traits, 238–241
dry matter production, 93
economic uses, 3–5
environmental sustainability, 356
epigenetic modulation of genomes and DNA methylation, 235–236
exemplar crops comparison to, 23–24
genomic tools application in breeding, 246–256
germplasm management, characterization, and conservation, 237–238
gross CO_2 assimilation, 88
GS for, 159–160, 263–269
height, 89
HI, 88
improving productivity, 236
leaf measurements, 92–93
marker-assisted selection to improving breeding efficiency, 241–243
molecular cytogenetics of, 232–235
monoculture, 396–397
omics technologies to oil palm yield improvement, 256–263
one-shot method of growth recording, 94, 95
parent selection, 89–91
post-genomics era, 243–245
process of photosynthesis, 87
ramets, 194
recording techniques, 92
secondary/complementary products, 5
seed companies, 298–301
selection for auxiliary traits, 89
selection for physiological traits in, 87
sustainability and climate change effects, 355–356
sustainability, 397–398
traits for parent selection, 91–92
unraveling genome, 238–241
World Trade in palm oil, 3
Oil palm genetic material evaluation, 38
bunch quality characters, 38–39
vegetative traits, 39
yield and components, 38
Oil Palm Production Simulator (OPRODSIM), 373
Oil palm resistance to drought
distribution and culture conditions, 101
methods of breeding drought-resistant crosses, 103
physiological impact of water deficit, 103–104
prospects, 104
symptoms and impact of drought, 101–102
water-deficit environments, 102–103
Oil Palm Simulator (OPSIM), 373
Oil palm tissue culture (OPTC), 192, 301, 307
clonal industry, 197
clonal propagation, 193–194
culture stages and durations, 198
cutting-edge molecular tools, 199
embryogenesis process, 199–201
liquid culture system, 201–203
post-genomics for in vitro cultures, 209–211
potential of organelle research in tissue culture, 208–209
refining in vitro protocols, 201
revolutionizing OPTC research, 197
somaclonal variation associated with micro-propagation, 203–208
Oil palm tissue culture labs (OPTCLs), 196–197
Oil Palm Tissue Culture Tracking System (OPTRACKS), 215
Oil palm yield modeling
APSIM-Oil Palm module, 374–375
crop models, 372–373
future use of crop models, 377–378
model results at Rompin, Malaysia, 375–376
oil palm crop models, 373–374
Oil to bunch ratio (O/B ratio), 38, 319
Oil to dry mesocarp ratio (O/DM ratio), 38, 319
Oil to mesocarp (O/M), 284
Oil/wet mesocarp ratio (O/WM ratio), 38
Oil yield (OY), 38, 65, 173–174, 285, 319
Oligonucleotide donor sequences, 369–370
O/M, *see* Oil to mesocarp

Omega-3 oils, 370
"Omics" technologies, 59, 229, 269
  biological classifications, 259
  increased balance of carbon flow, 260
  integrating multiple omics and genetic data in GS, 269
  malic acid/citric acid ratios, 261
  to oil palm yield improvement, 256
  in oil palm yield studies, 258–262
  pitfalls and challenges in, 262–263
  twenty-first century omics platforms in oil crop studies, 256–257
  yield enhancement, 263
One-shot method of growth recording, 94, 95
OPRODSIM, *see* Oil Palm Production Simulator
OPSIM, *see* Oil Palm Simulator
OPTC, *see* Oil palm tissue culture
OPTCLs, *see* Oil palm tissue culture labs
Optimal planting density, 90
OPTIMAS analysis software, 385
OPTRACKS, *see* Oil Palm Tissue Culture Tracking System
Original compact palm (OCP), 286
Ortet(s), 148
  availability, 311
  principles and procedures in selection, 307
  selection, 195–197
Oryza glaberrima (*O. glaberrima*), 24
Oryza sativa L, *see* Rice
O/WM ratio, *see* Oil/wet mesocarp ratio
OY, *see* Oil yield

P

Palmelit/cirad's breeding program, *see* Breeding programs
Palmitoyl thioesterase, 366
Palm kernel oil (PKO), 3, 9
Palm NF 32.3005, 27
Palm Oil Research Institute of Malaysia (PORIM), 2
Palm populations, resequencing of, 249–250
PALMSIM crop model, 373
Papua New Guinea calibration, 375
PAR, *see* Photosynthetically active radiation
Parent palm selection methods, 148
  multiple trait selection, 149–150
  single trait selection, 148–149
Parent selection, 89–91
  traits for parent selection, 91
  traits of D palms in Dura Lines and T palms in Tenera lines, 91
  traits of P parents, 92
Pathogen
  biology, 107–108
  effectors, 117–118
  genetics, 107–108
  toxins, 117–118
  variability, 109–110
Pathogenesis-related proteins (PR proteins), 210
Pattern recognition receptors (PRRs), 116
PBR, *see* Plant Breeder's Rights
PCR, *see* Polymerase chain reaction
PCS, *see* Petiole cross-section
Peanut (*Arachis hypogaea* L.), 21
Peat plantings, 361
PEG, *see* Polyethylene glycol
Percent fruit to bunch (F/B%), 65
Percent kernel to fruit (K/F%), 65
Percent mesocarp to fruit (M/F%), 65
Percent shell to fruit (S/F%), 65
Perennial crops, GS for, 159
Perennial tree crops, 192
"Perfect" markers, 59
Performance index (PI), 391–392
per hectare (p/ha), 286
Peru, *E. oleifera* seeds, 47
Pests, 105
Petiole cross-section (PCS), 39
Petroleum, 13
PGR, *see* Plant growth regulators
p/ha, *see* per hectare
Pharmaceuticals, 370
Pheno-copter, 386
Phenolic compounds, 308
Phenomics, 229, 256, 378–379
Phenomobiles, 386
Phenotype, 378–379
Phenotypic diversity, 47–48
Phenotypic QTL (pQTL), 257
Phenotypic traits
  fruit abscission, 98–99
  fruit ripening, 97–98
  tree architecture, 96–97
Phenotyping, 378–379
  geometric variables, 389
  light interception efficiency, 390
  methods, 229, 387
  Monteith equation, 388
  new challenges in oil palm, 386
  photosynthesis phenotyping for drought tolerance, 391–394
  plant architecture and light interception, 388–391
Photochemical reflectance index (PRI), 381–382
Photon Systems Instruments, 379
Photosynthesis, 88, 357
  phenotyping for drought tolerance, 391–394
  photosynthetic PI parameter, 392
  process, 87
Photosynthetically active radiation (PAR), 13, 87–88, 372
Photosystem II (PSII), 382

Physiological
age, 127–128
approaches, 358
disorders, 105
impact of water deficit, 103–104
Phytopathological principles, 105
Phytophthora, 117
P. palmivora, 105, 113, 120, 401–402
Phytosanitary measures in country of origin, 37–38
PI, *see* Performance index
Pisifera (P), 25, 60, 88, 154, 166–167
palms, 8, 89
traits of P parents, 92
PKO, *see* Palm kernel oil
Plant; *see also* Crop
crop physiological traits, 9–10
morphology and reproductive biology, 7–10
reproductive traits, 8–9
vegetative traits, 7–8
Plant Breeder's Rights (PBR), 314
Plant genetics, 58
estimates of within-family genetic variability, 76–79
estimation of genetic covariance, correlation, and correlated response to selection, 76
estimation of genetic variance and heritability, 71–75
estimation of genotype–environment interaction effects, 76
qualitative trait inheritance and applications in oil palm breeding, 60–63
quantitative trait inheritance and applications in oil palm breeding, 63–70
selection response and realized heritability, 75–76
Plant growth regulators (PGR), 306
Plant ideotype concept, 10
Plant mitochondrial genomes, 209
Plant tissue culture, 192
Plant uptake efficiency, 356
Plant variety protection (PVP), 312–313
issues with, 323–324
Malaysia, 322–323
other countries, 323
Plasticity of bunch weight, 70
Pollen
gardens, 301
morphology, 49–50
mother cells, 231
supply, 305
Pollinated bunch, 9
Poly-cross mating design, 73
Polyethylene glycol (PEG), 368
Polygalacturonase EgPG4, 98–99
Polygenic resistance, 107–108
Polymerase chain reaction (PCR), 200, 231
Polyploid, 21
Population hybrid varieties, 315
PORIM, *see* Palm Oil Research Institute of Malaysia
PORIM/MPOB transformation group, 365
Post-genomics era, 243–245
for in vitro cultures, 209–211
Potential yield concept, 10
Power analysis, 335–336
P phenotype and genotype, 61
pQTL, *see* Phenotypic QTL
Pre-reciprocal recurrent selection strategy, 169
Preheated seeds, 304
Premature bunch harvesting, 9
PRI, *see* Photochemical reflectance index
Primary ortets, 308, 309
Private plantation research companies, 2
Private seed companies, 298
Process inefficiency, 311
Proembryoids, 311
Progeny testing, 149, 247
Proteomics, 210, 243–245, 256
PR proteins, *see* Pathogenesis-related proteins
PRRs, *see* Pattern recognition receptors
PSII, *see* Photosystem II
Pudricion de la flecha (PF), *see* Spear rot
Pudricion del cogollo (PC), *see* Bud rot
Pudricion difusa de cogollo (PCD), *see* Diffused bud rot
Pulse-modulated fluorescence, 382
Putative brassinosteroid leucine-rich repeat receptor kinase (*EgBrRK*), 200–201
Putative cytokinin dehydrogenase (*EgCKX*), 200–201
PVP, *see* Plant variety protection
"Pyramiding" approach, 404, 405

Q

QTL, *see* Quantitative trait loci
Qualitative trait inheritance, 60–61
crown disease markers, 62–63
delayed fruit shedding with low lipase markers, 62
dwarfing markers, 62
long bunch stalk markers, 62
oil palm breeding applications, 61
shell marker, 62
virescens fruit marker, 62
Quantitative genetic theory and principles, 63–64, 65
Quantitative trait inheritance, 63
and applications in oil palm breeding, 63
combining ability and BV estimates, 67–68
estimation of genetic variance and heritability, 63–67

genetic covariance and correlation, 68
genotype–environment interaction, 68–70
Quantitative trait loci (QTL), 59, 173, 229, 285, 362
Quarantine procedures, 37–38

R

R2R3-MYB transcription factor, 98
Rachis length (RL), 38, 90, 91
Radiation-limiting equations, 372
Radiation-use efficiency (RUE), 359–360, 374
Rainfall, 10
Ramet(s), 148
crop mill OER, 318–319
market, 298–301
RAMs, *see* Randomly amplified microsatellites
Random amplified polymorphic DNA (RAPD), 48, 111, 238
Randomization, 329
Randomized complete block (RCB), 154
Randomized complete block design (RCBD), 38, 330
Randomly amplified microsatellites (RAMs), 111
Random model, 331
Random regression BLUP (RR-BLUP), 156, 269
R&D programs, 368
RAPD, *see* Random amplified polymorphic DNA
Rapeseed, 370
RCB, *see* Randomized complete block
RCBD, *see* Randomized complete block design
Reactive oxygen species (ROS), 129
Realized heritability, 75–76
Reciprocal recurrent genomic selection (RRGS), 252, 264, 266, 267
Reciprocal recurrent selection (RRS), 24, 59, 102, 145, 194, 247–248
Côte d'Ivoire, 170
implementation, 174
Indonesia, 170–171
population improvement in, 171
scheme, 169
Recloning efficiencies, 308–309
Recombinant inbred line (RIL), 146, 363
Recombinant Inbred Variety (RIV), 146
Recoverable yield, 95
Red green blue (RGB), 381
Reduced Height genes (*Rht* genes), 58, 89, 96
Reflectance measurements, 381
Regression approach, 71
Relative humidity (RH), 11
ReML, *see* Residual ML
REML method, *see* Restricted maximum likelihood method
Repeatability (R), 72
ANOVA for estimation of, 73
data structure, 72
Replication, 329
Reproductive traits, 8–9
Research Institute of Sumatra Planters Association (RISPA), 2
Resequencing of palm populations, 249–250
Residual ML (ReML), 75
Resistance durability, 107–108
Resistance genes (R genes), 107, 116–117
Resolvable design, 331
Resource use efficiency (R_eUE), 15, 362–363
breeding for, 356
light, 359–360
nutrients, 356–357
water, 357–359
Restricted maximum likelihood method (REML method), 155, 332
Restricted taxonomic functionality, 116
Restriction fragment-length polymorphism (RFLP), 42, 238
Restriction fragment-linked polymorphisms, 48–49
R_eUE, *see* Resource use efficiency
RFLP, *see* Restriction fragment-length polymorphism
RGB, *see* Red green blue
RH, *see* Relative humidity
Rht genes, *see Reduced Height* genes
Ribulose 1,5 bisphosphate carboxylase/oxygenase (Rubisco), 360
Rice (*Oryza sativa* L.), 249
Rice Annotation Project Database, 249
RIL, *see* Recombinant inbred line
RISPA, *see* Research Institute of Sumatra Planters Association
RIV, *see* Recombinant Inbred Variety
RL, *see* Rachis length
Root phenotyping methods, 383–385
Root system architecture traits (RSAT), 383
ROS, *see* Reactive oxygen species
RR-BLUP, *see* Random regression BLUP
RRGS, *see* Reciprocal recurrent genomic selection
RRS, *see* Reciprocal recurrent selection
RSAT, *see* Root system architecture traits
Rubber wood blocks (RWBs), 111
Rubin's method, 337
Rubisco, *see* Ribulose 1,5 bisphosphate carboxylase/oxygenase
RUE, *see* Radiation-use efficiency
RWBs, *see* Rubber wood blocks

S

saHS, *see* SapricHistosols
SAIN, *see* Sarana Inti Pratama, Indonesia
Salinity, 12–13
Sanger technology, 240

SapricHistosols (saHS), 375
Sarana Inti Pratama, Indonesia (SAIN), 175, 179
achieved and expected genetic improvements, 181
alpha mating design crosses, 180
Saturated vapor pressure (SVP), 11
SCA, *see* Specific combining ability
Sclerotinia sclerotiorum (*S. sclerotiorum*), 118
SE, *see* Somatic embryogenesis
S.E. Asian palm oil, 396
Secondary breeding program, 175, 177
Secreted in xylem effectors (SIX effectors), 118
SED, *see* Standard error of difference
Seed
establishment, 301
germination, 306
inventory, 306
moisture adjustment, 306
processing, 303–304, 305
seed-specific genes, 369–370
sorting and selection, 306
storage, 40
Seed certification, 312–313
Indonesia, 320
Malaysia, 319–320
other countries, 320
requirements, 321
Selection index (SI), 59, 149
Selection response, 75–76
Semi/biclonal D × P oil palm seed, 301
Semiclonal seed, 196
Senegal, *E. guineensis*, 35
Sensitivity analysis, 390
Sensor mobility, 386
Septoria, 108
Sequencing
hypomethylated genome of oil palm, 240–241
technology, 242–243
Serdang Avenue (S), 25
S/F%, *see* Percent shell to fruit
S/F ratio, *see* Shell to fruit ratio
S/G composition, *see* Syringyl/guaiacyl composition
SHELL gene, 60, 227
Shell marker, 62, 242
Shell to fruit ratio (S/F ratio), 38, 319
Shoot phenotyping methods, *see* Canopy phenotyping methods
Short sequence repeats (SSR), 173
Shovelomics, 384–385
SI, *see* Selection index
Sib selection, 148–149
Sierra Leone, *E. guineensis*, 35
Simple sequence repeat, 49
Simple sequence repeats (SSRs), 42, 228
analysis, 44
markers, 313
Simulations, GS studies based on, 263–264
Single-crosses model, 155–156
Single lens reflex (SLR), 383
Single nucleotide polymorphism (SNP), 42, 156, 228, 240, 285, 313
genotyping array for oil palm, 250
Single seed descent (SSD), 146
Single sequence repeats (SSRs), 111
Single trait selection, 148–149
Sire model, 154–155
SIRIM, *see* Standard and Industrial Research Institute of Malaysia
Site yield potential (SYP), 10, 13
SIX effectors, *see* Secreted in xylem effectors
SLIM-FaTT, 212
SLR, *see* Single lens reflex
SNOOP version, *see* Soil Nitrogen Overview—Oil Palm
SNP, *see* Single nucleotide polymorphism
Socfin (Soc), 25
breeding program, 168
breeding strategy, 172
Indonesia, 168
Soil(s), 11, 361
soil-environment × genotype interactions, 384
Soil and Water Assessment Tool (SWAT), 377
Soil Nitrogen Overview—Oil Palm (SNOOP version), 373
Solar radiation, 11
Somaclonal variation, 193
associated with micro-propagation, 203
epigenome, 205–208
flower development, 204–205
Somatic embryogenesis (SE), 192, 210
influence of hormones on SE, 200–201
molecular aspects of SE, 199–200
South America, *E. oleifera* genetic resources, 46–47
South East Asia, oil palm diseases, 120
SP, *see* Sungai Pancur
Spear rot, 112, 124
breeding for resistance and resistance expression, 113
screening for disease resistance, 113–115
Specific combination, *see* Specific combining ability (SCA)
Specific combining ability (SCA), 67, 145, 247, 301, 339
alpha (α) incomplete factorial mating design, 338–345
Split-plot design, 330
SSD, *see* Single seed descent
SSR, *see* Short sequence repeats
SSRs, *see* Simple sequence repeats; Single sequence repeats

Standard and Industrial Research Institute of Malaysia (SIRIM), 307, 313
Standard error of difference (SED), 331
Statistical models, 331–332, 339–345
Stearoyl-ACP, *see* Stearoyl-acyl carrier protein
Stearoyl-acyl carrier protein (Stearoyl-ACP), 366
Stress level, 392
Strong winds, 360–361
sui generis system, 313
Sumatra Bioscience, Indonesia (SUMBIO), 175, 179
  achieved and expected genetic improvements, 183
  breeding for ease of harvesting traits, 182
  comparative expected time-frames, 184
  SUMBIO's main breeding program, 182
SUMBIO, *see* Sumatra Bioscience, Indonesia
Sungai Pancur (SP), 25–26
Supply, nutrient balance approach, 14
*SureSawit*Shell, molecular diagnostic assay, 227, 242, 246
Sure Shell Kit™, 305
SVP, *see* Saturated vapor pressure
SWAT, *see* Soil and Water Assessment Tool
Synthetic seed, 215, 216
SYP, *see* Site yield potential
Syringyl/guaiacyl composition (S/G composition), 130
Systems biology approach, 209–211

T

TAG, *see* Triacylglyceride
TAL effectors, *see* Transcriptional activator like effectors
TALEN, *see* Transcription activator-like effector nucleases
Tandem culling-level selection, 149–150
Tanzania, *E. guineensis*, 34–35
TargetIng Local Lesions IN Genomes (TILLING), 362
TCA, *see* Tricarboxylic acid
T-derived lines, 89
T. diversifolia, see Mexican sunflower
TDM, *see* Trunk dry matter
TDM production, *see* Total dry matter production
Temperature, 11, 360
Tenera palms (T palms), 1, 60, 88, 91, 227
TEP, *see* Total economic product
TEs, *see* Transposable elements
Theobroma grandiflorum, see Cupucu
TILLING, *see* TargetIng Local Lesions IN Genomes
Tim Penilai dan Pelepas Varietas (TP2V), 313, 320
Tissue culture, 288
  potential of organelle research in, 208–209
  propagation, 205–206
Total dry matter production (TDM production), 9
Total economic product (TEP), 38
Total leaf surface per hectare, 88
Totipotency principle, 192
Toxins, pathogen, 117–118
TP2V, *see Tim Penilai dan Pelepas Varietas*
"Training set", 156
Transcription activator-like effector nucleases (TALEN), 117, 369
Transcriptional activator like effectors (TAL effectors), 117
Transcriptomics, 256
Transgene technology, 368
Transgenic approaches, 366, 370
Transgenic oil palm, 363
  achieving commercial success, 370–371
  current technical challenges, 367–369
  development, 364–365
  genome editing technologies, 371–372
  from GM crops, 371
  historical approaches to oil palm transformation, 366–367
  target traits, 365–366
  technologies for genome manipulation, 369–370
  transgenic crops, 363–364
Transitioning oil palm plantation, 396
  agroforestry, 397
  oil palm monoculture, 396–397
Transpiration, 14
Transposable elements (TEs), 206
Tree architecture, 96–97
Triacylglyceride (TAG), 365–366
Trial conduct, 336–337
Tricarboxylic acid (TCA), 258
Trichoderma harzianum (*T. harzianum*), 244
Triosephosphate isomerase, 259
Triticum aestivum L, *see* Bread wheat
Triticum tauchii (D) germplasm, 24
"Tropical oils", 365
True second cycle of selection, 174
Trunk diameter, 93–94
Trunk dry matter (TDM), 93
$T \times T/P$ trials, 333
Two-dimensional electrophoresis (2-DE), 210
Two-in-One MPOB Simple Impeller (2-in-1 MoSLIM), 212
Type-2 acyl-CoA:diacylglycerol acyltransferase (DGAT2), 366

U

Ulu Remis (UR), 25
Ulu Remis Ts (URT), 27, 147
Umbelluria californica (*U. californica*), 364

UN Consultative Group on International Agricultural Research (CGIAR), 22
United Plantations (UP), 2
United Plantations Berhad (UPB), 289
  E. oleifera × *E. guineensis* hybrids, 291–293
  E. oleifera germplasm in, 290–291
  interspecific hybrid improvement at, 289–293
  yield and oil yield performance, 291–293
Unit Pelaksana Teknis Daerah (UPTD), 320
Unraveling genome, 238
  ESTs projects, 238–239
  MAS, 238
  sequencing hypomethylated genome of oil palm, 240–241
  whole-genome sequencing, 241
UP, *see* United Plantations
UPB, *see* United Plantations Berhad
UPOV, *see* International Union for Protection of New Varieties of Plants
Upper stem rot (USR), 110
UPTD, *see Unit Pelaksana Teknis Daerah*
URT, *see* Ulu Remis Ts
USR, *see* Upper stem rot
UTP, *see* L'Union Tropicale de Plantation

V

Vapor pressure (VP), 11
Vapor pressure deficit (VPD), 11
Varieties reproducing families, 315
VDM production, *see* Vegetative dry matter production
Vegetative
  criteria, 170
  growth, 337
  propagation, 192
  traits, 7–8, 39
Vegetative dry matter production (VDM production), 9, 93, 94
Vigor score, 94
Virescens, 60, 98, 242
  fruit, 227, 402–403, 404
  fruit marker, 62
VIRESCENS gene, 98–99
VP, *see* Vapor pressure
VPD, *see* Vapor pressure deficit

W

WAA, *see* Weeks after anthesis
WAIFOR, *see* West African Institute for Oil Palm Research
WaNuLCAS, 374
Water-use efficiency (WUE), 358
Water, 14, 357–359
  water-deficit environments, 102–103
  water-limiting type of models, 372–373
Water deficit
  other traits, 104
  physiological impact of, 103
  root development, 104
Weeks after anthesis (WAA), 258
Weevil (*Elaeidobius kamerunicus*), 8, 302
West Africa (WA), 144, 166–167
West African Institute for Oil Palm Research (WAIFOR), 2
Wet heat method, 304
Whole-genome
  resequencing, 256
  sequencing, 241, 248–249
Wind, 11
WinRHIZO, 385
Within family
  genetic variability estimation, 76–79
  selection method, 149
World Trade in palm oil, 3
WUE, *see* Water-use efficiency

X

Xanthomonas oryzae (*X. oryzae*), 117

Y

Yangambi population (Ybi population), 26
Yield, 337; *see also* Harvestable yield
  and components, 38
  and oil yield performance, 291, 292, 293
  oil yield progress, 173–174
  potential, 10, 398–399
  stability, 403–404

Z

Zairean, introgression programs, 40
Zaire, *E. guineensis*, 34

Printed in the United States
by Baker & Taylor Publisher Services

Printed in the United States
by Baker & Taylor Publisher Services